T0211806

Quantum
Continuous
Variables

Quantum Continuous Variables

A Primer of Theoretical Methods

Alessio Serafini

Department of Physics & Astronomy
University College London

CRC Press
Taylor & Francis Group
Boca Raton London New York

CRC Press is an imprint of the
Taylor & Francis Group, an **informa** business

CRC Press
Taylor & Francis Group
6000 Broken Sound Parkway NW, Suite 300
Boca Raton, FL 33487-2742

First issued in paperback 2021

© 2017 by Taylor & Francis Group, LLC
CRC Press is an imprint of Taylor & Francis Group, an Informa business

No claim to original U.S. Government works

ISBN 13: 978-0-367-87026-3 (pbk)
ISBN 13: 978-1-4822-4634-6 (hbk)
ISBN 13: 978-1-315-11872-7 (ebk)

DOI: 10.1201/9781315118727

Library of Congress Cataloging-in-Publication Data

Names: Serafini, Alessio, author.
Title: Quantum continuous variables : a primer of theoretical methods /
Alessio Serafini.
Description: Boca Raton, FL : CRC Press, Taylor & Francis Group, [2017] |
Includes bibliographical references and index.
Identifiers: LCCN 2016058596| ISBN 9781482246346 (hardback ; alk. paper) |
ISBN 1482246341 (hardback ; alk. paper) | ISBN 9781482246353 (e-book) |
ISBN 148224635X (e-book)
Subjects: LCSH: Quantum computing.
Classification: LCC QA76.889 .S47 2017 | DDC 006.3/843--dc23
LC record available at https://lccn.loc.gov/2016058596

Visit the Taylor & Francis Web site at
http://www.taylorandfrancis.com

and the CRC Press Web site at
http://www.crcpress.com

Publisher's Note
The publisher has gone to great lengths to ensure the quality of this reprint but points out that some imperfections in the original copies may be apparent.

Contents

PART V **Technologies**

CHAPTER 8 ▪ Quantum information protocols with continuous variables 223

List of Problems

Preface

This is a book with no figures (but with great cover art), which is somewhat anomalous in dealing with a subject where optical schemes play such a major role. As well as betraying one of the author's idiosyncrasies, this is a deliberate choice: while the subject matter rests definitely within the domain of physics, and holds considerable potential for technological implications, the main emphasis in the treatment presented is in fact on the unifying power of the mathematical formalism. At the same time, this is certainly not a book on what one calls mathematical physics, whose subtleties I have quite on the contrary systematically steered clear of (at times even questionably!). The prominence of formal manipulations arose not from a commitment to mathematical detail, but rather from striving for generality within the framework set by physics and quantum information. This approach arguably implies a steeper initial learning curve, but does pay off once the general vision behind it reveals itself to the reader. Personal inclination, pedagogical practice, and the lack of a like-minded attempt in the literature have informed these choices, which I hope will prove beneficial to some.

London, December 2016 *A. Serafini (Serafozzi)*

Acknowledgements

Needless to say, a number of people have contributed, more or less directly, to the realisation of this endeavour. Fabrizio Illuminati and Silvio De Siena set me on this path, during my PhD at the University of Salerno. Since then, almost all interactions with collaborators, senior and junior, on the topic of continuous variables have helped me to form an appreciation of both wide thematics and specific methods that ended up being part of this monograph.

Given my relatively modest network, and the effort I made to contain references, I would like to mention them all, in order of appearance, lest they don't find any other credit in this work: Matteo Paris, Gerardo Adesso, Jens Eisert, Michael Wolf, Julien Laurat, Gaëlle Keller, Thomas Coudreau, Stefano Mancini, Sougato Bose, Mauro Paternostro, Myungshik Kim, Oscar Dahlsten, Martin Plenio, David Gross, Alex Retzker, Marcus Cramer, Masaki Owari, Eugene Polzik, Kasper Jensen, Thomas Fernholz, Stefano Pirandola, Zhen-Biao Yang, Sai-Yun Ye, Hulya Yadsan-Appleby, Marco Genoni, Daniel Burgarth, Dmitrii Shalashilin, Davide Girolami, Tommaso Tufarelli, Alessandro Ferraro, Howard Wiseman, Jinglei Zhang, James Millen, Peter Barker, Vittorio Giovannetti, Oscar Duarte, Uther Shackerley-Bennett, Alexander Pitchford, Ludovico Lami and Alberto Carlini (note I limited this list to people I had direct collaborative exchanges with, on the subject of continuous variables).

This book is thought out especially for students, was written with a constant eye to them, and is hence dedicated to all of my students. The PhD students I had the pleasure of supervising over the years (Hulya Yadsan-Appleby, Sai-Yun Ye, Michele Avalle, Uther Pendragon Franck Shackerley-Bennet and Sofia Qvarfort) have all helped to shape material that is part of it. My current students, Uther and Sofia, directly proofread and commented on portions of the manuscript; besides conducting an extensive error hunt, Uther also took unstintingly of his time to inquire into several facts and factoids relevant to the book. Théo Sépulchre, who spent a summer internship with us from ENS Lyon, was also of help, for his annotations to Chapter 2.

This textbook would not be the same without the incredibly inspiring Quantum Optics lectures I had the opportunity to teach at Scuola Normale Superiore in Pisa in the academic years 2013–14 and 2014–15. In particular, the students sitting the 2013–14 module (Alberto Biella, Giacomo De Palma, Ludovico Lami, Luca Rigovacca) all gave direct inputs to the manuscript (three of which will be explicitly acknowledged on the relevant pages). The support and participation of Ludovico Lami calls for a special mention: not only did he contribute two outstanding original mathematical proofs, but he also spent hours in front of a blackboard helping me to clear out specific issues.

Although more condensed, the Quantum Optics and Atoms lectures I have been teaching since 2014 within UCL's Centre for Doctoral Training in Delivering Quantum Technologies have also been key to adjusting my pedagogical approach and refining the manuscript. Among the CDT students, I should acknowledge Carlo Sparaciari, whose handwritten lecture notes were helpful in reconstructing two sections of the book from the lectures, and Abdulah Fawaz, who pointed me towards references on coherent control.

Vittorio Giovannetti was instrumental in setting this project in motion. His (and Rosario Fazio's) kind hospitality at Scuola Normale Superiore during a sabbatical break allowed me to lay plans for it, and also, crucially, gave me the opportunity to teach the fantastic classes mentioned above.

Marco Genoni has been my closest collaborator while this book was being conceived and written. Justice can hardly be done to his contribution. It will suffice to mention that the whole approach to open and monitored quantum systems contained in Chapter 6 would not be there if I hadn't been working with him (on that topic, pivotal discussions with Stefano Mancini, Howard Wiseman and Andrew Doherty should also be acknowledged). Nor would the study on quantum Fisher information found in Chapter 8, which was spurred by expertise Marco brought over from Matteo Paris's school in Milan.

Tommaso Tufarelli's proofreading of Chapter 9 has been prompt and invaluable. Gerardo Adesso, a longstanding collaborator, gave precious suggestions on content he would have liked to see covered, besides having been a most committed promoter of this project since its very inception. I do hope his enthusiastic support will not be proven unfounded. Andrea Mari was as quick as lightening in clarifying specific issues on quantum channel capacities. Mark Wilde took an interest in Chapter 5, and fed me with several valuable remarks. Discussions directly relevant to the book with Andrew Fisher and Sougato Bose should also be acknowledged.

Much help came, of course, from outside academia. The collaboration with Stefano Ronchi was extraordinarily interesting. I would like to thank him especially for compromising on some of his artistic tenets, a most difficult sacrifice, and meeting me (more than) halfway to produce what is quite possibly the part of this book I am most proud of. Eleonora Serafini contributed to the cover concept too, with some well-placed words of wisdom. Andrea Montanari freely offered to finalise the cover's layout, for which I am immensely grateful (and pleased with the quality of what he produced). Rebecca Davies and Francesca McGowan at CRC were wonderful publishers, who humoured my whims, accommodated my times and saw the project through.

Monica's pervasive care, understanding and most valuable advice trickled into the book: her support had a direct impact on it, from the very early stages of its composition. Viola is a constant source of inspiration, and often valiantly makes up for my lack thereof. Anna took care of much in my stead, thus freeing abundant time I could pour into this effort.

A final mention goes to Marcus Cramer, whose cultural integrity I have long admired.

I

Preliminaries

Introduction

CONTENTS

This chapter sets the stage for the book, serving as a prelude and introducing the mathematical notation and background.

1.1 WHAT IS THIS BOOK ABOUT?

This book caters to graduate students approaching the study of quantum continuous variables, or for researchers keen on gaining a first grounding in the theoretical methods to describe such systems, be it from the angle of quantum optics, of controlled quantum dynamics or, more widely, of quantum technologies. It is not meant to be, it should be noted right away, an exhaustive handbook collecting all advanced results in the field. Certain notable developments have been entirely overlooked, while other specific aspects, included to illustrate the application of the basic formalism, might have been magnified beyond their merit by the author's perspective.

Central to the treatment presented is the formalism of Gaussian states and operations, which arises naturally when one deals with linear and quadratic Hamiltonians. Gaussian states and systems have played a privileged role in quantum optics and field theories, essentially since the very early steps of such disciplines, due to their ease of theoretical description and major relevance to experimental practice. Over the last twenty years, such a privilege has carried over to quantum information science, where Gaussian states form the core of the continuous variable toolbox.

In quantum optics, Gaussian states are ubiquitous, because second-order Hamiltonians are typically dominant (the free Hamiltonian of the field is Gaussian, for one thing), and allow for remarkably accurate coherent manipulations in practical set-ups. In the field theoretical tradition, Gaussian dynamics are often mocked as 'quasi-free', referring to the fact that their diagonalisation is trivial, in a specific sense that will be clarified later on. This supposed triviality should not fool the reader though as, while Gaussian dynamics may essentially be dealt with entirely at the phase space level, the analysis of their quantum information properties, crucial to their exploitation as engineering resources, requires one to confront their Hilbert space description. The latter is often very challenging, and in fact its full understanding still presents several loose ends and open questions. So, even though Gaussian dynamics might well be exactly solvable with comparatively elementary tools, the properties of Gaussian states related to the Hilbert space and tensor product structures are far from being equally transparent.

In this regard, it is also worthwhile to comment here on what is arguably the most debated facet of Gaussian systems from the quantum information standpoint: that they can be mimicked with stochastic classical systems, and thus may grant no genuine quantum advantage. If one restricts entirely to Gaussian manipulations and measurements, this statement is technically true. However, it is crucial to emphasise that quantum systems, even if described by Gaussian states, do imply an underlying Hilbert space structure. Hence, if one is able to access non-Gaussian resources at any one stage of a given protocol, be it through non-Gaussian measurements or dynamical operations, then one can harness quantum advantages from them. To mention a well-known, exemplary case, let us point out that Gaussian entanglement may display quantum non-locality if tested with non-Gaussian measurements. The correlations and entropies of Gaussian states are thus worth investigating, from both a theoretical and a practical standpoint.

Furthermore, it should be stressed that Gaussian quantum systems are bound, due to the subsumed quantum structure on a Hilbert space, to fundamental bounds that systems in the classical regime do not have to comply with, which do have far-reaching consequences. This is gloriously illustrated by the security proof of Gaussian quantum key distribution, which will close the theoretical portion of the book. Let us also point out that there exist other widely applicable Gaussian tools, such as squeezed light for enhanced metrology[1] or quadratic couplings for ground state cooling, which are not diminished in the slightest by the classical analogy mentioned above.

The book will cover, in great depth and generality, Gaussian dynamics of

---

[1] It is perhaps worth noting that the quest for squeezed light was greatly spurred by the realisation that beating the vacuum noise limit would facilitate the detection of gravitational waves.

open and closed systems, also including measurements and continuous monitoring, which will be essentially carried out at the phase space level. It will also introduce the instruments to deal with the Hilbert space description of Gaussian systems, and go on to analyse their quantum correlations, their information and communication capacity, their application to quantum metrology, and their use in quantum key distribution.

While the Gaussian and symplectic frameworks will be central to our treatment, they will often serve as keys to access more general aspects of continuous variable quantum systems. Many forays into non-Gaussian continuous variable systems have thus been incorporated into the book, including the derivation of general master equations and stochastic master equations, entanglement distillation, as well as more self-standing subjects, such as entanglement criteria beyond the Gaussian case. The general formalism presented in the book, hinging on the Fourier–Weyl expansion of quantum states, should put the reader in a position to explore further ramifications of the subject matter outside the Gaussian regime. A number of techniques relevant to this aim have been illustrated in dealing with optimisation problems, such as teleportation thresholds and information capacities, that called for us to go beyond the mere Gaussian description in order to be tackled with adequate generality.

Notwithstanding the proclaimed weight of the formalism, this book also makes an effort to rely the theoretical methods, be they Gaussian or non-Gaussian, to systems of practical interest to the realisation of quantum technologies. Thus, optical set-ups are referred to repeatedly throughout the text, and a final chapter of the monograph is dedicated to an overview of physical systems where quantum continuous variables are embodied and a high degree of coherent control is achievable in the laboratory.

Quantum computation is the great absentee, in the sense that no attempt was made at a systematic treatment of approaches towards gate- or measurement-based ("one-way") quantum computation where continuous variables play a direct computational role. Somehow, the author was always wary of the 'analogue' element present in most such schemes. There are, however, several architectures where continuous variables play an important auxiliary role, typically as a coherent bus of quantum information: these are exemplified through a dedicated treatment of the Cirac–Zoller trapped ion quantum computer, to be found in Chapter 9. Besides, linear optical quantum computation, where some of the methods developed in the text play a prominent role, is included through a basic description of the seminal Knill–Laflamme–Milburn quantum computer (also in Chapter 9). An infinite number of alternative ways may be envisaged to digitalise quantum continuous variables, so there is certainly scope for the author's scepticism towards quantum computation with such systems to be proven spectacularly wrong.

1.1.1 Synopsis

After a cursory look at quantum mechanics (Chapter 2), the book introduces Gaussian states, rather unconventionally, as ground and thermal states of second-order Hamiltonians (Chapter 3).[2] The associated symplectic and Weyl groups will also emerge naturally, as Heisenberg evolutions of the canonical operators, and so will the normal mode decomposition, yielding eventually the spectrum of a generic Gaussian state as well as its entropy. The uncertainty principle, that constrains the set of quantum Gaussian states, will also be derived. The connection with the more general phase space formalism, dear to quantum optics, is then made by deriving the Fourier–Weyl relation that allows one to expand bounded operators in terms of Weyl operators, and by defining the characteristic function in the process (Chapter 4). These tools will link the Gaussian and phase space descriptions to the Hilbert space.

Next, the discrete transformations that preserve the Gaussian character of the input state will be described in great detail and generality, including open systems subject to quantum noise and conditional evolutions due to measurements (Chapter 5). Afterwards, the continuous-time versions of such Gaussian evolutions, termed diffusive for reasons that will be clear, shall be introduced, also including their monitored counterparts, giving rise to stochastic dynamics (Chapter 6). This approach will include an alternative derivation of Markovian master equations, using, so to speak, the comfortably solvable Gaussian dynamics as a probe.

Our coverage will then specialise to the mathematical problem of qualifying and quantifying the quantum entanglement of continuous variables, which will be dealt with in some detail (Chapter 7).

The theoretical methods developed in the book will finally be applied to four selected topics in the area of quantum technologies, namely, quantum teleportation, quantum channel capacities, quantum estimation, and quantum key distribution (Chapter 8). Last, but certainly not least, will come the *à posteriori* justification of all the theoretical methods developed by direct linking with the practical set-ups where they apply, ranging from quantum optics to bosonic atoms, through optomechanics, trapped ions, atomic ensembles, and integrated circuits (Chapter 9).

1.2 HOW TO USE THIS BOOK

This book assumes acquaintance with basic linear algebra (essentially, with the theory of linear operators and their diagonalisation) and quantum mechanics. Some slightly more specific notions concerning the singular value decomposition and Schur complements are reported in Section 1.3, where the

---

[2] We hope students and researchers with a background in physics may find this approach, whereby Gaussian states are first presented without any reference to the mathematical machinery of characteristic functions and quasi-probability distributions, more directly accessible and to their taste than the customary definitions adopted in quantum optics.

whole notation is also set, and a few more standard mathematical reminders are included. Some care should be taken in getting to terms with the simple and expedient, though slightly unusual, outer product notation fleshed out in detail in Section 1.3.3, since much of the book relies on it. Quantum mechanics is sketched out in Chapter 2, with the added purpose of clarifying further notation and adopted conventions. In general, the book aims at providing the reader with a self-contained treatment, in the sense that a postgraduate student with a standard university training in linear algebra and quantum mechanics, but no previous research experience, should ideally be able to access all of it, without reference to external sources. On certain, more specialised subjects, this was made possible by the inclusion of dedicated appendixes. Some compromises were in order, with notions which would have required too longwinded a digression: most notably, the asymptotic attainability of the Holevo bound will not be proven; likewise, the Hudson theorem will be invoked without formal statement or proof.

The core of the book (Chapters 3–6) unfolds like a coherent whole with a number of interwoven connections and dependencies between different parts. The basic theory and formalism set out in Chapters 3 and 4 are preparatory and essential to the whole remainder of the volume. An appreciation of Gaussian measurements, described in Chapter 5, is necessary to understand the continuously monitored dynamics addressed in Chapter 6. The foundations of the entanglement characterisation developed in Chapter 7, may be tackled without much reference to the preceding two chapters, although case studies taken from them will feature prominently. Chapter 8, being an applied synthesis of all the methods developed, is bound to contain frequent referencing to the previous chapters, although it was written in such a way that the reader already familiar with the basics of the formalism should be able to access it directly. The same is true of Chapter 7. Also note that, while an effort has been made to include all the elements necessary to the understanding of the material presented, the early part of the book (Chapters 3–6) is deliberately more pedagogical and detailed than the more specialised, later part (Chapters 7–8). Chapter 9 is to some extent self-standing, containing a discussion of the fundamental physics that justifies the mathematical description of certain systems as quantum continuous variables. The discussion of input-output interfaces, relating continua of modes to localised degrees of freedom, does however account for the basic correlation equations utilised in Chapter 6 to describe Markovian environments, which would otherwise be left as a somewhat arbitrary starting point.

References to external sources have been kept to a minimum. Although direct credit is given in the main text when a specific approach existing in the literature is followed closely, all of the references are deferred, for each chapter, to a conclusive Further Reading section, together with some words of commentary by the author. Further bibliography, typically on topics that are more peripheral to our main concerns, may be found in the appendices. It should be strongly emphasised that references have been chosen in strict relation to the approach taken and material treated, in the interest of complementing the book's coverage, of addressing side issues that cropped up but could not be properly developed, and of nurturing the reader's understanding of the wider repercussions of the subject matter in hand. Such choices were obviously distorted by the author's personal judgment and partial perspective towards the literature. This should not compromise, but perhaps rather strengthen, the book's intention of serving as a basic tool.

From Chapter 3 onward, outstanding statements and results are framed in titled boxes like this

---

Outstanding result. This is outstanding, ain't it?

While it may hinder at times the narrative flow, this has proven a very popular choice among students using my lecture notes, and has hence been maintained.

Some of the statements are followed by formal proofs, like this one:

Proof. That was outstanding as it stood out, and it deserves these vertical lines on both sides. □

Not all proofs in the text are laid out with vertical sidelines as above, only the most technical ones. This is meant to make it easy for the reader that is not interested in the details of a derivation to skip it altogether.

The book includes 70 solved problems, framed like this

Problem 1.1. (*Example of a problem*). This is a problem, ain't it?
Solution. Fortunately, this problem is solved by this solution.

Typically, problems found in the middle of a discourse are instrumental to it, and may be small parts of derivations which the reader is encouraged to tackle as an exercise; the solution is however always included right after each problem, so that the reader who is unwilling to fill in a derivation may nonetheless access full, consistent proofs and arguments. Other problems, typically found at the end of a section, are instead meant to be self-standing applications of the techniques introduced in the main text. Some of them are a pretext to include notable topics which would otherwise have found no other space in this textbook.

1.3 MATHEMATICAL NOTATION, CONVENTIONS AND BASIC FORMULAE

1.3.1 Hilbert spaces and operators

Much of the mathematics in the book will take place in the Hilbert space of square integrable functions $L^2(\mathbb{R}^n)$, with n finite. Because the parallel use of the (finite dimensional) phase space and the (infinite dimensional) Hilbert space notations may potentially raise confusion, all linear operators acting on Hilbert spaces will always wear a hat, in order to distinguish them from finite dimensional transformations, with the sole exception of density operators indicating quantum states, which will be typically denoted with ϱ (as customary), without a hat. Thus, $\hat{\mathbb{1}}$ will stand for the infinite dimensional identity operator on the Hilbert space, while $\mathbb{1}$ will stand for a finite dimensional identity matrix, with dimension d clear from context unless explicitly indicated through the notation $\mathbb{1}_d$. Note also that the identity 'superoperator' (mapping a linear operator into the same operator) will be denoted as $\mathcal{I}$.

The Dirac notation, whereby Hilbert space vectors will be denoted with kets, as in $|\psi\rangle$ and their dual with a bra, as in $\langle\psi|$, will be adopted. The notation $|\psi, \phi\rangle$ will stand for the tensor product of the vectors $|\psi\rangle$ and $|\phi\rangle$, belonging to different Hilbert spaces: $|\psi, \phi\rangle = |\psi\rangle \otimes |\phi\rangle$.

The abridged notation $(\hat{o} + \text{h.c.})$ will stand for $(\hat{o} + \hat{o}^\dagger)$, where $\hat{o}^\dagger$ is the Hermitian conjugate of the operator $\hat{o}$.

As customary in quantum mechanics, the notation $\langle\hat{o}\rangle$ for operator $\hat{o}$ will stand for the expectation value of $\hat{o}$ with respect to the quantum state ϱ of the relevant system, clear from context or unspecified: $\langle\hat{o}\rangle = \text{Tr}\,[\varrho\hat{o}]$.

1.3.2 Linear algebra and direct sums

Finite-dimensional vectors, which most often will belong to an even-dimensional phase space, will be denoted with bold letters, such as $\mathbf{v}$, with Euclidean and sesquilinear duals $\mathbf{v}^\mathsf{T}$ and $\mathbf{v}^\dagger$, respectively. They will thus always be kept distinct from Hilbert space vectors expressed in bra-ket notation. The Euclidean scalar product between two vectors is therefore written $\mathbf{v}^\mathsf{T}\mathbf{w} = \mathbf{v} \cdot \mathbf{w}$. The modulus of a real vector $|\mathbf{v}|$ is also understood to be its Euclidean norm: $|\mathbf{v}| = \sqrt{\mathbf{v}^\mathsf{T}\mathbf{v}}$. The entries of a vector $\mathbf{v}$ will be denoted with a light font, as v_j.

Linear operators and quadratic forms on finite dimensional spaces will be most often denoted with capital letters, such as A, with the exception of covariance matrices which, in compliance with the author's longstanding usage, will be denoted with the bold Greek letter $\boldsymbol{\sigma}$, up to labels, with entries σ_{jk} in light font. All finite dimensional linear operators not acting on Hilbert spaces are distinguished from Hilbert space operators by the lack of a hat (with the exception of quantum states, like ϱ, also without a hat). As already mentioned, we will refer to the $d \times d$ identity matrix with the symbol $\mathbb{1}_d$, while the generic $\mathbb{1}$ will denote the identity matrix when the dimension is deemed clear from the context. Occasionally, the set of $n \times n$ matrices on the field $\mathbb{F}$ will be denoted with $\mathcal{M}(n, \mathbb{F})$.

The diagonal matrix with entries $\{\lambda_1, \ldots, \lambda_d\}$ on the main diagonal and 0 elsewhere will be denoted with $\text{diag}(\lambda_1, \ldots, \lambda_d)$.

The standard two-dimensional Pauli matrices will also be useful:

$$\sigma_z = \begin{pmatrix} 1 & 0 \\ 0 & -1 \end{pmatrix}, \quad \sigma_x = \begin{pmatrix} 0 & 1 \\ 1 & 0 \end{pmatrix}, \quad \sigma_y = \begin{pmatrix} 0 & -i \\ i & 0 \end{pmatrix}. \tag{1.1}$$

$\mathrm{Tr}M$ and $\mathrm{Det}M$ stand for, respectively, the trace and determinant of the matrix M.

The direct sum notation $\oplus$ will be used often, especially when dealing with phase space variables: let us remind the reader that the direct sum of two square matrices M and N, of dimensions m and n respectively, is an $m + n$-dimensional square matrix with M and N as diagonal blocks and 0 entries elsewhere:

$$M \oplus N = \begin{pmatrix} M & 0_{m \times n} \\ 0_{n \times m} & N \end{pmatrix} , \tag{1.2}$$

where $0_{j \times k}$ stands for the null matrix with j rows and k columns. Henceforth, the generic symbol 0 will stand for a null matrix with dimension imposed from the context (as per the null blocks in the example above), including zero vectors or infinite dimensional operators if need be.

1.3.2.1 Singular value decomposition

Given an $m \times n$ matrix M on the complex field, one has

$$M = UDV , \tag{1.3}$$

where U and V are, respectively, $m \times m$ and $n \times n$ unitary matrices, while D is an $m \times n$ matrix with real, non-negative entries d_j on the main diagonal, and zeros elsewhere. Assuming $n \leq m$, the singular values $\{d_j\}$ are the positive square roots of the eigenvalues of $D^\dagger D$.

If $m = n$, then all the matrices involved are square matrices. If M is real with positive determinant, then U and V are real orthogonal transformations.

1.3.2.2 Schur complements

Although this monograph assumes previous familiarity with basic linear algebra, it seems worthwhile to recall the properties of Schur complements, which are slightly more exotic than other fundamental tools and will play a major part in analysing Gaussian measurements and entanglement.

In this section, let M be a Hermitian matrix partitioned into blocks as per

$$M = \begin{pmatrix} A & B \\ B^\dagger & C \end{pmatrix} , \tag{1.4}$$

with A invertible. Then, the Schur complement of M with respect to its principal submatrix A, denoted with M/A, is defined as

$$M/A = C - B^\dagger A^{-1} B . \tag{1.5}$$

The Schur complement, when well defined, offers a powerful condition equivalent to the strict positivity of the matrix M:

$$M > 0 \quad \Leftrightarrow \quad A > 0 \quad \text{and} \quad (C - B^\dagger A^{-1} B) > 0 . \tag{1.6}$$

The equivalence (1.6) is straightforward to prove by noting that $M > 0$ may be written as $\mathbf{v}^\dagger A\mathbf{v} + 2\mathbf{v}^\dagger B\mathbf{w} + \mathbf{w}^\dagger C\mathbf{w} > 0$ for all complex vectors $\mathbf{v}$ and $\mathbf{w}$ (with the same dimension as A and C respectively). For given $\mathbf{w}$, the latter is a convex second-order polynomial in $\mathbf{v}$ – convexity being a consequence of $A > 0$ – and it can be easily minimised by setting its gradient to 0. The same procedure to evaluate the minimum of a second-order polynomial of a single variable then leads to the matrix condition (1.6). Note also that $A > 0$ is clearly necessary for $M > 0$.

The equivalence (1.6) implies a less known variational characterisation of M/A, which will turn out to be useful: if $A > 0$, then

$$M/A = \sup\{\tilde{C} : \; M \geq 0 \oplus \tilde{C}\}, \tag{1.7}$$

where the supremum is taken with respect to the matrix partial order induced by positivity: $M \geq N \Leftrightarrow M - N \geq 0$ (the so-called 'Löwner' partial order). The equivalence between the definitions $M/A = C - B^\dagger A^{-1}B$ and (1.7) is a direct consequence of the positivity condition (1.6) applied to $M - (0 \oplus \tilde{C})$.

Schur complements are also central to the process of Gaussian elimination (whereby a matrix is decomposed in a product of block triangular matrices) which leads to the following expression for the inverse of a non-singular matrix M in terms of its submatrices:

$$M^{-1} = \begin{pmatrix} (A - BC^{-1}B^\dagger)^{-1} & A^{-1}B(B^\dagger A^{-1}B - C)^{-1} \\ (B^\dagger A^{-1}B - C)^{-1}B^\dagger A^{-1} & (C - B^\dagger A^{-1}B)^{-1} \end{pmatrix}. \tag{1.8}$$

This may also be verified by direct matrix multiplication, noting that

$$(A - BC^{-1}B^\dagger)^{-1}A + A^{-1}B(B^\dagger A^{-1}B - C)^{-1}B^\dagger = \mathbb{1}, \tag{1.9}$$

$$(A - BC^{-1}B^\dagger)^{-1}B + A^{-1}B(B^\dagger A^{-1}B - C)^{-1}C = 0, \tag{1.10}$$

which are both proven through a little algebra after having multiplied both sides by $M/C = (A - BC^{-1}B^\dagger)$.

1.3.3 Compact outer product notation

When not dealing with tensor products in Hilbert spaces, the outer product of two real vectors $\mathbf{a}$ and $\mathbf{b}$ is defined as customary according to standard matrix multiplication, and denoted by $\mathbf{ab}^\mathsf{T}$. In components: $(\mathbf{ab}^\mathsf{T})_{jk} = a_j b_k$.

Vectors and matrices of operators (where each entry is an operator acting on a Hilbert space) will be used extensively throughout the book. In general, for a vector of Hermitian operators $\hat{\mathbf{a}}$, one has $\hat{\mathbf{a}}\hat{\mathbf{a}}^\mathsf{T} \neq (\hat{\mathbf{a}}\hat{\mathbf{a}}^\mathsf{T})^\mathsf{T}$, as the operators in the vector might not commute with each other. It is hence useful to define the symmetrized and anti-symmetrized forms of such products, as

$$\{\hat{\mathbf{a}}, \hat{\mathbf{a}}^\mathsf{T}\} = \hat{\mathbf{a}}\hat{\mathbf{a}}^\mathsf{T} + (\hat{\mathbf{a}}\hat{\mathbf{a}}^\mathsf{T})^\mathsf{T}, \tag{1.11}$$

$$[\hat{\mathbf{a}}, \hat{\mathbf{a}}^\mathsf{T}] = \hat{\mathbf{a}}\hat{\mathbf{a}}^\mathsf{T} - (\hat{\mathbf{a}}\hat{\mathbf{a}}^\mathsf{T})^\mathsf{T}. \tag{1.12}$$

In components, one has

$$\left\{\hat{\mathbf{a}}, \hat{\mathbf{a}}^{\mathsf{T}}\right\}_{jk} = \hat{a}_j \hat{a}_k + \hat{a}_k \hat{a}_j \,, \tag{1.13}$$

$$\left[\hat{\mathbf{a}}, \hat{\mathbf{a}}^{\mathsf{T}}\right]_{jk} = \hat{a}_j \hat{a}_k - \hat{a}_k \hat{a}_j \,. \tag{1.14}$$

Also note that

$$\left\{\hat{\mathbf{a}}, \hat{\mathbf{a}}^{\mathsf{T}}\right\} + \left[\hat{\mathbf{a}}, \hat{\mathbf{a}}^{\mathsf{T}}\right] = 2\hat{\mathbf{a}}\hat{\mathbf{a}}^{\mathsf{T}} \,. \tag{1.15}$$

1.3.4   Delta functions

The Dirac delta function $\delta(x)$ admits a representation as

$$\delta(x) = \frac{1}{2\pi} \int_{-\infty}^{+\infty} e^{ipx} \, \mathrm{d}p \,, \tag{1.16}$$

which can be turned into the following relation for complex variables:

$$\frac{1}{4\pi^2} \int_{-\infty}^{+\infty} \int_{-\infty}^{+\infty} e^{ipx'} e^{-ixp'} \, \mathrm{d}x' \mathrm{d}p' = \frac{1}{\pi^2} \int_{\mathbb{C}} e^{\alpha\beta^* - \alpha^*\beta} \mathrm{d}^2\beta = \delta^2(\alpha) \,, \tag{1.17}$$

where the changes of variable $\sqrt{2}\alpha = (x + ip)$ and $\sqrt{2}\beta = (x' + ip')$ were applied, and $\delta^2(\alpha) = \delta\left(\mathrm{Re}(\alpha)\right) \delta\left(\mathrm{Im}(\alpha)\right)$.

The Heaviside θ function is defined as $\theta(x) = 0$ for $x < 0$, $\theta = 1$ for $x > 0$ and $\theta(x) = 1/2$ for $x = 0$, so that $\frac{\mathrm{d}}{\mathrm{d}x}\theta = \delta(x)$.

The discrete Kronecker delta δ_{jk} will be employed, with values 1 for $j = k$ and 0 for $j \neq k$.

1.3.5   Gaussian integrals

Gaussian integrals with many variables will occur very often throughout the book, most often over integration domains of even dimension (since, as we shall see, a phase space is a real space of dimension $2n$). Given a symmetric, real, positive definite $2n \times 2n$ matrix A, and a $2n$-dimensional vector $\mathbf{b}$, one has the equality:

$$\int_{\mathbb{R}^{2n}} \mathrm{d}\mathbf{r} \, e^{-\mathbf{r}^{\mathsf{T}} A \mathbf{r} + \mathbf{r}^{\mathsf{T}} \mathbf{b}} = \frac{\pi^n}{\sqrt{\mathrm{Det}\, A}} \, e^{\frac{1}{4} \mathbf{b}^{\mathsf{T}} A^{-1} \mathbf{b}} \,, \tag{1.18}$$

where the shorthand notation $\mathrm{d}\mathbf{r}$ indicates the product of differentials of the $2n$ integration variables that compose the vector $\mathbf{r}$: $\mathrm{d}\mathbf{r} = \mathrm{d}r_1 \ldots \mathrm{d}r_{2n}$. Eq. (1.18) may be proven by switching to the eigenbasis of A (which can always be done by an orthogonal transformation since A is symmetric, and decouples the integral), completing the squares at the exponent, and then inserting $\int_{\mathbb{R}} e^{-x^2} \, \mathrm{d}x = \sqrt{\pi}$. The determinant appears as the product of the eigenvalues of A. Note that Eq. (1.18) also holds for a complex $\mathbf{b}$.

1.3.6 Miscellanea

A dot will stand for time-derivative: $\dot{f} = \frac{\mathrm{d}f}{\mathrm{d}t}$. Partial derivatives will be often expressed in shortened notation: $\partial_x f = \frac{\partial f}{\partial x}$ for the multivariate function f.

We will make use of the Landau "little-o" symbol $o(\varepsilon)$, denoting an unspecified quantity such that $\lim_{\varepsilon \to 0} \frac{o(\varepsilon)}{\varepsilon} = 0$.

Summation signs labelled by a condition will assume the involved variables to extend over the whole interval of definition, as clear from context. For instance, if $j \in [1\ldots,n]$ and $k \in [1\ldots,n]$, then $\sum_{j<k} f(j,k) = \sum_{k=1}^{n} \sum_{j=1}^{k-1} f(j,k)$. Likewise, $\sum_{j \neq k} f(j,k)$ will stand for $\sum_{j=1}^{n} \left(\sum_{k=1}^{j-1} f(j,k) + \sum_{k=j+1}^{n} f(j,k) \right)$.

Also, we will at times make use of Einstein's summation convention, whereby repeated indexes in an expression are summed over the whole range of possible values: for instance, if $j \in [1\ldots,n]$, then $f_j g_j = \sum_{j=1}^{n} f_j g_j$.

The infinite sum of the geometric progression will be used several times:

$$\sum_{j=0}^{\infty} z^j = \frac{1}{1-z}, \quad \text{for } |z| < 1 ; \tag{1.19}$$

whence, after, respectively, one or two differentiations and further straightforward manipulation, one obtains the other useful infinite sums:

$$\sum_{j=0}^{\infty} n z^j = \frac{z}{(1-z)^2}, \quad \text{for } |z| < 1 , \tag{1.20}$$

$$\sum_{j=0}^{\infty} n^2 z^j = \frac{z(z+1)}{(1-z)^3}, \quad \text{for } |z| < 1 . \tag{1.21}$$

CHAPTER 2

Quantum mechanics: Instructions for use

CONTENTS

This book is not meant to contain a detailed analysis of the foundations of quantum mechanics. However, we felt an introduction to the general theory was required in order to make the presentation self-sufficient and clear the following chapters from possible ambiguities. We shall, wherever possible, avoid the intricacies of the mathematical description of infinite dimensional spaces, as well as the conceptual subtleties related to the interpretation of the formalism (although the approach taken will necessarily betray aspects of the author's stand on these matters), and content ourselves with a 'pragmatic' introduction to the theory, as the chapter's title suggests. Throughout the chapter, we will even occasionally refer to a hypothetical "user" of the theory as an artefact to elucidate relationships between the formalism and its operational interpretation.

Notions related to quantum non-locality and the information content of quantum states, such as entropies and the characterisation of quantum correlations (also known as entanglement), will be the subject of the second half of the chapter, as they are very relevant to the material covered in the book.

2.1 QUANTUM STATES AND QUANTUM MEASUREMENTS

Any quantum system is associated with a Hilbert space $\mathcal{H}$, that is a vector space on the complex field endowed with an inner product. We shall adopt Dirac's notation, whereby the inner product between the 'ket' vectors $|v\rangle$ and $|w\rangle$ is denoted with $\langle v|w\rangle$. The inner product is sesquilinear[1] in the sense that $\langle \alpha v|\beta w\rangle = \alpha^* \beta \langle v|w\rangle$, for all $\alpha, \beta \in \mathbb{C}$ and all $|v\rangle, |w\rangle \in \mathcal{H}$. If the dimension of $\mathcal{H}$ is infinite, the space is still supposed to admit a numerable basis of vectors, which we shall denote with $\{|j\rangle, j \in \mathbb{N}\}$, a requirement referred to as "separability", so that any vector $|v\rangle$ in $\mathcal{H}$ can be expressed as an infinite sum in such a basis: $|v\rangle = \sum_{j=0}^{\infty} v_j |j\rangle$. Unless otherwise noted, we shall assume the basis we work with to be orthonormal, such that it satisfies $\langle j|k\rangle = \delta_{jk}$.

A linear operator $\hat{A}$ on $\mathcal{H}$ is bounded if and only if $m \in \mathbb{R}$ exists such that $|\langle v|\hat{A}|v\rangle| \leq m \langle v|v\rangle \; \forall \; |v\rangle \in \mathcal{H}$. While in finite dimensions, where linear operators just correspond to finite dimensional matrices, any bounded operator $\hat{A}$ is also "trace class", i.e., it admits a well-defined and finite $\text{Tr}[\hat{A}] = \sum_j \langle j|\hat{A}|j\rangle$, in infinite dimensions this obviously does not hold (consider, for instance, the identity operator), so that one has to require specifically that an operator is trace class (which implies boundedness). We will also refer to a trace class operator $\hat{A}$ as 'normalised' if $\text{Tr}[\hat{A}] = 1$.

Besides, while in finite dimensions all linear operators associated with observables are represented by Hermitian matrices – such that $A_{jk} = A_{kj}^*$, if A is the matrix representing the operator – and admit a spectral decomposition (a basis of eigenvectors, as ensured by the spectral theorem), the stronger assumption of being 'self-adjoint'[2] must be made in infinite dimensions to state that an operator $\hat{A}$ admits a spectral decomposition: $\hat{A} = \sum a|a\rangle\langle a|$. Notice that the sum will typically be infinite and has been kept completely generic as it may also generalise to an integral with a given measure on a (see below).

As a further subtlety concerning infinite dimensional Hilbert spaces, note that the objects $|a\rangle$ and their dual $\langle a|$ might not even be proper, normalised vectors of $\mathcal{H}$, but only linear forms whose inner product $\langle a|v\rangle$ with any vector $|v\rangle \in \mathcal{H}$ is well defined. This is the case for the self-adjoint operators $\hat{x}$ and $\hat{p}$ that describe position and momentum of a quantum degree of freedom, and satisfy $[\hat{x}, \hat{p}] = i\hbar\hat{\mathbb{1}}$ which, as we shall see, may be represented on the Hilbert space L^2 of square integrable functions on the real line. For instance, $\hat{x}$ may be written as

$$\hat{x} = \int_{-\infty}^{+\infty} x|x\rangle\langle x|\,\mathrm{d}x \,, \tag{2.1}$$

where the $\{|x\rangle, x \in \mathbb{R}\}$ satisfy $\langle y|x\rangle = \delta(x - y)$.[3] The notation above means

---

[1] From Latin, "one and a half linear".

[2] A self-adjoint operator is Hermitian on a dense domain with respect to the topology induced, on $\mathcal{H}$, by the inner product.

[3] The fact that none of the self-adjoint operators fulfilling the canonical commutation relation $[\hat{x}, \hat{p}] = i\hbar\hat{\mathbb{1}}$ may admit any normalisable eigenvector $|v\rangle \in L^2$ is easy to show. Assume $\hat{x}|x\rangle = x|x\rangle$ with $\langle x|x\rangle = 1$, then $\hat{x}' = \hat{x} - x\hat{\mathbb{1}}$ would be another operator that

that, given a self-adjoint $\hat{A} = \sum_a a|a\rangle\langle a|$, all expectation values $\mathrm{Tr}\left(\hat{A}\hat{x}^n\right)$ may be evaluated as

$$\mathrm{Tr}\left(\hat{A}\hat{x}^n\right) = \int_{-\infty}^{+\infty} \sum_a x^n a|\langle a|x\rangle|^2 \, \mathrm{d}x, \tag{2.2}$$

where all the $\langle a|x\rangle$ are well defined. If $\hat{A}$ were a quantum state, the expression (2.2) would allow one to retrieve all the statistics related to measurements of the operator $\hat{x}$. Further specifics on the structure and usage of infinite dimensional Hilbert spaces are to be found in Section 2.4, where quantum continuous variable systems are introduced.

To complete our cursory survey of linear operators on Hilbert spaces, let us also remind the reader that a 'positive' operator $\hat{A}$ is one such that $\langle v|\hat{A}|v\rangle \geq 0$ for all $|v\rangle \in \mathcal{H}$.[4]

The set of quantum states $\mathcal{B}(\mathcal{H})$ is the set of normalised (i.e., trace class with trace 1), self-adjoint and positive linear operators on the Hilbert space $\mathcal{H}$. Any quantum state ϱ hence admits a spectral decomposition: $\varrho = \sum_j \lambda_j|v_j\rangle\langle v_j|$, where the sum may generalise to an integral, the $|v_j\rangle$ may be proper or improper vectors, and all the eigenvalues λ_j are positive and such that $\sum_j \lambda_j = 1$. Positivity and the normalisation requirement ensure that the theory allows a user to make consistent probabilistic predictions.

A quantum state is said to be 'pure' if it is the projector on a one-dimensional subspace of the Hilbert space, in which case one may write $\varrho = |\psi\rangle\langle\psi|$. This is equivalent to stating that a quantum state is pure if and only if all of its eigenvalues are 0 except for one which is 1. Analogously, pure states may be seen as the extreme points in the convex set of quantum states. Therefore, any ϱ admits an "ensemble" decomposition into a convex combination of pure states, $\varrho = \sum_\nu p_\nu|\nu\rangle\langle\nu|$, which is generally not unique, unless ϱ is itself a pure state and $\varrho = |\nu\rangle\langle\nu|$.

The operational interpretation of the mathematical object ϱ, representing the state of a quantum system, requires one to introduce a description for the act of measurement. Any quantum measurement is characterised by a positive operator-valued measure (POVM),[5] which is simply a set of linear operators $\{\hat{K}_\mu\}$ satisfying

$$\sum_\mu \hat{K}_\mu^\dagger \hat{K}_\mu = \hat{\mathbb{1}} \, . \tag{2.3}$$

Note that, once again, the summation over the label μ has been kept completely generic as it may refer to a complex or real space of any dimension

---

satisfies $[\hat{x}', \hat{p}] = i\hbar\hat{\mathbb{1}}$ and has a 0 eigenvalue with associated normalised eigenvector $|x\rangle$. Then, $\langle x|[\hat{x}', \hat{p}]|x\rangle = 0$, while $i\hbar\langle x|\hat{\mathbb{1}}|x\rangle = i\hbar$, in clear contradiction with the commutation relation.

[4] In the literature, the definition of a positive operator usually also implies self-adjointness. We kept the two notions distinct here for the sake of clarity.

[5] This, somewhat haughty, terminology is borrowed from the mathematical physics tradition, where POVMs were first considered.

and may also extend to an integral over some measure on μ. It is to the normalisation condition (2.3), realised by a sum of operators over the variable μ, that the designation POVM is hinting at.

The set of values μ corresponds to all the possible outcomes of the measurement process described by the POVM defined above. If the system that is being measured is prepared in state ϱ, each outcome μ is obtained with probability $p(\mu)$ given by the Born rule:

$$p(\mu) = \mathrm{Tr}\left(\varrho \hat{K}_\mu^\dagger \hat{K}_\mu\right) . \tag{2.4}$$

The formalism introduced so far assures that all the $p(\mu)$ are positive and add up to 1 if summed over all the possible outcomes μ. Eq. (2.4) contains the full operational characterisation of the quantum state ϱ, which is nothing but a mathematical means that allows a user of the theory to predict the probability of any possible outcome in any possible measurement process.

An idealised quantum measurement process, whereby the quantum state of the system is not destroyed or fundamentally altered – think, as a prototypical instance of such an alteration, of the detection of a photon in a photodetector, where the measured light field is converted into an electric signal – would leave the system in the 'conditional' state

$$\frac{\hat{K}_\mu \varrho \hat{K}_\mu^\dagger}{\mathrm{Tr}\left(\hat{K}_\mu \varrho \hat{K}_\mu^\dagger\right)} \tag{2.5}$$

after a measurement with outcome μ occurs. If instead the measurement is "coarse-grained", in the sense that the hypothetical user of the theory only knows that one of a set $\mathcal{M}$ of filtered outcomes was obtained, the updated quantum state would be

$$\frac{\sum_{\mu \in \mathcal{M}} \hat{K}_\mu \varrho \hat{K}_\mu^\dagger}{\mathrm{Tr}\left[\sum_{\mu \in \mathcal{M}} \hat{K}_\mu \varrho \hat{K}_\mu^\dagger\right]} , \tag{2.6}$$

where, as usual, the sum might generalise to an integral.

If the $\hat{K}_\mu$ are all projectors on one-dimensional spaces, so that $\hat{K}_\mu = |\mu\rangle\langle\mu|$, then the Born rule (2.4) becomes

$$p(\mu) = \langle\mu|\varrho|\mu\rangle , \tag{2.7}$$

and the conditional state is simply $|\mu\rangle\langle\mu|$ (von Neumann postulate). The measurement may then be associated with a self-adjoint observable $\hat{O} = \sum_\mu O_\mu |\mu\rangle\langle\mu|$. Note how the eigenvalues O_μ play an entirely conventional role since, on any given measurement process, they can each be rescaled at one's leisure up to degeneracies, as discussed below. The only aspect that matters operationally is maintaining the distinction between distinct eigenvalues. The Born rule then implies that the statistical moments of the operator $\hat{O}$ may all be evaluated as $\langle \hat{O}^n \rangle = \text{Tr} \left(\varrho \hat{O}^n \right)$. Throughout the book, the terminology $\langle \hat{O} \rangle$ will denote the average ('expectation') value of the measurement described by the self-adjoint operator $\hat{O}$ over the quantum state in question ϱ, which will always be clear from the context if not specified.

Projectors on subspaces of dimension greater than 1 describe 'degenerate' observables, where the measurement process is coarse-grained in a way that does not distinguish between the different one-dimensional states associated with a given value of O_μ. Then, $p(O) = \sum_{\mu:O_\mu=O}\langle \mu|\varrho|\mu \rangle$ and the conditional state associated with the outcome O is $\frac{1}{p(O)}\sum_{\mu:O_\mu=O}\langle \mu|\varrho|\mu \rangle|\mu \rangle\langle \mu|$. POVMs whose elements $\hat{K}_\mu$ are projectors are commonly referred to as projective or von Neumann measurements.

The Hilbert space of a composite quantum system comprising a system with Hilbert space $\mathcal{H}_A$ and a system with Hilbert space $\mathcal{H}_B$ is given by the tensor product $\mathcal{H}_{AB} = \mathcal{H}_A \otimes \mathcal{H}_B$. Given a quantum state $\varrho_{AB} \in \mathcal{B}(\mathcal{H}_{AB})$, the reduced, local state of system A is given by the partial trace ϱ_A:

$$\varrho_A = \text{Tr}_B\left(\varrho_{AB}\right) = \sum_j {}_B\langle j|\varrho_{AB}|j \rangle_B \,, \tag{2.8}$$

where $\{|j\rangle_B\}$ denotes an orthonormal basis of the Hilbert space $\mathcal{H}_B$.

It is well known, since Naimark's work, that the most general POVM $\{\hat{K}_\mu\}$ can be equivalently recast as a projective measurement in a larger Hilbert space, in the sense that, for any $\{\hat{K}_\mu\}$ satisfying Eq. (2.3), one has that a state $\varrho_E \in \mathcal{B}(\mathcal{H}_E)$ on an auxiliary Hilbert space $\mathcal{H}_E$ (the label E referring to a hypothetical "environment") and an orthonormal basis $\{|\mu\rangle\}$ of $\mathcal{H} \otimes \mathcal{H}_E$ exist, such that

$$\text{Tr}\left(\hat{K}_\mu \varrho \hat{K}_\mu^\dagger\right) = \langle \mu|(\varrho \otimes \varrho_E)|\mu \rangle \quad \forall \, \varrho \in \mathcal{B}(\mathcal{H}) \,. \tag{2.9}$$

2.2   CP-MAPS AND UNITARY TRANSFORMATIONS

Any map Φ between density operators that represents a deterministic physical dynamical process must be:

- Linear: $\Phi(\alpha\varrho + \beta\sigma) = \alpha\Phi(\varrho) + \beta\Phi(\sigma) \, \forall \alpha, \beta \in \mathbb{C}$ and all linear operators ϱ and σ. Although slightly stronger, this condition ensures that convex combinations of input states are mapped into the corresponding convex combinations of output states, which makes the probabilistic interpretation of the ensemble decomposition consistent with dynamical processes.

- Trace-preserving: $\text{Tr}\left[\Phi(\varrho)\right] = \text{Tr}\left[\varrho\right]$, so that the probabilities of outcomes of any measurement still add up to one after the dynamics.

- Positive: $\Phi(\varrho) \geq 0$ if $\varrho \geq 0$, so that the Born rule still predicts positive probabilities after the dynamical process.

- Completely positive: a more subtle requirement that amounts to asking that the map resulting from applying Φ on a subsystem of a composite system while leaving the other subsystem unchanged is still positive. In formulae: $\Phi \otimes \mathcal{I}(|\psi\rangle\langle\psi|) \geq 0 \quad \forall \, |\psi\rangle \in \mathcal{H}_A \otimes \mathcal{H}_B$, where Φ is assumed to act on linear operators on $\mathcal{H}_A$ while $\mathcal{I}$ denotes the identity superoperator[6] on $\mathcal{H}_B$. Clearly, complete positivity implies positivity, although here we have kept the two requirements distinct for the sake of clarity. Note also that the spectral decomposition ensures that a superoperator which is positive on all one-dimensional projectors is also positive on all quantum states.

Such deterministic physical quantum maps are known as completely positive (CP) maps or "quantum channels".

A general way of describing a CP-map is through its 'Kraus operators' $\hat{K}_\mu$:

$$\Phi(\varrho) = \sum_\mu \hat{K}_\mu \varrho \hat{K}_\mu^\dagger \, . \tag{2.10}$$

The operators $\hat{K}_\mu$ (where μ might be a continuous label, so that the sum above generalises to an integral) satisfy the same condition as the POVM that describes a measurement process: $\sum_\mu \hat{K}_\mu^\dagger \hat{K}_\mu = \hat{\mathbb{1}}$.

A special class of CP-maps is given by the action of unitary operations, which may be seen as CP-maps with a single Kraus operator, under which the input state ϱ is mapped into $\hat{U}\varrho\hat{U}^\dagger$, with $\hat{U} \in U(d)$, where d is the dimension of the Hilbert space. Notice that the unitarity condition $U^\dagger U = \hat{\mathbb{1}}$ is then equivalent to the CP-map condition (2.10). Unitary transformations play a role analogous to that of pure states within the set of quantum states: in particular, a unitary transformation U will send the generic pure state $|\psi\rangle\langle\psi|$ into $U|\psi\rangle\langle\psi|U^\dagger$, thus preserving its purity (as this notation is equivalent to stating that the vector $|\psi\rangle$ is mapped to the vector $U|\psi\rangle$).

An equivalent representation of completely positive maps may be given in terms of their 'Stinespring dilation': any CP-map is equivalent to a unitary operation $\hat{U}$ acting on an extended global system comprising the original system S and an environment E, upon partial tracing over E. In formulae, this amounts to stating that the most general Kraus operators $\{\hat{K}_\mu\}$, may be written as

$$\hat{K}_\mu = {}_E\langle\mu|\hat{U}|0\rangle_E \, , \tag{2.11}$$

where the $\{|\mu\rangle_E\}$ are a basis of the environmental Hilbert space. Observe that these $\{\hat{K}_\mu\}$ do form a resolution of the identity, because of the unitarity of $\hat{U}$.

Yet another equivalent representation of CP-maps is given by the Choi–Jamiolkowski isomorphism, whose general form is reviewed in Section 5.5.1, at the end of Chapter 5, where an infinite-dimensional version of the isomorphism will be adopted to describe Gaussian CP-maps.

Notice that the term "deterministic", employed above, is meant to exclude

---

[6] A 'super'-operator is an operator acting between linear operators on a Hilbert space.

filtering operations described by Eq. (2.6), realised by performing a quantum measurement and by keeping the output state only if a certain set of measurement outcomes was obtained. Such filtering maps are not linear, as the normalisation factor they involve also depends on ϱ. While, once a given outcome is post-selected, they may be construed as transforming the state with certainty under Eq. (2.6), we refer to them as "non-deterministic" in that they imply the heralded discarding of physical systems.

2.3 DYNAMICS: HAMILTONIANS AND MASTER EQUATIONS

The continuous-time version of the unitary operations and CP-maps introduced above is based on the notion of a self-adjoint Hamiltonian operator, that governs the dynamics of ideal, closed quantum systems according to the Schrödinger equation for the (pure) state vector $|\psi\rangle$:

$$\frac{\mathrm{d}}{\mathrm{d}t}|\psi\rangle = -i\hat{H}|\psi\rangle . \tag{2.12}$$

In the Heisenberg picture (where vectors stay put and operators evolve), this is equivalent to the Heisenberg equation for a generic operator $\hat{o}$:

$$\frac{\mathrm{d}}{\mathrm{d}t}\hat{o} = i[\hat{H}, \hat{o}] . \tag{2.13}$$

If the system's initial state ϱ is mixed, then the evolution equation is of Heisenberg form:

$$\frac{\mathrm{d}}{\mathrm{d}t}\varrho = -i[\hat{H}, \varrho] \tag{2.14}$$

[note however the relevant sign difference between (2.14) and (2.13)]. These equations yield unitary evolutions at any time t: $\varrho \mapsto \mathrm{e}^{-i\hat{H}t}\varrho\mathrm{e}^{i\hat{H}t}$ and $\hat{o} \mapsto \mathrm{e}^{i\hat{H}t}\hat{o}\mathrm{e}^{-i\hat{H}t}$.

Open quantum dynamics are modelled as a Hamiltonian, unitary dynamics on a larger global system, comprising an environment which is eventually traced out. Under certain dynamical assumptions[7] this leads to a time-local quantum master equation for the density operator ϱ, which is said to be of 'Lindblad form':

$$\dot{\varrho} = \sum_j \left[\hat{L}_j\varrho\hat{L}_j - \frac{1}{2}\left(\hat{L}_j^2\varrho + \varrho\hat{L}_j^2\right)\right] = \sum_j \mathcal{D}[\hat{L}_j](\varrho) , \tag{2.15}$$

---

[7]Essentially, small coupling with the environment (Born approximation) and humongous environmental recurrence times with respect to the system typical dynamical time scales, so that information, once spread into the vast environment, shall not flow back to the system (Markov approximation, which leads to memory-less dynamics without integral kernels). Let us note that the exact, general necessary and/or sufficient conditions under which master equations in the Lindblad form may be obtained are a complicated matter, that we will not concern ourselves with here.

where we implicitly defined the Lindblad superoperator $\mathcal{D}[\hat{L}_j]$. The equation above describes the most general time-local evolution (in the sense that it is a first-order differential equation, whose solution depends only on the state at a certain time) which gives rise to a completely positive and trace-preserving dynamics. We will derive master equations in Lindblad form in Chapter 6.

2.4   CONTINUOUS VARIABLES

The standard approach to quantising a non-relativistic dynamical degree of freedom proceeds by defining a pair of self-adjoint position and momentum operators $\hat{x}$ and $\hat{p}$ that satisfy

$$[\hat{x}, \hat{p}] = i\hbar \hat{\mathbb{1}} \,, \tag{2.16}$$

where $\hbar = \frac{h}{2\pi}$ is the reduced Planck constant. The commutation relation above is known as the canonical commutation relation (CCR), while $\hat{x}$ and $\hat{p}$ are referred to as a pair of 'canonical' operators, in analogy with the terminology of classical Hamiltonian dynamics, where commutators would be replaced with Poisson brackets. If natural units, where $\hbar = 1$, are adopted, the canonical commutation relation above may be written in terms of the so called ladder – or 'annihilation' and 'creation' – operators $\hat{a} = (\hat{x} + i\hat{p})/\sqrt{2}$ and $\hat{a}^\dagger = (\hat{x} - i\hat{p})/\sqrt{2}$, as

$$[\hat{a}, \hat{a}^\dagger] = \hat{\mathbb{1}} \,. \tag{2.17}$$

The non-Hermitian operators $\hat{a}$ and $\hat{a}^\dagger$ effect the annihilation and creation of a quantum of energy in the description of quantum harmonic oscillators. In the framework of quantum field theory (see below) their action instead represents the creation and destruction of particles.

The extension of the commutation relations above to n pairs of canonical variables – each labelled by an integer – reads, in natural units, $[\hat{x}_j, \hat{p}_k] = i\delta_{jk}$. Inspection reveals that the CCR algebra, defined by these commutation relations, does not allow for a representation through finite dimensional matrices. For instance, by taking the trace of the left- and right-hand side of Eq. (2.16) and assuming that the trace of a commutator vanishes, as is always the case in finite dimensions, one would get $\text{Tr}\,\hat{\mathbb{1}} = 0$, which is clearly false. However, infinite dimensional representations of the CCR algebra do exist. As well known from basic quantum mechanics, one can consider the space of square-integrable functions on the real space $L^2(\mathbb{R}^n)$, and define:

$$\hat{x}_j|f\rangle = x_j f(\mathbf{x}) \,, \tag{2.18}$$

$$\hat{p}_j|f\rangle = -i\frac{\mathrm{d}}{\mathrm{d}x_j}f(\mathbf{x}) \,, \quad \forall\,|f\rangle \equiv f(\mathbf{x}) \in L^2(\mathbb{R}^n) \,, \tag{2.19}$$

with $\mathbf{x} = (x_1, \dots, x_n)$. By virtue of the Stone–von Neumann theorem, all representations of the CCR algebra on a finite set of degrees of freedom are unitarily equivalent to the representation given by Eqs. (2.18) and (2.19). In this book, we shall not venture to consider continua of degrees of freedom, as would be necessary in quantum field theory, and will thus be content with this representation of the CCR algebra.[8]

The eigenstates of $\hat{x}_j$ (and $\hat{p}_j$) are not part of $L^2(\mathbb{R}^n)$, although we shall still indicate them in the Dirac notation as $|x_j\rangle$ and $|p_j\rangle$, by which we denote linear forms acting on $L^2(\mathbb{R}^n)$ (or on its dual), such that

$$\langle x'_j | f \rangle = \langle f | x'_j \rangle^* = f(x_1 \ldots, x_{j-1}, x'_j, x_{j+1}, \ldots, x_n) \in L^2(\mathbb{R}^{n-1}) \,, \quad (2.20)$$

$\forall \, |f\rangle \equiv f(\mathbf{x}) \in L^2(\mathbb{R}^n)$. For a trace-class operator $\hat{O}$, one can then write

$$\mathrm{Tr}\left(\hat{O}\right) = \int_{-\infty}^{+\infty} \langle x | \hat{O} | x \rangle \, \mathrm{d}x \,. \quad (2.21)$$

The operators $\hat{x}_j$ and $\hat{p}_j$ admit a spectral decomposition in terms of projectors on these improper eigenvectors. Their eigenvalues form a continuous set covering the whole real line, whence the terminology "quantum continuous variables" for systems described by pairs of canonical operators.

In quantum field theory, which allows one to include relativistic systems with varying number of particles through the formalism of second quantization, the commutation relation above applies to pairs of bosonic field operators $\hat{x}$ and $\hat{p}$. A prominent example of such a quantum field is the electromagnetic one, where $\hat{x}$ and $\hat{p}$ are the quantum counterpart of the magnetic and electric fields along one polarisation direction. This is the reason why systems described by quantum continuous variables are often referred to as "bosonic" systems in the literature. Clearly, all the formal results that apply to first quantized systems also apply to bosonic quantum fields in second quantization.

Let us briefly mention that our treatment could be adapted to fermionic fields by replacing the commutator in the left-hand side of Eq. (2.17) with an anti-commutator, and by imposing additional anti-commutator prescriptions on multiple degrees of freedom, summed up as $\{\hat{a}_j, \hat{a}_k^\dagger\} = \delta_{jk} \hat{\mathbb{1}}$ and $\{\hat{a}_j, \hat{a}_k\} = 0$. The resulting algebra, which describes systems abiding by the fermionic statistics, admits however a finite dimensional representation whenever a finite number of degrees of freedom is considered. We shall therefore not include fermionic fields in our main coverage, as the specific traits that characterise quantum continuous variables do not apply to them. Nonetheless, an interesting parallel between the methods presented and fermionc fields is drawn in Appendix A.

Notice that, in the following, we will treat the mathematics of quantum continuous variables in the much more convenient natural units ($\hbar = 1$).

---

[8] The essential mathematical distinction between quantum optics and quantum field theory is precisely the fact that the former is concerned with a finite number of degrees of freedom, while the latter deals with a continuum thereof. In Chapter 9, we shall come back to this issue and, in a sense, bridge between quantum field theory and quantum optics by analysing the quantization of the electromagnetic field in full detail.

We will however always introduce units explicitly when considering concrete examples of such systems.

2.5 ENTROPIES

Given a statistical ensemble X defined by a random classical variable x and its associated probability p_x, the Shannon entropy $S(X) = -\sum_x p_x \log_2(p_x)$ gives the number of bits (classical two-level systems) per letter needed to completely specify x in the asymptotic limit of infinitely long strings. Thus, $S(X)$ is an operationally well-defined indicator of how uncertain the random variable x is or, conversely, of how much information each letter in the string that uses the alphabet X carries.

In the same vein, the information content of a quantum state ϱ may be quantified rigorously by determining, in the asymptotic limit of an infinite ensemble of physical systems, how many two-dimensional quantum systems (*qubits*) per system are needed to represent ϱ faithfully. This amounts to taking the base-2 logarithm of the dimension of the typical subspace spanned by ϱ, which is sufficient to represent the state arbitrarily well, as only states belonging to such a subspace will occur in the asymptotic limit. It is easy to show that such a quantity is given by the classical Shannon entropy $S(\lambda_j) = -\sum_j \lambda_j \log_2 \lambda_j$ of the classical probability distribution defined by the set $\{\lambda_j\}$ of eigenvalues of ϱ.[9] The quantity $S(\lambda_j)$ is also known as the von Neumann entropy of the quantum state, and may be expressed equivalently as

$$S_V(\varrho) = -\text{Tr}\left(\varrho \log_2 \varrho\right) . \tag{2.22}$$

In a Hilbert space of dimension d, one has $0 \leq S_V(\varrho) \leq \log_2 d$. A quantum state is pure, i.e., its density matrix is the projector on a one-dimensional subspace, if and only if its entropy is zero, while the maximum entropy state is always given by the normalised state proportional to the identity operator, which is clearly a legitimate quantum state.

Thus, the von Neumann entropy characterises the entropic content of a quantum state, in the sense that high values of the entropy indicate that the direction of the state in the Hilbert space is highly ambiguous, whereas lower values correspond to a quantum state whose direction in the Hilbert space is comparatively predictable. It is common to refer to a state with higher von Neumann entropy as "more mixed" than a state with lower entropy.

The evaluation of the von Neumann entropy defined by Eq. (2.22) requires the full spectrum of the quantum state and is at times cumbersome to carry out. It is hence convenient to introduce another quantifier of mixedness, termed "purity" and often indicated with $\mu(\varrho)$. The purity is given by

$$\mu(\varrho) = \text{Tr}\left(\varrho^2\right) . \tag{2.23}$$

---

[9] Notice that in the continuous variable case the spectrum of ϱ could be continuous and the summation may have to be replaced by an integral over a probability density.

The purity is related to the notion of linear entropy, which we shall refrain from introducing here, along with a whole family of generalised entropies that will not be necessary in our treatment. Although it does in general yield a different ordering of quantum states with respect to the von Neumann entropy, the extreme values of the purity are still achieved by pure states, which are the only ones for which $\mu(\varrho) = 1$, and by states proportional to the identity, for which $\mu(\varrho) = \frac{1}{d}$. In infinite dimensions, the purity hence takes values between 0 and 1. While not imbued with a clear operational significance, the purity of a quantum state is, as we shall see, a very expedient means to check whether a quantum state is pure or not, as it is in general much easier to assess analytically.

In analogy with the classical mutual information $I(X;Y) = S(X)+S(Y)-S(X,Y)$, where $S(X,Y)$ is the entropy of the joint ensemble (X,Y), that quantifies the correlations between two statistical ensembles X and Y as the reduction in the number of bits per letter needed to specify one of the variables once the other is known, one may define the quantum mutual information $I(\varrho_{AB}) = S_V(\varrho_A) + S_V(\varrho_B) - S_V(\varrho_{AB})$ in terms of the global and local von Neumann entropies of the state ϱ_{AB}. The quantity $I(\varrho_{AB})$ is always positive, which is equivalent to stating that the von Neumann entropy is subadditive, and quantifies the total correlations contained in the bipartite state ϱ_{AB}. Besides, the von Neumann entropy S_V is *strictly* concave: $S_V(p\varrho_1 + (1 - p)\varrho_2) \geq pS_V(\varrho_1) + (1 - p)S_V(\varrho_2)$ for all density operators ϱ_1, ϱ_2 (this just amounts to stating that S_V grows under mixing) and $S_V(p\varrho_1 + (1 - p)\varrho_2) = pS_V(\varrho_1) + (1 - p)S_V(\varrho_2)$ if and only if $\varrho_1 = \varrho_2$.

Another notable property of the von Neumann entropy is the so called 'strong subadditivity': let $\varrho_{ABC} \in \mathcal{B}(\mathcal{H}_A \otimes \mathcal{H}_B \otimes \mathcal{H}_C)$, then

$$S_V(\varrho_{ABC}) + S_V(\varrho_A) \leq S_V(\varrho_{AB}) + S_V(\varrho_{AC}), \qquad (2.24)$$

where ϱ_A, ϱ_{AB} and ϱ_{AC} are the local density operators, obtained by partial tracing.

2.6   QUANTUM ENTANGLEMENT

As already mentioned, a composite quantum system comprising n sub-systems associated with distinct Hilbert spaces $\{\mathcal{H}_j, \ j \in [1,\ldots,n]\}$ is described by quantum states defined on the tensor product Hilbert space $\bigotimes_{j=1}^{n} \mathcal{H}_j$. The tensor product structure implies the existence of quantum states of composite systems that cannot be generated by the separate owning parties through local operations and classical communication (LOCC) alone. States with this property are called entangled states, and play a privileged role as a resource in quantum computation and quantum communication. Hereafter, we will restrict to the notion of bipartite entanglement, which is assessed across a certain split $\mathcal{H} = \mathcal{H}_A \otimes \mathcal{H}_B$ of the global Hilbert space $\mathcal{H}$, as the tensor product of the two local spaces $\mathcal{H}_A$ and $\mathcal{H}_B$ (distinct from the more general multipartite case, where LOCC defined with respect to a partition into a generic number of local spaces are taken into account).

Entanglement is best defined through the opposite notion of separability: a quantum state of a bipartite system $\varrho \in \mathcal{B}(\mathcal{H}_A \otimes \mathcal{H}_B)$ is said to be separable if and only if there exist local states $\varrho_{A,j} \in \mathcal{B}(\mathcal{H}_A)$ and $\varrho_{B,j} \in \mathcal{B}(\mathcal{H}_B)$, with $j \in [1, \ldots, m]$ and probabilities p_j, with $\sum_{j=1}^{m} p_j = 1$, such that

$$\varrho = \sum_{j=1}^{m} p_j \left(\varrho_{A,j} \otimes \varrho_{B,j} \right) . \tag{2.25}$$

It is clear that a quantum state of the form above can be created by the two parties A and B by coordinating via classical communication the creation of the local states $\varrho_{A,j}$ and $\varrho_{B,j}$, dictated by the sampling of the probability distribution p_j. Conversely, a quantum state is entangled if and only if it is not separable.

Quantum entanglement, which the founding fathers of the theory used to discuss as a "spooky action at distance", has by now acquired a well-established role as a resource for quantum communication, computation and metrology. Substantial efforts have hence gone to characterise it, in both qualitative and quantitative terms, mainly aimed at answering the questions of whether a given state is entangled or not and, if it is, by how much, in some sense to be defined. In the remainder of this section, we will briefly review very basic results in this area which will be relevant to the following of the book, without any attempt at being exhaustive or very detailed.

2.6.1 Entanglement of pure quantum states

It is easy to convince oneself that a globally pure state $|\psi\rangle \in \mathcal{H}_A \otimes \mathcal{H}_B$ may be written in the form (2.25) if and only if it is the tensor product of two local pure states: $|\psi\rangle = |\psi\rangle_A \otimes |\psi\rangle_B$, with $|\psi\rangle_A \in \mathcal{H}_A$ and $|\psi\rangle_B \in \mathcal{H}_B$. If that is the case, then the local, reduced states of subsystems A and B are pure as well, being given by $|\psi\rangle_A$ and $|\psi\rangle_B$ respectively. On the other hand, if the local states are pure, then the global state is also pure and just given by their tensor product. It follows that a globally pure state is separable if and only if the von Neumann entropy of its reduced states is zero (or, equivalently, that a globally pure state is entangled if and only if the entropy of its local reduced states is not zero).

More generally, given any pure quantum state $|\psi\rangle \in \mathcal{H}_A \otimes \mathcal{H}_B$, orthonormal local bases $\{|j\rangle_A\}$ and $\{|j\rangle_B\}$ of $\mathcal{H}_A$ and $\mathcal{H}_B$ may be defined such that one has

$$|\psi\rangle = \sum_j d_j |j\rangle_A \otimes |j\rangle_B , \tag{2.26}$$

where the quantities d_j may always be made real and positive by a proper choice of phases, and are normalised: $\sum_j d_j^2 = 1$. The particular form (2.26) is known as the Schmidt decomposition, where the $\{d_j\}$ are the Schmidt coefficients of the bipartite pure state $\{|\psi\rangle\langle\psi|\}$. Note that the real and positive Schmidt coefficients associated to a pure quantum state are always unique. Evaluating the partial trace on a pure state in Schmidt decomposition yields local density matrices ϱ_A and ϱ_B directly in their diagonal form. Namely, $\varrho_A = \sum_j d_j^2 |j\rangle_{AA}\langle j|$ and, analogously, $\varrho_B = \sum_j d_j^2 |j\rangle_{BB}\langle j|$. The spectra of the two reduced states of bipartite pure states are hence identical, and so are their local entropies. As a consequence, a globally pure state is separable if and only if it admits only one nonzero Schmidt coefficient, equal to 1 (in which case the Schmidt decomposition above trivially reduces to a product of two local pure states).

In quantitative terms, the amount of the entanglement contained in a pure bipartite state can be evaluated by the von Neumann entropy of the reduced state. Note that, as highlighted in the discussion of the Schmidt coefficients above, this quantity does not depend on which of the two parties is traced out. The von Neumann entropy of the reduced state, the so called "entropy of entanglement", has an operational interpretation in terms of distillable entanglement: It represents the maximum ratio m/n between how many 'singlets' m (i.e., states of the form $(|0\rangle \otimes |1\rangle - |1\rangle \otimes |0\rangle)/\sqrt{2}$, or equivalent under local unitary transformations) can be obtained by LOCC from n copies of the original state, in the asymptotic limit of infinite n. Note that the distillable entanglement is upper bounded by $\log_2(d_-)$, where d_- is the smaller of the dimensions of $\mathcal{H}_A$ and $\mathcal{H}_B$, which may well be infinite if both constituent spaces are infinite dimensional.

2.6.2   Partial transposition and logarithmic negativity

The entanglement of mixed quantum states is much more difficult to characterise. In fact, even in the bipartite case, there exists no clear-cut criterion to decide whether a given quantum state, in a generic Hilbert space of arbitrary dimension, is entangled or not. Obviously, if the global state is mixed, the reduced local state can be mixed without any implication on the correlations between the two parties in question. The local entropy is thus not an indicator of entanglement and different, more sophisticated techniques are in order.

Among the number of diverse approaches developed to address the question of mixed state entanglement, we will privilege partial transposition, which is rather revealing and will be very relevant to the subject matter of the remainder of the book. The roots of this treatment lie in the simple observation that a superoperator which is positive, but not completely positive (i.e., whose tensor product with the identity superoperator on a composite Hilbert space is not positive), will still result in a positive operator if applied on a separable state of the form (2.25). This is immediately evident by inspection. Therefore, if such a positive but not completely positive map is applied on a subsystem of a bipartite quantum state and the output operator is found to be negative, then the original state must be entangled (because it cannot be separable).

The non-positivity of the output of a local application of a positive but not completely positive map is a sufficient condition for entanglement (conversely, the positivity of the same output is a necessary condition for separability).

A positive but not completely positive map that serves a key role in the characterisation of quantum entanglement is transposition, in any given basis. Besides being a very expedient – in the sense of being comparatively easy to evaluate, given the linearity of the map – entanglement criterion, the connection between the negativity of the partially transposed state and quantum correlations runs deeper. In fact, it turns out that states with positive partial transposition, while possibly entangled under the definition above, cannot be distilled into singlets by LOCC at all. Besides, the negativity of the partial transposition is also necessary for entanglement in small bipartite Hilbert spaces, of dimension 2×2 and 2×3. Furthermore, this criterion plays a key role in the entanglement theory of the so called Gaussian states, which will be central to our discussion of continuous variables in the following chapters.

The positivity of the partial transposition criterion ('PPT') is also related to a much used quantifier of quantum entanglement, the so called logarithmic negativity $E_{\mathcal{N}}$. As is the case for the associated criterion, the logarithmic negativity is relatively accessible computationally: if $\varrho_{AB} \in \mathcal{B}(\mathcal{H}_A \otimes \mathcal{H}_B)$ is a bipartite state, and we denote by $\tilde{\varrho}_{AB}$ its partial transposition (the operator obtained by transposition with respect to any local basis, in only one of the two subsystems),[10] then one has

$$E_{\mathcal{N}} = \log_2 \|\tilde{\varrho}\|_1 , \tag{2.27}$$

where $\|\hat{o}\|$ denotes the trace norm of operator $\hat{o}$, i.e., the sum of the absolute values of its eigenvalues, if diagonalisable. Note that partial transposition cannot change the trace of an operator, so that the sum of the eigenvalues of a partially transposed state $\tilde{\varrho}$ is still 1. The quantity $\|\tilde{\varrho}\|_1$ may hence be different from – and in particular larger than – 1 if and only if $\tilde{\varrho}$ has negative eigenvalues. It follows that $E_{\mathcal{N}}$ is equal to 0 for all states with positive partial transpose, and larger than 0 for all states which violate the PPT criterion.

The adoption of the logarithmic negativity as an entanglement quantifier is somewhat unsatisfactory, in that there exist entangled states, known as "bound entangled" states because their entanglement cannot be distilled, that have positive partial transpose and hence zero logarithmic negativity. Even so, the logarithmic negativity is still a consistent entanglement monotone – a quantity that does not increase under local operations and classical communication – and it does provide one with an upper bound on the distillable entanglement. The base 2 appearing in the log of Eq. (7.40) is needed for the logarithmic negativity to represent such an upper value.

---

[10] Notice that neither the PPT criterion nor the value of the logarithmic negativity depend on which subsystem is transposed, since the two partial transpositions are related by a global transposition, that is a positive map. Also, the choice of local basis is irrelevant too, since any two local bases are related by local unitary transformations.

2.7 FURTHER READING

Most of the traditional textbooks on quantum mechanics take the historic route, which typically attempts to justify the abstraction of the formalism by presenting the involved amount of experimental evidence the founding fathers were confronted with. Although some of the old classics are extremely deep and certainly worth reading, we would argue such a 'defensive' approach ends up being typically confusing (to students and scholars alike), and would rather advocate one where the formalism is laid out in a more systematic fashion. An exception to such a tradition is, to some extent, Sakurai's classic book, which we would still recommend as a basic reference on quantum mechanics [74]. A more modern take on the subject, coming from a quantum computation perspective and setting out from finite dimensional Hilbert spaces (stripped of the technical difficulties of infinite dimensions) is Schumacher and Westmoreland's book [77]. A very inspired introduction to "modern" quantum mechanics and its connection to information theory (as well as, more specifically, to quantum communication and computation) is the set of lecture notes by John Preskill, which has been serving the quantum information community since its early days, and has been recently updated [71]. Notably, such lecture notes also include a very useful account of classical information theory. While written when the field was still relatively young, Nielsen and Chuang's textbook is still a very useful introductory reference to quantum information and computation [64], where a proof of the strong subadditivity of the von Neumann entropy, first shown in [59], may also be found.

II

Foundations

CHAPTER 3

Gaussian states of continuous variable systems

CONTENTS

Quantum continuous variable systems are quantum systems obeying the canonical commutation relations. As we have seen in the previous chapter, such systems always require the adoption of infinite dimensional Hilbert spaces, even when a finite number of degrees of freedom is considered. Besides making their dynamics potentially intractable, this fact also implies that the study of properties related to the information and correlation content of continuous variable systems turns out to be particularly difficult.[1]

A widely known way to make continuous variable systems more tractable is

---

[1] As we saw in Section 2.6.2 with the case of the negativity of partial transposition as an entanglement criterion, which is necessary only up to dimension 6, the entanglement properties of quantum states do depend on the dimension of the constituent Hilbert spaces. The complete characterisation of bipartite entanglement in products of Hilbert spaces with infi-

to consider the restriction to Gaussian states. Already well trodden in the older quantum optics literature, the study of Gaussian systems has more recently become a central theme of continuous variable quantum information. If open quantum systems, coupled to an external environment, are considered, the Gaussian regime applies when the system couples linearly to its environment and when system and environment are governed by Hamiltonians which are at most quadratic in the canonical operators (or, equivalently, when the overall Hamiltonian of system and environment is at most quadratic).

These conditions may seem rather restrictive but, as we shall point out in the course of the book, are actually ordinarily met by a vast number of existing experimental set-ups in the areas of quantum optics, trapped ions, optomechanics, atomic ensembles and certain superconducting degrees of freedom (all of these platforms are reviewed in Chapter 9). What we will refer to as Gaussian measurements – POVMs that preserve the Gaussian character of initial states – are also customarily carried out with comparatively high efficiency, while the description of quantum noise can be incorporated seamlessly in the Gaussian picture.

What is perhaps even more important to note is that the restriction to the Gaussian realm, while not exhaustive of the full wealth of dynamics and situations allowed in the whole Hilbert space, still lets one include most of the processes relevant to quantum technologies. Squeezing, whereby certain canonical quadratures have uncertainties below the vacuum state noise, and can hence be used in precision measurements, quantum entanglement, and cooling, where the entropy is drained out of a system in order to reach a pure quantum state, and initialise information protocols or low noise experiments, may all be described within the Gaussian regime. Within the quantum information discourse, the Gaussian restriction has been often criticised on the grounds that no real genuinely quantum effects can ever be observed without leaving it at some point, since the statistics obtained from Gaussian states through Gaussian measurements can be reproduced by classical probability distributions. While technically correct, the reader should be reminded that such critiques often disregard the fact that the object described by a Gaussian state does entail a genuinely quantum description in a Hilbert space. Hence, for instance, the preparation of a pure Gaussian quantum state is the preparation of a pure vector of the Hilbert space. Likewise, a highly squeezed Gaussian state prepared for a metrological protocol is a state with very low noise in a certain physical observable, regardless of whether it can be mimicked by a classical distribution or not. The testing and exploitation of quantum non-locality (stronger than classical correlations that imply the presence of quantum entanglement) with Gaussian states does instead require a departure from Gaussian measurements. However, as we will see, certain non-Gaussian measurements are customarily implemented with current tech-

---

nite dimensions would apply to any product of Hilbert spaces, regardless of dimensionality, and is hence bound to be impervious.

nology, and it is hence still relevant to study in detail the creation of such a resource for Gaussian states.

This chapter will introduce Gaussian states from a somewhat unusual angle: as ground and thermal states of at most second-order Hamiltonians. Our treatment will highlight the privileged role played by quadratic canonical (*i.e.*, 'symplectic') transformations in this context, which we will take care to describe in full generality, and will hinge on the reduction of Gaussian states into their normal modes, which allows for their systematic diagonalisation. We will then proceed to explore an equivalent characterisation in terms of statistical moments of the canonical operators, and to introduce the implications of non-commutativity on such statistics in the form of the Robertson–Schrödinger uncertainty relation.

3.1 CANONICAL COMMUTATION RELATIONS

The canonical commutation relations (CCR) are central to the standard approach to the quantization of continuous systems, be they motional degrees of freedom of non-relativistic particles ("first" quantization) or bosonic quantum fields ("second" quantization). Given a finite set of degrees of freedom represented by pairs of self-adjoint canonical operators $\hat{x}_j$ and $\hat{p}_j$, for $j = 1, \ldots, n$, the CCR read

$$[\hat{x}_j, \hat{p}_k] = i\delta_{jk}\hbar . \tag{3.1}$$

It is now convenient to introduce a real, canonical, anti-symmetric form Ω (also known as the 'symplectic form', for reasons that will become clear in the following), given by the direct sum of identical 2×2 blocks:

$$\Omega = \bigoplus_{j=1}^{n} \Omega_1 , \quad \text{with} \quad \Omega_1 = \begin{pmatrix} 0 & 1 \\ -1 & 0 \end{pmatrix} . \tag{3.2}$$

Note that the identity operator should be understood on the right-hand side of Eq. (3.1): the commutators of pairs of canonical operators are proportional to "*c*-numbers",[2] and can hence be represented by a complex-valued, rather than operator-valued, matrix $i\Omega$, as follows.

Canonical Commutation Relations (CCR). By defining the vector of canonical operators $\hat{\mathbf{r}} = (\hat{x}_1, \hat{p}_1 \ldots \hat{x}_n, \hat{p}_n)^\mathsf{T}$, Eq. (3.1) can be expediently recast as the following geometric, label-free, expression

$$[\hat{\mathbf{r}}, \hat{\mathbf{r}}^\mathsf{T}] = i\Omega , \tag{3.3}$$

where the commutator of row and column vectors of operators should be taken as an outer product (see Section 1.3.3).

Henceforth, we shall set $\hbar = 1$ and only reinstate it explicitly in dealing

---

[2] A fancy word for the identity operator times a complex number.

with practical cases. Borrowing from the optical and field-theoretical termi-
nologies, canonical degrees of freedom are also referred to as 'modes'.

Throughout the book, we will adopt the expedient convention whereby the
symbol Ω without a label will stand for the anti-symmetric form of the right
dimension given the matrix it multiplies to the left or right. When beneficial
for clarity, we shall instead use the labelled symbol Ω_k to indicate the $2k \times 2k$-
dimensional symplectic form of k modes. This arrangement will come in handy
when dealing with composite systems, comprising subsystems with possibly
different numbers of modes. Notice also that $\Omega = -\Omega^\mathsf{T}$ and $\Omega^2 = -\mathbb{1}_{2n}$, where
$\mathbb{1}_{2n}$ is the $2n \times 2n$ identity matrix. Also, Ω is a real orthogonal transformation:
$\Omega^\mathsf{T}\Omega = -\Omega^2 = \mathbb{1}_{2n}$.

Re-ordering the canonical operators as $\hat{\mathbf{s}} = (\hat{x}_1, \ldots \hat{x}_n, \hat{p}_1, \ldots, \hat{p}_n)^\mathsf{T}$ (which
will be occasionally referred to as 'xp-order') yields the following equivalent
expression of the CCR:

$$[\hat{\mathbf{s}}, \hat{\mathbf{s}}^\mathsf{T}] = iJ, \quad \text{with} \quad J = \begin{pmatrix} 0_n & \mathbb{1}_n \\ -\mathbb{1}_n & 0_n \end{pmatrix}, \tag{3.4}$$

where $\mathbb{1}_n$ and 0_n are, respectively, the $n \times n$ identity and null matrices and
J is the symplectic form in the re-ordered basis. In the interest of clarity, we
will denote phase space vectors with $\mathbf{s}$, instead of $\mathbf{r}$, whenever this choice of
basis is adopted in the book.

Another relevant, equivalent form to express the CCR is given by consid-
ering bosonic annihilation and creation operators $\hat{a}_j$ and $\hat{a}_j^\dagger$, defined as

$$\hat{a}_j = \frac{\hat{x}_j + i\hat{p}_j}{\sqrt{2}}. \tag{3.5}$$

It is easy to see that the vector of annihilation and creation operators $\hat{\mathbf{a}} = (\hat{a}_1, \hat{a}_1^\dagger, \ldots, \hat{a}_n, \hat{a}_n^\dagger)^\mathsf{T}$ is related to $\hat{\mathbf{r}}$ by the unitary transformation $\bar{U}_n$, given by

$$\bar{U} = \bigoplus_{j=1}^n \bar{u}, \quad \text{with} \quad \bar{u} = \frac{1}{\sqrt{2}} \begin{pmatrix} 1 & i \\ 1 & -i \end{pmatrix}, \tag{3.6}$$

so that the CCR may be equivalently recast as

$$[\hat{\mathbf{a}}, \hat{\mathbf{a}}^\dagger] = [\bar{U}\hat{\mathbf{r}}, \hat{\mathbf{r}}^\dagger \bar{U}^\dagger] = \bar{U}[\hat{\mathbf{r}}, \hat{\mathbf{r}}^\dagger]\bar{U}^\dagger = i\bar{U}\Omega\bar{U}^\dagger = \Sigma = \bigoplus_{j=1}^n \sigma_z, \tag{3.7}$$

where σ_z is defined as the standard z Pauli matrix:

$$\sigma_z = \begin{pmatrix} 1 & 0 \\ 0 & -1 \end{pmatrix}. \tag{3.8}$$

Note that the adjoint of a vector of operators has been implicitly defined
as the vector obtained by transposing the original one and conjugating each

of its operator entries, e.g., $\hat{\boldsymbol{a}}^\dagger = (\hat{a}_1^\dagger, \hat{a}_1, \ldots, \hat{a}_n^\dagger, \hat{a}_n)$. Also, $\hat{\mathbf{r}}^\mathsf{T} = \hat{\mathbf{r}}^\dagger$, since all of its entries are Hermitian operators.

Often, and especially in the mathematical physics literature, the CCR are expressed by exponentiating the canonical operators, which has the advantage of making the operators involved bounded:[3]

$$e^{i(\hat{x}_j - \hat{p}_k)} = e^{i\hat{x}_j} e^{-i\hat{p}_k} e^{-\frac{i}{2}\delta_{jk}} = e^{-i\hat{p}_k} e^{i\hat{x}_j} e^{\frac{i}{2}\delta_{jk}} . \tag{3.9}$$

The final phase factors in the previous equation imply the non-commutativity of position and momentum shifts, and are a typical signature of quantum mechanics. Eq. (3.9) may be generalised to consider arbitrary shift operators, also known as Weyl operators in the case of the CCR algebra:

$$\begin{aligned}
e^{i(\mathbf{r}_1 + \mathbf{r}_2)^\mathsf{T} \Omega \hat{\mathbf{r}}} &= e^{i\mathbf{r}_1^\mathsf{T} \Omega \hat{\mathbf{r}}} e^{i\mathbf{r}_2^\mathsf{T} \Omega \hat{\mathbf{r}}} e^{[\mathbf{r}_1^\mathsf{T} \Omega \hat{\mathbf{r}}, \mathbf{r}_2^\mathsf{T} \Omega \hat{\mathbf{r}}]/2} = e^{i\mathbf{r}_1^\mathsf{T} \Omega \hat{\mathbf{r}}} e^{i\mathbf{r}_2^\mathsf{T} \Omega \hat{\mathbf{r}}} e^{-[\mathbf{r}_1^\mathsf{T} \Omega \hat{\mathbf{r}}, \hat{\mathbf{r}}^\mathsf{T} \Omega \mathbf{r}_2]/2} \\
&= e^{i\mathbf{r}_1^\mathsf{T} \Omega \hat{\mathbf{r}}} e^{i\mathbf{r}_2^\mathsf{T} \Omega \hat{\mathbf{r}}} e^{-\mathbf{r}_1^\mathsf{T} \Omega [\hat{\mathbf{r}}, \hat{\mathbf{r}}^\mathsf{T}] \Omega \mathbf{r}_2/2} = e^{i\mathbf{r}_1^\mathsf{T} \Omega \hat{\mathbf{r}}} e^{i\mathbf{r}_2^\mathsf{T} \Omega \hat{\mathbf{r}}} e^{-i\mathbf{r}_1^\mathsf{T} \Omega^3 \mathbf{r}_2/2} \qquad (3.10) \\
&= e^{i\mathbf{r}_1^\mathsf{T} \Omega \hat{\mathbf{r}}} e^{i\mathbf{r}_2^\mathsf{T} \Omega \hat{\mathbf{r}}} e^{i\mathbf{r}_1^\mathsf{T} \Omega \mathbf{r}_2/2} = e^{i\mathbf{r}_2^\mathsf{T} \Omega \hat{\mathbf{r}}} e^{i\mathbf{r}_1^\mathsf{T} \Omega \hat{\mathbf{r}}} e^{-i\mathbf{r}_1^\mathsf{T} \Omega \mathbf{r}_2/2} , \quad \forall \, \mathbf{r}_1, \mathbf{r}_2 \in \mathbb{R}^{2n} ,
\end{aligned}$$

where we have used the anti-symmetry of Ω and the property $\Omega^2 = -\mathbb{1}$. Therefore, one has:

Composition of Weyl operators. Let $\hat{D}_\mathbf{r} = e^{i\mathbf{r}^\mathsf{T} \Omega \hat{\mathbf{r}}}$, then

$$\hat{D}_{\mathbf{r}_1 + \mathbf{r}_2} = \hat{D}_{\mathbf{r}_1} \hat{D}_{\mathbf{r}_2} e^{i\mathbf{r}_1^\mathsf{T} \Omega \mathbf{r}_2/2} . \tag{3.11}$$

This equation, central to the construction of the general formalism of quantum optics, will find frequent application throughout the book.

3.2 QUADRATIC HAMILTONIANS AND GAUSSIAN STATES

The object of this section is to introduce the set of Gaussian states of a continuous variable system of n degrees of freedom, and establish their general parametrisation. This will be done in terms of second-order Hamiltonians, whose analysis will also naturally lead to the diagonal form of a generic Gaussian state.

Throughout this book we refer, somewhat loosely, to a 'second-order' Hamiltonian as a Hamiltonian which can be expressed as a polynomial of order two in the canonical operators. In terms of the vector of operators $\hat{\mathbf{r}}$

---

[3]The equivalence between this expression and Eq. (3.1) is a straightforward consequence of the following well-known corollary of the Baker–Campbell–Hausdorff formula:

$$e^{\hat{A} + \hat{B}} = e^{\hat{A}} e^{\hat{B}} e^{-[\hat{A}, \hat{B}]/2} ,$$

which holds whenever $[\hat{A}, \hat{B}]$ is *central*, that is, when it commutes with both $\hat{A}$ and $\hat{B}$ so that $\hat{A}$, $\hat{B}$ and their commutator form a closed algebra.

defined above, the most general second-order Hamiltonian operator $\hat{H}$ reads, up to an irrelevant additive constant:

$$\hat{H} = \frac{1}{2}\hat{\mathbf{r}}^{\mathsf{T}} H \hat{\mathbf{r}} + \hat{\mathbf{r}}^{\mathsf{T}} \mathbf{r} \, , \tag{3.12}$$

where $\mathbf{r}$ is a $2n$-dimensional real vector and H a symmetric matrix, known as the Hamiltonian matrix and not to be confused with the Hamiltonian operator $\hat{H}$.[4] The matrix H can be assumed to be symmetric since any anti-symmetric component in it would just add a term proportional to the identity operator, because of the CCR, and would thus amount to adding a constant to the Hamiltonian. Note that the set of second-order Hamiltonians, as defined above, include also all strictly quadratic Hamiltonians, where the linear term $\mathbf{r}$ is set to zero, as well as all linear Hamiltonians, where the quadratic term H is set to zero. The convenience of the factor $\frac{1}{2}$ in Eq. (3.12) will become clear shortly.

The modelling of quantum dynamics through second-order Hamiltonians is very common when higher-order terms are inconspicuous and negligible, as is often the case for quantum light fields. Besides, second-order Hamiltonians represent a consistent approximation in other situations of great interest for experiments, such as ion traps, optomechanical systems, nanomechanical oscillators, and several other systems which will be reviewed in Chapter 9. Systems governed by such Hamiltonians are also commonly referred to as 'harmonic'. Up to interactions, the 'free', local Hamiltonian of a quantum oscillator, $\hat{x}^2 + \hat{p}^2$ in rescaled units, is obviously quadratic.

The diagonalisation of any second-order Hamiltonian is a rather straightforward mathematical routine. Because, as we shall see, such a diagonalisation rests on identifying degrees of freedom that are decoupled from each other, systems governed by second-order Hamiltonians are referred to as "quasi-free" in the quantum field theory literature. Notwithstanding the ease with which their dynamics is solved, such systems still offer, as already mentioned above, a very rich scenario for quantum information theory, where the standard methods used for the analysis of second-order Hamiltonians prove to be powerful tools.

Let us now define the set of Gaussian states as *all the ground and thermal states of second-order Hamiltonians with positive definite Hamiltonian matrix* $H > 0$. The restriction to positive definite Hamiltonian matrices corresponds

---

[4] In certain strands of the mathematical physics literature, a Hamiltonian matrix is actually one which may be written as ΩH, whose significance in this context will become clear in the following. We shall not adhere to such a terminology.

to considering 'stable' systems – *i.e.*, Hamiltonian operators bounded from below – and makes our definition consistent.[5]

Definition of a Gaussian state. Any Gaussian state ϱ_G may be written as

$$\varrho_G = \frac{e^{-\beta \hat{H}}}{\mathrm{Tr}\left[e^{-\beta \hat{H}}\right]} , \qquad (3.13)$$

with $\beta \in \mathbb{R}^+$ and $\hat{H}$ defined by $\mathbf{r}$ and H as in Eq. (3.12), including the limiting instance

$$\varrho_G = \lim_{\beta \to \infty} \frac{e^{-\beta \hat{H}}}{\mathrm{Tr}\left[e^{-\beta \hat{H}}\right]} . \qquad (3.14)$$

Clearly, all states of the form (3.13) for finite β are by construction mixed states, while all pure Gaussian states are described by Eq. (3.14).

By the definition above, Gaussian states have been parametrized through the Hamiltonian matrix H, the vector $\mathbf{r}$, whose meaning will become clear shortly, and the parameter β, which is intended to mimic a notation well established in thermodynamics where it would indicate the inverse temperature (up to the Boltzmann constant).

3.2.1 Displacement operators

The overarching aim of this and the following sections is to obtain the diagonal form of the states (3.13) and (3.14). First off, notice that, upon re-defining $\bar{\mathbf{r}} = -H^{-1}\mathbf{r}$ (always possible, as any positive definite H is invertible), and up to an irrelevant constant, the Hamiltonian $\hat{H}$ is equivalent to

$$\hat{H}' = \frac{1}{2}(\hat{\mathbf{r}} - \bar{\mathbf{r}})^\mathsf{T} H (\hat{\mathbf{r}} - \bar{\mathbf{r}}) , \qquad (3.15)$$

which is an alternative form of the most general second-order Hamiltonian with positive definite H, where the real vector $\bar{\mathbf{r}}$ merely shifts the vector of operators $\hat{\mathbf{r}}$. We can prove that this shift is equivalent to the action of a unitary operator by considering the action of a Weyl operator, introduced in

---

[5]This is easily proven, since the spectra of canonical operators comprise the whole real line, so that it is clear from Eq. (3.12) that, if H were not positive, one would have at least a canonical operator $\hat{q}$ (associated with a phase space direction $\mathbf{q}$ such that $\mathbf{q}^\mathsf{T} H \mathbf{q} \not> 0$) with improper eigenvectors $\{|q\rangle, q \in \mathbb{R}\}$ such that the function $\langle q|\hat{H}|q\rangle$ is unbounded from below, contradicting the stability of $\hat{H}$. Conversely, a Hamiltonian matrix $H > 0$ always implies that $\hat{H}$ is bounded from below.

Eq. (3.11), on the vector of operators $\hat{\mathbf{r}}$. We intend to prove the following relation:

Action of Weyl operators on canonical operators. Let $\hat{D}_{\bar{\mathbf{r}}} = e^{i\bar{\mathbf{r}}^\mathsf{T}\Omega\hat{\mathbf{r}}}$, then

$$\hat{D}_{-\bar{\mathbf{r}}}\,\hat{\mathbf{r}}\,\hat{D}_{\bar{\mathbf{r}}} = \hat{\mathbf{r}} - \bar{\mathbf{r}}\,, \tag{3.16}$$

where it is understood that the same Weyl operator acts on all entries of the vector $\hat{\mathbf{r}}$. Note that $\hat{D}_{-\bar{\mathbf{r}}} = \hat{D}_{\bar{\mathbf{r}}}^\dagger$.

Proof. To this aim, let us define the vector of operators $\hat{\mathbf{f}}(\bar{\mathbf{r}}) = e^{-i\bar{\mathbf{r}}^\mathsf{T}\Omega\hat{\mathbf{r}}}\hat{\mathbf{r}}e^{i\bar{\mathbf{r}}^\mathsf{T}\Omega\hat{\mathbf{r}}}$, for which one has $\hat{\mathbf{f}}(0) = \hat{\mathbf{r}}$, as well as

$$\partial_{r'_j}\hat{f}_k\Big|_{\bar{\mathbf{r}}=0} = \left(\partial_{r'_j}e^{-i\sum_{lm}r'_l\Omega_{lm}\hat{r}_m}\hat{r}_k e^{i\sum_{st}r'_s\Omega_{st}\hat{r}_t}\right)\Big|_{\bar{\mathbf{r}}=0}$$

$$= -i\sum_m \Omega_{jm}[\hat{r}_m,\hat{r}_k] = \sum_m \Omega_{jm}\Omega_{mk} = -\delta_{jk}\,,$$

while all the higher-order derivatives of $\hat{f}_j$ are obviously zero. Hence, all the derivatives in zero of the smooth operator-valued function $\hat{\mathbf{f}}(\bar{\mathbf{r}})$ coincide with the derivatives of the function $\hat{\mathbf{r}} - \bar{\mathbf{r}}$, which proves Eq. (3.16).[6] □

Because of Eq. (3.16), the Weyl operators are also known as shift or displacement operators, typically in the quantum optics literature.

Inserting Eq. (3.16) into (3.15) yields

$$\hat{H}' = \frac{1}{2}(\hat{\mathbf{r}} - \bar{\mathbf{r}})^\mathsf{T}H(\hat{\mathbf{r}} - \bar{\mathbf{r}}) = \frac{1}{2}\hat{D}_{-\bar{\mathbf{r}}}\hat{\mathbf{r}}^\mathsf{T}H\hat{\mathbf{r}}\hat{D}_{\bar{\mathbf{r}}} \tag{3.17}$$

(observe that, in the notation in use here, the quadratic form H only represents a set of coefficients with no operatorial content, such that operators may slide through it as in the right-hand side of Eq. (3.17) above). Up to first-order displacement operators, one can hence set the vector $\bar{\mathbf{r}}$ in the second-order Hamiltonian $\hat{H}'$ to zero. The effect of the purely quadratic part $\frac{1}{2}\hat{\mathbf{r}}^\mathsf{T}H\hat{\mathbf{r}}$ can be understood by considering the transformations it induces on the vector of operators $\hat{\mathbf{r}}$ in the Heisenberg picture, which will be the focus of the next section.

---

[6]The same conclusion could have been reached by applying the well-known Baker–Campbell–Hausdorff relationship

$$e^{\hat{X}}\hat{Y}e^{-\hat{X}} = \hat{Y} + [\hat{X},\hat{Y}] + \frac{1}{2!}[\hat{X},[\hat{X},\hat{Y}]] + \frac{1}{3!}[\hat{X},[\hat{X},[\hat{X},\hat{Y}]]] + \cdots\,.$$

3.2.2 The symplectic group of linear canonical transformations

Let us now consider the Heisenberg evolution of the vector of operators $\hat{\mathbf{r}}$ under the dynamics governed by the Hamiltonian $\hat{H} = \frac{1}{2}\hat{\mathbf{r}}^{\mathsf{T}}H\hat{\mathbf{r}}$. One has

$$
\dot{\hat{r}}_j = \frac{i}{2}[\hat{H}, \hat{r}_j] = \frac{i}{2}\sum_{kl}[\hat{r}_k H_{kl}\hat{r}_l, \hat{r}_j]
$$

$$
= \frac{i}{2}\sum_{kl}H_{kl}\left(\hat{r}_k[\hat{r}_l, \hat{r}_j] + [\hat{r}_k, \hat{r}_j]\hat{r}_l\right) = \sum_{kl}\Omega_{jk}H_{kl}\hat{r}_l , \tag{3.18}
$$

which can be recast in vector form as

$$
\dot{\hat{\mathbf{r}}} = \Omega H \hat{\mathbf{r}} . \tag{3.19}
$$

The solution to the differential equation (3.19) is straightforward and given by $\hat{\mathbf{r}}(t) = e^{\Omega H t}\hat{\mathbf{r}}(0)$. Since it represents the action of a unitary operation, the transformation $e^{\Omega H t}$ must preserve the CCR when applied on the vector $\hat{\mathbf{r}}$, that is

$$
i\Omega = [\hat{\mathbf{r}}, \hat{\mathbf{r}}^{\mathsf{T}}] = [e^{\Omega H t}\hat{\mathbf{r}}, \hat{\mathbf{r}}^{\mathsf{T}}(e^{\Omega H t})^{\mathsf{T}}] = e^{\Omega H t}[\hat{\mathbf{r}}, \hat{\mathbf{r}}^{\mathsf{T}}](e^{\Omega H t})^{\mathsf{T}} = ie^{\Omega H t}\Omega(e^{\Omega H t})^{\mathsf{T}} . \tag{3.20}
$$

The transformation $e^{\Omega H t}$ must hence preserve the canonical anti-symmetric form Ω when acting by congruence.[7] This can be restated by claiming that $e^{\Omega H t}$ belongs to the group of linear canonical transformations, well-known from classical Hamiltonian mechanics. This group is also known as the real symplectic group in dimension $2n$, denoted by $Sp_{2n,\mathbb{R}}$ (let us remind the reader that H and Ω are $2n \times 2n$ matrices). The quadratic form Ω, which encodes the commutation relations in our formalism, is also known as the symplectic form, and the symplectic group is defined as the set of transformations that preserve Ω under congruence:

$$
S \in Sp_{2n,\mathbb{R}} \quad \Leftrightarrow \quad S\Omega S^{\mathsf{T}} = \Omega . \tag{3.21}
$$

The group character of such a set is ascertained by noting that its elements must be invertible because their determinant cannot be zero by Binet's theorem, and that $S^{-1}\Omega S^{-1\mathsf{T}} = \Omega$ (the inclusion of the identity matrix and of any product of two elements are obvious).

Since Eq. (3.19) is the Heisenberg equation for the operator vector $\hat{\mathbf{r}}$ under the quadratic Hamiltonian $\frac{1}{2}\hat{\mathbf{r}}^{\mathsf{T}}H\hat{\mathbf{r}}$, its solution must be given by the action of the unitary operator generated by such a Hamiltonian on $\hat{\mathbf{r}}$, that is

$$
e^{\frac{i}{2}\hat{\mathbf{r}}^{\mathsf{T}}H\hat{\mathbf{r}}}\hat{\mathbf{r}}e^{-\frac{i}{2}\hat{\mathbf{r}}^{\mathsf{T}}H\hat{\mathbf{r}}} = e^{\Omega H t}\hat{\mathbf{r}} . \tag{3.22}
$$

---

[7]We will refer to the matrix A acting "by congruence" on the quadratic form B to indicate the transformation $B \mapsto ABA^{\mathsf{T}}$. The invertible matrix A will instead be said to act "by similarity" on B when it transforms it according to $B \mapsto ABA^{-1}$.

It is then expedient to introduce the shorthand notation $\hat{S}_H$ for operators with purely quadratic generators:

$$\hat{S}_H = e^{\frac{i}{2}\hat{r}^{\mathsf{T}} H \hat{r}}, \tag{3.23}$$

such that Eq. (3.22) can be recast compactly as follows:

Action of quadratic Hamiltonians on canonical operators. Let $\hat{S}_H = e^{i\frac{1}{2}\hat{r}^{\mathsf{T}} H \hat{r}}$, then

$$\hat{S}_H \hat{r} \hat{S}_H^\dagger = S_H \hat{r}, \tag{3.24}$$

where $S_H = e^{\Omega H} \in Sp_{2n,\mathbb{R}}$.[8]

Notice that, in Eq. (3.24), the time variable t has just been absorbed in the symmetric matrix H. The relationship (3.24) will be used extensively throughout the book.

The prominent role of symplectic transformations in characterising Gaussian states will become clear in the next section. The more mathematically oriented reader may find more detailed information about the properties of the symplectic group, especially in regard to decompositions which prove particularly useful when dealing with problems in physics, in Appendix B.

3.2.3 Normal modes

The normal mode decomposition, whereby a positive definite quadratic form is split into 'decoupled' degrees of freedom, is instrumental in diagonalising second-order Hamiltonians, and will represent one of the methodological cornerstones of the material covered in this book. This technique, well established since the early days of classical mechanics, can be summarised in the following statement:

Normal mode decomposition. Given a $2n \times 2n$ positive definite real matrix M, there exists a symplectic transformation $S \in Sp_{2n,\mathbb{R}}$ such that

$$SMS^{\mathsf{T}} = D \quad \text{with} \quad D = \mathrm{diag}(d_1, d_1, \ldots, d_n, d_n), \tag{3.25}$$

with $d_j \in \mathbb{R}^+ \ \forall \ j \in [1, \ldots, n]$.

---

[8] This equation may be seen as a consequence of the fact that the correspondence between symplectic transformations and unitary operators generated by purely second-order Hamiltonians constitutes a projective representation of $Sp_{2n,\mathbb{R}}$, that is a representation up to transformation-dependent phase factors. It turns out that a mapping can be defined where such phase factors are always either -1 or $+1$: Technically, one thus achieves a faithful representation of the *metaplectic* group, a double cover of the symplectic group whose exact definition would go beyond the scope of this book. It may however be useful to add that this construction is, somewhat loosely, also referred to as the "metaplectic" representation in the literature.

Proof. Since M is invertible and with strictly positive eigenvalues, a set of real matrices S satisfying Eq. (3.25) may be constructed as $S = D^{1/2}OM^{-1/2}$, for all $O \in O(2n)$. We have to show that a choice of the orthogonal transformation O exists such that this matrix is symplectic, which is equivalent to

$$D^{1/2}OM^{-1/2}\Omega M^{-1/2}O^{\mathsf{T}}D^{1/2} = \Omega \,, \tag{3.26}$$

where we have made use of the symmetry of M and D. Now, the matrix $\Omega' = M^{-1/2}\Omega M^{-1/2}$ is clearly anti-symmetric, and has full rank (since Ω and $M^{-1/2}$ have both full rank: $M^{-1/2}\Omega M^{-1/2}\mathbf{v} = 0 \Leftrightarrow \mathbf{v} = 0$, this just corresponds to stating that $\mathrm{Det}\Omega' = \mathrm{Det}\Omega/\mathrm{Det}M \neq 0$) and for any $2n \times 2n$ real anti-symmetric matrix there exists an orthogonal transformation $O \in O(2n)$ which puts it in a decoupled canonical form:

$$O\Omega'O^{\mathsf{T}} = \bigoplus_{j=1}^{n} d_j^{-1}\Omega_1 \,, \tag{3.27}$$

where Ω_1 is the 2×2 antisymmetric block defined in Eq. (3.2) and $d_j \in \mathbb{R}$ $\forall\, j \in [1, \ldots, n]$.[9] Note that the quantities $\{d_j\}$ are certainly different from zero because Ω' is full rank. Hence, one has:

$$D^{1/2}OM^{-1/2}\Omega M^{-1/2}O^{\mathsf{T}}D^{1/2} = D^{1/2}O\Omega'O^{\mathsf{T}}D^{1/2} = \bigoplus_{j=1}^{n} d_j d_j^{-1}\Omega_1 = \Omega\,,$$
$$\tag{3.28}$$

where D has been set to $\mathrm{diag}(d_1, d_1, \ldots, d_n, d_n)$, as anticipated in Eq. (3.25), thus proving the theorem. Note that the quantities d_j must be strictly positive because M is strictly positive. □

The normal mode decomposition above is often referred to as the 'Williamson's theorem', although it is only a consequence of the theorem, in the sense of being a specific instance thereof. Williamson's general organisation of all quadratic Hamiltonians was accomplished surprisingly recently, in the twentieth century, while the notion of normal mode decomposition,

---

[9] The canonical decomposition of anti-symmetric matrices follows from the diagonalisability of symmetric ones: let A be a real, anti-symmetric, $2n \times 2n$ matrix (even dimension is imposed to fix ideas and because it applies to our case), then A^2 is symmetric and can be diagonalised as per $OA^2O^{\mathsf{T}} = B$, with B diagonal and $O \in O(2n)$. Consider then a generic eigenvector $\mathbf{e}_1$ of A^2, with eigenvalue $d_1^2 \in \mathbb{R}$. The vector $\mathbf{e}'_1 = A\mathbf{e}_1/d_1$ is clearly orthogonal to $\mathbf{e}_1$, because A is antisymmetric: $\mathbf{e}_1^{\mathsf{T}}A\mathbf{e}_1 = 0$. Let $\mathbf{v}$ be a generic vector in the linear subspace orthogonal to the space spanned by $\mathbf{e}_1$ and $\mathbf{e}'_1$, then one has

$$\mathbf{v}^{\mathsf{T}}A\mathbf{e}_1 = \mathbf{v}^{\mathsf{T}}\mathbf{e}'_1 d_1 = 0 \quad \text{and} \quad \mathbf{v}^{\mathsf{T}}A\mathbf{e}'_1 = \mathbf{v}^{\mathsf{T}}A^2\mathbf{e}_1/d_1 = 0 \,,$$

as $\mathbf{e}_1$ is an eigenvalue of A^2 by hypothesis. The equation above shows that any choice of orthogonal basis including $\mathbf{e}_1$ and $\mathbf{e}'_1$ would result in A acting as a diagonal block $d_1\Omega_1$ in the subspace spanned by $\mathbf{e}_1$ and $\mathbf{e}'_1$. Iterating this argument leads to the canonical decomposition of Eq. (3.27).

which Huygens and Galileo were probably already aware of, was developed mathematically by Lagrange.

The symplectic transformation S that turns a matrix M into its normal form is determined by the linear transformation L that diagonalises the matrix $i\Omega M$ (where the conventional factor i is included because the eigenvalues of ΩM are purely imaginary). This can be seen by taking the converse of the normal mode decomposition (3.25): $M = S^{-1}DS^{\mathsf{T}-1}$, and noticing that

$$i\Omega M = i\Omega S^{-1}DS^{\mathsf{T}-1} = iS^{\mathsf{T}}\Omega DS^{\mathsf{T}-1} = S^{\mathsf{T}}\bar{U}^{\dagger}\left(\bigoplus_{j=1}^{n} d_j\sigma_z\right)\bar{U}S^{\mathsf{T}-1}, \quad (3.29)$$

where $\bar{U} = \bigoplus_{j=1}^{n}\bar{u}$, and

$$\bar{u} = \frac{1}{\sqrt{2}}\begin{pmatrix} 1 & i \\ 1 & -i \end{pmatrix} \quad (3.30)$$

is the transformation that diagonalises Ω_1: $\bar{u}\Omega_1\bar{u}^{\dagger} = -i\sigma_z$ (note that this is the very same unitary transformation that appeared in Eq. (3.7) to relate ladder and canonical operators). Eq. (3.29) shows that $i\Omega M$ is always diagonalisable for positive definite M and that, if L is the matrix that diagonalises $i\Omega M$ by similarity (such that $L\Omega ML^{-1}$ is diagonal), then one has $S = \bar{U}^{\dagger}L^{\dagger-1}$ (where we accounted for the fact that transposing a real matrix is the same as conjugating it) for the symplectic transformation S that decomposes M in normal modes by congruence.

Eq. (3.29) also implies that the n quantities $\{d_j, j \in [1, \ldots, n]\}$ are the absolute values of the eigenvalues of $i\Omega M$ (which come in pairs of equal modulus and opposite sign). The d_j's are referred to as the symplectic eigenvalues of the positive definite matrix M, while the normal mode decomposition is also known as "symplectic diagonalisation".

Note that, since it preserves the CCR, a symplectic transformation acting on the vector of canonical operators corresponds to defining new canonical degrees of freedom which are linearly related to the original ones. As we shall see in detail in the next section, the normal mode decomposition of a $2n \times 2n$ positive definite second-order Hamiltonian defines a new set of degrees of freedom which are dynamically decoupled under such a Hamiltonian, and oscillate like free harmonic oscillators. In this instance, the n symplectic eigenvalues correspond to the frequencies of the normal modes (also known as "eigenfrequencies" or "normal" frequencies). It is perhaps worthwhile remarking that the normal mode decomposition derived above applies to *all* stable second-order Hamiltonians, and is thus much more general with respect to the standard, orthogonal diagonalisation of a symmetric potential matrix V one may be used to. That the latter is just a particular case of the statement above may be seen by adopting the re-ordering of the vector of operators $\hat{\mathbf{s}} = (\hat{x}_1, \ldots, \hat{x}_n, \hat{p}_1, \ldots, \hat{p}_n)^{\mathsf{T}}$, with symplectic form J as in Eq. (3.4). In this basis, any matrix $S = R \oplus R$ with $R \in O(n)$ (the same rotation acting on the

x's and the p's) is symplectic, as it is easy to show that it preserves J. Now, if the Hamiltonian matrix is assumed to be $V \oplus \mathbb{1}$, where the identity represents the kinetic energy contribution of momenta (which is the case only if *all degrees of freedom are associated with the same mass*), then the symplectic $R \oplus R$, where R is the orthogonal transformation that diagonalises V, defines a new set of decoupled canonical coordinates (as R diagonalises both V and $\mathbb{1}$), and thus allows one to solve the problem. This is usually how the dynamics of coupled harmonic oscillators is dealt with in elementary treatments. Notice however that the new variables, as defined above, will typically not be the normal modes, as the redefined x_j and p_j contributions to the Hamiltonian, while decoupled, will be unbalanced. An explicit example of a normal mode decomposition is given later in Problem 3.4.

Let us emphasise once again that the normal mode decomposition only holds for positive definite matrices, in that a real symplectic S doing the job could only be constructed as above for such matrices.[10] Non-definite cases, notably including the free particle Hamiltonian $\hat{p}^2$, are the actual object of the Williamson theorem, which classifies the symplectic reduction of all possible quadratic Hamiltonians, including pathological cases where ΩH is not diagonalisable. Notwithstanding their mathematical interest, these instances will be disregarded here.

> **Problem** 3.1. (*A pathological Hamiltonian*). Show that the single-particle Hamiltonian $2\hat{p}^2$ does not admit a normal mode decomposition.
>
> *Solution.* The Hamiltonian operator $2\hat{p}^2$ corresponds to the single-mode Hamiltonian matrix $H_f = \mathrm{diag}(0,1)$, whence
>
> $$i\Omega H_f = i \begin{pmatrix} 0 & 1 \\ 0 & 0 \end{pmatrix}, \tag{3.31}$$
>
> which is an upper triangular Jordan block, and cannot be diagonalised. Therefore, the symplectic eigenvalues of H do not exist.

3.2.4 Normal mode decomposition of a Gaussian state

Let us now go back to our definition of the set of Gaussian states given by Eqs. (3.12), (3.13) and (3.14), in terms of a positive definite Hamiltonian matrix H, a vector of displacements $\mathbf{r}$ and an inverse temperature β. We have already seen how to reduce the vector parameter $\mathbf{r}$ to the action of a unitary Weyl operator. The normal mode decomposition puts us in a position to analyse the role of H as well. In fact, because of the theorem proven in the

---

[10] Trivially, the decomposition extends to negative definite matrices too, which are equivalent to positive definite up to a minus sign.

previous section, one can write

$$H = S^{(H)\mathsf{T}} \left(\bigoplus_{j=1}^{n} \omega_j \mathbb{1}_2 \right) S^{(H)} \quad \text{for} \quad S^{(H)} \in Sp_{2n,\mathbb{R}} , \qquad (3.32)$$

where $S^{(H)}$ is the transpose and inverse of the (not necessarily unique) symplectic transformation that puts H in normal form acting by congruence, which always exists because $H > 0$ by hypothesis, and the ω_j's are the symplectic eigenvalues of H (the frequencies of its normal modes). Eq. (3.32) leads to

$$\hat{\mathbf{r}}^{\mathsf{T}} H \hat{\mathbf{r}} = \hat{\mathbf{r}}^{\mathsf{T}} S^{(H)\mathsf{T}} \left(\bigoplus_{j=1}^{n} \omega_j \mathbb{1}_2 \right) S^{(H)} \hat{\mathbf{r}} . \qquad (3.33)$$

But our introduction of the symplectic group was based upon the equivalence between the action of a symplectic transformation on the vector of canonical operators, as in $S^{(H)}\hat{\mathbf{r}}$, and the action of a unitary operator generated by a second-order Hamiltonian. In point of fact, any symplectic transformation S may be decomposed into the product of two matrix exponentials.[11] Therefore, a pair of symmetric matrices A and B must exist such that

$$S^{(H)}\hat{\mathbf{r}} = \hat{S}_A \hat{S}_B \hat{\mathbf{r}} \hat{S}_B^\dagger \hat{S}_A^\dagger , \qquad (3.34)$$

where the operators $\hat{S}_A$ and $\hat{S}_B$, defined as per Eq. (3.23), act by similarity on each entry of $\hat{\mathbf{r}}$. The matrices A and B are not necessarily unique, and satisfy the equality $S^{(H)} = e^{\Omega A} e^{\Omega B}$, where $S^{(H)}$ was defined above as the inverse and transpose of the symplectic transformation that brings H into normal form. Although in practice it may not be easy to determine A and B analytically, such matrices always exist for any given Hamiltonian matrix H. Notice that, in the interest of rigour, we were forced to introduce two symplectic generators, ΩA and ΩB, because there exist symplectic transformations which require the product of two matrix exponentials to be expressed. However, we can now simplify our notation by defining the unitary transformation $\hat{S} = \hat{S}_A \hat{S}_B$. Inserting Eq. (3.34) into (3.33) yields

$$\hat{\mathbf{r}}^{\mathsf{T}} H \hat{\mathbf{r}} = \hat{S} \hat{\mathbf{r}}^{\mathsf{T}} \left(\bigoplus_{j=1}^{n} \omega_j \mathbb{1}_2 \right) \hat{\mathbf{r}} \hat{S}^\dagger = \hat{S} \left(\sum_{j=1}^{n} \omega_j \left(\hat{x}_j^2 + \hat{p}_j^2 \right) \right) \hat{S}^\dagger . \qquad (3.35)$$

We have thus shown that *every second-order Hamiltonian with zero displacement and positive definite Hamiltonian matrix is unitarily equivalent to the*

---

[11] Any element of a *compact* Lie group – e.g., the unitary group in any finite dimension – may be written as a single matrix exponential. This is equivalent to stating that the exponential map from the algebra of generators into the group is surjective for such groups. However, the symplectic group $Sp_{2n,\mathbb{R}}$ is not compact, and it turns out that some of its elements may only be obtained as the product of two exponentials of generators. We will not elaborate further on this issue, which is rather technical and would call for an extensive and detailed mathematical analysis of the properties of $Sp_{2n,\mathbb{R}}$.

Hamiltonian of a set of free, non-interacting harmonic oscillators. From now on, let us denote the free Hamiltonian of mode j with frequency ω_j by the shorthand notation $\hat{H}_{\omega_j}$:

$$\hat{H}_{\omega_j} = \frac{\omega_j}{2}\left(\hat{x}_j^2 + \hat{p}_j^2\right) . \qquad (3.36)$$

Putting together Eqs. (3.17) and (3.35), the most general second-order Hamiltonian $\hat{H}$ of Eq. (3.15) with positive definite Hamiltonian matrix H may be recast in the form:

$$\hat{H} = \frac{1}{2}(\hat{\mathbf{r}} - \bar{\mathbf{r}})^{\mathsf{T}} H(\hat{\mathbf{r}} - \bar{\mathbf{r}}) = \hat{D}_{-\bar{\mathbf{r}}}\hat{S}\left(\sum_{j=1}^{n} \hat{H}_{\omega_j}\right)\hat{S}^{\dagger}\hat{D}_{\bar{\mathbf{r}}}, \qquad (3.37)$$

$H = e^{\Omega A}e^{\Omega B} \bigoplus_{j=1}^{n} \omega_j \mathbb{1}_2 e^{-B\Omega}e^{-A\Omega}$, where $\{\omega_j, j \in [1, \ldots, n]\}$ is the set of (doubly-degenerate) eigenvalues of $|i\Omega H|$.[12]

The same unitary equivalence carries over to Gaussian states, whose operator form ϱ_G, as defined in Eqs. (3.13) and (3.14), only depends on the second-order Hamiltonian $\hat{H}$. Hence, we have derived the most general expression for a Gaussian state ϱ_G as:

$$\varrho_G = \hat{D}_{-\bar{\mathbf{r}}}\hat{S}\frac{\left(\bigotimes_{j=1}^{n} e^{-\beta\hat{H}_{\omega_j}}\right)}{\prod_{j=1}^{n} \mathrm{Tr}\left[e^{-\beta\hat{H}_{\omega_j}}\right]}\hat{S}^{\dagger}\hat{D}_{\bar{\mathbf{r}}}, \quad \beta > 0 . \qquad (3.38)$$

One then just needs to put the free Hamiltonian $\hat{H}_{\omega_j}$ in diagonal form in order to obtain the spectrum of any Gaussian state.

Notice how the direct sum of Hamiltonians Eq. (3.35), in the 'phase space' picture set by the vector of canonical operators $\hat{\mathbf{r}}$, turned into a tensor product in the Hilbert space representation. This trait, due to the linearity of the operations acting on $\hat{\mathbf{r}}$, will be an important ingredient in much of our treatment of Gaussian states and their information properties.

3.2.5 The Fock basis

As we just saw, the diagonal form of the free oscillator's Hamiltonian $\hat{H}_{\omega_j}$ of Eq. (3.36) determines the spectrum of any Gaussian state. Obtaining such a diagonal form is one of the very first notions dealt with in most basic quantum

---

[12]Note that the $2n^2 + n$ real parameters contained in the $2n \times 2n$ real, symmetric matrix H, have been transferred into the symplectic matrix S, which symplectically diagonalises H, plus the set of n symplectic eigenvalues ω_j's. The number of free parameters has not changed though, as the transformation performing the symplectic diagonalisation of any positive definite matrix is in general ambiguous due to the invariance of the Hamiltonian matrix $\mathbb{1}_2$ under local rotations, which are symplectic. Hence, the number of parameters is still $2n^2 + n$ (a generic symmetric matrix) minus n (due to the invariance that was just mentioned) plus n (the number of symplectic eigenvalues), which is consistent with the previous counting.

mechanics courses. Nonetheless, it is so relevant to our theoretical framework that we will take the opportunity to concisely recall it here.

The Hamiltonian $\hat{H}_{\omega_j}$ may be recast as $\hat{H}_{\omega_j} = \omega_j \left(\hat{a}_j^\dagger \hat{a}_j + \frac{1}{2} \right)$ in terms of the annihilation and creation operators defined in Eq. (3.5). Because of the CCR, it can be easily shown that if $|\lambda\rangle_j$ is an eigenstate of $\hat{H}_{\omega_j}$ with generic eigenvalue $\omega_j \lambda$, then $\hat{a}_j |\lambda\rangle_j$ is an eigenvector too, with eigenvalue $\omega_j (\lambda - 1)$. But since the spectrum of $\hat{H}_{\omega_j}$ must be bounded from below, because it is manifestly a positive operator,[13] this implies that a state $|0\rangle_j$ must exist such that $\hat{a}_j |0\rangle_j = 0$. Such a state is referred to as the vacuum state, and it is the ground state of $\hat{H}_{\omega_j}$, with eigenvalue $1/2$. It is then easy to show that all other eigenvectors of $\hat{H}_{\omega_j}$ may be obtained by m repeated applications of the creation operator $\hat{a}_j^\dagger$ on $|0\rangle_j$, and that they have eigenvalues $\omega_j \left(m + \frac{1}{2} \right)$, with $m \in \mathbb{N}$. The normalised eigenstates of $\hat{H}_{\omega_j}$ are known as Fock, or number states, and will be denoted with $\{|m\rangle_j, m \in \mathbb{N}\}$. The operator $\hat{a}_j^\dagger \hat{a}_j$ is known as the number operator. Let us summarise the action of creation and annihilation operators in the Fock basis:

$$\hat{a}_j |m\rangle_j = \sqrt{m} |m - 1\rangle_j \,, \tag{3.39}$$

$$\hat{a}_j^\dagger |m\rangle_j = \sqrt{m + 1} |m + 1\rangle_j \,, \tag{3.40}$$

$$\hat{a}_j^\dagger \hat{a}_j |m\rangle_j = m |m\rangle_j \,. \tag{3.41}$$

For a bosonic quantum field, in the second quantization picture, the number state $|m\rangle_j$ represents the presence of m particles (excitations of the field) in mode j.

We can now make use of Eq. (3.38) to determine the spectrum of a generic Gaussian state ϱ_G and express it in the Fock basis. The normalisation factor is promptly evaluated:

$$\frac{1}{\text{Tr}\left[e^{-\beta \hat{H}_{\omega_j}} \right]} = \frac{1}{e^{-\frac{\beta \omega_j}{2}} \sum_{m=0}^\infty e^{-\beta \omega_j m}} = e^{\frac{\beta \omega_j}{2}} - e^{-\frac{\beta \omega_j}{2}}, \tag{3.42}$$

so that

$$\varrho_G(\beta) = \left(\prod_{j=1}^n (1 - e^{-\beta \omega_j}) \right) \hat{D}_{\bar{\mathbf{r}}}^\dagger \hat{S} \left(\bigotimes_{j=1}^n \left(\sum_{m=0}^\infty e^{-\beta \omega_j m} |m\rangle_{jj}\langle m| \right) \right) \hat{S}^\dagger \hat{D}_{\bar{\mathbf{r}}} \,. \tag{3.43}$$

The limit of pure states, $\beta \mapsto \infty$, is particularly simple and instructive in this representation:

$$\lim_{\beta \to \infty} \varrho_G(\beta) = \hat{D}_{\bar{\mathbf{r}}}^\dagger \hat{S} |0\rangle \langle 0| \hat{S}^\dagger \hat{D}_{\bar{\mathbf{r}}} \,, \tag{3.44}$$

---

[13] Any operator that can be written as $\hat{o}^\dagger \hat{o}$, for a linear operator $\hat{o}$, is necessarily positive semi-definite.

where the shorthand notation $|0\rangle = \bigotimes_{j=1}^{n} |0\rangle_j$ has been introduced to represent the vacuum of the whole field. *All pure Gaussian states are obtained by applying unitary operations generated by second-order Hamiltonians on the vacuum state.*

Note that the positive parameter β is technically redundant, as it might have been absorbed in the Hamiltonian matrix H and, in particular, in its normal frequencies ω_j, as apparent from Eq. (3.43). We have preferred to render it explicit in our treatment because it allows for a very clear definition of the set of pure Gaussian states, and because it relates our formalism to a physical interpretation: the Gaussian state with parameters H, $\bar{\mathbf{r}}$ and β is the equilibrium state of a system with local second-order Hamiltonian $\hat{H}$ after thermalisation with a reservoir at rescaled temperature $1/\beta$. In the light of this notation, it should be particularly evident that the quest for ground state cooling in a harmonic potential is essentially the quest for the preparation of a pure Gaussian state. Another remarkable consequence of Eq. (3.43) is that Gaussian states have either rank 1, if they are pure, or infinite. Given the stringent restriction on the form of its spectrum, no Gaussian state with finite rank greater than 1 may exist.

In the next section, we will move on to yet another equivalent parametrisation of a generic Gaussian state, and a more direct rendition of the spectrum of a Gaussian state in terms of observable parameters will be derived.

3.3 STATISTICAL MOMENTS OF A GAUSSIAN STATE AND THE COVARIANCE MATRIX

We intend now to derive another, more expedient, parametrisation of a Gaussian state, which we shall adopt for the remainder of the book. Let us then drop the dependence on the second-order Hamiltonian $\hat{H}$ and assume instead the parametrisation of the most general n-mode Gaussian state, given by Eq. (3.43), in terms of a generic symplectic transformation S, represented in the Hilbert space by $\hat{S}$, of an arbitrary displacement operator $\hat{D}_{\bar{\mathbf{r}}}$, and of a set of n strictly positive real numbers ξ_j (each replacing $\beta\omega_j$ above). The limit of pure states may still be taken by sending all the ξ_j to infinity. Then one has, for the most general Gaussian state ϱ_G:

$$\varrho_G = \left(\prod_{j=1}^{n} \left(1 - e^{-\xi_j} \right) \right) \hat{D}_{\bar{\mathbf{r}}}^{\dagger} \hat{S}^{\dagger} \left(\bigotimes_{j=1}^{n} \left(\sum_{m=0}^{\infty} e^{-\xi_j m} |m\rangle_{jj}\langle m| \right) \right) \hat{S} \hat{D}_{\bar{\mathbf{r}}} . \quad (3.45)$$

Also recall that, by virtue of Eqs. (3.16) and (3.34), one has

$$\hat{D}_{\bar{\mathbf{r}}} \hat{\mathbf{r}} \hat{D}_{\bar{\mathbf{r}}}^{\dagger} = \hat{\mathbf{r}} + \bar{\mathbf{r}} , \quad (3.46)$$

$$\hat{S} \hat{\mathbf{r}} \hat{S}^{\dagger} = S\hat{\mathbf{r}} , \quad (3.47)$$

which express the fact that Weyl operators and unitary operators generated by Hamiltonians of order two projectively represent, respectively, the abelian group of translations in dimension $2n$ and the real symplectic group. Notice that we used $\hat{D}_{-\bar{\mathbf{r}}} = \hat{D}_{\bar{\mathbf{r}}}^{\dagger}$.

Let us now evaluate the expectation value of $\hat{\mathbf{r}}$ for the Gaussian state ϱ_G:

$$\mathrm{Tr}\left[\varrho_G \hat{\mathbf{r}}\right] = \prod_{j=1}^{n}\left(1 - e^{-\xi_j}\right)\mathrm{Tr}\left[\left(\bigotimes_{j=1}^{n}\left(\sum_{m=0}^{\infty}e^{-\xi_j m}|m\rangle_{jj}\langle m|\right)\right)\hat{S}\hat{D}_{\bar{\mathbf{r}}}\hat{\mathbf{r}}\hat{D}_{\bar{\mathbf{r}}}^{\dagger}\hat{S}^{\dagger}\right]$$

$$= \prod_{j=1}^{n}\left(1 - e^{-\xi_j}\right)\mathrm{Tr}\left[\bigotimes_{j=1}^{n}\left(\sum_{m=0}^{\infty}e^{-\xi_j m}|m\rangle_{jj}\langle m|\right)\left(S\hat{\mathbf{r}} + \bar{\mathbf{r}}\right)\right]$$

$$= \prod_{j=1}^{n}\left(1 - e^{-\xi_j}\right)\mathrm{Tr}\left[\bigotimes_{j=1}^{n}\left(\sum_{m=0}^{\infty}e^{-\xi_j m}|m\rangle_{jj}\langle m|\right)\bar{\mathbf{r}}\right] = \bar{\mathbf{r}}, \qquad (3.48)$$

where we used the fact that the expectation value of any linear combination of canonical operators vanishes when calculated on a state which is diagonal in the Fock basis. This can be understood by inspecting Eqs. (3.39) and (3.40), and keeping in mind that a linear combination of $\hat{x}_j$'s and $\hat{p}_j$'s is a linear combination of a_j's and $\hat{a}_j^{\dagger}$'s. The vector parameter $\bar{\mathbf{r}}$ is then just the vector of expectation values of the canonical operators on the state ϱ_G, which could be determined by performing measurements of positions and momenta on non-relativistic particles described by such operators. This will also be referred to as the vector of first (statistical) moments.

Let us then move on to consider second statistical moments of canonical operators on our state. In particular, let us consider the second moments in their symmetrised version, which we shall group together in the 'covariance matrix' $\boldsymbol{\sigma}$, defined below in terms of the outer product notation of Section 1.3.3:

$$\boldsymbol{\sigma} = \mathrm{Tr}\left[\left\{(\hat{\mathbf{r}} - \bar{\mathbf{r}}), (\hat{\mathbf{r}} - \bar{\mathbf{r}})^{\mathsf{T}}\right\}\varrho_G\right] = \prod_{j=1}^{n}\left(1 - e^{-\xi_j}\right)$$

$$\times \mathrm{Tr}\left[\left(\bigotimes_{j=1}^{n}\left(\sum_{m=0}^{\infty}e^{-\xi_j m}|m\rangle_{jj}\langle m|\right)\right)\hat{S}\hat{D}_{\bar{\mathbf{r}}}\left\{(\hat{\mathbf{r}} - \bar{\mathbf{r}}), (\hat{\mathbf{r}} - \bar{\mathbf{r}})^{\mathsf{T}}\right\}\hat{D}_{\bar{\mathbf{r}}}^{\dagger}\hat{S}^{\dagger}\right]$$

$$= \prod_{j=1}^{n}\left(1 - e^{-\xi_j}\right)\mathrm{Tr}\left[\left(\bigotimes_{j=1}^{n}\left(\sum_{m=0}^{\infty}e^{-\xi_j m}|m\rangle_{jj}\langle m|\right)\right)\hat{S}\left\{\hat{\mathbf{r}}, \hat{\mathbf{r}}^{\mathsf{T}}\right\}\hat{S}^{\dagger}\right]$$

$$= \prod_{j=1}^{n}\left(1 - e^{-\xi_j}\right)\mathrm{Tr}\left[\left(\bigotimes_{j=1}^{n}\left(\sum_{m=0}^{\infty}e^{-\xi_j m}|m\rangle_{jj}\langle m|\right)\right)\left\{S\hat{\mathbf{r}}, \hat{\mathbf{r}}^{\mathsf{T}}S^{\mathsf{T}}\right\}\right]$$

$$= \prod_{j=1}^{n}\left(1 - e^{-\xi_j}\right)S\,\mathrm{Tr}\left[\left(\bigotimes_{j=1}^{n}\left(\sum_{m=0}^{\infty}e^{-\xi_j m}|m\rangle_{jj}\langle m|\right)\right)\left\{\hat{\mathbf{r}}, \hat{\mathbf{r}}^{\mathsf{T}}\right\}\right]S^{\mathsf{T}}.$$

$$(3.49)$$

Thanks to the use we made of group representations, we were able to bring the unitary operators outside the evaluation of the expectation value, which is now reduced to determining the trace in the last expression above. That is a rather straightforward task. First of all, notice that any expectation value involving two canonical operators pertaining to different modes is zero, because the state left inside the trace is diagonal in the Fock basis (this is the same argument by which we showed that the expectation values of linear functions of canonical operators are zero for such states). We are left with the task of evaluating the expectation values of the operators

$$2\hat{x}_j^2 = 2\hat{a}_j^\dagger \hat{a}_j + 1 + \hat{a}_j^2 + \hat{a}_j^{\dagger 2} \,, \tag{3.50}$$

$$2\hat{p}_j^2 = 2\hat{a}_j^\dagger \hat{a}_j + 1 - \hat{a}_j^2 - \hat{a}_j^{\dagger 2} \,, \tag{3.51}$$

$$\hat{x}_j \hat{p}_j + \hat{p}_j \hat{x}_j = i(\hat{a}_j^{\dagger 2} - \hat{a}_j^2) \,, \tag{3.52}$$

which have been expressed as functions of creation and annihilation operators, for the local state $\sum_{m=0}^{\infty} e^{-\xi_j m} |m\rangle_{jj}\langle m|$. Only terms with the same number of $\hat{a}_j$ and $\hat{a}_j^\dagger$ contribute to the expectation value, because the state we are considering is diagonal in the Fock basis. The only operator that does contribute, besides the identity, is the number operator $\hat{a}_j^\dagger \hat{a}_j$, for which one finds

$$\langle \hat{a}_j^\dagger \hat{a}_j \rangle = \left(1 - e^{-\xi_j}\right) \mathrm{Tr} \left[\sum_{m=0}^{\infty} e^{-\xi_j m} |m\rangle_{jj}\langle m| \hat{a}_j^\dagger \hat{a}_j \right] \tag{3.53}$$

$$= \left(1 - e^{-\xi_j}\right) \sum_{m=0}^{\infty} e^{-\xi_j m} m = \frac{e^{-\xi_j}}{1 - e^{-\xi_j}} \,, \tag{3.54}$$

so that

$$2\langle \hat{x}_j^2 \rangle = 2\langle \hat{p}_j^2 \rangle = \frac{2e^{-\xi_j}}{1 - e^{-\xi_j}} + 1 = \frac{1 + e^{-\xi_j}}{1 - e^{-\xi_j}} \,. \tag{3.55}$$

Hence, for the expectation values of the anti-commutators entering Eq. (3.49), one has

$$\prod_{j=1}^{n} \left(1 - e^{-\xi_j}\right) \mathrm{Tr} \left[\left(\bigotimes_{j=1}^{n} \left(\sum_{m=0}^{\infty} e^{-\xi_j m} |m\rangle_{jj}\langle m| \right) \right) \{\hat{\mathbf{r}}, \hat{\mathbf{r}}^\mathsf{T}\} \right] = \bigoplus_{j=1}^{n} \frac{1 + e^{-\xi_j}}{1 - e^{-\xi_j}} \mathbb{1}_2 \,. \tag{3.56}$$

Upon defining

$$\nu_j = \frac{1 + e^{-\xi_j}}{1 - e^{-\xi_j}} \geq 1 \,, \tag{3.57}$$

we can write the matrix of the second statistical moments of canonical operators of any Gaussian state as follows:

Covariance matrix of a generic Gaussian state. The covariance matrix (CM) of the most general Gaussian state can be written as

$$\boldsymbol{\sigma} = S \left(\bigoplus_{j=1}^{n} \nu_j \mathbb{1}_2 \right) S^{\mathsf{T}} , \qquad (3.58)$$

with $\nu_j \geq 1$ [see Eq. (3.57)] and $S \in Sp_{2n,\mathbb{R}}$.

Eq. (3.58), formally analogous to (3.25), is the normal mode decomposition of the CM $\boldsymbol{\sigma}$, with ν_j as its symplectic eigenvalues which, let us remind the reader, are nothing but the absolute values of the eigenvalues of $i\Omega\boldsymbol{\sigma}$, which come in pairs of opposite sign $\mp\nu_j$.

By definition, the covariance matrix $\boldsymbol{\sigma}$ and the vector of first moments $\bar{\mathbf{r}}$ are directly accessible in practice as the first and second statistical moments of measurements of the canonical operators. Because the covariance matrix contains all the information about S and the symplectic eigenvalues $\{\nu_j\}$, our analysis can be summarised by the following characterisation of a Gaussian state ϱ_G:

Parametrisation of a generic Gaussian state. A Gaussian state ϱ_G is completely determined by its vector of first moments $\bar{\mathbf{r}}$ and covariance matrix $\boldsymbol{\sigma}$, defined as

$$\bar{\mathbf{r}} = \mathrm{Tr}[\varrho_G \hat{\mathbf{r}}] \quad \text{and} \quad \boldsymbol{\sigma} = \mathrm{Tr}\big[\varrho_G \{(\hat{\mathbf{r}} - \bar{\mathbf{r}}), (\hat{\mathbf{r}} - \bar{\mathbf{r}})^{\mathsf{T}}\}\big] . \qquad (3.59)$$

The conditions that the covariance matrix $\boldsymbol{\sigma}$ of a physical state must fulfil will be determined in the next section. This parametrisation is in obvious analogy with a classical Gaussian probability distribution, which is also completely determined by first- and second-order moments. In situations where first moments are irrelevant (which will be explicitly pointed out), we will let covariance matrices alone represent Gaussian states.

The spectrum of the most general Gaussian state ϱ_G can also be characterised:

Spectrum of a generic Gaussian state. Let ν_j be the n symplectic eigenvalues of the n-mode Gaussian state ϱ_G, then one has

$$\varrho_G = \hat{D}_{\bar{\mathbf{r}}}^{\dagger} \hat{S}^{\dagger} \left(\bigotimes_{j=1}^{n} \left(\sum_{m=0}^{\infty} \frac{2}{\nu_j + 1} \left(\frac{\nu_j - 1}{\nu_j + 1} \right)^m |m\rangle_{jj}\langle m| \right) \right) \hat{S} \hat{D}_{\bar{\mathbf{r}}} . \qquad (3.60)$$

Observe that the quadratic unitary $\hat{S}$ acts on the diagonal form of the state in Eq. (3.60) as $\hat{S}^\dagger \cdot \hat{S}$, while the corresponding symplectic transformation would act on the covariance matrix of the diagonal state as $S \cdot S^\mathsf{T}$. The conjugation of $\hat{S}$ occurring here is analogous to the conjugation that relates the Schrödinger and Heisenberg pictures, reflected in the sign difference between Eq. (2.13) and Eq. (2.14). This issue is illustrated explicitly in a specific case in the problem that follows.

Problem 3.2. (*Squeezed state in the Fock basis*). Prove that the single-mode, Gaussian 'squeezed state' $|z\rangle\langle z|$, with covariance matrix $\mathrm{diag}(z, 1/z)$ for $z > 0$ and zero first moments, may be written in the Fock basis as

$$|z\rangle = \sqrt{\frac{2\sqrt{z}}{z+1}} \sum_{m=0}^{\infty} \left(\frac{z-1}{2(z+1)}\right)^m \frac{\sqrt{(2m)!}}{m!} |2m\rangle . \qquad (3.61)$$

Solution. Let $z = e^{2r}$, so that the expression above may be recast as

$$|z\rangle = \frac{1}{\sqrt{\cosh(r)}} \sum_{m=0}^{\infty} \left(\frac{\tanh(r)}{2}\right)^m \frac{\sqrt{(2m)!}}{m!} |2m\rangle . \qquad (3.62)$$

Notice now that the covariance matrix $\mathrm{diag}(e^{2r}, e^{-2r})$ is the result of acting on the vacuum with the symplectic transformation $\mathrm{diag}(e^r, e^{-r})$ (also known as 'single-mode squeezing'). Clearly, $\mathrm{diag}(e^r, e^{-r}) = e^{r\sigma_z} = e^{r\Omega_1\sigma_x}$, in terms of the Pauli matrices σ_z and σ_x. We have thus determined the quadratic Hamiltonian $\hat{H}_{sq}$ that generates the squeezing operator: $\hat{H}_{sq} = \frac{1}{2}\hat{\mathbf{r}}^\mathsf{T}\sigma_x\hat{\mathbf{r}} = \hat{x}\hat{p} = \frac{1}{2i}(\hat{a}^2 - \hat{a}^{\dagger 2})$ (see Chapter 5 for a taxonomy of quadratic Hamiltonian generators). The state $|z\rangle$ may thus also be written as

$$|z\rangle = e^{-ir\hat{H}_{sq}} = e^{\frac{r}{2}(\hat{a}^{\dagger 2} - \hat{a}^2)}|0\rangle , \qquad (3.63)$$

where the minus sign at the exponent reflects the conjugate $\hat{S}^\dagger$ acting on states from the left as per Eq. (3.60). In order to prove (3.62), we shall now assume it as an *ansatz*, and show that it is consistent with the derivative $\partial_r|z\rangle$ evaluated from Eq. (3.63), which we know to hold. Since the equality for the derivative will hold for all r, this argument will prove the identity. One has

$$\sqrt{\cosh(r)}\partial_r|z\rangle = \left(\frac{\hat{a}^{\dagger 2} - \hat{a}^2}{2}\right) \sum_{m=0}^{\infty} \left(\frac{\tanh(r)}{2}\right)^m \frac{\sqrt{(2m)!}}{m!} |2m\rangle$$

$$= \sum_{m=0}^{\infty} \left(\frac{\tanh(r)}{2}\right)^{m-1} \frac{\sqrt{(2m)!}}{m!} \left[\frac{2m - \tanh(r)^2(2m+1)}{4}\right] |2m\rangle , \qquad (3.64)$$

which is indeed the same as

$$\sqrt{\cosh(r)}\partial_r|z\rangle = \sqrt{\cosh(r)}\,\partial_r\left[\frac{1}{\sqrt{\cosh(r)}}\sum_{m=0}^{\infty}\left(\frac{\tanh(r)}{2}\right)^m\frac{\sqrt{(2m)!}}{m!}|2m\rangle\right]$$

$$= \sum_{m=0}^{\infty}\left(\frac{\tanh(r)}{2}\right)^{m-1}\frac{\sqrt{(2m)!}}{m!}\left[\frac{m}{2\cosh(r)^2} - \frac{\tanh(r)^2}{4}\right]|2m\rangle.$$

$$(3.65)$$

Note that r might also be taken complex, which corresponds to rotating the optical phase of the squeezing (*i.e.*, to letting the squeezed state go through a phase shifter, which will be introduced in Chapter 5).

Problem 3.3. (*Null eigenstate of the canonical operators*). Prove that, in the limits $z \to 0$ and $z \to \infty$, the state $|z\rangle$ of the previous problem tends to an eigenstate of $\hat{x}$ and $\hat{p}$ associated with the eigenvalue 0.

Solution. Up to normalisation, in the limit $z \to 0$, $|z\rangle$ tends to the unnormalised vector $\sum_{m=0}^{\infty}\left(-\frac{1}{2}\right)^m\frac{\sqrt{(2m)!}}{m!}|2m\rangle$. Applying $\sqrt{2}\hat{x} = (\hat{a} + \hat{a}^\dagger)$ on such a Fock basis expansion yields

$$\sum_{m=0}^{\infty}\left(-\frac{1}{2}\right)^m\left[-\frac{\sqrt{(2m+2)!}}{2(m+1)!}\sqrt{2m+2} + \frac{\sqrt{(2m)!}}{m!}\sqrt{2m+1}\right]|2m+1\rangle = 0.$$

$$(3.66)$$

An analogous cancellation takes place, for $\sqrt{2}i\hat{p} = (\hat{a} - \hat{a}^\dagger)$, in the limit $z \to \infty$, where the state tends to $\sum_{m=0}^{\infty}\left(\frac{1}{2}\right)^m\frac{\sqrt{(2m)!}}{m!}|2m\rangle$.

Problem 3.4. (*Translationally invariant harmonic chain*). In order to demonstrate the technique of symplectic diagonalisation in the context of solving a dynamical model, and as a minimalist foray into the complex world of many-body physics, let us consider here the simple example of a closed, translationally invariant harmonic chain with nearest neighbour interactions. In properly rescaled units, the Hamiltonian operator of n such oscillators is given by:

$$\hat{H} = \sum_{j=1}^{n}\left(\omega\hat{x}_j^2 + \hat{p}_j^2 + 2\chi\hat{x}_{j-1}\hat{x}_j\right), \qquad (3.67)$$

where j labels the oscillators on the ring, and it is understood that $\hat{x}_0 = \hat{x}_n$ (which entails exact translational invariance on a closed ring, without boundary effects). Show that the condition $\omega > 2\chi > 0$ ensures that $\hat{H}$ is bounded from below, and determine the spectrum of $\hat{H}$ under that condition.

Solution. The Hamiltonian $\hat{H}$ is obviously quadratic and may be rewritten, using vector of operators $\hat{s} = (\hat{x}_1, \ldots, \hat{x}_n, \hat{p}_1, \ldots, \hat{p}_n)^\mathsf{T}$ – it is convenient here to reorder the operators with the symplectic form J as in Eq. (3.4), as

$$\hat{H} = \hat{s}^\mathsf{T}(H_0 + \chi V)\hat{s} \,, \tag{3.68}$$

for

$$H_0 = \begin{pmatrix} \omega \mathbb{1}_n & 0 \\ 0 & \mathbb{1}_n \end{pmatrix}, \quad V = \begin{pmatrix} 0 & 1 & 0 & \cdots & 0 & 1 \\ 1 & 0 & 1 & 0 & \cdots & 0 \\ 0 & 1 & 0 & \ddots & 0 & \vdots \\ \vdots & 0 & \ddots & \ddots & \ddots & 0 \\ 0 & \vdots & 0 & \ddots & 0 & 1 \\ 1 & 0 & \cdots & 0 & 1 & 0 \end{pmatrix} \oplus 0,$$

$$\tag{3.69}$$

where 0 stands for the $n \times n$ null matrix. As already noted in the main text, under this ordering, any matrix $S = R \oplus R$ with $R \in O(n)$ (the same rotation acting on the x's and the p's) is symplectic, since it is straightforward to show that it preserves the symplectic form J. Moreover, such an S leaves H_0 unchanged. Clearly, one such matrix diagonalises the real and symmetric V. Once the diagonal form of V is determined, the symplectic diagonalisation of the whole Hamiltonian matrix, and hence the diagonalisation of the associated Hamiltonian operator, will be straightforward to achieve.

As was discussed, a quadratic Hamiltonian is bounded from below if and only if its associated Hamiltonian matrix is strictly positive. We will be able to assess this criterion by determining the spectrum of H. In order to determine eigenvalues and eigenvectors of V, notice that such a matrix is translationally invariant, and hence must commute with the shift operator T that describes the translation by one site along the ring:

$$T = \begin{pmatrix} 0 & 0 & \cdots & 0 & 1 \\ 1 & 0 & 0 & \cdots & 0 \\ 0 & 1 & \ddots & 0 & \vdots \\ \vdots & \ddots & \ddots & \ddots & 0 \\ 0 & \cdots & 0 & 1 & 0 \end{pmatrix}. \tag{3.70}$$

The eigenvectors $\{\mathbf{t}_j\}$ and eigenvalues $\{\tau_j\}$ of T are easily verified to be given by the discrete Fourier series vectors:

$$
\mathbf{t}_j = \begin{pmatrix} 1 \\ e^{ij\frac{2\pi}{n}} \\ \vdots \\ e^{ikj\frac{2\pi}{n}} \\ \vdots \\ e^{i(n-1)j\frac{2\pi}{n}} \end{pmatrix} , \quad \tau_j = e^{-ij\frac{2\pi}{n}} \quad \text{for} \quad j \in [0, \dots, n-1]. \quad (3.71)
$$

These must also be eigenvectors of V, with real eigenvalues $\{v_j\}$: in fact, it is plain to see that $V\mathbf{t}_j = 2\cos(j\frac{2\pi}{n})\mathbf{t}_j = v_j\mathbf{t}_j$. Notice that $v_j = v_{n-j}$: such a degeneracy allows one to linearly combine the degenerate pairs of $\mathbf{t}_j$'s in order to select a real, orthonormal eigenbasis for V, which must clearly exist, and which we shall term the 'Fourier modes'. Thus, $V = 2RDR^{\mathsf{T}}$, with $D = \text{diag}\left(1, \cos(\frac{2\pi}{n}), \dots, \cos((n-1)\frac{2\pi}{n})\right)$ and $R \in O(n)$.

The symplectic $S = R \oplus R$ diagonalises the full Hamiltonian matrix, as per $H = S((\omega\mathbb{1} + 2\chi D) \oplus \mathbb{1})S^{\mathsf{T}}$. Hence, the eigenvalues of H are strictly positive if and only if $\omega > -2\chi\cos(j\frac{2\pi}{n})$ for all integers j: for even n, this corresponds to $\omega > 2\chi$; for odd n, $\cos(j\frac{2\pi}{n})$ never equals -1 and a slightly weaker condition depending on n could be imposed. However, the condition $\omega > 2\chi$ is always sufficient for stability. This condition is instrumental in the following, as it ensures the matrix H can be symplectically diagonalised. Let us briefly remark that it may be possible to proceed with the symplectic diagonalisation and then determination of the spectrum along the same lines in more general cases, although the checks and provisos to apply would be rather tedious. We will see a case of this sort in Chapter 8, when discussing teleportation thresholds, where Fourier modes will somewhat unexpectedly come to the rescue.

Inspection of the diagonal form of H also shows that the j-th Fourier mode (i.e., the pair of variables (x_j, p_j) in the new basis), is associated with a diagonal Hamiltonian matrix $\text{diag}(\omega + 2\chi\cos(j\frac{2\pi}{n}), 1)$, whose associated symplectic eigenvalue is just

$$
\nu_j = \sqrt{\omega + 2\chi\cos\left(j\frac{2\pi}{n}\right)} \qquad (3.72)
$$

(the symplectic eigenvalue associated with a quadratic form with a single decoupled mode is just the square of the determinant). Inserting such an expression into Eq. 3.60, yields the full spectrum of $\hat{H}$, given by

$$\lambda_{m_1,\dots,m_n} = \prod_{j=1}^{n} \frac{2}{\nu_j + 1} \left(\frac{\nu_j - 1}{\nu_j + 1} \right)^{m_j}, \quad \text{for} \quad m_j \in \mathbb{N}, \qquad (3.73)$$

with $\{\nu_j\}$ given by Eq. (3.72). The diagonalisation of the harmonic chain's Hamiltonian is thus complete.

3.4 THE UNCERTAINTY PRINCIPLE

Not all symmetric matrices belong to the set of covariances of a quantum state. The non-commutativity of the canonical operators, along with the probabilistic interpretation of the quantum state, impose specific constraints on the variances and covariances of such observables that go under the name of uncertainty principles. We will derive here a geometric uncertainty relation which turns out to be necessary and sufficient for σ to be the covariance matrix of a Gaussian state. In order to set the scene for such an iconic, distinctively quantum, feature, let us mention that the covariance matrix σ of a multivariate, classical probability distribution over continuous variables must satisfy the sole condition $\sigma > 0$. As is well-known, more stringent restrictions apply to quantum states.

Given a – not necessarily Gaussian – quantum state ϱ on a system of n modes, with vector of first moments $\bar{\mathbf{r}}$, let us define the matrix τ as

$$\tau = 2 \text{Tr} \left[\varrho (\hat{\mathbf{r}} - \bar{\mathbf{r}})(\hat{\mathbf{r}} - \bar{\mathbf{r}})^{\mathsf{T}} \right] , \qquad (3.74)$$

where $\hat{\mathbf{o}} \hat{\mathbf{o}}^{\mathsf{T}}$ for the vector of operators $\hat{\mathbf{o}}$ is to be taken in the outer product sense (see Section 1.3.3). One has

$$\tau = \text{Tr} \left[\varrho \{ (\hat{\mathbf{r}} - \bar{\mathbf{r}}), (\hat{\mathbf{r}} - \bar{\mathbf{r}})^{\mathsf{T}} \} + [\hat{\mathbf{r}}, \hat{\mathbf{r}}^{\mathsf{T}}] \right] = \sigma + i\Omega . \qquad (3.75)$$

Now, define the operator $\hat{O} = \sqrt{2} \mathbf{y}^{\dagger} (\hat{\mathbf{r}} - \bar{\mathbf{r}})$ for a generic $\mathbf{y} \in \mathbb{C}^{2n}$. Because of the positivity of the density matrix ϱ, one has

$$0 \le \text{Tr} \left[\varrho \hat{O} \hat{O}^{\dagger} \right] = \mathbf{y}^{\dagger} \tau \mathbf{y} = \mathbf{y}^{\dagger} (\sigma + i\Omega) \mathbf{y}, \quad \forall \, \mathbf{y} \in \mathbb{C}^{2n} . \qquad (3.76)$$

That is

Robertson–Schrödinger uncertainty relation. Let σ be the covariance matrix of a quantum state ϱ. Then

$$\sigma + i\Omega \ge 0 . \qquad (3.77)$$

This relationship is manifestly invariant under symplectic transformations, since $\Omega = S\Omega S^\mathsf{T}$. It can therefore be expressed in terms of the symplectic eigenvalues, which are the n independent symplectic invariant quantities of a $2n$-mode covariance matrix. By writing (3.77) for the normal mode form of σ one gets

$$\nu + i\Omega = \bigoplus_{j=1}^{n} \begin{pmatrix} \nu_j & i \\ -i & \nu_j \end{pmatrix} \geq 0 , \tag{3.78}$$

which is equivalent to

$$\nu_j \geq 1 , \quad j \in [1, \ldots, n] . \tag{3.79}$$

Taking the transposition of the uncertainty relation (3.77) does not affect the positivity of the matrix and gives the equivalent formulation $\sigma - i\Omega \geq 0$ (since $\Omega^\mathsf{T} = -\Omega$).

It may also be shown that (3.77) directly implies that $\sigma > 0$, and hence that any σ pertaining to a quantum state can be symplectically diagonalised, so that (3.79) is completely equivalent to (3.77). The positive semi-definiteness $\sigma \geq 0$ is obvious from (3.77) since $\mathbf{v}^\mathsf{T}\Omega\mathbf{v} = 0$ for all $\mathbf{v} \in \mathbb{R}^{2n}$. The more restrictive condition $\sigma > 0$ can be established by noting that, if a vector $\mathbf{v}_0$ existed such that $\mathbf{v}_0^\mathsf{T}\sigma\mathbf{v}_0 = 0$, then one could always identify a vector $\mathbf{v}_1$ such that $\mathbf{v}_0^\mathsf{T}\Omega\mathbf{v}_1 = b \neq 0$ and define a vector $\mathbf{w} = ix\mathbf{v}_0 + \mathbf{v}_1$ for $x \in \mathbb{R}$ for which

$$\mathbf{w}^\dagger(\sigma + i\Omega)\mathbf{w} = \mathbf{v}_1^\mathsf{T}\sigma\mathbf{v}_1 + 2bx . \tag{3.80}$$

A choice of x then would always exist such that $\mathbf{w}^\dagger(\sigma + i\Omega)\mathbf{w}$ is negative, thus violating the relationship (3.77).

The uncertainty relations above stem directly from the canonical commutation relations and from the positivity of the quantum state ϱ: given the CCR, the inequality (3.77) – or, equivalently, (3.79) – are necessary for the operator ϱ to be positive semi-definite. If the operator ϱ is Gaussian, then such inequalities are not only necessary but also sufficient for its positivity. In fact, in our constructive definition of the set of Gaussian states, the relationship (3.79) arose naturally as a sufficient condition on the symplectic eigenvalues. Hence, we can claim that the uncertainty relation (3.77), which is equivalent to (3.79), is necessary and sufficient for σ to represent the covariance matrix of a Gaussian state.

At times, depending on the problem at hand, it may be more expedient to express symplectically invariant relationships through a set of symplectic invariants other than the symplectic eigenvalues. Recall that the symplectic eigenvalues ν_j's associated with a Gaussian state with covariance matrix σ are, up to a sign, the eigenvalues of $i\Omega\sigma$. Because of the properties of Ω, and of the symmetry of σ, the latter come in pairs of alternate sign, forming the set $\{-\nu_j, +\nu_j, j \in [1, \ldots, n]\}$. Hence, the characteristic polynomial of $i\Omega\sigma$, $C(x)$, satisfies

$$C(x) = \prod_{j=1}^{n} [(x - \nu_j)(x + \nu_j)] = \prod_{j=1}^{n} (x^2 - \nu_j^2) . \tag{3.81}$$

Note that the latest equation implies that the characteristic polynomial $C(x)$ is even (all the coefficients of odd powers of x must vanish). The nonzero coefficients of the characteristic polynomials $C(x)$ form a set of n independent symplectic invariants alternative to $\{\nu_j, j \in [1, \ldots, n]\}$. It is customary in the literature to define such invariants $\{\Delta_{j,n}, j \in [0, \ldots, n]\}$ by setting

$$C(x) = \sum_{j=0}^{n} x^{2j} (-1)^{j+n} \Delta_{j,n} , \qquad (3.82)$$

such that, by definition, $\Delta_{n,n} = 1$ for all n and the independent invariants are n and not $n+1$, as it should be. This yields immediately a necessary condition on the new invariants from the uncertainty relation (3.79). The latter may in fact be inserted in (3.81) to obtain $(-1)^n C(1) \geq 0$, whence

$$\sum_{j=0}^{n} (-1)^j \Delta_{j,n} \geq 0 . \qquad (3.83)$$

Notice that one has $\Delta_{0,n} = \prod_{j=1}^{n} \nu_j^2 = \mathrm{Det}\,\sigma$ for all n.

For $n = 1$, the inequality (3.83) reduces to $\mathrm{Det}\,\sigma \geq 1$ which, along with the additional requirement $\sigma > 0$ is yet another rendition of the Robertson–Schrödinger relation (3.77) for single-mode systems:

Uncertainty relation for single-mode systems. The 2×2 matrix σ is the covariance matrix of a quantum state of a single degree of freedom if and only if

$$\mathrm{Det}\,\sigma \geq 1 \quad \text{and} \quad \sigma > 0 . \qquad (3.84)$$

The independent positivity condition $\sigma > 0$ must always be added since quadratic functions of the symplectic eigenvalues, like the $\Delta_{j,n}$, are clearly the same for σ and its opposite $-\sigma$. Notice that the inequality (3.84) reads $\sigma_{11}\sigma_{22} - \sigma_{12}^2 \geq 1$ and hence implies $\sigma_{11}\sigma_{22} \geq 1$. By setting $\sigma_{11} = 2\Delta\hat{x}^2$ and $\sigma_{22} = 2\Delta\hat{p}^2$, one recovers the standard (Heisenberg's) formulation of the uncertainty principle: $4\Delta\hat{x}^2\Delta\hat{p}^2 \geq 1 = |\langle[\hat{x}, \hat{p}]\rangle|^2$, which is subsumed here as a necessary condition.

It is easy to see from Eqs. (3.81) and (3.82) that $\Delta_{1,n} = \sum_{j=1}^{n} \nu_j^2$ for all n. For two-mode states, $\Delta_{1,n}$ is sometimes referred to as the 'seralian' invariant and denoted with the shorthand notation Δ. A complete characterisation of the uncertainty relation for two-mode states is given by (see Problem 3.5):

Uncertainty relation for two-mode systems. The 4×4 matrix σ is the covariance matrix of a quantum state of two modes if and only if

$$\mathrm{Det}\,\sigma - \Delta + 1 \geq 0 , \qquad (3.85)$$

$$\Delta^2 \geq 4\,\mathrm{Det}\,\sigma , \qquad (3.86)$$

$$\sigma > 0 . \qquad (3.87)$$

We will take up the description of two-mode systems in terms of symplectic invariants again in Chapter 7, where it will turn out to be useful in order to characterise the entanglement of such systems.

Problem 3.5. (*Two-mode uncertainty relation*). Prove that the Robertson–Schrödinger relation for two-mode states is equivalent to the relations (3.85–3.87).

Solution. The two symplectic invariants for a two-mode state, the determinant and the seralian, fulfil $\mathrm{Det}\boldsymbol{\sigma} = \nu_-^2\nu_+^2$ and $\Delta = (\nu_-^2 + \nu_+^2)$ respectively, where the two symplectic eigenvalues have been denoted with ν_- and ν_+. Hence, the symplectic eigenvalues may be expressed in terms of the symplectic invariants as follows:

$$\nu_{\mp}^2 = \frac{\Delta \mp \sqrt{\Delta^2 - 4\mathrm{Det}\boldsymbol{\sigma}}}{2}. \qquad (3.88)$$

Now, Eq. (3.86) is necessary and sufficient for the symplectic eigenvalues to be real. Once reality is established, the condition $\nu_-^2 \geq 1$, on the smaller of the two eigenvalues, ensures that $\nu_{\mp}^2 \geq 1$ holds for both eigenvalues, giving $\Delta - 2 \geq \sqrt{\Delta^2 - 4\mathrm{Det}\boldsymbol{\sigma}} \Rightarrow 4 - 4\Delta \geq -4\mathrm{Det}\boldsymbol{\sigma}$, which is (3.85). Note that we could square the inequality in the last step, since both terms were certainly positive: in fact, inserting (3.86) in (3.85) and inferring $\mathrm{Det}\boldsymbol{\sigma} \geq 0$ from (3.87) yields $\mathrm{Det}\boldsymbol{\sigma} \geq 1$ and hence $\Delta \geq 2$ from (3.86). The positivity condition (3.87) must be added ad hoc, as discussed above in the one-mode instance.

Notice that the inequality (3.85) is the two-mode version of the general uncertainty relation (3.83) introduced above.

3.5 PURITY AND ENTROPIES OF GAUSSIAN STATES

The general diagonalisation (3.60) allows one to evaluate, or at least to give a compact expression for, any unitarily invariant functional of a Gaussian state ϱ_G with covariance matrix $\boldsymbol{\sigma}$. Clearly, all such functions will depend solely on the symplectic eigenvalues of the covariance matrix $\boldsymbol{\sigma}$ obtained, let us remind the reader once more, as the pairwise degenerate eigenvalues of the matrix $|\Omega\boldsymbol{\sigma}|$.

The purity $\mu(\varrho_G) = \mathrm{Tr}\left(\varrho_G^2\right)$, typically the most expedient indicator of mixedness of a quantum state (see Section 2.5), is particularly straightforward to express for an n-mode state. Such a quantity is in fact multiplicative under tensor products, so that (3.60) yields

$$\mathrm{Tr}\left(\varrho_G^2\right) = \prod_{j=1}^{n}\left[\frac{4}{(\nu_j+1)^2}\sum_{m=0}^{\infty}\left(\frac{\nu_j-1}{\nu_j+1}\right)^{2m}\right] = \prod_{j=1}^{n}\frac{1}{\nu_j}. \qquad (3.89)$$

Now, using the normal mode decomposition of $\boldsymbol{\sigma}$ (3.58), and recalling that $\mathrm{Det}S = 1$, as proven in Appendix B, one obtains $\prod_{j=1}^{n} \nu_j^2 = \mathrm{Det}\boldsymbol{\sigma}$, which can be inserted in (3.89) to yield:

Purity of a Gaussian state. One has

$$\mu(\varrho_G) = \mathrm{Tr}\left(\varrho_G^2\right) = \frac{1}{\sqrt{\mathrm{Det}\boldsymbol{\sigma}}}, \qquad (3.90)$$

where $\boldsymbol{\sigma}$ is the covariance matrix of the Gaussian state ϱ_G.

In particular, it follows that *a Gaussian state is pure if and only if its covariance matrix has determinant 1.* Equivalently, a Gaussian state is pure if and only if all the symplectic eigenvalues of its covariance matrix $\boldsymbol{\sigma}$ are equal to 1, which implies that the normal mode decomposition of $\boldsymbol{\sigma}$ is simply $\boldsymbol{\sigma} = SS^\mathsf{T}$ for some $S \in Sp_{2n,\mathbb{R}}$. Yet another equivalent condition for the purity of a Gaussian state is $-\Omega S \Omega S = \mathbb{1}$ (obtained by squaring the matrix $i\Omega S$).

More generally, it is often interesting to determine unitarily invariant norms of a density matrix ϱ, defined in general as the 'Schatten p-norms' $\|\varrho\|_p = (\mathrm{Tr}\,[\varrho^p])^{1/p}$ for $p \geq 1$. It is worth noting that the 2-norm, which determines the purity of the state, is also referred to as the Frobenius or Hilbert–Schmidt norm, while the limits $p \to 1$ and $p \to \infty$ define, respectively, the trace and operator norms. For Gaussian states, all of these quantities may be determined in terms of the symplectic eigenvalues of $\boldsymbol{\sigma}$, through Eq. (3.60).

In turn, the p-norms lead to the definition of generalised entropies, such as the manifestly non-additive Bastiaans–Tsallis entropies $T_p(\varrho) = (1 - \|\varrho\|_p^p)/(p - 1)$ for $p > 1$ (which include the so-called linear entropy $T_2(\varrho) = 1 - \mu(\varrho)$), or the Rényi entropies $S_p(\varrho) = \log_2 \|\varrho\|_p^p/(1 - p)$ for $p > 1$, which are instead all additive when evaluated on tensor product states. It is an interesting exercise to prove that

$$\lim_{p \to 1+} S_p(\varrho) = -\mathrm{Tr}\,[\varrho \log_2 \varrho] = S_V(\varrho), \qquad (3.91)$$

where the S_V is the von Neumann entropy that, as discussed in Section 2.5, quantifies operationally the quantum information content of the state.

The von Neumann entropy of a Gaussian state is given by:

Entropy of a Gaussian state. Let $\boldsymbol{\sigma}$ be the covariance matrix of an n-mode Gaussian state ϱ_G. Then

$$S_V(\varrho_G) = -\mathrm{Tr}[\varrho_G \log_2(\varrho_G)] = \sum_{j=1}^{n} s_V(\nu_j), \qquad (3.92)$$

where $\{\nu_j, j \in [1, \ldots, n]\}$ are the symplectic eigenvalues of $\boldsymbol{\sigma}$ and

$$s_V(x) = \frac{x+1}{2} \log_2\left(\frac{x+1}{2}\right) - \frac{x-1}{2} \log_2\left(\frac{x-1}{2}\right). \qquad (3.93)$$

It is worth emphasising here that single-mode Gaussian states have only one independent unitarily invariant quantity: the symplectic eigenvalue ν_1. All the entropies defined above are increasing functions of $\nu_1 \geq 1$ (and they all vanish for $\nu_1 = 1$). Hence, all such entropies induce the same hierarchy of mixedness on the set of single-mode Gaussian states, a fact which is very specific to this set of states and certainly not true for more general quantum states, not even for multimode Gaussian states.

Problem 3.6. (*Entropy of Gaussian states*). Prove Eq. (3.92).

Solution. By noting that the von Neumann entropy of tensor products of quantum states is additive, and using the diagonal form of the Gaussian state ϱ_G from Eq. (3.60), one can write

$$
\begin{aligned}
S_V(\varrho_G) &= \sum_{j=1}^{n} \left[\log_2\left(\frac{\nu_j+1}{2}\right) - \frac{2\log_2\left(\frac{\nu_j-1}{\nu_j+1}\right)}{\nu_j+1} \sum_{m=0}^{\infty} m\left(\frac{\nu_j-1}{\nu_j+1}\right)^m \right] \\
&= \sum_{j=1}^{n} \left[\log_2\left(\frac{\nu_j+1}{2}\right) - \log_2\left(\frac{\nu_j-1}{\nu_j+1}\right)\left(\frac{\nu_j-1}{2}\right) \right] \\
&= \sum_{j=1}^{n} \left[\left(\frac{\nu_j+1}{2}\right)\log_2\left(\frac{\nu_j+1}{2}\right) - \left(\frac{\nu_j-1}{2}\right)\log_2\left(\frac{\nu_j-1}{2}\right) \right],
\end{aligned}
$$
(3.94)

where we used $\sum_{m=0}^{\infty} mz^m = z/(1-z)^2$ for $|z| < 1$, which may be derived by differentiating the geometric series $\sum_{m=0}^{\infty} z^m = 1/(1-z)$.

Problem 3.7. (*p-norms of Gaussian states*). Show that, for a Gaussian state ϱ_G with covariance matrix $\boldsymbol{\sigma}$, one has $\mathrm{Tr}(\varrho_G^p) = \prod_{j=1}^{n} g_p(\nu_j)$, where the $\{\nu_j\}$ are the n symplectic eigenvalues of $\boldsymbol{\sigma}$ and $g_p = \frac{2^p}{(x+1)^p-(x-1)^p}$. Hence, insert the expression into the definition of the Rényi entropy and take the limit $p \to 1+$ to prove Eq. (3.92).

Solution. The multiplicativity of the p-norm of tensor products and the diagonalisation (3.60) of the state ϱ_G allows one to compute

$$
\mathrm{Tr}(\varrho_G^p) = \prod_{j=1}^{n} \left[\frac{2^p}{(\nu_j+1)^p} \sum_{m=0}^{\infty} \left(\frac{\nu_j-1}{\nu_j+1}\right)^{pm} \right] = \prod_{j=1}^{n} g_p(\nu_j), \quad (3.95)
$$

by just summing the geometric series $\sum_{m=0}^{\infty} z^m = 1/(1-z)$ for $|z| < 1$.

The associated, additive, Rényi entropy S_p is therefore,

$$S_p(\varrho_G) = \sum_{j=1}^{n} \left[\frac{1}{(1-p)} \log_2 \left(\frac{2^p}{(\nu_j + 1)^p - (\nu_j - 1)^p} \right) \right] \qquad (3.96)$$

for $p > 1$. In the limit $p \to 1+$, both numerator and denominator inside the square brackets tend to 0, so that one has to resort to L'Hôpital's rule to get

$$\lim_{p \to 1+} -\log_2(2) + \frac{(\nu_j + 1)^p \log_2(\nu_j + 1) - (\nu_j - 1)^p \log_2(\nu_j - 1)}{[(\nu_j + 1)^p - (\nu_j - 1)^p]} = s_V(\nu_j),$$
$$(3.97)$$

which proves the identity

$$\lim_{p \to 1+} S_p(\varrho_G) = S_V(\varrho_G) . \qquad (3.98)$$

3.6   FURTHER READING

The paper [3] is a very useful resource on the symplectic group, which has served the continuous variable quantum information community since its very early steps. Concerning the group and its application to quantum mechanics, one should also be aware of de Gosson's work, in particular of the textbook [22], which will be of direct relevance to Chapter 5 too.

The compact derivation of the normal mode decomposition reported here is due to Simon, Srinivasan and Chaturvedi [81]. Williamson's original paper [87], is summarised in the notorious, and rather intense, Appendix C to Arnold's classic textbook [2].

Broader mathematical context to the formalism we adopted here is provided by the body of research on harmonic analysis, well captured by the textbook [28], which contains detailed coverage of the metaplectic representation and the related Stone–von Neumann theorem.

Phase space methods

CONTENTS

The language of statistical moments and covariance matrices we introduced in the previous chapter offers a compact and efficient formalism to deal with Gaussian states. A much more general approach may be taken to describe any quantum state in a setting which is reminiscent of classical phase space. This is the framework of quantum characteristic functions and quasi-probability distributions, which goes back to seminal work by Wigner on quantum corrections to classical statistical mechanics, and bloomed in the sixties with the rise of theoretical quantum optics and the emergence of a general unifying picture.

Conceptually, the phase space description of quantum states hinges on the completeness of the set of displacement operators, which we shall prove in the form of the Fourier–Weyl relation between density matrices and characteristic functions. This relationship will constitute the bridge between phase space and Hilbert space descriptions, which will be useful in several applications in the context of quantum information. It will be convenient to handle most of the proofs and mathematical arguments concerning characteristic functions and quasi-probability distributions on a single mode of the system. Because displacement operators of multimode systems are just tensor products of local displacement operators, the extension of the formalism to systems with many degrees of freedom will be straightforward. Nevertheless, we shall always take care of explicitly linking the single-mode formulae which will appear in this chapter to the general multimode description adopted in the previous one.

4.1 COHERENT STATES

The coherent state $|\alpha\rangle$ is the eigenvector of the operator $\hat{a} = (\hat{x} + i\hat{p})/\sqrt{2}$ with eigenvalue α. If one defines $\alpha = (x + ip)/\sqrt{2}$, $\mathbf{r} = (x, p)^\mathsf{T}$, $\hat{\mathbf{r}} = (\hat{x}, \hat{p})^\mathsf{T}$ and the operator $\hat{D}_\alpha$ as

$$\hat{D}_\alpha = \hat{D}_{-\mathbf{r}} = \mathrm{e}^{-i\mathbf{r}^\mathsf{T}\Omega\hat{\mathbf{r}}} = \mathrm{e}^{\alpha\hat{a}^\dagger - \alpha^*\hat{a}}, \tag{4.1}$$

in keeping with the convention adopted in Chapter 3, one finds $\hat{D}_\alpha^\dagger \hat{a} \hat{D}_\alpha = \hat{a} + \alpha$. The eigenstate $|\alpha\rangle$ is then easily determined:

Eigenvectors of the annihilation operator. One has

$$\hat{a}\hat{D}_\alpha|0\rangle = \alpha\hat{D}_\alpha|0\rangle . \tag{4.2}$$

In fact:

$$\hat{a}\hat{D}_\alpha|0\rangle = \hat{D}_\alpha\hat{D}_\alpha^\dagger\hat{a}\hat{D}_\alpha|0\rangle = \hat{D}_\alpha(\hat{a} + \alpha)|0\rangle = \alpha\hat{D}_\alpha|0\rangle . \tag{4.3}$$

The expression of $|\alpha\rangle$ in the Fock basis is also readily established, and reads:

Coherent states in the Fock basis. Let $|\alpha\rangle$ be a coherent state such that $a|\alpha\rangle = \alpha|\alpha\rangle$, then

$$|\alpha\rangle = \mathrm{e}^{-\frac{|\alpha|^2}{2}} \sum_{m=0}^{\infty} \frac{\alpha^m}{\sqrt{m!}}|m\rangle . \tag{4.4}$$

Problem 4.1. (*Coherent states in the Fock basis*). Prove Eq. (4.4).

Solution. Assume $|\alpha\rangle = \sum_{m=0}^{\infty} c_m|m\rangle$ in the Fock basis $\{|m\rangle, m \in \mathbb{N}\}$, then the eigenvalue equation reads

$$\hat{a}|\alpha\rangle = \sum_{m=0}^{\infty} c_m\sqrt{m}|m-1\rangle = \sum_{m=0}^{\infty} c_{m+1}\sqrt{m+1}|m\rangle = \alpha\sum_{m=0}^{\infty} c_m|m\rangle \tag{4.5}$$

which, because of the linear independence of the basis vectors, must be satisfied separately by all coefficients in the sum: $c_{m+1} = \alpha c_m/\sqrt{m+1}$, which is possible if and only if $c_m = c\alpha^m/\sqrt{m!}$ for some normalisation factor c. The latter may be determined by noticing that $\sum_{m=0}^{\infty} \frac{|\alpha|^2}{m!} = \mathrm{e}^{-|\alpha|^2}$, which yields the normalised coherent state of Eq. (4.4).

Two further relationships we will use throughout the chapter are the expression of the CCR for Weyl operators with complex variables

$$\hat{D}_\alpha\hat{D}_\beta = \mathrm{e}^{\frac{1}{2}(\alpha\beta^* - \alpha^*\beta)}\hat{D}_{\alpha+\beta} \tag{4.6}$$

and the overlap between two coherent states:

$$\langle\beta|\alpha\rangle = \langle 0|\hat{D}_{-\beta}\hat{D}_\alpha|0\rangle = \langle 0|\hat{D}_{\alpha-\beta}|0\rangle\,e^{\frac{1}{2}(\alpha\beta^*-\alpha^*\beta)}$$
$$= \langle 0|\alpha-\beta\rangle\,e^{\frac{1}{2}(\alpha\beta^*-\alpha^*\beta)} = e^{-\frac{1}{2}|\alpha-\beta|^2}\,e^{\frac{1}{2}(\alpha\beta^*-\alpha^*\beta)}\,. \tag{4.7}$$

While not orthogonal, the coherent states form a complete set, in that the identity operator can be represented as a weighted integral of projectors on coherent states:

Coherent states resolution of the identity. One has

$$\frac{1}{\pi}\int_{\mathbb{C}}|\alpha\rangle\langle\alpha|\,d^2\alpha = \hat{\mathbb{1}}\,. \tag{4.8}$$

Proof. This property can be verified by using the Fock basis decomposition (4.4):

$$\frac{1}{\pi}\int_{\mathbb{C}}d^2\alpha\,|\alpha\rangle\langle\alpha| = \frac{1}{\pi}\sum_{m,n=0}^{\infty}\int_{\mathbb{C}}d^2\alpha\,\frac{\alpha^m\alpha^{*n}}{\sqrt{m!n!}}\,e^{-|\alpha|^2}|m\rangle\langle n|$$

$$= \frac{1}{\pi}\sum_{m,n=0}^{\infty}\int_0^{\infty}d\rho\int_0^{2\pi}d\varphi\,e^{i(m-n)\varphi}\,\frac{e^{-\rho^2}\rho^{m+n+1}}{\sqrt{m!n!}}|m\rangle\langle n| \tag{4.9}$$

$$= \sum_{m=0}^{\infty}2\int_0^{\infty}d\rho\,\frac{e^{-\rho^2}\rho^{2m+1}}{m!}|m\rangle\langle m| = \sum_{m=0}^{\infty}|m\rangle\langle m| = \hat{\mathbb{1}}\,,$$

where we used $\int_0^{2\pi}e^{i(m-n)\varphi}\,d\varphi = 2\pi\delta_{mn}$ and $\int_0^{\infty}e^{-\rho^2}\rho^{2m+1}\,d\rho = \frac{m!}{2}$. □

Since they are not orthogonal and yet resolve the identity as an integral of one-dimensional projectors, it may be shown that the coherent states form actually an *overcomplete* set, that is a set which is still complete after removal of any one element. Eq. (4.9) implies that the trace of any trace-class operator $\hat{O}$ can be determined as an integral over coherent states:

$$\text{Tr}\left[\hat{O}\right] = \sum_{m=0}^{\infty}\langle m|\hat{O}|m\rangle = \frac{1}{\pi}\int_{\mathbb{C}}d^2\alpha\sum_{m=0}^{\infty}\langle m|\alpha\rangle\langle\alpha|\hat{O}|m\rangle$$

$$= \frac{1}{\pi}\int_{\mathbb{C}}d^2\alpha\sum_{m=0}^{\infty}\langle\alpha|\hat{O}|m\rangle\langle m|\alpha\rangle = \frac{1}{\pi}\int_{\mathbb{C}}d^2\alpha\,\langle\alpha|\hat{O}|\alpha\rangle\,. \tag{4.10}$$

In terms of the language we developed in the previous chapter, it is immediately clear from the definition (4.2) that a coherent state $|\alpha\rangle$ is a Gaussian state with covariance matrix $\boldsymbol{\sigma} = \mathbb{1}$ and first moments $x = \sqrt{2}\,\text{Re}(\alpha)$ and $p = \sqrt{2}\,\text{Im}(\alpha)$ (recall that $\hat{D}_\alpha = \hat{D}_{\mathbf{r}}^\dagger$ with the definitions above for x and p).

4.2   THE FOURIER–WEYL RELATION

We shall now establish a direct connection between density matrices on $L^2(\mathbb{R}^n)$ and functions of $2n$ variables, which stems from the fact that the displacement operators $\hat{D}_{\mathbf{r}}$ form an orthogonal complete set on the space of operators on $L^2(\mathbb{R}^n)$ with respect to the Hilbert–Schmidt scalar product.[1] For simplicity, we will prove the statement explicitly for a single mode, adopting the complex single-mode notation $\hat{D}_\gamma$ for the Weyl operators. As anticipated above, the extension of such a result to the multimode case is straightforward.

Fourier–Weyl relation. Given a bounded operator $\hat{O}$ on the Hilbert space of one bosonic mode, one has

$$\hat{O} = \frac{1}{\pi} \int_{\mathbb{C}} d^2\alpha \, \mathrm{Tr}\left[\hat{D}_\alpha \hat{O}\right] \hat{D}_{-\alpha} . \qquad (4.11)$$

Proof. First note that, due to the decomposition of the identity (4.9), any bounded operator $\hat{O}$ for which $\langle\alpha|\hat{O}|\beta\rangle$ is well defined may be decomposed as follows in terms of coherent states:

$$\hat{O} = \frac{1}{\pi^2} \int_{\mathbb{C}^2} d\alpha \, d\beta \, \langle\alpha|\hat{O}|\beta\rangle \, |\alpha\rangle\langle\beta| , \qquad (4.12)$$

so that one just needs to prove that the operator $|\alpha\rangle\langle\beta|$ can be expanded in terms of displacement operators to extend the proof to all bounded operators. We intend to show that

$$|\alpha\rangle\langle\beta| = \frac{1}{\pi} \int_{\mathbb{C}} d^2\gamma \, \mathrm{Tr}\left[|\alpha\rangle\langle\beta|\hat{D}_\gamma\right] \hat{D}_\gamma^\dagger . \qquad (4.13)$$

Eq. (4.13) is equivalent to

$$
\begin{aligned}
|0\rangle\langle0| &= \frac{1}{\pi} \int_{\mathbb{C}} d^2\gamma \, \mathrm{Tr}\left[|\alpha\rangle\langle\beta|\hat{D}_\gamma\right] \hat{D}_{-\alpha}\hat{D}_{-\gamma}\hat{D}_\beta \\
&= \frac{1}{\pi} \int_{\mathbb{C}} d^2\gamma \, \mathrm{Tr}\left[|\alpha\rangle\langle\beta - \gamma|\right] e^{\frac{1}{2}(\gamma\beta^* - \gamma^*\beta)} \hat{D}_{-\alpha}\hat{D}_{-\gamma}\hat{D}_\beta \\
&= \frac{1}{\pi} \int_{\mathbb{C}} d^2\gamma \, \langle\beta - \gamma|\alpha\rangle e^{\frac{1}{2}(\gamma\beta^* - \gamma^*\beta)} \hat{D}_{-\alpha}\hat{D}_{-\gamma}\hat{D}_\beta \\
&= \frac{1}{\pi} \int_{\mathbb{C}} d^2\gamma \, e^{-\frac{1}{2}|\beta-\alpha-\gamma|^2} \hat{D}_{\beta-\alpha-\gamma} = \frac{1}{\pi} \int_{\mathbb{C}} d^2\gamma \, e^{-\frac{1}{2}|\gamma|^2} \hat{D}_\gamma ,
\end{aligned}
$$

where we used Eqs. (4.6) and (4.7). We are thus left with having to prove

---

[1] The Hilbert–Schmidt scalar product between two operators $\hat{A}$ and $\hat{B}$ is defined as $\mathrm{Tr}(\hat{A}^\dagger \hat{B})$.

the following relationship:

$$|0\rangle\langle 0| = \frac{1}{\pi} \int_{\mathbb{C}} d^2\gamma \, e^{-\frac{1}{2}|\gamma|^2} \hat{D}_\gamma \,. \tag{4.14}$$

In order to do that, let us apply the operator on the right-hand side on the Fock basis vector $|m\rangle$:

$$
\begin{aligned}
\frac{1}{\pi} \int_{\mathbb{C}} d^2\gamma \, e^{-\frac{1}{2}|\gamma|^2} \hat{D}_\gamma \, |m\rangle &= \frac{1}{\pi} \int_{\mathbb{C}} d^2\gamma \, e^{-\frac{1}{2}|\gamma|^2} \hat{D}_\gamma \frac{\hat{a}^{\dagger m}}{\sqrt{m!}} |0\rangle \\
&= \frac{1}{\pi} \int_{\mathbb{C}} d^2\gamma \, e^{-\frac{1}{2}|\gamma|^2} \frac{(\hat{a}^\dagger - \gamma^*)^m}{\sqrt{m!}} |\gamma\rangle \\
&= \int_{\mathbb{C}} \frac{d^2\gamma}{\pi} \, e^{-|\gamma|^2} \frac{(\hat{a}^\dagger - \gamma^*)^m}{\sqrt{m!}} \sum_{n=0}^{\infty} \frac{\gamma^n}{\sqrt{n!}} |n\rangle \\
&= \sum_{n=0}^{\infty} \sum_{j=0}^{m} \int_{\mathbb{C}} \frac{d^2\gamma}{\pi} \, e^{-|\gamma|^2} \binom{m}{j} \frac{(-\gamma)^{*j} \gamma^n}{\sqrt{m!n!}} \hat{a}^{\dagger(m-j)} |n\rangle \\
&= \sum_{j=0}^{m} \binom{m}{j} (-1)^j |m\rangle = \delta_{m0}|0\rangle \,, \tag{4.15}
\end{aligned}
$$

where we inserted the following integral, already employed in (4.9):[2]

$$\frac{1}{\pi} \int_{\mathbb{C}} d^2\gamma \, e^{-|\gamma|^2} \gamma^{*j}\gamma^n = n! \, \delta_{jn} \,. \tag{4.16}$$

Eq. (4.15) is equivalent to (4.14) and hence to (4.13). □

Whilst it seemed convenient to adopt the somewhat generic terminology "Fourier–Weyl relation" for the mathematical result above, it should be noted that, in harmonic analysis, the expansion in the set of Weyl operators is usually referred to as the "Fourier–Weyl transform".

> **Characteristic function.** Given a state ϱ, we are thus led to define $\chi(\alpha) = \mathrm{Tr}[\hat{D}_\alpha \varrho]$ such that
> $$\varrho = \frac{1}{\pi} \int_{\mathbb{C}} d^2\alpha \, \chi(\alpha)\hat{D}_{-\alpha} \,. \tag{4.17}$$

As we will see in the next section, the function $\chi(\alpha)$ is known as the symmetrically ordered characteristic function associated to the quantum state ϱ. Clearly, complete knowledge of $\chi(\alpha)$ provides one with complete information about ϱ.

For a system of n modes, in the notation of the previous chapter, the

---

[2] Also note that, in the very last step of Eq. (4.15), $\sum_{j=0}^{m} \binom{m}{j}(-1)^j = (1-1)^m$ for $m > 0$ (and 1 otherwise).

Fourier–Weyl relation we just derived reads[3]

$$\varrho = \frac{1}{(2\pi)^n} \int_{\mathbb{R}^{2n}} d\mathbf{r}\, \chi(\mathbf{r}) \hat{D}_{\mathbf{r}} , \qquad (4.18)$$

with $d\mathbf{r} = dx_1 dp_1 \ldots dx_n dp_n$ and

$$\chi(\mathbf{r}) = \text{Tr}\left[\hat{D}_{-\mathbf{r}}\varrho\right] . \qquad (4.19)$$

Notice the change of sign due to the convention in defining $\hat{D}_\alpha$ set by Eq. (4.1). The additional factor $(1/2)^n$ is due to the change of measure associated with the change of variables $\alpha = \frac{x + ip}{\sqrt{2}}$.

Applying Eq. (4.11) to the operator $\hat{D}_\beta$ itself leads to the orthogonality of the Weyl operators with respect to the Hilbert–Schmidt inner product:

Orthogonality of Weyl operators. Let $\hat{D}_\alpha = e^{\alpha \hat{a}^\dagger - \alpha^* \hat{a}} = e^{-i\mathbf{r}^T \Omega \hat{\mathbf{r}}} = \hat{D}_{-\mathbf{r}}$, then

$$\text{Tr}\left[\hat{D}_\alpha \hat{D}_{-\beta}\right] = \pi \delta^2(\alpha - \beta) , \qquad (4.20)$$

or, with $2n$ real variables,

$$\text{Tr}\left[\hat{D}_{\mathbf{r}} \hat{D}_{-\mathbf{s}}\right] = (2\pi)^n \delta^{2n}(\mathbf{r} - \mathbf{s}) . \qquad (4.21)$$

4.3 CHARACTERISTIC FUNCTIONS AND QUASI-PROBABILITY DISTRIBUTIONS

The Fourier–Weyl relation gives rise to more than one equivalent representation of a quantum state ϱ. For a system comprising a single mode, one can introduce the real parameter $s \in [-1, +1]$ and define the corresponding s-ordered characteristic function $\chi_s(\alpha)$ as

$$\chi_s(\alpha) = \text{Tr}\left[\hat{D}_\alpha \varrho\right] e^{\frac{s}{2}|\alpha|^2} . \qquad (4.22)$$

The significance of the exponential factor becomes manifest if one considers the evaluation of expectation values of ordered products of ladder operators. By recalling that $\hat{D}_\alpha = e^{\alpha \hat{a}^\dagger} e^{-\alpha^* \hat{a}} e^{-\frac{1}{2}|\alpha|^2} = e^{-\alpha^* \hat{a}} e^{\alpha \hat{a}^\dagger} e^{+\frac{1}{2}|\alpha|^2}$ and inserting such expressions in the definition of the s-ordered characteristic function, one

---

[3]The multimode relation follows straightforwardly from the single-mode one because a basis of operators acting on the tensor product Hilbert space $L^2(\mathbb{R}^2)^{\otimes n}$ may be constructed as $\bigotimes_{j=1}^n \hat{D}_{\mathbf{r}_j} = \hat{D}_{\oplus_{j=1}^n \mathbf{r}_j} = \hat{D}_{\mathbf{r}}$, for all $\mathbf{r}_j \in \mathbb{R}^2$, whose direct sums form all $\mathbf{r} \in \mathbb{R}^{2n}$.

obtains

$$\langle \hat{a}^{\dagger m} \hat{a}^n \rangle_1 \equiv \mathrm{Tr} \left[\hat{a}^{\dagger m} \hat{a}^n \varrho \right] = \left(\frac{\partial}{\partial \alpha} \right)^m \left(-\frac{\partial}{\partial \alpha^*} \right)^n \chi_1(\alpha) \Big|_{\alpha=0} , \qquad (4.23)$$

$$\langle \hat{a}^{\dagger m} \hat{a}^n \rangle_0 = \left(\frac{\partial}{\partial \alpha} \right)^m \left(-\frac{\partial}{\partial \alpha^*} \right)^n \chi_0(\alpha) \Big|_{\alpha=0} , \qquad (4.24)$$

$$\langle \hat{a}^{\dagger m} \hat{a}^n \rangle_{-1} \equiv \mathrm{Tr} \left[\hat{a}^n \hat{a}^{\dagger m} \varrho \right] = \left(\frac{\partial}{\partial \alpha} \right)^m \left(-\frac{\partial}{\partial \alpha^*} \right)^n \chi_{-1}(\alpha) \Big|_{\alpha=0} , \qquad (4.25)$$

where the normally and anti-normally ordered products of field operators $\langle \hat{a}^{\dagger m} \hat{a}^n \rangle_{\pm 1}$ were implicitly defined as the expectation values of a product of m $\hat{a}^\dagger$ and n $\hat{a}$ where all the creation operators are on the left and all the annihilation operators are on the right (normal ordering) or *vice versa* (anti-normal ordering). The symmetric ordering $\langle \hat{a}^{\dagger m} \hat{a}^n \rangle_0$ is instead given the normalised sum of products of m $a^\dagger$ and n a in all possible orders (which is therefore symmetric under the exchange of $\hat{a}$ and $\hat{a}^\dagger$).[4] Differentiating the s-ordered characteristic function in $\alpha = 0$ allows one to retrieve the expectation value of s-ordered products of creation and annihilation operators, with $s = 1$, $s = 0$ and $s = -1$ corresponding to normal, symmetric and anti-normal ordering respectively. This property carries over to the multimode case.

Throughout the book, we shall refer to the function χ_0 as "the" characteristic function, omitting the "symmetrically ordered" specification as well as the label 0.

Notice also that the following relationship holds for all values of s:

$$\chi_s(0) = \mathrm{Tr} \left[\varrho \right] = 1 . \qquad (4.26)$$

4.3.1 General properties of the characteristic function

The description of states through their characteristic function turns out to be instrumental to a number of achievements, several of which will feature in this book. Let us therefore see how the properties of a quantum state ϱ are reflected by the associated symmetric characteristic function $\chi(\mathbf{r}) = \mathrm{Tr}(\hat{D}_{-\mathbf{r}} \varrho)$.

First, notice that $\chi(\mathbf{r})$ is continuous. Also, one has $\chi(0) = \mathrm{Tr}(\hat{D}_0 \varrho) = \mathrm{Tr} \varrho = 1$ (as $\hat{D}_0 = \hat{\mathbb{1}}$, by definition): normalisation is simply equivalent to the fact that the characteristic function evaluated at the origin is one:

$$\chi(0) = 1 \qquad (4.27)$$

(in compliance with our general notation, the symbol 0 stands here for the zero vector in generic dimension).

---

[4] For instance,

$$\langle a^{\dagger 2} a \rangle_0 = \mathrm{Tr} \left[\frac{1}{3} (a^{\dagger 2} a + a^\dagger a a^\dagger + a a^{\dagger 2}) \varrho \right] .$$

From $\varrho = \varrho^\dagger$, one gets $\chi(\mathbf{r}) = \text{Tr}(\varrho \hat{D}_{-\mathbf{r}}) = \text{Tr}(\varrho^\dagger \hat{D}_\mathbf{r})^* = \chi^*(-\mathbf{r})$ (where $\text{Tr}(A^\dagger)^* = \text{Tr}(A)$ and $\hat{D}^\dagger_{-\mathbf{r}} = \hat{D}_\mathbf{r}$ have been utilised): Hermiticity is equivalent to

$$\chi(\mathbf{r}) = \chi^*(-\mathbf{r}) . \tag{4.28}$$

Also, the probabilistic interpretation of quantum states imposes $\varrho \geq 0$. This condition is slightly more delicate to describe for characteristic functions and we shall justify it heuristically. One can exploit the fact that an operator is positive if and only if $\langle v|\varrho|v\rangle \geq 0$ for all $|v\rangle \in L^2(\mathbf{R}^n)$. Since any vector can be expanded in the Fock basis, one may infer that, for ϱ to be positive, it is sufficient – and clearly necessary – that $\text{Tr}(\varrho \hat{O}^\dagger_j \hat{O}_j) \geq 0$ for all operators $\hat{O}_j = |0\rangle\langle j|$ in terms of the Fock vectors $|0\rangle$ and $|j\rangle$. Now, the operators $\hat{O}_j$ are clearly bounded, and hence admit a Fourier–Weyl decomposition, that is, they can always be expressed in the basis set of Weyl operators $\hat{D}_\mathbf{r}$ (orthonormal with respect to the Hilbert–Schmidt product, as we saw above).

Now, let $\mathcal{R} = \{\mathbf{r}_j : \mathbf{r}_j \in \mathbb{R}^{2n}, j \in [1, \ldots, j_{max}]\}$ be a generic set of j_{max} vectors. In light of the above, the positivity of ϱ is equivalent to the positivity of all the quantities of the form $\text{Tr}(\varrho \hat{R}^\dagger_\mathbf{c} \hat{R}_\mathbf{c})$, with $\hat{R}_\mathbf{c} = \sum_{j=1}^{j_{max}} c_j \hat{D}_{\mathbf{r}_j}$, where $\mathbf{c} \in \mathbb{C}^{j_{max}}$ is a generic vector of complex coefficients (we are thus encompassing all the operators $\hat{R}_\mathbf{c}$ which admit a decomposition in terms of Weyl operators). One has

$$\text{Tr}(\varrho \hat{R}^\dagger_\mathbf{c} \hat{R}_\mathbf{c}) = \sum_{j,k=1}^{j_{max}} c^*_j c_k \text{Tr}\left[\varrho \hat{D}^\dagger_{\mathbf{r}_j} \hat{D}_{\mathbf{r}_k}\right] \geq 0 . \tag{4.29}$$

Hence, the positivity of the density matrix ϱ is equivalent to the positivity of the matrix $D_{jk} \equiv \text{Tr}\left[\varrho \hat{D}^\dagger_{\mathbf{r}_j} \hat{D}_{\mathbf{r}_k}\right]$: $D \geq 0 \ \forall \mathcal{R} \in \mathcal{M}_{2n,j_{max}}$ (a set of j_{max} $2n$-dimensional vectors can be parameterised as a real $2n \times j_{max}$ matrix). Let us now translate this condition in a corresponding requirement on the characteristic function:

$$\text{Tr}\left[\varrho \hat{D}^\dagger_{\mathbf{r}_j} \hat{D}_{\mathbf{r}_k}\right] = \frac{1}{(2\pi)^n} \int_{\mathbb{R}^{2n}} d\mathbf{r} \, \chi(\mathbf{r}) \text{Tr}\left(\hat{D}_\mathbf{r} \hat{D}_{-\mathbf{r}_j} \hat{D}_{\mathbf{r}_k}\right) = \chi(\mathbf{r}_j - \mathbf{r}_k) e^{i\mathbf{r}_k^\mathsf{T} \Omega \mathbf{r}_j/2} , \tag{4.30}$$

where (3.11) and (4.21) were applied. Summarising, $\varrho \geq 0$ if and only if the matrix

$$D_{jk} = \chi(\mathbf{r}_j - \mathbf{r}_k) e^{i\mathbf{r}_k^\mathsf{T} \Omega \mathbf{r}_j/2} \tag{4.31}$$

is positive semi-definite for all sets of vectors $\mathcal{R} = \{\mathbf{r}_j : \mathbf{r}_j \in \mathbb{R}^{2n}, j \in [1, \ldots, j_{max}]\}$. Often, this prescription, along with $\chi(0) = 1$, is referred to as the condition for a function to be the characteristic of a quantum state.[5] Significantly, setting $\Omega = 0$ in Eq. (4.31) gives a necessary and sufficient condition for a continuous function $\chi(\mathbf{r})$ to be a *classical* characteristic function, i.e., the Fourier transform of a multivariate classical probability distribution.

---

[5] This is the case if one, as is usually done, subsumes the property of Hermiticity in the definition of "positive semi-definite". Then, if one sets $\mathbf{r}_1 = 0$ and $\mathbf{r}_2 = \mathbf{r}$ the positivity (and hence Hermiticity) of D implies $\chi^*(\mathbf{r}) = \chi(-\mathbf{r})$.

4.3.2 Quasi-probability distributions

By taking the complex Fourier transform of $\chi_s(\alpha)$, one may define the s-ordered "quasi-probability" distribution $W_s(\alpha)$:

$$W_s(\alpha) = \frac{1}{\pi^2} \int_{\mathbb{C}} d^2\beta \, e^{(\alpha\beta^* - \alpha^*\beta)} \chi_s(\beta) \, , \tag{4.32}$$

which is normalised:

$$\int_{\mathbb{C}} d^2\alpha \, W_s(\alpha) = \chi_s(0) = 1 \tag{4.33}$$

and has the property that

$$\int_{\mathbb{C}} d^2\alpha \, \alpha^{*m} \alpha^n W_s(\alpha) = \frac{1}{\pi^2} \int_{\mathbb{C}^2} d^2\alpha \, d^2\beta \left(\left(-\frac{\partial}{\partial\beta} \right)^m \left(\frac{\partial}{\partial\beta^*} \right)^n e^{(\alpha\beta^* - \alpha^*\beta)} \right) \chi_s(\beta)$$

$$= \int_{\mathbb{C}} d^2\beta \left(\left(-\frac{\partial}{\partial\beta} \right)^m \left(\frac{\partial}{\partial\beta^*} \right)^n \delta^2(\beta) \right) \chi_s(\beta)$$

$$= \left(\frac{\partial}{\partial\beta} \right)^m \left(-\frac{\partial}{\partial\beta^*} \right)^n \chi_s(\beta) \Big|_{\beta=0} = \langle a^{\dagger m} a^n \rangle_s \, , \tag{4.34}$$

where we applied the representation of the derivative of the delta function given by the integral over α. Eqs. (4.33) and (4.34) justify the terminology 'quasi-probability' distributions, with the "quasi-" there to remind one that the quantity $W_s(\alpha)$ is in general not positive, and in fact it may not even be a proper function at all, as we shall see in what follows.

Although they allow for a simple, unified treatment, the quasi-probability distributions for different values of s emerged historically at different times in different contexts, responding to different demands.

For $s = 1$, the quantity $W_1(\alpha)$ is the celebrated Glauber–Sudarshan P-representation, and is commonly denoted with $P(\alpha)$. One has the remarkable property:

Glauber–Sudarshan P-representation of a quantum state. Let $P(\alpha) = W_1(\alpha)$ be the Fourier transform of the normally ordered characteristic function associated with the quantum state ϱ. Then,

$$\varrho = \int_{\mathbb{C}} d^2\alpha \, P(\alpha) |\alpha\rangle\langle\alpha| \, . \tag{4.35}$$

Proof. The P-representation identity is easily proven by applying the operator $\hat{D}_\alpha$ by similarity on both sides of Eq. (4.14), getting

$$|\alpha\rangle\langle\alpha| = \frac{1}{\pi} \int_{\mathbb{C}} d^2\gamma \, e^{-\frac{1}{2}|\gamma|^2} e^{(\alpha\gamma^* - \alpha^*\gamma)} \hat{D}_{-\gamma} \, , \tag{4.36}$$

and then by inserting this expression in the right-hand side of Eq. (4.35), to obtain

$$\int_{\mathbb{C}} d^2\alpha\, P(\alpha)|\alpha\rangle\langle\alpha| = \frac{1}{\pi}\int_{\mathbb{C}} d^2\alpha \int_{\mathbb{C}} d^2\gamma\, P(\alpha)e^{-\frac{1}{2}|\gamma|^2}e^{(\alpha\gamma^*-\alpha^*\gamma)}\hat{D}_{-\gamma}$$

(4.37)

$$= \int_{\mathbb{C}} \frac{d^2\gamma}{\pi}\chi_1(\gamma)e^{-\frac{1}{2}|\gamma|^2}\hat{D}_{-\gamma} = \int_{\mathbb{C}} \frac{d^2\gamma}{\pi}\chi_0(\gamma)\hat{D}_{-\gamma} = \varrho,$$

where we solved the integral over α through the inverse Fourier transform relation

$$\chi_1(\gamma) = \int_{\mathbb{C}} d^2\alpha\, e^{(\alpha\gamma^*-\alpha^*\gamma)}P(\alpha)\,, \tag{4.38}$$

and applied the Fourier–Weyl relation (4.13) in the last identity. □

Due to their over-completeness, coherent states allow for a diagonal decomposition of any density matrix. However, this decomposition is given in terms of the quasi-probability $P(\alpha) = W_1(\alpha)$, which may not be a distribution or a proper function. For instance, the P-representation of a coherent state $|\alpha\rangle\langle\alpha|$ is the delta function $\delta^2(\gamma-\alpha)$, while the P-representation of a number state may only be expressed in terms of derivatives of the delta function.

States with a P-representation "not more singular" than a delta function, including coherent states, are often referred to as 'classical' states. We will apply this definition to Gaussian states later on, in Problem 4.2, as it bears a remarkable consequence on their separability properties, as we shall see in Chapter 7. We shall, however, not dwell on the general inspection of this notion any further.

The Fourier–Weyl relation allows one to shed light on the nature of the anti-symmetrically ordered quasi-probability function $W_{-1}(\alpha)$ too. Such a function is commonly referred to as the 'Husimi' Q-function, and denoted as $Q(\alpha)$. One has

$$\frac{1}{\pi}\langle\alpha|\varrho|\alpha\rangle = \frac{1}{\pi^2}\int_{\mathbb{C}} d^2\beta\,\chi_0(\beta)\langle\alpha|\hat{D}_{-\beta}|\alpha\rangle = \frac{1}{\pi^2}\int_{\mathbb{C}} d^2\beta\, e^{\frac{1}{2}(\alpha\beta^*-\alpha^*\beta)}\chi_0(\beta)\langle\alpha|\alpha-\beta\rangle$$

$$= \frac{1}{\pi^2}\int_{\mathbb{C}} d^2\beta\, e^{\alpha\beta^*-\alpha^*\beta}\chi_0(\beta)e^{-\frac{|\beta|^2}{2}} = W_{-1}(\alpha) = Q(\alpha)\,, \tag{4.39}$$

whence:

Husimi Q-function. Let $Q(\alpha) = W_{-1}(\alpha)$ be the Fourier transform of the anti-normally ordered characteristic function associated with the quantum state ϱ. Then,

$$Q(\alpha) = \frac{1}{\pi}\langle\alpha|\varrho|\alpha\rangle\,, \tag{4.40}$$

where $|\alpha\rangle$ is the coherent state with amplitude α.

The Q-function is therefore always positive and does not diverge. On account of these properties, it found wide application in the study of quantum dynamical systems, whenever a well-defined probability distribution is desirable. As we shall see in Chapter 5, the value of the Q-function at a point α represents the probability that the heterodyne measurement of the system yields the outcome α (heterodyne measurements give complex outcomes).

As for the symmetrically ordered $W_0(\alpha)$, it was historically the first quasi-probability to be introduced, by Wigner, and is called the Wigner function. In the remainder of the book, we shall denote it as $W(\alpha)$, omitting the subscript. Likewise, χ will represent the symmetrically ordered characteristic function χ_0. It is instructive to express the integral (4.32), that defines the Wigner function, in terms of real variables through the identifications $\alpha = \frac{x+ip}{\sqrt{2}}$ and $\beta = \frac{x'+ip'}{\sqrt{2}}$, obtaining

$$
\begin{aligned}
W(x,p) &= \frac{1}{\pi^2} \int_{\mathbb{R}} \int_{\mathbb{R}} \frac{\mathrm{d}x'\,\mathrm{d}p'}{2}\, e^{i(px'-xp')}\chi_0(x',p') \\
&= \frac{1}{2\pi^2} \int_{\mathbb{R}} \int_{\mathbb{R}} \mathrm{d}x'\,\mathrm{d}p'\, e^{i(px'-xp')} \int_{\mathbb{R}} \mathrm{d}q\, \langle q|\hat{D}_{-\frac{x'}{2}}\varrho\hat{D}_{-\frac{x'}{2}}|q\rangle \\
&= \frac{1}{2\pi^2} \int_{\mathbb{R}^3} \mathrm{d}q\,\mathrm{d}x'\,\mathrm{d}p'\, e^{ipx'}e^{ip'(q-x)}\langle q+\tfrac{x'}{2}|\varrho|q-\tfrac{x'}{2}\rangle \qquad (4.41) \\
&= \frac{1}{\pi} \int_{\mathbb{R}} \mathrm{d}x'\, e^{ipx'}\langle x+\tfrac{x'}{2}|\varrho|x-\tfrac{x'}{2}\rangle = \frac{2}{\pi} \int_{\mathbb{R}} \mathrm{d}x'\, e^{i2px'}\langle x+x'|\varrho|x-x'\rangle,
\end{aligned}
$$

where we expressed the trace, as well as the final result, in terms of the improper quadrature eigenvalues $|q\rangle$, as per Eq. (2.21), and $\hat{D}_{-\frac{x'}{2}} = e^{-i\frac{x'}{2}\hat{p}}e^{i\frac{p'}{2}\hat{x}}e^{i\frac{p'x'}{8}} = e^{i\frac{p'}{2}\hat{x}}e^{-i\frac{x'}{2}\hat{p}}e^{-i\frac{p'x'}{8}}$ (after ordering via Baker-Campbell-Hausdorff). We also applied the representation of the delta function (1.16).

Operational interpretation of the Wigner function. Eq. (4.41) implies that

$$
\frac{1}{2} \int_{-\infty}^{+\infty} \mathrm{d}p\, W(x,p) = \langle x|\varrho|x\rangle. \qquad (4.42)
$$

Up to a factor $\frac{1}{2}$, the integral of the Wigner function over a certain phase space quadrature gives the probability of measuring the conjugate quadrature. This statement is clearly phase invariant, in the specific sense that, if one defines the generalised canonical operators $\hat{x}_\theta = \cos\theta\,\hat{x} - \sin\theta\,\hat{p}$ and associated canonical variables x_θ, one has

$$
\frac{1}{2} \int_{-\infty}^{+\infty} \mathrm{d}x_{\theta-\frac{\pi}{2}}\, W(x,p) = \langle x_\theta|\varrho|x_\theta\rangle, \qquad (4.43)
$$

since $[\hat{x}_\theta, \hat{x}_{\theta-\frac{\pi}{2}}] = i$.

The marginal of the Wigner function along any phase space direction is hence, up to a factor $\frac{1}{2}$, a positive probability distribution which describes the statistics of quadrature measurements. Such measurements, corresponding to position or momentum measurements for a material particle in the cases $\theta = 0, -\frac{\pi}{2}$, are implemented through 'homodyne' detection in quantum optics (see Chapter 5).

4.4 CHARACTERISTIC FUNCTION OF A GAUSSIAN STATE

In order to link the phase space approach back to the material we introduced in the previous chapter, let us now determine the (symmetrically ordered) characteristic function χ_G of the most general Gaussian state, which was parametrised in Eq. (3.45) in terms of a covariance matrix $\boldsymbol{\sigma} = S\nu S^{\mathsf{T}}$ (with normal form $\boldsymbol{\nu}$) and a vector of first moments which was denoted with $\bar{\mathbf{r}}$. We will start off with the general case with any number of variables, as in Eq. (4.19), and then reduce it to the single-mode problem, for which we will adopt more convenient complex variables.

By inserting Eq. (3.45) into Eq. (4.19) one gets

$$
\begin{aligned}
\chi_G(\mathbf{r}) &= \left(\prod_{j=1}^{n} \left(1 - \mathrm{e}^{-\xi_j}\right) \right) \mathrm{Tr} \left[\left(\bigotimes_{j=1}^{n} \left(\sum_{m=0}^{\infty} \mathrm{e}^{-\xi_j m} |m\rangle_{jj}\langle m| \right) \right) \hat{S}\hat{D}_{\bar{\mathbf{r}}} D_{-\mathbf{r}} \hat{D}_{\bar{\mathbf{r}}}^\dagger \hat{S}^\dagger \right] \\
&= \left(\prod_{j=1}^{n} \left(1 - \mathrm{e}^{-\xi_j}\right) \right) \mathrm{Tr} \left[\left(\bigotimes_{j=1}^{n} \left(\sum_{m=0}^{\infty} \mathrm{e}^{-\xi_j m} |m\rangle_{jj}\langle m| \right) \right) D_{S^{-1}\mathbf{r}} \right] \mathrm{e}^{i\bar{\mathbf{r}}^\mathsf{T}\Omega\mathbf{r}},
\end{aligned}
$$

(4.44)

where we applied Eq. (3.11) and the projective representation of the symplectic group through the identity $\hat{S}\hat{D}_{\mathbf{r}}\hat{S}^\dagger = \mathrm{e}^{i\mathbf{r}^\mathsf{T}\Omega S\hat{\mathbf{r}}} = \mathrm{e}^{i\mathbf{r}^\mathsf{T} S^{\mathsf{T}-1}\Omega\hat{\mathbf{r}}} = \hat{D}_{S^{-1}\mathbf{r}}$. Since the displacement operators are tensor products of local operators, the problem of determining χ_G has been reduced to finding the characteristic function of the single mode operator $\sum_{m=0}^{\infty} \mathrm{e}^{-\xi_j m} |m\rangle\langle m|$, where $|m\rangle$ is a Fock state. In fact, the characteristic function of a tensor product of operators is the product of their individual characteristic functions, so the Gaussian characteristic function we are after may be determined by taking the product of such single-mode characteristic functions, multiplying it by the phase factor we determined above, and applying the transformation S^{-1} on the variable $\mathbf{r}$.

Switching temporarily to complex variables is expedient to evaluate the single-mode characteristic function $\chi^{(1)}(\alpha) = \mathrm{Tr}\left[\sum_{m=0}^{\infty} \mathrm{e}^{-\xi_j m} |m\rangle\langle m|\hat{D}_\alpha \right]$:

$$
\begin{aligned}
\chi^{(1)}(\alpha) &= \frac{1}{\pi} \int_{\mathbb{C}} \mathrm{d}^2\gamma \sum_{m=0}^{\infty} \mathrm{e}^{-\xi_j m} \langle\gamma|m\rangle\langle m|\hat{D}_\alpha|\gamma\rangle \\
&= \frac{1}{\pi} \int_{\mathbb{C}} \mathrm{d}^2\gamma\, \mathrm{e}^{-\frac{1}{2}(|\gamma|^2+|\alpha+\gamma|^2+\alpha^*\gamma-\alpha\gamma^*)} \sum_{m=0}^{\infty} \frac{\left(\gamma^*(\alpha+\gamma)\mathrm{e}^{-\xi_j}\right)^m}{m!} \\
&= \frac{\mathrm{e}^{-\frac{|\alpha|^2}{2}}}{\pi} \int_{\mathbb{C}} \mathrm{d}^2\gamma\, \mathrm{e}^{-|\gamma|^2\left(1-\mathrm{e}^{-\xi_j}\right)} \mathrm{e}^{\alpha\gamma^* \mathrm{e}^{-\xi_j}-\alpha^*\gamma}.
\end{aligned}
$$

(4.45)

The latter is a Gaussian integral, which can be reduced to a particularly simple form by setting $\gamma = (x + iy)/\sqrt{1 - e^{-\xi_j}}$ and yields, upon use of Eq. (1.18)

$$\chi^{(1)}(\alpha) = \text{Tr}\left[\sum_{m=0}^{\infty} e^{-\xi_j m}|m\rangle\langle m|\hat{D}_\alpha\right] = \frac{e^{-\frac{|\alpha|^2}{2}\left(\frac{1+e^{-\xi_j}}{1-e^{-\xi_j}}\right)}}{1 - e^{-\xi_j}} = \frac{e^{-\frac{|\alpha|^2}{2}\nu_j}}{1 - e^{-\xi_j}}, \quad (4.46)$$

where we have inserted the symplectic eigenvalue of the covariance matrix σ according to Eq. (3.57). Combining Eqs. (4.44) and (4.46) leads to the following expression for the characteristic function of a multimode Gaussian state, after the substitution $\alpha = (x_j + ip_j)/\sqrt{2}$ for each different j:

$$\chi_G(\mathbf{r}) = e^{-\frac{1}{4}\mathbf{r}^T S^{-1T}(\oplus_{j=1}^n \nu_j \mathbb{1}_2)S^{-1}\mathbf{r}}e^{i\mathbf{r}^T\Omega^T\bar{\mathbf{r}}} = e^{-\frac{1}{4}\mathbf{r}^T S^{-1T}\Omega^T(\oplus_{j=1}^n \nu_j \mathbb{1}_2)\Omega S^{-1}\mathbf{r}}e^{i\mathbf{r}^T\Omega^T\bar{\mathbf{r}}}$$

$$= e^{-\frac{1}{4}\mathbf{r}^T\Omega^T S(\oplus_{j=1}^n \nu_j \mathbb{1}_2)S^T\Omega\mathbf{r}}e^{i\mathbf{r}^T\Omega^T\bar{\mathbf{r}}} = e^{-\frac{1}{4}\mathbf{r}^T\Omega^T\sigma\Omega\mathbf{r}}e^{i\mathbf{r}^T\Omega^T\bar{\mathbf{r}}}, \quad (4.47)$$

where we exploited the invariance of $(\oplus_{j=1}^n \nu_j \mathbb{1}_2)$ under the action of Ω by congruence (the latter represents a product of local rotations on 2×2 subspaces, and thus preserves local identity matrices, and also repeatedly employed $S\Omega S^T = \Omega$.

Characteristic function of a Gaussian state. Summarising, we have found that the characteristic function χ_G of a Gaussian state with covariance matrix σ and vector of first moments $\bar{\mathbf{r}}$ is given by:

$$\chi_G(\mathbf{r}) = e^{-\frac{1}{4}\mathbf{r}^T\Omega^T\sigma\Omega\mathbf{r}}e^{i\mathbf{r}^T\Omega^T\bar{\mathbf{r}}}. \quad (4.48)$$

Notice how the vector variable $\mathbf{r}$ always enters this expression after multiplication by the symplectic form Ω. Summing up, the full Fourier–Weyl description of a Gaussian state ϱ_G reads

$$\varrho_G = \frac{1}{(2\pi)^n}\int_{\mathbb{R}^{2n}}\mathrm{d}\mathbf{r}\, e^{-\frac{1}{4}\mathbf{r}^T\Omega^T\sigma\Omega\mathbf{r}+i\mathbf{r}^T\Omega^T\bar{\mathbf{r}}}\hat{D}_\mathbf{r} = \frac{1}{(2\pi)^n}\int_{\mathbb{R}^{2n}}\mathrm{d}\tilde{\mathbf{r}}\, e^{-\frac{1}{4}\tilde{\mathbf{r}}^T\sigma\tilde{\mathbf{r}}+i\tilde{\mathbf{r}}^T\bar{\mathbf{r}}}\hat{D}_{\Omega^T\tilde{\mathbf{r}}}, \quad (4.49)$$

where we have also reported the more compact notation in terms of the variables $\tilde{\mathbf{r}} = \Omega\mathbf{r}$. The latter is often employed in the literature, and we shall also adopt it when expedient, although we preferred to introduce first the notation in terms of the variable $\mathbf{r}$ as it relates directly to phase space displacements as per Eq. (3.16).

We are thus led to yet another general characterisation of Gaussian states, as the quantum states with a Gaussian characteristic function. The Wigner function W_G of a Gaussian state, obtained by taking the complex Fourier transform of Eq. (4.48), is readily evaluated by applying (1.18), obtaining

$$W_G(\mathbf{r}) = \frac{1}{2^n\pi^{2n}}\int_{\mathbb{R}^{2n}}\mathrm{d}\mathbf{r}'\, e^{-\frac{1}{4}\mathbf{r}'^T\Omega^T\sigma\Omega\mathbf{r}'}e^{i\mathbf{r}'^T\Omega^T(\bar{\mathbf{r}}-\mathbf{r})}$$

$$= \frac{2^n}{\pi^n\sqrt{\text{Det}\,\sigma}}e^{-(\mathbf{r}-\bar{\mathbf{r}})^T\sigma^{-1}(\mathbf{r}-\bar{\mathbf{r}})}. \quad (4.50)$$

The latter is, with respect to the measure $\frac{d\mathbf{r}}{2^n}$, a Gaussian probability distribution centred in $\bar{\mathbf{r}}$ and with covariances described by $\boldsymbol{\sigma}$.

As we saw above, the marginal Wigner function along any phase space direction describes the probability distribution of the quantum measurement of the associated quadrature operator. Hence, as far as Gaussian states are concerned, the Wigner function provides one with a local, 'realistic' model to describe quadrature measurements. *If one restricts to quadrature measurements*, such systems may be mimicked by multivariate classical Gaussian distributions and will never show any signature of quantum non-locality, such as a violation of Bell or CHSH inequalities. Clearly, quantum Gaussian states admit, besides the phase space description akin to classical distributions, an underlying Hilbert space description: general quantum measurements on the Hilbert space do allow for quantum non-locality to become manifest with Gaussian states. It is however clear from the analogy with classical distributions that, whilst Gaussian states are entirely described by first and second moments of their Wigner distributions, any analysis regarding genuinely quantum features, such as quantum entanglement, requires one to look beyond the phase space formalism.

Inserting the relation (4.49) into $\mathrm{Tr}(\varrho_1\varrho_2)$ for two Gaussian states ϱ_1 and ϱ_2, and using Eqs. (4.21) and (1.18), leads to a first notable and consequential finding that illustrates remarkably well the usefulness of the Fourier–Weyl description, and which will be used repeatedly throughout the book:

Overlap (Hilbert–Schmidt product) of two Gaussian states. Given two n-mode Gaussian states ϱ_1 and ϱ_2, with covariance matrices $\boldsymbol{\sigma}_1$ and $\boldsymbol{\sigma}_2$ and vectors of first moments $\mathbf{r}_1$ and $\mathbf{r}_2$, one has:

$$\mathrm{Tr}(\varrho_1\varrho_2) = \frac{2^n}{\sqrt{\mathrm{Det}(\boldsymbol{\sigma}_1 + \boldsymbol{\sigma}_2)}}\, e^{-(\mathbf{r}_1 - \mathbf{r}_2)^{\mathsf{T}}(\boldsymbol{\sigma}_1 + \boldsymbol{\sigma}_2)^{-1}(\mathbf{r}_1 - \mathbf{r}_2)}\,. \qquad (4.51)$$

Proof. The formula above is proven by adopting the Fourier–Weyl representation of a Gaussian state (4.49)

$$\varrho_j = \frac{1}{(2\pi)^n}\int_{\mathbb{R}^{2n}} d\tilde{\mathbf{r}}\, e^{-\frac{1}{4}\tilde{\mathbf{r}}^{\mathsf{T}}\boldsymbol{\sigma}_j\tilde{\mathbf{r}} + i\tilde{\mathbf{r}}^{\mathsf{T}}\mathbf{r}_j}\, \hat{D}_{\Omega^{\mathsf{T}}\tilde{\mathbf{r}}} \quad \text{for} \quad j = 1, 2\,, \qquad (4.52)$$

along with (4.21) and (1.18). $\qquad\qquad\Box$

Problem 4.2. (*Classicality of Gaussian states*). Show that a Gaussian state is classical, in the sense that it does admit a P-representation which is a positive, proper function, if and only if the smallest eigenvalue of its covariance matrix $\boldsymbol{\sigma}$ is greater than or equal to 1.

Solution. Adopting $2n$ real variables and inserting the symmetric

Gaussian characteristic function (4.48) into the definition of W_{-1} (4.32) yields

$$W_{-1}(\mathbf{r}) = \frac{1}{(2\pi^2)^n} \int_{\mathbb{R}^{2n}} d\mathbf{r}' e^{i\mathbf{r}'^\mathsf{T}\Omega\mathbf{r}} e^{\frac{1}{4}\mathbf{r}'^\mathsf{T}\mathbf{r}'} e^{-\frac{1}{4}\mathbf{r}'^\mathsf{T}\Omega^\mathsf{T}\boldsymbol{\sigma}\Omega\mathbf{r}' + i\mathbf{r}'^\mathsf{T}\Omega^\mathsf{T}\bar{\mathbf{r}}} . \quad (4.53)$$

Clearly, this Gaussian integral results in a well-defined, proper function of $\mathbf{r}$, as per Eq. (1.18), if and only if the quadratic form in the integration variable $\mathbf{r}'$ at the exponent is positive definite, which is in fact the case if and only if $\Omega^\mathsf{T}\boldsymbol{\sigma}\Omega - \mathbb{1} > 0 \Leftrightarrow \boldsymbol{\sigma} > \mathbb{1}$. The latter condition is precisely the same as stating that the smallest eigenvalue of $\boldsymbol{\sigma}$ is strictly greater than 1. Notice that for a coherent state, for which $\boldsymbol{\sigma} = \mathbb{1}$, the P-representation above reduces to the representation of a delta function. We shall not need any more detail concerning non-classicality than this basic remark in our future endeavours.

Problem 4.3. (*Schrödinger cat state*). Determine the characteristic and Wigner functions of the normalised pure state $|\psi\rangle = \frac{|\alpha\rangle + |-\alpha\rangle}{\sqrt{2 + 2e^{-2|\alpha|^2}}}$, where $|\alpha\rangle$ and $|-\alpha\rangle$ are coherent states.

Solution. Notice that the normalisation of $|\psi\rangle$ can be verified through Eq. (4.7). The density matrix associated with the cat state reads $\varrho = |\psi\rangle\langle\psi| = \frac{1}{2\eta}(|\alpha\rangle\langle\alpha| + |-\alpha\rangle\langle-\alpha| + |-\alpha\rangle\langle\alpha| + |\alpha\rangle\langle-\alpha|)$, where $\eta = (1 + e^{-2|\alpha|^2})$ was defined. Since the mapping that relates ϱ to its characteristic function is manifestly linear, the first two terms will just contribute the characteristic functions of the Gaussian coherent states $|\alpha\rangle$ and $|-\alpha\rangle$ (properly weighted). All that is left to determine is the 'coherent', off-diagonal contribution $|\alpha\rangle\langle-\alpha|$, given by

$$\mathrm{Tr}\left(|\alpha\rangle\langle-\alpha|\hat{D}_\gamma\right) = \mathrm{Tr}\left(\hat{D}_\alpha|0\rangle\langle0|\hat{D}_\alpha\hat{D}_\gamma\right) = \langle0|\hat{D}_\alpha\hat{D}_\gamma\hat{D}_\alpha|0\rangle$$

$$= \langle0|\hat{D}_{2\alpha+\gamma}|0\rangle = \langle0|2\alpha+\gamma\rangle = e^{-\frac{1}{2}(2\alpha^*+\gamma^*)(2\alpha+\gamma)}$$

$$= e^{-2|\alpha|^2 - \frac{1}{2}|\gamma|^2 - (\alpha^*\gamma + \alpha\gamma^*)} , \quad (4.54)$$

where we used Eqs. (4.6), (4.7) and $\hat{D}_\alpha^\dagger = \hat{D}_{-\alpha}$. The conjugate term $|-\alpha\rangle\langle\alpha|$ will give a contribution of the same form, but with α replaced by $-\alpha$. As for the characteristic functions of the Gaussian states $|\alpha\rangle$ and $|-\alpha\rangle$, they can be derived from the generic Gaussian expression (4.48) by substituting $\gamma = \frac{x+ip}{\sqrt{2}}$ and $\alpha = \frac{\bar{x}+i\bar{p}}{\sqrt{2}}$ or $\alpha = -\frac{\bar{x}+i\bar{p}}{\sqrt{2}}$, respectively. The complete characteristic function $\chi(\gamma)$ is hence:

$$\chi(\gamma) = \frac{1}{2\eta}e^{-\frac{1}{2}|\gamma|^2}\left(e^{\gamma\alpha^* - \gamma^*\alpha} + e^{-\gamma\alpha^* + \gamma^*\alpha} + 2e^{-2|\alpha|^2}\cosh(\alpha^*\gamma + \alpha\gamma^*)\right) . \quad (4.55)$$

Note that the state was normalised correctly, as $\chi(0) = 1$. Alternately, in terms of the vectors $\mathbf{r} = (x, p)^{\mathsf{T}}$ and $\mathbf{r}' = \sqrt{2}(\mathrm{Re}(\alpha), \mathrm{Im}(\alpha))^{\mathsf{T}}$:

$$\chi(\mathbf{r}) = \frac{1}{2\eta} e^{-\frac{1}{4}\mathbf{r}^{\mathsf{T}}\mathbf{r}} \left(e^{i\mathbf{r}^{\mathsf{T}}\Omega^{\mathsf{T}}\mathbf{r}'} + e^{-i\mathbf{r}^{\mathsf{T}}\Omega^{\mathsf{T}}\mathbf{r}'} + 2e^{-|\mathbf{r}'|^2} \cosh(\mathbf{r}^{\mathsf{T}}\mathbf{r}') \right) . \quad (4.56)$$

The corresponding Wigner function $W(\mathbf{r})$ is then obtained by Fourier transformation, as in Eq. (4.50):

$$W(\mathbf{r}) = \frac{1}{\eta\pi} \left(e^{-(\mathbf{r}-\mathbf{r}')^{\mathsf{T}}(\mathbf{r}-\mathbf{r}')} + e^{-(\mathbf{r}+\mathbf{r}')^{\mathsf{T}}(\mathbf{r}+\mathbf{r}')} + 2e^{-\mathbf{r}^{\mathsf{T}}\mathbf{r}} \cos(2\mathbf{r}^{\mathsf{T}}\Omega\mathbf{r}') \right) .$$
$$(4.57)$$

The function $W(\mathbf{r})$ is not necessarily positive: the cosine term implies ripples in phase space that are a distinctive signature of quantum coherence. Although the marginal distribution must always be positive, since the interpretation (4.42) in terms of measurement probabilities is general, the multivariate function $W(\mathbf{r})$ cannot be thought of as a probability distribution. In Chapter 7, we shall encounter an entangled version of this state, and see that this impossibility has significant consequences for the nonlocality of such states.

Non-Gaussian states of this form, which involve the coherent superposition of two states with different first moments, are commonly referred to as Schrödinger cat states, in homage to the notorious dead and/or alive cat hypothesised by the pioneer of quantum mechanics. They are much renowned for their potential, typically as probe states, in the investigation of fundamental effects, such as decoherence, and in more general applications to quantum sensing. In Problem 9.7, we will also touch on a scheme for the preparation of such states in practice.

4.5   FURTHER READING

An insightful account of the phase space formalism is contained in Barnett and Radmore's classic textbook [5], which complements well our comparatively scant rendition. Other coverages with more of an eye to quantum optical implementations are to be found in [75, 83].

Dynamics

Gaussian operations

CONTENTS

Gaussian operations, as one usually refers to CP-maps that send Gaussian states into Gaussian states, are apt to model a wide range of situations of practical interest and may often be controlled and applied on demand in the laboratory, in a variety of set-ups (that will be the subject matter of Chapter 9). Furthermore, they can be classified in a hierarchy of compact, powerful and far-reaching theoretical descriptions, which will be introduced by degrees and discussed in the current chapter. Note that here we are dealing with discrete quantum operations, rather than dynamical equations, which will be the subject of the next chapter.

We shall start from Gaussian unitary operations, corresponding to the symplectic group already introduced in Chapter 3, and will also clarify how other basic manipulations such as tensoring and partial tracing carry over to the Gaussian picture. Then, we shall broaden our view to include all deterministic (trace-preserving) maps resulting from Gaussian unitary dilations, which are often known as the set of "bosonic Gaussian channels" and encompass open dynamics subject to noise and 'decoherence' (the loss of quantum coherence due to the interaction with an environment). Next, we will move on to consider POVMs corresponding to the well-known homodyne and heterodyne detection schemes which, when acting on a subsystem, preserve the Gaussian character of the initial state both when the system is filtered by recording the measurement outcomes, and when acting unconditionally (that is, when the outcome of the measurement is not recorded). Our description will be further enlarged to the wider class of 'general-dyne' detections, and we will also introduce methods to account for imperfect detections. The ability to accommodate noise in the analytical description, often daunting with discrete variable and, more generally, outside the Gaussian regime, is probably one of the greatest strengths of the formalism under discussion. Finally, we shall introduce the Gaussian version of the Choi–Jamiolkowski isomorphism, and attain a unified description of all the Gaussian maps previously defined.

Throughout our exposition, we will refer to quantum optical systems to exemplify the practical relevance of the operations encountered.

5.1   GAUSSIAN UNITARY TRANSFORMATIONS

In view of the discussion contained in Chapter 3, it should be intuitively clear that a unitary transformation sends all Gaussian states into Gaussian states if and only if it is generated by a second-order Hamiltonian, comprised in general of a linear and a quadratic term in the canonical operators. Recovering the notation of Chapter 3, one may write the most general such Hamiltonian for n modes as in Eq. (3.17):

$$\hat{H} = \frac{1}{2}\hat{D}_{-\bar{\mathbf{r}}}\,\hat{\mathbf{r}}^\mathsf{T} H \hat{\mathbf{r}}\,\hat{D}_{\bar{\mathbf{r}}}\,, \tag{5.1}$$

where $\hat{D}_{\bar{\mathbf{r}}} = \mathrm{e}^{i\bar{\mathbf{r}}^\mathsf{T}\Omega\hat{\mathbf{r}}}$ is a Weyl operator with displacement $\bar{\mathbf{r}}$ and H is a $2n \times 2n$ symmetric matrix of Hamiltonian couplings. The equation above may be exponentiated to obtain a relationship between unitary operations:

$$\mathrm{e}^{i\hat{H}} = \hat{D}_{-\bar{\mathbf{r}}}\,\mathrm{e}^{\frac{i}{2}\hat{\mathbf{r}}^\mathsf{T} H \hat{\mathbf{r}}}\,\hat{D}_{\bar{\mathbf{r}}}\,. \tag{5.2}$$

We know that the central term on the right-hand side, a unitary generated by a purely quadratic Hamiltonian, corresponds to the symplectic transformation

$S = e^{\Omega H}$ acting linearly on the vector of canonical operators, and will therefore be denoted with $\hat{S} = e^{\frac{i}{2}\hat{r}^{\mathsf{T}} H \hat{r}}$. In formulae, S and $\hat{S}$ are related by Eq. (3.24), which is worth recalling here: $\hat{S}\hat{r}\hat{S}^{\dagger} = S\hat{r}$, and which can be carried over to Weyl operators to obtain $\hat{S}e^{i\bar{r}^{\mathsf{T}}\Omega\hat{r}}\hat{S}^{\dagger} = e^{i\bar{r}^{\mathsf{T}}\Omega S\hat{r}} = e^{i\bar{r}^{\mathsf{T}}S^{\mathsf{T}-1}\Omega\hat{r}} = \hat{D}_{S^{-1}\bar{r}}$, where the symplecticity of S was employed in transferring its action to the other side of Ω. Whence:

Composition of quadratic and displacement operators. Let $\hat{D}_{\bar{r}}$ be a Weyl operator and $\hat{S}$ a purely quadratic unitary operation related to the symplectic transformation S as per Eq. (3.24). Then one has

$$\hat{S}\hat{D}_{\bar{r}} = \hat{D}_{S^{-1}\bar{r}}\hat{S} . \tag{5.3}$$

Note that the following equivalent relation also holds true for the inverse transformations:

$$\hat{S}^{\dagger}\hat{D}_{\bar{r}} = \hat{D}_{S\bar{r}}\hat{S}^{\dagger} . \tag{5.4}$$

This simple prescription to re-order quadratic operators and linear displacements allows one to recast the most general second-order unitary transformation of Eq. (5.2) as

$$e^{i\hat{H}} = \hat{D}_{-\bar{r}}\hat{D}_{S^{-1}\bar{r}}\hat{S} = e^{i\bar{r}^{\mathsf{T}}\Omega S^{-1}\bar{r}/2}\hat{D}_{(S^{-1}-1)\bar{r}}\hat{S}, \tag{5.5}$$

where the Weyl representation of the CCR (3.11) was inserted. Up to an irrelevant overall phase, any second-order Hamiltonian can hence be decomposed in a sequence of Weyl displacement operators generated by linear Hamiltonians and Gaussian unitary operations generated by purely quadratic Hamiltonians. This argument, along with the repeated application of the relation (5.3) to the form of the most general Gaussian state (3.60), puts on firm, rigorous ground the intuition that a unitary operation is Gaussian if and only if it is generated by a second-order Hamiltonian.

Once untwined, the two classes of Gaussian unitary operations may be treated separately, as will be the case in the remainder of this section.

5.1.1 Linear displacements

Unitary operators generated by linear Hamiltonians form the class of Weyl operators, already introduced as $\hat{D}_{\mathbf{r}} = e^{i\mathbf{r}^{\mathsf{T}}\Omega\hat{r}}$ for $\mathbf{r} \in \mathbb{R}^{2n}$ and also known as 'linear displacements'. It is plain to see from Eq. (3.16) that one such operator acts as follows on the first moment vector $\bar{\mathbf{r}} = \mathrm{Tr}[\varrho_{G}\hat{r}]$ and covariance matrix $\boldsymbol{\sigma} = \mathrm{Tr}[\varrho\{(\hat{r} - \bar{r}), (\hat{r} - \bar{r})^{\mathsf{T}}\}]$ of a generic state ϱ. It is worth recalling here that a Gaussian state ϱ_{G} is uniquely determined by the first and second moments, so that the mapping below completely characterises the evolution of Gaussian states under displacement operators.

Linear displacements on statistical moments. The unitary action $\hat{D}_{\mathbf{r}}^{\dagger} \varrho \hat{D}_{\mathbf{r}}$ of a linear displacement $\hat{D}_{\mathbf{r}}$ maps the initial first moments $\bar{\mathbf{r}}$ and second moments $\boldsymbol{\sigma}$ of the state ϱ according to

$$\bar{\mathbf{r}} \mapsto \bar{\mathbf{r}} + \mathbf{r} , \tag{5.6}$$

$$\boldsymbol{\sigma} \mapsto \boldsymbol{\sigma} . \tag{5.7}$$

First moments may hence be arbitrarily adjusted by unitary displacement operators. Therefore, no properties depending only on the spectrum of the state, such as entropies, may depend on first moments (note that this could have already been inferred from Eq. (3.60), where the independence of the spectrum from the first moments is manifest). Besides, generic Weyl operators acting on a multimode system are by construction tensor products of Weyl operators, each acting independently on the Hilbert space of their mode. Hence, no properties invariant under local unitary operations, such as the mutual information and all entanglement measures, may depend on the first moments. First moments may then be disregarded, and are typically set to zero, in the analysis of all such properties, which allows for a substantial simplification. Information is still contained in the first moments though, and they do in fact play a prominent role in certain scenarios, such as the design and implementation of quantum communication or metrological protocols, as we shall see in Chapter 8.

In practice, controlled linear operations may be implemented by driving optical cavities with laser light in a strong coherent state (see Chapter 9 for more details on the description of driven systems) or by letting the system interact with a classical current, which enacts a linear Hamiltonian. More generally, first-order operations may be realised by coupling the system's canonical operators quadratically to external canonical variables, and preparing the latter in coherent states with high optical amplitude (i.e., with $|\alpha| \gg 1$ in our notation), where the canonical operators may be approximated by their average values and the Hamiltonian becomes effectively linear rather than quadratic.[1]

5.1.2 Symplectic transformations

The effect of an operation $\hat{S} = e^{\frac{i}{2}\hat{\mathbf{r}}^{\mathsf{T}} H \hat{\mathbf{r}}}$, generated by a purely quadratic Hamiltonian, on the parameters $\bar{\mathbf{r}}$ and $\boldsymbol{\sigma}$, and hence on generic Gaussian states, is easily inferred from Eq. (3.24) and from the definitions $\bar{\mathbf{r}} = \text{Tr}[\varrho_G \hat{\mathbf{r}}]$ and

---

[1]This "classical" approximation will be discussed in more detail in Section 5.4.1, for the description of homodyne detection.

$\boldsymbol{\sigma} = \mathrm{Tr}[\{(\hat{\mathbf{r}} - \bar{\mathbf{r}}), (\hat{\mathbf{r}} - \bar{\mathbf{r}})^{\mathsf{T}}\}]$. One thus obtains the following characterisation of symplectic actions.

Symplectic transformations on statistical moments. The unitary action $\hat{S}^{\dagger} \varrho \hat{S}$ of a quadratic operator $\hat{S} = \mathrm{e}^{\frac{i}{2}\hat{\mathbf{r}}^{\mathsf{T}} H \hat{\mathbf{r}}}$ maps the initial first moments $\bar{\mathbf{r}}$ and second moments $\boldsymbol{\sigma}$ of the state ϱ according to

$$\bar{\mathbf{r}} \mapsto S\bar{\mathbf{r}} \,, \tag{5.8}$$

$$\boldsymbol{\sigma} \mapsto S\boldsymbol{\sigma}S^{\mathsf{T}} \,, \tag{5.9}$$

where $S = \mathrm{e}^{\Omega H}$.

Symplectic transformations – which, let us remind the reader, are defined as the linear transformations that satisfy $S\Omega S^{\mathsf{T}} = \Omega$, and thus preserve the anti-symmetric symplectic form Ω under congruence – admit a number of useful decompositions. Among them, the most consequential in the context of physics is arguably the so called "Euler" (or "Bloch–Messiah") decomposition, whose traditional name refers to a faint analogy with the Euler decomposition of orthogonal transformations, but which is actually nothing but the singular value decomposition of a symplectic matrix. The latter takes a very specific form due to the strong constraints imposed by the symplectic condition and yields the decomposition of a symplectic transformation S into compact and non-compact symplectic transformations, as per:

Singular value decomposition of symplectic transformations. Any symplectic matrix $S \in Sp_{2n,\mathbb{R}}$ can be decomposed as

$$S = O_1 Z O_2 \,, \tag{5.10}$$

with $O_1, O_2 \in Sp_{2n,\mathbb{R}} \cap SO(2n)$ and

$$Z = \bigoplus_{j=1}^{n} \begin{pmatrix} z_j & 0 \\ 0 & z_j^{-1} \end{pmatrix} \,. \tag{5.11}$$

The proof of this statement may be found in Section B.2 of Appendix B, where several other properties of the symplectic group are derived and discussed in some detail. There, it is also proven that the compact symplectic subgroup $Sp_{2n,\mathbb{R}} \cap SO(2n)$ is isomorphic to $U(n)$, through a direct construction given by Eq. (B.19), that allows one to parametrise any orthogonal symplectic in terms of a unitary matrix. This property will come in handy below.

5.1.2.1 Passive transformations: Phase shifters and beam splitters

The transformations O_1 and O_2, belonging to $SO(2n)$ as well as $Sp_{2n,\mathbb{R}}$, preserve the identity, which is typically identified with the 'free' Hamiltonian matrix of a continuous variable system, since it corresponds to an equally weighted sum of squares of all the canonical operators: $\hat{\mathbf{r}}^{\mathsf{T}}\hat{\mathbf{r}}$ in our vector notation. Hence, such transformations are also referred to as "energy-preserving" or "passive" transformations. In the quantum optics laboratory, they model passive elements, essentially beam splitters (semi-reflective mirrors that mix up two modes) and phase shifters (dielectric plates that rotate the optical phase of a travelling electromagnetic wave with respect to a given reference). More precisely, a phase shifter is simply a local rotation

$$R_\varphi = \begin{pmatrix} \cos\varphi & \sin\varphi \\ -\sin\varphi & \cos\varphi \end{pmatrix}, \tag{5.12}$$

where φ represents the phase shift of the polarisation vector, while a beam splitter is described by an orthogonal on two modes mixing x's and p's in the same way:

$$R_\theta = \begin{pmatrix} \cos\theta & 0 & \sin\theta & 0 \\ 0 & \cos\theta & 0 & \sin\theta \\ -\sin\theta & 0 & \cos\theta & 0 \\ 0 & -\sin\theta & 0 & \cos\theta \end{pmatrix}, \tag{5.13}$$

where $(\cos\theta)^2$ is the transmittivity of the semi-reflective mirror.

Note that, if one switches to complex variables $\alpha_j = (x_j + ip_j)/\sqrt{2}$ for $j \in [1,\ldots,n]$, n being the total number of modes, a phase shifter just multiplies α_1 by a phase $e^{i\varphi}$, while a beam splitter rotates α_1 and α_2. Hence, by employing the univocal parametrisation (B.19) of orthogonal symplectic transformations in terms of unitary matrices, and by observing that any unitary can be decomposed into a product of diagonal phase multiplications and 2×2 real rotations,[2] one may conclude that:

Decomposition of passive transformations. *Any passive symplectic transformation may be decomposed as a product of phase shifters and beam splitters.*

It is revealing, and of crucial importance in applications, to derive the Hamiltonian that generates beam splitter and phase shifter transformations. This will also offer the opportunity to review the Gaussian formalism developed in Chapter 3. Let us start with phase shifts: a symplectic generator is always equal to the symplectic form Ω times a symmetric Hamiltonian matrix φH_{PS}; hence, assuming the transformation R_φ may be obtained as a single exponentiation, one has $R_\varphi = e^{\Omega H_{PS}\varphi}$, where the explicit real parameter φ at the exponential allows one to reproduce the whole one-parameter set of transformations. In practice, assuming a dimensionless Hamiltonian matrix, φ would be the product of time and interaction strength (time had been previously implicitly absorbed in the Hamiltonian matrix for mathematical

---

[2]In turn, this is a consequence of the fact that any two-dimensional unitary transformation U_2 may be written as $U_2 = \mathrm{diag}(e^{i\varphi_1}, e^{i\varphi_2})R_\varphi\mathrm{diag}(e^{i\varphi_3}, e^{-i\varphi_3})$, and that any n-dimensional unitary U is the composition of two-dimensional unitaries.

convenience). In principle, one could now take the matrix logarithm of R_φ to obtain H_{PS} but, in this simple 2×2 case, it is straightforward to see that the choice $\Omega H_{PS}\varphi = \Omega\varphi$ will do.[3] Therefore, the Hamiltonian matrix equal to the identity generates phase shifters. At the Hilbert space level, this corresponds to the Hamiltonian operator given by

$$\hat{H}_{PS} = \frac{1}{2}\hat{\mathbf{r}}^\mathsf{T} H_{PS}\hat{\mathbf{r}} = \frac{1}{2}\left(\hat{x}_1^2 + \hat{p}_1^2\right) . \tag{5.14}$$

The "free", harmonic oscillator Hamiltonian is thus responsible for phase shifters: most natural systems are subject to such a Hamiltonian at all times, and hence rotate in the phase space. The effect of the always-on free Hamiltonian is however typically disregarded by moving to the rotating frame (see Chapter 9 for details), and phase shifters are enacted by dielectric slabs or other means that vary the optical path length of the field, and thus speed up or slow down its rotation with respect to the time-dependent phase reference.

Let us now investigate the Hamiltonian generating the beam splitter transformation: it is apparent from Eq. (5.13) that such a transformation acts as two identical two-dimensional rotations on separate sets of phase space variables, so that the problem of identifying the generator may be reduced to the same 2×2 question that was tackled in the phase shifter's case. By analogy, the generating Hamiltonian must hence satisfy

$$\Omega H_{BS} = \begin{pmatrix} 0 & 0 & 1 & 0 \\ 0 & 0 & 0 & 1 \\ -1 & 0 & 0 & 0 \\ 0 & -1 & 0 & 0 \end{pmatrix} \Rightarrow H_{BS} = \begin{pmatrix} 0 & 0 & 0 & -1 \\ 0 & 0 & 1 & 0 \\ 0 & 1 & 0 & 0 \\ -1 & 0 & 0 & 0 \end{pmatrix}, \tag{5.15}$$

which corresponds to the operator

$$\hat{H}_{BS} = \frac{1}{2}\hat{\mathbf{r}}^\mathsf{T} H_{BS}\hat{\mathbf{r}} = \hat{p}_1\hat{x}_2 - \hat{x}_1\hat{p}_2 = i(\hat{a}_1^\dagger\hat{a}_2 - \hat{a}_1\hat{a}_2^\dagger). \tag{5.16}$$

This coupling swaps excitations between the two modes and is what takes place in semi-reflective mirrors. One has $R_\theta = e^{\Omega H_{BS}\theta}$, with the parameter θ given in practice by the product of interaction time and coupling strength. Typically, in quantum optics, beam splitters, like phase shifters, are not always-on elements but rather act as a whole on incoming travelling waves, and are thus often described as discrete operations rather than Hamiltonians. If $\theta = \pi/4$, one realises a so called 50:50, or "balanced", beam splitter, where half of the light goes through and the other half is reflected.

---

[3] The exponentiation of a 2×2 anti-symmetric block $\Omega\varphi = \begin{pmatrix} 0 & \varphi \\ -\varphi & 0 \end{pmatrix}$ yields

$$e^{\Omega\varphi} = \sum_{j=0}^{\infty} \frac{\Omega^j\varphi^j}{j!} = \sum_{j\,\text{even}} \frac{\Omega^j\varphi^j}{j!} + \sum_{j\,\text{odd}} \frac{\Omega^j\varphi^j}{j!} = \cos\varphi\mathbb{1} + \sin\varphi\Omega = \begin{pmatrix} \cos\varphi & \sin\varphi \\ -\sin\varphi & \cos\varphi \end{pmatrix},$$

where we used $\Omega^2 = -\mathbb{1}$.

5.1.2.2 Squeezing transformations

Going back to the singular value decomposition (5.10), it is clear that it can be interpreted by stating that any symplectic operation, and hence any purely quadratic operation at the Hilbert space level, can be decomposed as the product of passive operations and local squeezing operations, where by a local squeezing S_{SQ} we intend a symplectic acting on a single mode by contracting a canonical variable and expanding the conjugate one:

$$S_{SQ} = \begin{pmatrix} z_j & 0 \\ 0 & z_j^{-1} \end{pmatrix}. \tag{5.17}$$

As already shown in Problem 3.2 (and re-iterated in Problem 5.1 below), these transformations are generated by the Hamiltonian operator $\hat{x}_1\hat{p}_1 = i(\hat{a}_1^{\dagger 2} - \hat{a}_1^2)$. Such a Hamiltonian operator is not positive, since it corresponds to a Hamiltonian matrix which is not positive, and would thus not be thermodynamically stable, in the sense that its thermal Gibbs state is not even defined. In practice, such Hamiltonians are obtained as approximations of stable Hamiltonians acting on the relevant degrees of freedom plus ancillary ones, in regimes where the ancillae may be eliminated from the effective quantum dynamics. For instance, in quantum optics, single-mode squeezing operations are obtained by employing nonlinear crystals pumped by an accessory laser field with annihilation operator $\hat{b}$, in set-ups which are referred to as *degenerate* parametric oscillators or parametric amplifiers (depending on whether the light field is in a cavity or a travelling wave). The original coupling Hamiltonian is proportional to $i(\hat{a}_1^{\dagger 2}\hat{b} - \hat{a}_1^2\hat{b}^\dagger)$ (where the squared $\hat{a}_1^2$ and $\hat{a}_1^{\dagger 2}$ are a signature of the 'degenerate' nature of the parametric process), but the strong laser field operators $\hat{b}$ and $\hat{b}^\dagger$ may be replaced by c-numbers β and β^*, yielding the desired effective Hamiltonian $i|\beta|(\hat{a}_1^{\dagger 2} - \hat{a}_1^2)$, up to a phase factor (see Section 5.4.1 for a justification of such replacements).[4] Squeezing operations thus require energy, provided by the external laser, as reflected by the fact that they do not commute with the free Hamiltonian $\hat{\mathbf{r}}^\mathsf{T}\hat{\mathbf{r}}$.

As a further relevant piece of analysis, let us point out that, by combining the singular value decomposition and the discussion in the previous section, we have constructively shown that:

Decomposition of symplectic transformations. *Any symplectic transformation, on any number of modes, may be decomposed as a product of single-mode squeezers and phase shifters and two-mode beam splitters.*

---

[4] Note that the degrees of freedom of the nonlinear crystal, briefly mentioned above, have already been eliminated due to their fast dissipation rates in order to obtain the degenerate parametric coupling used here. Such a standard elimination method, termed 'adiabatic elimination', where the dynamics of quickly dissipating degrees of freedom is slave to that of less noisy ones, will not be discussed here.

Problem 5.1. (*Generator of single-mode squeezing*). Show that the single-mode squeezing transformations (5.17) are generated by the Hamiltonian operator $\hat{x}_1\hat{p}_1$.

Solution. One has $\hat{x}_1\hat{p}_1 = \frac{1}{2}\hat{\mathbf{r}}^\mathsf{T}\sigma_x\hat{\mathbf{r}}$, in terms of the Pauli matrix σ_x, whence one can determine the symplectic generator $\Omega\sigma_x = \sigma_z$. It is clear that $S_{SQ} = \mathrm{e}^{\sigma_z r}$, where r would in practice equal the product of interaction time and interaction strength of the degenerate parametric process.

Problem 5.2. (*More squeezing*). Determine the class of single-mode symplectic transformations generated by the Hamiltonian $\frac{1}{2}(\hat{p}^2 - \hat{x}^2)$.

Solution. The Hamiltonian under scrutiny may be written as $-\frac{1}{2}\hat{\mathbf{r}}^\mathsf{T}\sigma_z\hat{\mathbf{r}}$, where σ_z is the z Pauli matrix. This generates the class of transformations $\mathrm{e}^{-\Omega\sigma_z r} = \mathrm{e}^{\sigma_x r}$, in terms of the Pauli matrix σ_x, whose exponentiation is easily handled by observing that $\sigma_x^2 = \mathbb{1}$. One thus gets $\mathrm{e}^{\sigma_x r} = \sum_{j \text{ even}} \frac{r^j}{j!}\mathbb{1} + \sum_{j \text{ odd}} \frac{r^j}{j!}\sigma_x = \cosh(r)\mathbb{1} + \sinh(r)\sigma_x$. This is another form of single-mode squeezing which, it is plain to see, may be obtained by rotating the matrix S_{SQ} by a phase shifter with phase $\varphi = \pi/4$. Along with the generators of the phase shifters R_φ and S_{SQ}, this generator forms a basis of the three-dimensional symplectic algebra associated with $Sp_{2,\mathbb{R}}$.

Problem 5.3. (*Symmetric generators*). Let us consider a symplectic transformation $S = \mathrm{e}^{\Omega H}$ where the generator ΩH is symmetric, and thus has real eigenvalues. Show that any such Hamiltonian matrix different from zero cannot be positive.

Solution. The symmetry of ΩH implies that Ω and H anti-commute: $(\Omega H)^\mathsf{T} = -H\Omega = \Omega H$ (since $H = H^\mathsf{T}$ and $\Omega^\mathsf{T} = -\Omega$). Now, assume $\mathbf{v}^\mathsf{T} H\mathbf{v} > 0$ for some real vector $\mathbf{v}$, then one has

$$\mathbf{v}^\mathsf{T}\Omega^\mathsf{T} H\Omega\mathbf{v} = -\mathbf{v}^\mathsf{T} H\mathbf{v} < 0$$

by hypothesis. Thus, H cannot be positive nor negative definite, since for any vector on which the quadratic form H is positive one can identify an orthogonal vector on which it is negative.

Another pervasive class of symplectic transformations are the so called two-mode squeezing transformations S_r, whose symplectic matrix reads

$$S_r = \begin{pmatrix} \cosh r & 0 & \sinh r & 0 \\ 0 & \cosh r & 0 & -\sinh r \\ \sinh r & 0 & \cosh r & 0 \\ 0 & -\sinh r & 0 & \cosh r \end{pmatrix}, \qquad (5.18)$$

and is generated by the Hamiltonian operator $i(\hat{a}_1^\dagger \hat{a}_2^\dagger - \hat{a}_1 \hat{a}_2)$. This Hamiltonian is also not positive: in practice it is realised through the process of *non-degenerate* parametric down-conversion, where the system modes are coupled through a crystal to a laser field with annihilation operator $\hat{b}$ via the coupling Hamiltonian $i(\hat{a}_1^\dagger \hat{a}_2^\dagger \hat{b} - \hat{a}_1 \hat{a}_2 \hat{b}^\dagger)$, and the laser mode operators may be replaced by their expectation values, as in the single-mode, degenerate case discussed above.

Problem 5.4. (*Two-mode squeezing Hamiltonian*). Show that the two-mode squeezing transformations S_r are generated by the Hamiltonian $\hat{H}_{TB} = i(\hat{a}_1^\dagger \hat{a}_2^\dagger - \hat{a}_1 \hat{a}_2)$.

Solution. Let us switch to quadrature operators and write

$$\hat{H}_{TB} = i(\hat{a}_1^\dagger \hat{a}_2^\dagger - \hat{a}_1 \hat{a}_2) = \frac{1}{2} \hat{\mathbf{r}}^\mathsf{T} H_{TB} \hat{\mathbf{r}}, \qquad (5.19)$$

with

$$H_{TB} = \begin{pmatrix} 0 & 0 & 0 & 1 \\ 0 & 0 & 1 & 0 \\ 0 & 1 & 0 & 0 \\ 1 & 0 & 0 & 0 \end{pmatrix}. \qquad (5.20)$$

The corresponding class of symplectic transformations is then obtained as $S_r = e^{\Omega H_{TB} r}$, which yields Eq. (5.18).

The application of the transformation S_r on the two-mode vacuum state yields a pure Gaussian state with null first moments and covariance matrix $\sigma_r = S_r \mathbb{1}_4 S_r^\mathsf{T}$:

$$\sigma_r = S_r S_r^\mathsf{T} = \begin{pmatrix} \cosh(2r) & 0 & \sinh(2r) & 0 \\ 0 & \cosh(2r) & 0 & -\sinh(2r) \\ \sinh(2r) & 0 & \cosh(2r) & 0 \\ 0 & -\sinh(2r) & 0 & \cosh(2r) \end{pmatrix}. \qquad (5.21)$$

This state is clearly entangled for all $|r| > 0$, since the purity of the local state is just given by $1/\cosh(2r)$, with local von Neumann entropy that grows indefinitely in the limit $r \to \infty$ (see the next section for a formal clarification about partial traces of Gaussian states). It is, in fact, the celebrated 'two-mode squeezed state' with squeezing parameter r, also known in optics under the evocative byname of "twin-beam". As stated above, this quantum state represents the ideal, noiseless, output of degenerate parametric down-conversion processes, a standard way to produce optical continuous variable entanglement.

At the Hilbert space level, the Gaussian state in question is given by $|\psi_r\rangle = e^{-i\hat{H}_{TBr}}|0,0\rangle = e^{(\hat{a}_1^\dagger \hat{a}_2^\dagger - \hat{a}_1 \hat{a}_2)r}|0,0\rangle$. The Schmidt decomposition of this entangled state may also be determined, by observing that

$$|\psi_r\rangle = e^{(\hat{a}_1^\dagger \hat{a}_2^\dagger - \hat{a}_1 \hat{a}_2)r}|0,0\rangle = \frac{1}{\cosh r} \sum_{j=0}^{\infty} \tanh(r)^j |j,j\rangle , \qquad (5.22)$$

where the notation $|j,j\rangle$ stands for a tensor product of Fock states. Eq. (5.22) can be proven by differentiating the left- and right-hand sides with respect to r, and then inserting the original equation into the equation for the differentials, obtaining

$$\sum_{j=0}^{\infty} \tanh(r)^j (\hat{a}_1^\dagger \hat{a}_2^\dagger - \hat{a}_1 \hat{a}_2)|j,j\rangle = \sum_{j=0}^{\infty} \left(j \tanh(r)^{j-1} - (j+1)\tanh(r)^{j+1} \right)|j,j\rangle ,$$

$$(5.23)$$

which holds for all values of r. As $r \to \infty$, the state $|\psi_r\rangle$ tends to an evenly weighted superposition of tensor products of Fock states and hence approaches a maximally entangled state. In fact, in this limit the state $|r\rangle$ tends to the common (improper) eigenvector of the commuting quadrature operators $(\hat{x}_1 - \hat{x}_2)$ and $(\hat{p}_1 + \hat{p}_2)$ with eigenvalue 0 (as can be seen directly by applying such operators on the state, see also Section 5.4.2, where this argument finds a very direct application). Such an improper state, central to the seminal Einstein–Podolski–Rosen *gedanken* experiment concerning the reality and completeness of quantum mechanics, is thus approximated arbitrarily well by two-mode squeezed states.

It will also be expedient to our future efforts to define the unnormalised Hilbert space vector

$$|\xi\rangle = \lim_{r \to \infty} \cosh(r)|\psi_r\rangle = \sum_{j=0}^{\infty} |j,j\rangle . \qquad (5.24)$$

Two-mode squeezed states feature prominently in the theoretical description of Gaussian operations, and will appear again later on in the present chapter.

Problem 5.5. (*EPR states*). Show that $|\psi_r\rangle$ tends to the zero eigenvector of $(\hat{x}_1 - \hat{x}_2)$ and $(\hat{p}_1 + \hat{p}_2)$ in the limit $r \to \infty$, and to the zero eigenvector of $(\hat{x}_1 + \hat{x}_2)$ and $(\hat{p}_1 - \hat{p}_2)$ for $r \to -\infty$.

Solution. For $r \to +\infty$, $|\psi_r\rangle$ tends, up to normalisation, to the vector

$|\xi\rangle$ of Eq. (5.24). One has

$$(\hat{a}_1 - \hat{a}_2^\dagger + \hat{a}_1^\dagger - \hat{a}_2)\sum_{j=0}^{\infty}|j,j\rangle = \sum_{j=0}^{\infty}(\sqrt{j+1} - \sqrt{j+1})|j,j+1\rangle$$

$$+ \sum_{j=0}^{\infty}(\sqrt{j+1} - \sqrt{j+1})|j+1,j\rangle = 0,$$

(5.25)

$$(\hat{a}_1 - \hat{a}_2^\dagger - \hat{a}_1^\dagger + \hat{a}_2)\sum_{j=0}^{\infty}|j,j\rangle = \sum_{j=0}^{\infty}(\sqrt{j+1} - \sqrt{j+1})|j,j+1\rangle$$

$$- \sum_{j=0}^{\infty}(\sqrt{j+1} - \sqrt{j+1})|j+1,j\rangle = 0.$$

(5.26)

Observing that $\sqrt{2}(\hat{x}_1 - \hat{x}_2) = (\hat{a}_1 + \hat{a}_1^\dagger - \hat{a}_2 - \hat{a}_2^\dagger)$ and $\sqrt{2}i(\hat{p}_1 + \hat{p}_2) = (\hat{a}_1 - \hat{a}_1^\dagger + \hat{a}_2 - \hat{a}_2^\dagger)$ proves that this state is the common null eigenstate we were looking for.

As $r \to -\infty$, $|\psi_r\rangle$ tends, up to normalisation, to the vector $\sum_{j=0}^{\infty}(-1)^j|j,j\rangle$ (since $\tanh(r)^j = (-1)^j \tanh(|r|)$). The same cancellation shown above then occurs for $\sqrt{2}(\hat{x}_1 + \hat{x}_2) = (\hat{a}_1 + \hat{a}_1^\dagger + \hat{a}_2 + \hat{a}_2^\dagger)$ and $\sqrt{2}i(\hat{p}_1 - \hat{p}_2) = (\hat{a}_1 - \hat{a}_1^\dagger - \hat{a}_2 + \hat{a}_2^\dagger)$.

These are hence indeed the Einstein–Podolski–Rosen states, where the measurement of one local quadrature fixes the value of the other one. The full set of common eigenvectors of these pairs of commuting operators will be determined in Section 5.4.2.

Problem 5.6. (*Structure of pure Gaussian states*). Prove that the covariance matrix $\boldsymbol{\sigma}_p$ of a pure Gaussian state may be written as

$$\boldsymbol{\sigma}_p = ODO^\mathsf{T},$$

(5.27)

where O is passive: $O \in Sp_{2n,\mathbb{R}} \cap SO(2n)$ and $D = \mathrm{diag}(d_1, 1/d_1, d_2, 1/d_2, \ldots, d_n, 1/d_n)$, with $d_j > 0 \; \forall \; j \in [1,\ldots,n]$.

Solution. A pure Gaussian state is characterised by having all symplectic eigenvalues equal to 1. That is, by the fact that the normal form of its covariance matrix is the identity matrix $\mathbb{1}_{2n}$. Hence, $\boldsymbol{\sigma}_p = SS^\mathsf{T}$ for some symplectic transformation S. The statement above is then just a consequence of applying the singular value decomposition (5.10) to the covariance matrix $\boldsymbol{\sigma}_p = SS^\mathsf{T} = O_1 Z^2 O_1^\mathsf{T}$. Note that this implies that all single-mode covariance matrices belonging to pure states are just rotated squeezed states, and entirely specified by 2 real parameters. In general, the covariance matrix of an n-mode pure Gaussian state is characterised by $n^2 + n$ independent parameters (since $Sp_{2n,\mathbb{R}} \cap SO(2n) \sim U(n)$, which has n^2 free parameters).

Problem 5.7. (*Structure of single-mode Gaussian states*). Prove that the covariance matrix σ_1 of a single-mode Gaussian state may be written as

$$\sigma_1 = \nu O D O^{\mathsf{T}} , \qquad (5.28)$$

where R_φ is a phase shifter transformation as per Eq. (5.12), $D = \mathrm{diag}(d, 1/d)$, with $d > 0$, and $\nu \geq 1$.

Solution. This is a straightforward consequence of the singular value decomposition (5.10) applied to the normal mode decomposition $\sigma_1 = \nu S S^{\mathsf{T}}$, with $D = Z^2$ (note also that the set of phase shifters represents all single-mode orthogonal symplectic transformations).

5.2 TENSOR PRODUCTS AND PARTIAL TRACES OF GAUSSIAN STATES

Appending and discarding systems are also basic manipulations which preserve the Gaussian character of the states involved. Their description in the Gaussian domain may be readily inferred from the characteristic function formalism.

Given two Gaussian states ϱ_A and ϱ_B of any number of modes m and n, with first moments $\bar{\mathbf{r}}_A$ and $\bar{\mathbf{r}}_B$ and covariance matrices σ_A and σ_B, one has, using Eq. (4.49),

$$
\begin{aligned}
\varrho_A \otimes \varrho_B &= \int_{\mathbb{R}^{2(m+n)}} d\mathbf{r} \, \frac{e^{-\frac{1}{4}\mathbf{r}_A^{\mathsf{T}}\Omega^{\mathsf{T}}\sigma_A\Omega\mathbf{r}_A + \mathbf{r}_B^{\mathsf{T}}\Omega^{\mathsf{T}}\sigma_B\Omega\mathbf{r}_B + i\mathbf{r}_A^{\mathsf{T}}\Omega^{\mathsf{T}}\bar{\mathbf{r}}_A + i\mathbf{r}_B^{\mathsf{T}}\Omega^{\mathsf{T}}\bar{\mathbf{r}}_B}}{(2\pi)^{m+n}} \hat{D}_{\mathbf{r}_A} \otimes \hat{D}_{\mathbf{r}_B} \\
&= \frac{1}{(2\pi)^{m+n}} \int_{\mathbb{R}^{2(m+n)}} d\mathbf{r} \, e^{-\frac{1}{4}\mathbf{r}^{\mathsf{T}}\Omega^{\mathsf{T}}\sigma\Omega\mathbf{r} + i\mathbf{r}^{\mathsf{T}}\Omega^{\mathsf{T}}\bar{\mathbf{r}}} \hat{D}_{\mathbf{r}} , \qquad (5.29)
\end{aligned}
$$

where $\mathbf{r} = \mathbf{r}_A \oplus \mathbf{r}_B$, $\bar{\mathbf{r}} = \bar{\mathbf{r}}_A \oplus \bar{\mathbf{r}}_B$, $\sigma = \sigma_A \oplus \sigma_B$ and we have used the fact that the tensor product of two Weyl operators is a Weyl operator too. Hence, the tensor product of two Gaussian states is clearly still a Gaussian state, characterised as follows.

Tensor products of Gaussian states. The global vector of first moments $\bar{\mathbf{r}}$ and covariance matrix $\boldsymbol{\sigma}$ of the tensor product of two Gaussian states ϱ_A and ϱ_B, with first moments $\bar{\mathbf{r}}_A$ and $\bar{\mathbf{r}}_B$ and covariance matrices $\boldsymbol{\sigma}_A$ and $\boldsymbol{\sigma}_B$, are given by

$$\bar{\mathbf{r}} = \begin{pmatrix} \bar{\mathbf{r}}_A \\ \bar{\mathbf{r}}_B \end{pmatrix} = \bar{\mathbf{r}}_A \oplus \bar{\mathbf{r}}_B , \tag{5.30}$$

$$\boldsymbol{\sigma} = \begin{pmatrix} \boldsymbol{\sigma}_A & 0 \\ 0 & \boldsymbol{\sigma}_B \end{pmatrix} = \boldsymbol{\sigma}_A \oplus \boldsymbol{\sigma}_B . \tag{5.31}$$

Note that the composition properties above can also be derived directly from the definitions of first and second statistical moments of the states.

Conversely, the partial trace of a Gaussian state over subsystems A and B is evaluated by pinching out the relevant elements of the first-moments vector and covariance matrix:

Partial trace of a Gaussian state. Let

$$\bar{\mathbf{r}} = \begin{pmatrix} \bar{\mathbf{r}}_A \\ \bar{\mathbf{r}}_B \end{pmatrix} \quad \text{and} \quad \boldsymbol{\sigma} = \begin{pmatrix} \boldsymbol{\sigma}_A & \boldsymbol{\sigma}_{AB} \\ \boldsymbol{\sigma}_{AB}^{\mathsf{T}} & \boldsymbol{\sigma}_B \end{pmatrix} \tag{5.32}$$

be the first-moments vector and the covariance matrix of a Gaussian state ϱ_{AB} describing the subsystems A and B with, respectively, m and n modes (the dimension of the vectors and submatrices in the equations above is understood, e.g., the correlation block $\boldsymbol{\sigma}_{AB}$ is a $2m \times 2n$ matrix). Then, the state $\mathrm{Tr}_B[\varrho_{AB}]$ is a Gaussian state with a vector of first moments $\bar{\mathbf{r}}_A$ and a covariance matrix $\boldsymbol{\sigma}_A$.

Problem 5.8. (*Gaussian partial traces*). Prove the statement above.
Solution. Starting from Eq. (4.49) for a general $m + n$-mode Gaussian state ϱ_G

$$\varrho_G = \frac{1}{(2\pi)^{m+n}} \int_{\mathbb{R}^{2(m+n)}} \mathrm{d}\mathbf{r}\, e^{-\frac{1}{4}\mathbf{r}^{\mathsf{T}}\Omega^{\mathsf{T}}\boldsymbol{\sigma}\Omega\mathbf{r} + i\mathbf{r}^{\mathsf{T}}\Omega^{\mathsf{T}}\bar{\mathbf{r}}} \hat{D}_{\mathbf{r}_A} \otimes \hat{D}_{\mathbf{r}_B} , \tag{5.33}$$

with $\boldsymbol{\sigma}$ and $\bar{\mathbf{r}}$ as in Eq. (5.32) and $\mathbf{r} = \mathbf{r}_A \oplus \mathbf{r}_B$, and applying $\mathrm{Tr}_B \hat{D}_{\mathbf{r}_B} = (2\pi)^n \delta^{2n}(\mathbf{r}_B)$, one obtains

$$\mathrm{Tr}_B \varrho_G = \frac{1}{(2\pi)^m} \int_{\mathbb{R}^{2m}} \mathrm{d}\mathbf{r}_A\, e^{-\frac{1}{4}\mathbf{r}_A^{\mathsf{T}}\Omega^{\mathsf{T}}\boldsymbol{\sigma}_A\Omega\mathbf{r}_A + i\mathbf{r}_A^{\mathsf{T}}\Omega^{\mathsf{T}}\bar{\mathbf{r}}_A} \hat{D}_{\mathbf{r}_A} , \tag{5.34}$$

which proves the statement.

The agility with which tensor products and partial traces are dealt with is where the Gaussian formalism really shines. Such operations, which are certainly not equally accessible at the general Hilbert space level, are in fact key to the modelling of open systems, which we shall endeavour in the next section. As is often the case, this ease of description is a mixed blessing, since it reflects the features that make Gaussian quantum systems mimicable with classical means.

5.3   DETERMINISTIC GAUSSIAN CP-MAPS

The open dynamics resulting from considering an ancillary system – an 'environment' – in an initial Gaussian state, coupling such an ancilla to the system of interest through a quadratic Hamiltonian, and finally tracing out the ancilla preserves, as we have seen, the Gaussian character of the initial state. The sequence of operations described above gives rise to a class of Gaussian CP-maps, or "channels", capable of describing noise and decoherence. Such maps do not involve the probabilistic element associated with the outcome of a measurement, and we shall emphasise this aspect by adopting the slightly unorthodox term of *deterministic* Gaussian completely positive maps, in order to distinguish them from maps involving the conditioning of a quantum state due to measurements. In mathematical terms, deterministic maps preserve the trace of any input density operator, and are therefore commonly known as trace-preserving CP-maps. Yet another designation for this set of Gaussian quantum maps, often utilised in the quantum channels literature, is that of "bosonic Gaussian channels".

Note that a map described operationally in terms of a classical probability distribution depending on an inaccessible, auxiliary classical variable, but involving no act of measurement (which will crop up in the following), is still deterministic according to this terminology as it acts deterministically on an input density operator.

Deterministic Gaussian CP-maps are succinctly and exhaustively characterised as follows:

Complete characterisation of deterministic Gaussian CP-maps.
Given an initial Gaussian state of n modes with vector of first moments $\bar{\mathbf{r}}$ and covariance matrix σ, the evolution due to a deterministic Gaussian CP-map is completely described by two $2n \times 2n$ real matrices X and Y, which act as follows on the statistical moments:

$$\bar{\mathbf{r}} \mapsto X\bar{\mathbf{r}}\,, \tag{5.35}$$

$$\sigma \mapsto X\sigma X^\mathsf{T} + Y\,. \tag{5.36}$$

The matrices X and Y must be such that

$$Y + i\Omega \geq iX\Omega X^\mathsf{T}\,. \tag{5.37}$$

Eq. (5.37) ensures that enough additive noise, represented by Y, is acting for the final state to satisfy the uncertainty relation (3.77). Conversely, any X and Y satisfying the inequality (5.37) correspond to an open Gaussian dynamics.

An equivalent characterisation, but readily applicable to generic states, which we shall derive in Section 5.3.1 analysing the action of these CP-maps on Weyl operators, is the following:

General action of a Gaussian CP-map. Let $\chi_\varrho(\mathbf{r})$ be the characteristic function associated with the quantum state ϱ. The set of deterministic Gaussian CP-maps $\{\Phi\}$ is defined by the mapping:

$$\chi_\varrho(\mathbf{r}) \mapsto \chi_{\Phi(\varrho)}(\mathbf{r}) = e^{-\frac{1}{4}\mathbf{r}^\mathsf{T}\Omega^\mathsf{T} Y \Omega \mathbf{r}} \chi_\varrho(\Omega^\mathsf{T} X^\mathsf{T} \Omega \mathbf{r}) , \qquad (5.38)$$

for all X and Y satisfying (5.37).

Showing that any open Gaussian dynamics derived from a quadratic interaction and partial tracing over a Gaussian environment results in a map of the form (5.36) is straightforward and instructive. Let us assume that a system of n bosonic modes interacts with m environmental modes. The symplectic matrix describing the joint evolution of system and environment may be split into four submatrices, two of which, A and D, describe the internal evolution of system and environment, while the other two, B and C, issue from the quadratic coupling between system and environment:

$$S = \begin{pmatrix} A & B \\ C & D \end{pmatrix} . \qquad (5.39)$$

Notice that the most general quadratic coupling is contained in this description. Since S is symplectic, the sub-matrices A, B, C and D satisfy the following matrix equality

$$S\Omega S^\mathsf{T} = \begin{pmatrix} A\Omega_n A^\mathsf{T} + B\Omega_m B^\mathsf{T} & A\Omega_n C^\mathsf{T} + B\Omega_m D^\mathsf{T} \\ C\Omega_n A^\mathsf{T} + D\Omega_m B^\mathsf{T} & C\Omega_n C^\mathsf{T} + D\Omega_m D^\mathsf{T} \end{pmatrix} = \begin{pmatrix} \Omega_n & 0 \\ 0 & \Omega_m \end{pmatrix} , \qquad (5.40)$$

where $\Omega = \Omega_n \oplus \Omega_m$, and Ω_k is a symplectic form of k degrees of freedom.

The action of the CP-map is obtained by tracing out the environmental degrees of freedom after the action of S on the global state, i.e., by considering the diagonal block of $S(\sigma \oplus \sigma_E)S^\mathsf{T}$ pertaining to the system, where σ_E is the initial CM of the environment. The matrix σ_E is only constrained by the physicality condition (3.77). This evaluation yields the evolution of the covariance matrix σ as in Eq. (5.36), with

$$X = A \quad \text{and} \quad Y = B\sigma_E B^\mathsf{T} . \qquad (5.41)$$

The uncertainty principle (3.77) on the CM of the environment, $\sigma_E + i\Omega_m \geq 0$, implies, for any $n \times m$ matrix B,

$$B\sigma_E B^\mathsf{T} + iB\Omega_m B^\mathsf{T} \geq 0 . \qquad (5.42)$$

Because of Eq. (5.40) above, one has $B\Omega_m B^{\mathsf{T}} = \Omega_n - A\Omega_n A^{\mathsf{T}}$, which can be inserted into the previous expression to get

$$B\boldsymbol{\sigma}_E B^{\mathsf{T}} + i\Omega_n - iA\Omega_n A^{\mathsf{T}} \geq 0 , \qquad (5.43)$$

which is indeed identical to the relationship (5.37) between the matrices X and Y, thus showing that Eqs. (5.35)–(5.37) are necessary for CP-maps with a symplectic Stinespring dilation.

The proof of the converse statement, that any pair of matrices X and Y fulfilling the Inequality (5.37) corresponds to a deterministic Gaussian CP-map, is rather more technical in nature. Let us then indulge in some redundancy and report this claim here as a self-standing statement, followed by a formal proof.[5]

Sufficiency of the Gaussian CP-map description. Given $2n \times 2n$ X and Y fulfilling inequality (5.37), one may find a Gaussian Stinespring dilation which acts on a Gaussian state as per Eqs. (5.35) and (5.36). Moreover, the initial state of the environment may always be chosen to be the vacuum state – with covariance matrix $\mathbb{1}$ and zero first moments – in a space of $2n$ modes.

Proof. As will be apparent *à posteriori*, restricting to the case $m = 2n$ (number of environmental modes equal to twice the number of system modes) and $\boldsymbol{\sigma}_E = \mathbb{1}$ (zero temperature environment) will suffice to reproduce all of the CP-maps in question. Such assumptions will be hence made in what follows.

Eq. (5.41) shows that, given X and Y, one may reproduce the Gaussian CP-map through a symplectic reduction if and only if $A = X$. The choice of B allows instead for some freedom: by setting $\boldsymbol{\sigma}_E = \mathbb{1}$ and fixing the dimension of B as discussed above one may write

$$B = \sqrt{Y}O , \qquad (5.44)$$

with O is a $2n \times 4n$ real matrix with orthonormal rows. Note that, since $i\Omega - iX\Omega X^{\mathsf{T}}$ is anti-symmetric and hence yields no contribution if contracted with a real vector, Inequality (5.37) implies $Y \geq 0$, ensuring the existence of $\sqrt{Y}$. We must then find an O such that A and B satisfy the symplectic condition (5.40), i.e., such that

$$X\Omega_n X^{\mathsf{T}} + \sqrt{Y}O\,\Omega_{2n}O^{\mathsf{T}}\sqrt{Y} = \Omega_n . \qquad (5.45)$$

We can now restrict to the case of a strictly positive Y,[6] so that the matrix $Y^{1/2}$ may be assumed to be invertible, and the condition (5.45)

---

[5]This argument is courtesy of Ludovico Lami.

may be recast as

$$O\Omega_{2n}O^{\mathsf{T}} = Y^{-1/2}\left(\Omega_n - X\Omega_n X^{\mathsf{T}}\right)Y^{-1/2}. \tag{5.46}$$

The next step in the demonstration is showing that, for any X and Y, an O exists such that Eq. (5.46) is fulfilled.

The anti-symmetric $Y^{-1/2}\left(\Omega - X\Omega X^{\mathsf{T}}\right)Y^{-1/2}$ may be brought, through the action by similarity of an orthogonal $R \in O(2n)$, to the canonical form:

$$RY^{-1/2}\left(\Omega - X\Omega X^{\mathsf{T}}\right)Y^{-1/2}R^{\mathsf{T}} = \left(\bigoplus_{j=1}^{n} d_j \Omega_1\right) \tag{5.47}$$

(see the footnote on page 43). Notice now that the condition (5.37), when applied on such a canonical form, implies $d_j \leq 1 \; \forall j$.

On the other hand, each of the blocks $d_j\Omega_1$ may be obtained, on the left-hand side of Eq. (5.46), by applying, on the 4×4 matrix Ω_2, the 2×4 matrix O chosen as follows:

$$O = \begin{pmatrix} \cos\theta_j & 0 & -\sin\theta_j & 0 \\ 0 & \cos\theta_j & 0 & \sin\theta_j \end{pmatrix}, \tag{5.48}$$

so that, by setting $\sin(2\theta_j) = d_j$, one has

$$OO^{\mathsf{T}} = \begin{pmatrix} 1 & 0 \\ 0 & 1 \end{pmatrix} \quad \text{and} \quad O\begin{pmatrix} \Omega_1 & 0 \\ 0 & \Omega_1 \end{pmatrix}O^{\mathsf{T}} = \begin{pmatrix} 0 & d_j \\ -d_j & 0 \end{pmatrix}. \tag{5.49}$$

We have thus proven that, for all real X and Y satisfying (5.37), there exist real A and B such that $X = A$, $Y = BB^{\mathsf{T}}$ and $A\Omega_n A^{\mathsf{T}} + B\Omega_{2n}B^{\mathsf{T}} = \Omega_n$. The latter matrix equation can be recast by stating that the $2n$ vectors $\mathbf{v}_j$ forming the rows of the matrix $(A \; B)$ verify $\mathbf{v}_j \Omega \mathbf{v}_k^{\mathsf{T}} = \Omega_{jk}$, where $\Omega = \begin{pmatrix} \Omega_n & 0 \\ 0 & \Omega_{2n} \end{pmatrix}$ is the symplectic form of the bipartite system. It is left to show that matrices C and D exist such that $S = \begin{pmatrix} A & B \\ C & D \end{pmatrix}$ is symplectic, which is equivalent to stating that one can extend such vectors $\mathbf{v}_j$ to a global symplectic basis for Ω.

This can always be accomplished, as for any $2N \times 2N$ anti-symmetric matrix Ω, any set of $2s$ row vectors $\mathbf{v}_j$ such that $\mathbf{v}_j \Omega \mathbf{v}_k^{\mathsf{T}} = (\Omega_s)_{jk}$ for $j, k \in [1, \ldots, 2s]$, where Ω_s is the $2s \times 2s$ standard symplectic form of s modes, can be completed to a basis of $2N$ linearly independent vectors such that $\mathbf{v}_j \Omega \mathbf{v}_k^{\mathsf{T}} = (\Omega_N)_{jk}$ for $j, k \in [1, \ldots, 2N]$. In turn, this is the case since the first $2s$ vectors may be completed to a linearly independent basis and then orthogonalised with respect to the symplectic product according to the mapping $\mathbf{v}_j \mapsto \mathbf{v}_j - \sum_{k,l \leq 2s}(\Omega_s)_{kl}\mathbf{v}_j\Omega\mathbf{v}_l^{\mathsf{T}}\mathbf{v}_k$, for all $j \in [2s+1, \ldots, 2N]$. One then just needs to rotate the added orthogonal vectors to bring the antisymmetric matrix of their symplectic products in the canonical form (again, as per the footnote on page 43) and then rescale them to achieve a symplectic basis.

The proof that each pair of X and Y that verify (5.37) correspond to a deterministic Gaussian CP-map, resulting from the reduction of a symplectic dynamics on a larger space, is thus complete.

We have also shown that, as stated, such a reduction may always be constructed by considering an environment of $2n$ degrees of freedom (where n are the degrees of freedom of the system) initially in a pure state (as the CM $\sigma_E = \mathbb{1}$ has determinant 1 and is hence associated with a pure Gaussian state). □

Let us note that, as should be clear from the constructive definition of this set of channels, symplectic transformations are encompassed in this description: if $X \in Sp_{2n,\mathbb{R}}$, then $X\Omega X^\mathsf{T} = \Omega$ and the condition (5.37) reduces to $Y \geq 0$, which allows for the case $Y = 0$ corresponding to a noiseless (unitary) symplectic transformation. The case $X = \mathbb{1}$ and $Y \geq 0$ is very interesting too, as we will see shortly.

It is also worth noting that the output of a deterministic Gaussian CP-map on subsystem A of a multipartite system in a Gaussian state with first and second moments

$$\bar{\mathbf{r}}_{in} = \begin{pmatrix} \bar{\mathbf{r}}_A \\ \bar{\mathbf{r}}_B \end{pmatrix} \quad , \quad \sigma_{in} = \begin{pmatrix} \sigma_A & \sigma_{AB} \\ \sigma_{AB}^\mathsf{T} & \sigma_B \end{pmatrix} \tag{5.52}$$

is the Gaussian state with first and second moments

$$\bar{\mathbf{r}}_{out} = \begin{pmatrix} X\bar{\mathbf{r}}_A \\ \bar{\mathbf{r}}_B \end{pmatrix} \quad , \quad \sigma_{out} = \begin{pmatrix} X\sigma_A X^\mathsf{T} + Y & X\sigma_{AB} \\ \sigma_{AB}^\mathsf{T} X^\mathsf{T} & \sigma_B \end{pmatrix} . \tag{5.53}$$

A formal proof of this statement may be derived by adopting the Fourier-Weyl relation and the action of the CP-map on Weyl operators, which we will determine in Section 5.3.1 below (see Problem 5.9).

---

[6] This can be done without loss of generality by observing that, if one shows an O can be found for each $Y > 0$ such that (5.45) is fulfilled, then, for any $Y \geq 0$, one has that for all $\varepsilon > 0$ there exists an O_ε such that

$$X\Omega_n X^\mathsf{T} + \sqrt{Y + \varepsilon\mathbb{1}} \, O_\varepsilon \, \Omega_{2n} \, O_\varepsilon^\mathsf{T} \sqrt{Y + \varepsilon\mathbb{1}} = \Omega_n \tag{5.50}$$

(true since $Y + \varepsilon\mathbb{1} > 0$). As the set of matrices O's with orthogonal rows is compact, since it may inherit the topology of the compact set $O(4n)$, in the limit $\varepsilon \to 0$ there exists a converging subsequence of O_ε whose limit O_0 is also contained in the set of possible O's. Then one can apply the $\varepsilon \to 0$ limit on that subsequence to the equation above and obtain

$$X\Omega_n X^\mathsf{T} + \sqrt{Y} \, O_0 \, \Omega_{2n} \, O_0^\mathsf{T} \sqrt{Y} = \Omega_n . \tag{5.51}$$

We can hence restrict the remainder of the proof to the case of a positive definite Y.

5.3.1 Dual Gaussian CP-maps and action on Weyl operators

Let us now discuss in some detail the characterisation of the 'dual' of a Gaussian CP-map, which will turn out to be useful later on in describing noisy measurements, and explore some of its consequences. Given a generic CP-map Φ, its dual CP-map Φ^* is defined by the following relation:

$$\text{Tr}\left[\hat{A}\Phi^*(\hat{B})\right] = \text{Tr}\left[\Phi(\hat{A})\hat{B}\right] , \qquad (5.54)$$

for all bounded operators $\hat{A}$ and $\hat{B}$.

The dual of a trace-preserving Gaussian CP-map is also a Gaussian superoperator – in the sense that it preserves the Gaussian character of the input characteristic function – but is not necessarily trace-preserving (it is however, unital,[7] as always the case for the dual of a trace-preserving map).

In what follows, we denote by Φ^* the dual of a Gaussian CP-map Φ whose action is determined through Eqs. (5.35) and (5.36) by two matrices X, Y. Let us first understand how Φ^* acts on Weyl operators:

Dual of a deterministic Gaussian CP-map. Let Φ be the deterministic Gaussian CP-map parametrised by the matrices X and Y. Then, the dual map Φ^* acts on a Weyl operator $\hat{D}_{\Omega\mathbf{r}}$ as follows:

$$\Phi^*(\hat{D}_{\Omega\mathbf{r}}) = \hat{D}_{\Omega X^\mathsf{T}\mathbf{r}}\, e^{-\frac{1}{4}\mathbf{r}^\mathsf{T}Y\mathbf{r}} . \qquad (5.55)$$

Proof. The Stinespring representation of Φ allows us to write, for an arbitrary operator $\hat{A}$ (see the statement about the sufficiency of the Gaussian CP-map description on page 99)

$$\Phi(\hat{A}) = \text{Tr}_2\left[\hat{S}^\dagger\,\hat{A}\otimes|0\rangle\langle0|\,\hat{S}\right] , \qquad (5.56)$$

where $|0\rangle\langle0|$ is the vacuum state of an environment, labelled with 2, and $\hat{S}$ is a unitary related to a symplectic transformation S by

$$\hat{S}\,\hat{\mathbf{R}}\,\hat{S}^\dagger = S\,\hat{\mathbf{R}} , \qquad (5.57)$$

$$\hat{S}\,\hat{D}_{\mathbf{R}}\hat{S}^\dagger = \hat{D}_{S^{-1}\mathbf{R}} , \qquad (5.58)$$

where $\hat{\mathbf{R}}$ contains all the canonical operators of the bipartite system (including the environment), while $\mathbf{R}$ is the associated real vector which parametrises a global displacement operator [(Eq. (5.58) corresponds to Eq. (5.3)]. Moreover, we know from Eqs. (5.41) and (5.44) that S is rep-

---

[7]A unital CP-map is one that preserves the identity operator.

resented in block form as

$$S = \begin{pmatrix} X & \sqrt{Y}\,O \\ * & * \end{pmatrix}, \tag{5.59}$$

where O is an orthogonal transformation, while $*$ stands for generic, undefined matrix blocks. Now, by the very definition of dual map, we have, denoting with 1 and 2 the two subsystems in play:

$$\begin{aligned} \mathrm{Tr}_1\,[\hat{A}\,\Phi^*(\hat{B})] &= \mathrm{Tr}_1\,[\Phi(\hat{A})\hat{B}] = \mathrm{Tr}_{1,2}\,[\hat{S}^\dagger(\hat{A}\otimes|0\rangle\langle 0|)\hat{S}(\hat{B}\otimes\hat{\mathbb{1}})] \\ &= \mathrm{Tr}_1\,[\hat{A}\,\mathrm{Tr}_2[(\hat{\mathbb{1}}\otimes|0\rangle\langle 0|)\hat{S}(\hat{B}\otimes\hat{\mathbb{1}})\hat{S}^\dagger]]\,, \end{aligned} \tag{5.60}$$

for any operator $\hat{B}$. Since $\hat{A}$ is generic, this means that, for arbitrary $\hat{B}$,

$$\Phi^*(\hat{B}) = \mathrm{Tr}_2\,[(\hat{\mathbb{1}}\otimes|0\rangle\langle 0|)\hat{S}(\hat{B}\otimes\hat{\mathbb{1}})\hat{S}^\dagger]\,. \tag{5.61}$$

Applying this formula with $\hat{B} = \hat{D}_{\Omega\mathbf{r}}$ and inserting Eqs. (5.58) and (5.59), gives

$$\begin{aligned} \Phi^*(\hat{D}_{\Omega\mathbf{r}}) &= \mathrm{Tr}_2\,[(\hat{\mathbb{1}}\otimes|0\rangle\langle 0|)\hat{S}(\hat{D}_{\Omega\mathbf{r}}\otimes\hat{\mathbb{1}})\,\hat{S}^\dagger] \\ &= \mathrm{Tr}_2\,[(\hat{\mathbb{1}}\otimes|0\rangle\langle 0|)\hat{S}\hat{D}_{\Omega(\mathbf{r}\oplus 0)}\,\hat{S}^\dagger] = \mathrm{Tr}_2\,[(\hat{\mathbb{1}}\otimes|0\rangle\langle 0|)\hat{D}_{\Omega S^\mathsf{T}(\mathbf{r}\oplus 0)}] \\ &= \mathrm{Tr}_2\,[(\hat{\mathbb{1}}\otimes|0\rangle\langle 0|)\hat{D}_{(\Omega X^\mathsf{T}\mathbf{r}\,\oplus\,\Omega O^\mathsf{T}\sqrt{Y}\mathbf{r})}] \\ &= \hat{D}_{\Omega X^\mathsf{T}\mathbf{r}}\,\mathrm{Tr}\,[|0\rangle\langle 0|\,\hat{D}_{\Omega O^\mathsf{T}\sqrt{Y}\mathbf{r}}] = \\ &= \hat{D}_{\Omega X^\mathsf{T}\mathbf{r}}\,\chi_{|0\rangle\langle 0|}(-\Omega O^\mathsf{T}\sqrt{Y}\mathbf{r}) = \hat{D}_{\Omega X^\mathsf{T}\mathbf{r}}\,e^{-\frac{1}{4}\mathbf{r}^\mathsf{T}Y\mathbf{r}}\,, \end{aligned} \tag{5.62}$$

where we used $\hat{D}_0 = \hat{\mathbb{1}}$, $S^{-1}\Omega = \Omega S^\mathsf{T}$ and identified the characteristic function of the vacuum $\chi_{|0\rangle\langle 0|}(\mathbf{r}) = e^{-\frac{1}{4}|\mathbf{r}|^2}$. □

The relationship (5.55) completely determines the dual CP-map Φ^*, since the Weyl operators serve as a basis in operator space. For invertible X (i.e., $\mathrm{Det}\,X \neq 0$), to which we shall restrict for simplicity, the action of Φ^* on a generic Gaussian state ϱ_G is in fact also easily determined by applying the equation above to the Fourier-Weyl expansion of ϱ_G, given by Eq. (4.49):

$$\begin{aligned} \Phi^*(\varrho_G) &= \frac{1}{(2\pi)^n}\int_{\mathbb{R}^{2n}} e^{-\frac{1}{4}\tilde{\mathbf{r}}^\mathsf{T}\sigma\tilde{\mathbf{r}}+i\tilde{\mathbf{r}}^\mathsf{T}\bar{\mathbf{r}}}\,\Phi^*\big(\hat{D}_{\Omega\tilde{\mathbf{r}}}\big)\,\mathrm{d}\tilde{\mathbf{r}} \\ &= \frac{1}{(2\pi)^n}\int_{\mathbb{R}^{2n}} e^{-\frac{1}{4}\tilde{\mathbf{r}}^\mathsf{T}(\sigma+Y)\tilde{\mathbf{r}}+i\tilde{\mathbf{r}}^\mathsf{T}\bar{\mathbf{r}}}\,\hat{D}_{\Omega X^\mathsf{T}\tilde{\mathbf{r}}}\,\mathrm{d}\tilde{\mathbf{r}} \\ &= \frac{1}{(2\pi)^n|\mathrm{Det}\,X|}\int_{\mathbb{R}^{2n}} e^{-\frac{1}{4}\tilde{\mathbf{r}}^\mathsf{T}X^{-1}(\sigma+Y)X^{-1\mathsf{T}}\tilde{\mathbf{r}}+i\tilde{\mathbf{r}}^\mathsf{T}X^{-1}\bar{\mathbf{r}}}\,\hat{D}_{\Omega\tilde{\mathbf{r}}}\,\mathrm{d}\tilde{\mathbf{r}}\,, \tag{5.63} \end{aligned}$$

leading to:

Gaussian action of dual CP-maps. The dual of a Gaussian CP-map parametrised by X and Y, with X taken to be invertible, corresponds to a Gaussian CP-map with associated matrices X^* and Y^* given by

$$X^* = X^{-1} , \tag{5.64}$$

$$Y^* = X^{-1}YX^{-1\mathsf{T}} , \tag{5.65}$$

as well as a factor $1/|\mathrm{Det}X|$ multiplying the input density operator.

Note that $\mathrm{Tr}\left[\Phi^*\left(\varrho_G\right)\right] = \frac{1}{|\mathrm{Det}X|}$. The channel Φ^* is hence a Gaussian completely positive map, but not a trace-preserving one unless $|\mathrm{Det}X| = 1$. Since a quantum channel is trace-preserving if and only if its dual is unital, it follows that *a trace-preserving Gaussian CP-map Φ with X invertible is unital if and only if $|\mathrm{Det}X| = 1$*.

The transformation of a generic characteristic function $\chi(\mathbf{r})$ under the original CP-map Φ also follows directly from Eq. (5.55), by noting that

$$\mathrm{Tr}[\Phi(\varrho)\hat{D}_{-\mathbf{r}}] = \mathrm{Tr}[\varrho\Phi^*(\hat{D}_{-\mathbf{r}})] = e^{-\frac{1}{4}\mathbf{r}^{\mathsf{T}}\Omega^{\mathsf{T}}Y\Omega\mathbf{r}}\, \mathrm{Tr}[\varrho\hat{D}_{-\Omega^{\mathsf{T}}X^{\mathsf{T}}\Omega\mathbf{r}}] , \tag{5.66}$$

leading to Eq. (5.38), which has thus been derived independently.

Finally, we would like to establish the action of the original Gaussian CP-map Φ on Weyl operators, which is straightforward for invertible X, in light of the above. One may in fact assume the *ansatz* $\Phi(\hat{D}_{\Omega\mathbf{r}}) = c\,e^{-\frac{1}{4}\mathbf{r}^{\mathsf{T}}Y^*\mathbf{r}}\hat{D}_{\Omega X^*\mathbf{r}}$ for some $c \in \mathbb{R}$ and $2n \times 2n$ matrices X^* and Y^*, and apply Eq. (4.21) to write

$$e^{-\frac{1}{4}\bar{\mathbf{r}}^{\mathsf{T}}Y\bar{\mathbf{r}}}\delta^{2n}(\mathbf{r} + X^{\mathsf{T}}\bar{\mathbf{r}}) = \frac{e^{-\frac{1}{4}\bar{\mathbf{r}}^{\mathsf{T}}Y\bar{\mathbf{r}}}\mathrm{Tr}[\hat{D}_{\Omega\mathbf{r}}\hat{D}_{\Omega X^{\mathsf{T}}\bar{\mathbf{r}}}]}{(2\pi)^n} = \frac{\mathrm{Tr}[\hat{D}_{\Omega\mathbf{r}}\Phi^*(\hat{D}_{\Omega\bar{\mathbf{r}}})]}{(2\pi)^n}$$

$$= \frac{\mathrm{Tr}[\Phi(\hat{D}_{\Omega\mathbf{r}})\hat{D}_{\Omega\bar{\mathbf{r}}}]}{(2\pi)^n} = c\,e^{-\frac{1}{4}\mathbf{r}^{\mathsf{T}}Y^*\mathbf{r}}\delta^{2n}(X^*\mathbf{r} + \bar{\mathbf{r}}) , \tag{5.67}$$

which is verified for

$$c = \frac{1}{|\mathrm{Det}X|} , \quad X^* = X^{-1} , \quad Y^* = X^{-1}YX^{-1\mathsf{T}} . \tag{5.68}$$

This situation can be summarised by stating:

Action of Gaussian CP-maps on Weyl operators. A Gaussian CP-map characterised by X and Y, with X taken to be invertible, acts as follows on a Weyl operator:

$$\Phi(\hat{D}_{\Omega\mathbf{r}}) = \frac{e^{-\frac{1}{4}\mathbf{r}^{\mathsf{T}}X^{-1}YX^{-1\mathsf{T}}\mathbf{r}}}{|\mathrm{Det}X|}\hat{D}_{\Omega X^{-1\mathsf{T}}\mathbf{r}} . \tag{5.69}$$

Note that, since the dual map is unique, the constructive argument above ensures that, for invertible X, the dual map of a Gaussian CP-map is a Gaussian superoperator too, in the sense that it maps an operator with Gaussian characteristic function into another such operator. However, the dual map may not preserve the normalisation of quantum states. Also notice that the relationship (5.37) for X and Y is also satisfied by X^* and Y^*.

Problem 5.9. (*Partial action of Gaussian CP-maps*). Use the action of a deterministic Gaussian CP-map Φ on a Weyl operator to prove that the output of such a map, when acting on a subsystem of a Gaussian state, is given by Eq. (5.53).

Solution. Assuming subsystem A comprises m modes and subsystem B comprises n modes, the Fourier-Weyl relation (4.49) for the composite input state ϱ_{in} reads

$$\varrho_{in} = \int_{\mathbb{R}^{m+n}} \mathrm{d}\tilde{\mathbf{r}}_A \mathrm{d}\tilde{\mathbf{r}}_B \frac{e^{-\frac{1}{4}\left(\tilde{\mathbf{r}}_A^\mathsf{T}\sigma_A\tilde{\mathbf{r}}_A + 2\tilde{\mathbf{r}}_A^\mathsf{T}\sigma_{AB}\tilde{\mathbf{r}}_B + \tilde{\mathbf{r}}_B^\mathsf{T}\sigma_B\tilde{\mathbf{r}}_B\right) + i\left(\tilde{\mathbf{r}}_A^\mathsf{T}\bar{\mathbf{r}}_A + \tilde{\mathbf{r}}_B^\mathsf{T}\bar{\mathbf{r}}_B\right)}}{(2\pi)^{m+n}} \times$$
$$\times \hat{D}_{\Omega^\mathsf{T}\tilde{\mathbf{r}}_A} \otimes \hat{D}_{\Omega^\mathsf{T}\tilde{\mathbf{r}}_B} \tag{5.70}$$

where it was convenient to adopt the integration variables $\tilde{\mathbf{r}}_{A,B} = \Omega \mathbf{r}_{A,B}$, and we have exploited the fact that a composite, global Weyl operator is just the tensor product of local ones. We can then apply the Eq. (5.69), to write

$$\Phi(\varrho_{in}) = \int_{\mathbb{R}^{m+n}} \mathrm{d}\tilde{\mathbf{r}}_A \mathrm{d}\tilde{\mathbf{r}}_B \, \Phi(\hat{D}_{\Omega^\mathsf{T}\tilde{\mathbf{r}}_A}) \otimes \hat{D}_{\Omega^\mathsf{T}\tilde{\mathbf{r}}_B} \times$$
$$\times \frac{e^{-\frac{1}{4}\left(\tilde{\mathbf{r}}_A^\mathsf{T}\sigma_A\tilde{\mathbf{r}}_A + 2\tilde{\mathbf{r}}_A^\mathsf{T}\sigma_{AB}\tilde{\mathbf{r}}_B + \tilde{\mathbf{r}}_B^\mathsf{T}\sigma_B\tilde{\mathbf{r}}_B\right) + i\left(\tilde{\mathbf{r}}_A^\mathsf{T}\bar{\mathbf{r}}_A + \tilde{\mathbf{r}}_B^\mathsf{T}\bar{\mathbf{r}}_B\right)}}{(2\pi)^{m+n}}$$

$$= \int_{\mathbb{R}^{m+n}} \mathrm{d}\tilde{\mathbf{r}}_A \mathrm{d}\tilde{\mathbf{r}}_B \, \hat{D}_{\Omega^\mathsf{T}X^{-1\mathsf{T}}\tilde{\mathbf{r}}_A} \otimes \hat{D}_{\Omega^\mathsf{T}\tilde{\mathbf{r}}_B} e^{-\tilde{\mathbf{r}}_A^\mathsf{T}X^{-1}YX^{-1\mathsf{T}}\tilde{\mathbf{r}}_A} \times$$
$$\times \frac{e^{-\frac{1}{4}\left(\tilde{\mathbf{r}}_A^\mathsf{T}\sigma_A\tilde{\mathbf{r}}_A + 2\tilde{\mathbf{r}}_A^\mathsf{T}\sigma_{AB}\tilde{\mathbf{r}}_B + \tilde{\mathbf{r}}_B^\mathsf{T}\sigma_B\tilde{\mathbf{r}}_B\right) + i\left(\tilde{\mathbf{r}}_A^\mathsf{T}\bar{\mathbf{r}}_A + \tilde{\mathbf{r}}_B^\mathsf{T}\bar{\mathbf{r}}_B\right)}}{(2\pi)^{m+n}|\mathrm{Det}X|}$$

$$= \int_{\mathbb{R}^{m+n}} \mathrm{d}\tilde{\mathbf{r}}_A \mathrm{d}\tilde{\mathbf{r}}_B \, \hat{D}_{\Omega^\mathsf{T}\tilde{\mathbf{r}}_A} \otimes \hat{D}_{\Omega^\mathsf{T}\tilde{\mathbf{r}}_B} e^{-\tilde{\mathbf{r}}_A^\mathsf{T}Y\tilde{\mathbf{r}}_A} \times$$
$$\times \frac{e^{-\frac{1}{4}\left(\tilde{\mathbf{r}}_A^\mathsf{T}X\sigma_AX^\mathsf{T}\tilde{\mathbf{r}}_A + 2\tilde{\mathbf{r}}_A^\mathsf{T}X\sigma_{AB}\tilde{\mathbf{r}}_B + \tilde{\mathbf{r}}_B^\mathsf{T}\sigma_B\tilde{\mathbf{r}}_B\right) + i\left(\tilde{\mathbf{r}}_A^\mathsf{T}X\bar{\mathbf{r}}_A + \tilde{\mathbf{r}}_B^\mathsf{T}\bar{\mathbf{r}}_B\right)}}{(2\pi)^{m+n}}, \tag{5.71}$$

which does correspond to the output Gaussian state (5.53). The last step was just a change of variables to $X^{-1\mathsf{T}}\tilde{\mathbf{r}}_A$.

Before proceeding to the description of Gaussian quantum measurements, it is worthwhile to dwell on the properties of certain specific classes of deterministic Gaussian CP-maps, in view of their theoretical impact (which will find much illustration in the remainder of the book) and relevance to notable practical situations.

5.3.2 Classical mixing

Let us first consider the subset of channels for which $X = \mathbb{1}$: the inequality (5.37) then reduces to $Y \geq 0$: the only effect of the map is to add a positive semi-definite matrix Y to the covariance matrix of the input state. This type of channel has an interesting and consequential alternate interpretation to the unitary dilation picture that was adopted in the definition of the set of deterministic CP-maps.

Let us first assume $Y > 0$ (which implies $\mathrm{Det}Y > 0$), and consider the transformation that acts on a state by displacing it through a displacement operator $\hat{D}_{\mathbf{r}'}$ whose parameter $\mathbf{r}'$ is distributed according to the multivariate Gaussian probability distribution $e^{-\mathbf{r}'^\mathsf{T}Y^{-1}\mathbf{r}'}/(\pi^n\sqrt{\mathrm{Det}Y})$. The action of this map Φ_Y on an input state ϱ is defined as

$$\Phi_Y(\varrho) = \int_{\mathbb{R}^{2n}} \mathrm{d}\mathbf{r}' \frac{e^{-\mathbf{r}'^\mathsf{T}Y^{-1}\mathbf{r}'}}{\pi^n\sqrt{\mathrm{Det}Y}} \hat{D}_{\mathbf{r}'}\varrho\hat{D}_{\mathbf{r}'}^\dagger . \tag{5.72}$$

If the input is a Gaussian state ϱ_G with first moments $\bar{\mathbf{r}}$ and covariance matrix $\boldsymbol{\sigma}$, the effct of the map may be readily expressed in the characteristic function formalism:

$$\Phi_Y(\varrho_G) = \int_{\mathbb{R}^{4n}} \mathrm{d}\mathbf{r}'\mathrm{d}\mathbf{r} \frac{e^{-\mathbf{r}'^\mathsf{T}Y^{-1}\mathbf{r}' - \frac{1}{4}\mathbf{r}^\mathsf{T}\Omega^\mathsf{T}\boldsymbol{\sigma}\Omega\mathbf{r} + i\mathbf{r}^\mathsf{T}\Omega^\mathsf{T}\bar{\mathbf{r}}}}{(2\pi)^n\pi^n\sqrt{\mathrm{Det}Y}} \hat{D}_{\mathbf{r}'}\hat{D}_{\mathbf{r}}\hat{D}_{-\mathbf{r}'} \tag{5.73}$$

$$= \int_{\mathbb{R}^{4n}} \mathrm{d}\mathbf{r}'\mathrm{d}\mathbf{r} \frac{e^{-\mathbf{r}'^\mathsf{T}Y^{-1}\mathbf{r}' - \frac{1}{4}\mathbf{r}^\mathsf{T}\Omega^\mathsf{T}\boldsymbol{\sigma}\Omega\mathbf{r} + i\mathbf{r}^\mathsf{T}\Omega^\mathsf{T}\bar{\mathbf{r}} - i\mathbf{r}'^\mathsf{T}\Omega\mathbf{r}}}{(2\pi)^n\pi^n\sqrt{\mathrm{Det}Y}} \hat{D}_{\mathbf{r}} \tag{5.74}$$

$$= \frac{1}{(2\pi)^n} \int_{\mathbb{R}^{2n}} \mathrm{d}\mathbf{r}\, e^{-\frac{1}{4}\mathbf{r}^\mathsf{T}\Omega^\mathsf{T}(\boldsymbol{\sigma}+Y)\Omega\mathbf{r} + i\mathbf{r}^\mathsf{T}\Omega^\mathsf{T}\bar{\mathbf{r}}} \hat{D}_{\mathbf{r}} , \tag{5.75}$$

where the additional phase in (5.74) comes from the CCR (3.11) while the Gaussian integration in $\mathrm{d}\mathbf{r}'$ was carried out applying (1.18). Note that such an integration amounted to evaluating the Fourier transform of a multivariate Gaussian. As apparent from (5.75), random displacements distributed according to a classical Gaussian distribution with covariance matrix Y (which was in fact the covariance matrix of the classical displacement distribution) have the sole effect of adding the matrix Y to the input covariance matrix. The mathematical argument above may be summarised as:

Classical mixing. The CP-map

$$\Phi_Y(\varrho) = \int_{\mathbb{R}^{2n}} d\mathbf{r}' \frac{e^{-\mathbf{r}'^\mathsf{T} Y^{-1} \mathbf{r}'}}{\pi^n \sqrt{\mathrm{Det}Y}} \hat{D}_{\mathbf{r}'} \varrho \hat{D}_{\mathbf{r}'}^\dagger, \tag{5.76}$$

with $Y > 0$, corresponds to the Gaussian CP-map with $X = \mathbb{1}$ and Y as given.

Above, we have assumed $Y > 0$ in order to simplify the notation (so that the distribution could be normalised only in terms of $\mathrm{Det}Y$). The assumption of strict positivity is however not necessary, and it is very simple to convince oneself that any $Y \geq 0$ may be added to the original covariance matrix through a properly normalised classical mixing (by simply avoiding integration along the direction spanned by the null eigenvectors of Y).

Problem 5.10. (*Gaussian states as randomly displaced pure states*). Show that any Gaussian state may be written as the Gaussian classical mixing of a pure Gaussian state.

Solution. By virtue of the normal mode decomposition (3.58), the covariance matrix $\boldsymbol{\sigma}$ of a generic, n-mode Gaussian state may be written as $S\boldsymbol{\nu}S^\mathsf{T}$, where S is a symplectic transformation and $\boldsymbol{\nu} = \bigoplus_{j=1}^n (\nu_j \mathbb{1}_2)$, where the $\nu_j \geq 1$ for $j \in [1, \ldots, n]$ are the symplectic eigenvalues of $\boldsymbol{\sigma}$. We saw that a Gaussian state is pure if and only if $\boldsymbol{\nu} = \mathbb{1}$, so we can define a pure Gaussian state with covariance matrix $\boldsymbol{\sigma}_p = SS^\mathsf{T}$ and first moments equal to the target state to be reproduced. Then, one has $\boldsymbol{\sigma} = \Phi_{\boldsymbol{\sigma}-\boldsymbol{\sigma}_p}(\boldsymbol{\sigma}_p)$, which one may write since $\boldsymbol{\sigma} - \boldsymbol{\sigma}_p = S(\boldsymbol{\nu}-\mathbb{1})S^\mathsf{T} \geq 0$, and proves the statement.

5.3.3 Losses, attenuators and amplifiers

Let us now turn to CP-maps apt to describe the effect of quantum noise and decoherence, stemming from the coupling with an environment. In this section, alongside the general multimode definitions in the main text, we will include a number of significant specific properties of the single-mode versions of the channels in the form of sets of – often linked together – problems with associated solutions.

The interaction with a thermal environment of a single mode is represented, quite simply, by the Gaussian channel obtained by mixing the input state at a beam splitter with another mode in a Gaussian state with covariance matrix $(2N + 1)\mathbb{1}$, with $N \geq 0$, and null first moments. The latter is precisely the Gibbs thermal state associated to the free Hamiltonian, with mean number of thermal excitations equal to N.[8] We shall refer to this set of channels as attenuator channels.

If $N = 0$, the channel is also at times referred to as a 'pure-loss' channel, and describes the effect on the system of a zero-temperature bath. Since at room temperature and optical frequencies $N \ll 1$, the zero-temperature instance describes remarkably well the transmission of a signal undergoing losses along an optical fibre.

To keep our future notation more compact, it is expedient to re-parametrise this class of channels in terms of the 'thermal noise' $n_{th} = (2N + 1) \geq 1$. In the formalism of deterministic Gaussian CP-maps, attenuator channels admit the following characterisation:

Attenuator channels. An attenuator channel $\mathcal{E}_\theta^{n_{th}}$ is a deterministic Gaussian CP-map with

$$X = \cos\theta \mathbb{1}_2 \quad \text{and} \quad Y = (\sin\theta)^2 n_{th} \mathbb{1}_2 , \tag{5.77}$$

for $\theta \in [0, 2\pi[$ and $n_{th} \geq 1$.

The terminology "attenuators" refers to the reduction of the first moments' amplitude $\bar{\mathbf{r}}$ of an input state to $\cos\theta \, \bar{\mathbf{r}}$. In the case of pure loss channels ($n_{th} = 1$), also referred to as quantum limited attenuators, and input coherent states, this is done without adding any noise to the coherent state signal, in the sense that its covariance matrix stays unchanged (equal to the identity), and the input signal $|\alpha\rangle$ is mapped into $|\cos\theta \, \alpha\rangle$.

Problem 5.11. (*Attenuator channels*). Show that the class of attenuator channels, as defined above in terms of a beam splitter transformation on system and environment, is characterised by $X = \cos\theta \mathbb{1}_2$ and $Y = (\sin\theta)^2 n_{th} \mathbb{1}_2$.

Solution. It will be convenient to write down the beam splitter operation R_θ of Eq. (5.13) in terms of 2×2 matrix blocks:

$$R_\theta = \begin{pmatrix} \cos\theta \mathbb{1}_2 & \sin\theta \mathbb{1}_2 \\ -\sin\theta \mathbb{1}_2 & \cos\theta \mathbb{1}_2 \end{pmatrix} . \tag{5.78}$$

The action of the beam splitter on the covariance matrix on the tensor product of the input state with generic covariance matrix σ and the

---

[8]The parameter N may be connected to the inverse temperature β (with dimensions of energy) and the frequency ω of the environmental mode by the Bose statistics: $N = 1/(e^{\beta\hbar\omega} - 1)$.

environment with covariance matrix $n_{th}\mathbb{1}_2$ is represented as

$$R_\theta \left[\boldsymbol{\sigma} \oplus (n_{th}\mathbb{1}_2) \right] R_\theta^{\mathsf{T}} =$$

$$\begin{pmatrix} \cos\theta\mathbb{1}_2 & \sin\theta\mathbb{1}_2 \\ -\sin\theta\mathbb{1}_2 & \cos\theta\mathbb{1}_2 \end{pmatrix} \begin{pmatrix} \boldsymbol{\sigma} & 0 \\ 0 & n_{th}\mathbb{1}_2 \end{pmatrix} \begin{pmatrix} \cos\theta\mathbb{1}_2 & -\sin\theta\mathbb{1}_2 \\ \sin\theta\mathbb{1}_2 & \cos\theta\mathbb{1}_2 \end{pmatrix} =$$

$$\begin{pmatrix} \cos\theta\boldsymbol{\sigma} & \sin\theta n_{th}\mathbb{1}_2 \\ -\sin\theta\boldsymbol{\sigma} & \cos\theta n_{th}\mathbb{1}_2 \end{pmatrix} \begin{pmatrix} \cos\theta\mathbb{1}_2 & -\sin\theta\mathbb{1}_2 \\ \sin\theta\mathbb{1}_2 & \cos\theta\mathbb{1}_2 \end{pmatrix} =$$

$$\begin{pmatrix} n_{th}(\cos\theta)^2\boldsymbol{\sigma} + (\sin\theta)^2 n_{th}\mathbb{1}_2 & \sin\theta\cos\theta(n_{th}\mathbb{1}_2 - \boldsymbol{\sigma}) \\ \sin\theta\cos\theta(n_{th}\mathbb{1}_2 - \boldsymbol{\sigma}) & (\sin\theta)^2\boldsymbol{\sigma} + (\cos\theta)^2 n_{th}\mathbb{1}_2 \end{pmatrix} .$$

(5.79)

The partial trace over the environment of the final covariance matrix of Eq. (5.79) is just obtained by pinching the upper left 2×2 block, which corresponds to the system variables. The initial covariance matrix of the system $\boldsymbol{\sigma}$ is hence mapped to $(\cos\theta)^2\boldsymbol{\sigma} + (\sin\theta)^2 n_{th}\mathbb{1}_2$. The initial first moments $\bar{\mathbf{r}}\oplus 0$ (where, as customary, $0 = (0,0)^{\mathsf{T}}$ is the null vector of the environment's first moments) are transformed linearly by the beam splitter into $(\cos\theta\bar{\mathbf{r}}, -\sin\theta\bar{\mathbf{r}})^{\mathsf{T}}$. Again, partial tracing corresponds to just keeping the first moments pertaining to the system, which are hence $\cos\theta\bar{\mathbf{r}}$. Overall, this dynamics just corresponds to the action of the deterministic Gaussian channel with $X = \cos\theta\mathbb{1}_2$ and $Y = (\sin\theta)^2 n_{th}\mathbb{1}_2$, as stated.

Problem 5.12. (*Pure attenuator's outputs*). Show that if the output state of a quantum limited attenuator $\mathcal{E}_\theta^1$ with $\theta \neq \pi/2$ is pure:

$$\mathcal{E}_\theta^1(\varrho) = |\psi\rangle\langle\psi| , \quad \theta \neq \frac{\pi}{2} , \tag{5.80}$$

then the input ϱ must be a coherent state: $\varrho = \hat{D}_{\mathbf{r}}^\dagger|0\rangle\langle 0|\hat{D}_{\mathbf{r}}$.

Solution. Although several different paths may be taken, we shall tackle this problem through the formalism of the characteristic function, and by considering the Stinespring dilation of the attenuator channel as the mixing with the vacuum state at a beam splitter. The pure loss attenuator maps a characteristic function $\chi(\mathbf{r})$ as per $\chi(\mathbf{r}) \mapsto e^{-\frac{\sin\theta^2}{4}\mathbf{r}^{\mathsf{T}}\mathbf{r}}\chi(\cos\theta\,\mathbf{r})$ [see Eq. (5.38)]. Hence, aside from the trivial case $\theta = \pi/2$ where any state is mapped into the vacuum and any input yields a pure output, any two different characteristic functions will be mapped to different characteristic functions, and the same goes for the corresponding states. Hence, due to linearity, the output may be pure only if the input is pure too: $\mathcal{E}_\theta^1(\sum_j p_j|\psi_j\rangle\langle\psi_j|) = \sum_j p_j \mathcal{E}_\theta^1(|\psi_j\rangle\langle\psi_j|) = |\psi\rangle\langle\psi| \Leftrightarrow |\psi_j\rangle = |\psi_{in}\rangle \; \forall j$ (in this inference, we are using the fact that pure states are extremal in the convex set of density operators, and therefore cannot be expressed as the convex combination of two or more different states).

Let now $\hat{R}_\theta$ be the unitary corresponding to the beam splitter, that acts on the system and ancillary quadrature operators (pertaining to the vacuum mode that interacts with the system through the beam splitter) as the symplectic matrix R_θ of Eq. (5.13), and hence on the characteristic function of the composite system as per $\chi(\mathbf{r}) \mapsto \chi(R_\theta^{-1}\mathbf{r})$, where $\mathbf{r} = (\mathbf{r}_A \oplus \mathbf{r}_B)$ with $\mathbf{r}_A, \mathbf{r}_B \in \mathbb{R}^2$ describes the phase space variables of system and ancilla, respectively. The condition for the pure loss channel to have a pure output reads

$$\hat{R}_\theta^\dagger(|\psi_{in}\rangle \otimes |0\rangle) = |\psi\rangle \otimes |\psi_B\rangle \tag{5.81}$$

for some pure state of the ancilla $|\psi_B\rangle$ (for the global output to be pure, both its partial traces must be pure too). Eq. (5.81) can be recast in terms of characteristic functions as

$$\chi_{in}(\cos\theta\mathbf{r}_A - \sin\theta\mathbf{r}_B)e^{-\frac{1}{4}|\cos\theta\mathbf{r}_B+\sin\theta\mathbf{r}_A|^2} = \chi(\mathbf{r}_A)\chi_B(\mathbf{r}_B) \tag{5.82}$$

where, on the left-hand side, we have acted with the beam splitter on the initial characteristic function $\chi_{in}(\mathbf{r}_A)e^{-\frac{1}{4}|\mathbf{r}_B|^2}$. The characteristic function formalism enjoys a very expedient partial tracing, consisting in setting the variables traced upon to 0. Doing this alternatively for $\mathbf{r}_B$ and $\mathbf{r}_A$ yields (recall that any characteristic function evaluated at the origin is 1):

$$\chi(\mathbf{r}_A) = \chi_{in}(\cos\theta\mathbf{r}_A)e^{-\frac{1}{4}|\sin\theta\mathbf{r}_A|^2}, \tag{5.83}$$

$$\chi_B(\mathbf{r}_B) = \chi_{in}(-\sin\theta\mathbf{r}_B)e^{-\frac{1}{4}|\cos\theta\mathbf{r}_B|^2}, \tag{5.84}$$

which can be re-inserted in (5.82) to obtain, after the change of variables $\cos\theta\,\mathbf{r}_A \mapsto \mathbf{r}_A$ and $(-\sin\theta)\mathbf{r}_B \mapsto \mathbf{r}_B$,

$$\chi_{in}(\mathbf{r}_A)\chi_{in}(\mathbf{r}_B) = \chi_{in}(\mathbf{r}_A + \mathbf{r}_B)e^{\frac{1}{2}\mathbf{r}_A\cdot\mathbf{r}_B}. \tag{5.85}$$

Let us now switch to normally ordered characteristic functions, which we shall denote here with $\zeta_{in}(\mathbf{r}) = e^{\frac{1}{4}|\mathbf{r}|^2}\chi_{in}(\mathbf{r})$. If inserted in the expression above, this definition gives:

$$\zeta_{in}(\mathbf{r}_A)\zeta_{in}(\mathbf{r}_B) = \zeta_{in}(\mathbf{r}_A + \mathbf{r}_B). \tag{5.86}$$

The function $\zeta_{in}(\mathbf{r})$ is continuous and such that $\zeta_{in}(-\mathbf{r}) = \zeta_{in}(\mathbf{r})^*$. The only such function that satisfies Eq. (5.86) is a complex exponential, which can be written without loss of generality as $e^{i\mathbf{r}_A^\mathsf{T}\Omega^\mathsf{T}\mathbf{r}}$, for some $\mathbf{r} \in \mathbb{R}^2$. Going back to the symmetrically ordered characteristic function, one gets $\chi(\mathbf{r}_A) = e^{-\frac{1}{4}|\mathbf{r}_A|^2 + i\mathbf{r}_A^\mathsf{T}\Omega^\mathsf{T}\mathbf{r}}$, which is the characteristic function of the coherent state $\hat{D}_\mathbf{r}^\dagger|0\rangle$.

Note that the fact that any coherent state should yield a pure output state is an immediate consequence of the fact that the identity covariance matrix is obviously preserved by the channel. We have however shown that no other state enjoys such a property. Besides, the output corresponding to the input coherent state $\hat{D}_\mathbf{r}|0\rangle$ is also a coherent state, with reduced amplitude $\cos\theta\mathbf{r}$. This property, presented here as a curiosity, will be momentous in Chapter 8, when dealing with the information capacity of Gaussian CP-maps.

The converse, so to speak, of the attenuator channels is the class of amplifier channels, defined as the ones resulting from the interaction with a thermal environment – with covariance matrix $n_\mathsf{th}\mathbb{1}_2$ for $n_\mathsf{th} \geq 1$, and null first moments – through a two-mode squeezing, rather than a beam splitting, transformation. Amplifier channels admit the following characterisation:

Amplifier channels. An amplifier channel $\mathcal{A}_r^{n_\mathsf{th}}$ is a deterministic Gaussian CP-map with

$$X = \cosh r \mathbb{1}_2 \quad \text{and} \quad Y = (\sinh r)^2 n_\mathsf{th}\mathbb{1}_2 , \qquad (5.87)$$

for $r \in [0, \infty[$ and $n_\mathsf{th} \geq 1$.

In contrast with the attenuator class, the terminology 'amplifier' refers to the enhancement of the amplitude α of an input coherent state $|\alpha\rangle$ to $(\cosh r)\alpha$. If $n_\mathsf{th} = 1$, the channel is also known as a 'quantum limited amplifier', which enhances the optical signal of the input states whilst adding minimum noise, in the specific sense that the matrix inequality (5.37) is saturated for such a channel.

Problem 5.13. (*Amplifier channels*). Show that the class of amplifier channels, as defined above in terms of a beam splitter transformation on system and environment, is characterised by $X = \cosh r \mathbb{1}_2$ and $Y = (\sinh r)^2 n_\mathsf{th}\mathbb{1}_2$.

Solution. This proceeds exactly along the lines of the previous problem with the only difference that the beam splitter R_θ is replaced by the two-mode squeezing transformation S_r of Eq. (5.18)

$$S_r = \begin{pmatrix} \cosh r \mathbb{1}_2 & \sinh r \sigma_z \\ \sinh r \sigma_z & \cosh r \mathbb{1}_2 \end{pmatrix} , \tag{5.88}$$

where σ_z is the standard Pauli z matrix. The only property needed to achieve the form of X and Y as above is $\sigma_z^2 = \mathbb{1}_2$.

Problem 5.14. (*Quantum limited amplifiers*). Show that a quantum limited amplifier $\mathcal{A}_r^1$ partially saturates the inequality (5.37), in the sense that there exists one phase space direction along which the matrix inequality is an equality.

Solution. Inserting $X = \cosh r \mathbb{1}_2$ and $Y = (\sinh r)^2 \mathbb{1}_2$ into inequality (5.37) yields the following matrix inequality:

$$\begin{pmatrix} (\sinh r)^2 & -i(\sinh r)^2 \\ i(\sinh r)^2 & (\sinh r)^2 \end{pmatrix} \geq 0 . \tag{5.89}$$

The Hermitian matrix on the left-hand side has determinant 0, which means that one of its two eigenvalues is always 0, thus proving partial saturation of the inequality. The trace of the matrix would also equal 0 if $r = 0$: in such a case the inequality would be completely saturated but the channel would reduce to the trivial identity channel, which grants no amplification whatsoever.

Problem 5.15. (*Impossibility of fully quantum limited amplifiers*). Prove that no deterministic single-mode Gaussian channel that acts on all input vectors of first moments $\bar{\mathbf{r}}$ as an amplifier: $\bar{\mathbf{r}} \mapsto a\bar{\mathbf{r}} \; \forall \bar{\mathbf{r}} \in \mathbb{R}^2$, with $a > 1$, may ever completely saturate the inequality (5.37).

Solution. The amplifier condition, holding for all input $\bar{\mathbf{r}}$, implies that X is proportional to the identity: $X = a\mathbb{1}_2$, as can be seen from Eq. (5.35). In order for the inequality (5.37) to be completely saturated, both the real and imaginary parts of the left- and right-hand sides must equate, which implies $Y = 0$ and $\Omega = X\Omega X^\mathsf{T}$. But it is straightforward to see that only the trivial case $a = 1$ would allow X to be proportional to the identity and symplectic at the same time: no symplectic, 'noiseless' amplifiers may exist. Notice that this would not be true if one were interested in amplifying only along a certain phase space direction: a symplectic squeezing operation would then achieve the aim.

Problem 5.16. (*Phase-covariant channels*). Show that any phase-covariant single-mode Gaussian channel, i.e., any single-mode Gaussian channel Φ such that $\Phi(\hat{R}\varrho\hat{R}^\dagger) = \hat{R}\Phi(\varrho)\hat{R}^\dagger$ for all unitary $\hat{R}$ corresponding to a phase shifter transformation, is unitarily equivalent to a channel with $X = x\mathbb{1}_2$ and $Y = y\mathbb{1}$, for real x and y such that

$$y \geq |x^2 - 1| . \tag{5.90}$$

Solution. Since a bosonic Gaussian channel is entirely characterised by its action on Gaussian states, it is sufficient to consider a generic input Gaussian state with covariance matrix $\boldsymbol{\sigma}$ [the action of the channel may then be carried over to generic states through the relation (5.69)]. Denoting with R_φ all possible two-dimensional rotations, as given by Eq. (5.12), phase-covariance amounts to the fact that the channel, whose X and Y are still undetermined, commutes with the action of a phase shifter:

$$XR_\varphi\boldsymbol{\sigma}R_\varphi^\mathsf{T}X^\mathsf{T}+Y = R_\varphi X\boldsymbol{\sigma}X^\mathsf{T}R_\varphi^\mathsf{T}+R_\varphi YR_\varphi^\mathsf{T} \quad \forall\,\boldsymbol{\sigma}: \boldsymbol{\sigma}+i\Omega \geq 0 \tag{5.91}$$

and for all $\varphi \in [0, 2\pi[$. As the equality above must hold for all $\boldsymbol{\sigma}$, it is clear that the linear and constant parts of the equation above must equate separately and independently. This immediately implies that Y must be invariant under any rotation, which means it must be proportional to the identity. As for X, one is left with

$$(R_\varphi^\mathsf{T}XR_\varphi)\boldsymbol{\sigma}(R_\varphi^\mathsf{T}X^\mathsf{T}R_\varphi) = X\boldsymbol{\sigma}X^\mathsf{T}, \tag{5.92}$$

where one can insert $\boldsymbol{\sigma} = \mathbb{1}$ to get $R_\varphi^\mathsf{T}XX^\mathsf{T}R_\varphi = XX^\mathsf{T}$, which implies the necessary condition $XX^\mathsf{T} = \mathbb{1}$. Thus, it must be $X = xQ$ for some transformation $Q \in O(2)$ (beware: not necessarily in $SO(2)$, as the determinant of X can be minus or plus 1). Two-dimensional orthogonal transformations come in only two flavours: either $Q = R_\theta$ or $Q = \sigma_z R_\vartheta$, where σ_z is the Pauli z matrix and $R_\vartheta \in SO(2)$ for $\vartheta \in [0, 2\pi[$. Now, a major simplification to our task comes from the abelian nature of $SO(2)$, whereby $R_\varphi R_\vartheta = R(\varphi + \vartheta)$. If $Q = R_\vartheta$ is a special orthogonal – that is, a phase shifter – then Eq. (5.92) is always satisfied, for all $\boldsymbol{\sigma}$ and R_φ, as the abelian relation above implies that X is unchanged under any rotation acting by congruence (recall that $R_\varphi^\mathsf{T} = R_{-\varphi}$). If $Q = \sigma_z R_\vartheta$, then Eq. (5.92) turns into

$$R_{-\varphi}\sigma_z R_{\vartheta+\varphi}\boldsymbol{\sigma}R_{-\vartheta-\varphi}\sigma_z R_\varphi = \sigma_z R_{\vartheta+2\varphi}\boldsymbol{\sigma}R_{-\vartheta-2\varphi}\sigma_z = \sigma_z R_\vartheta\boldsymbol{\sigma}R_{-\vartheta}\sigma_z , \tag{5.93}$$

where we used the property $R_\varphi\sigma_z = \sigma_z R_{-\varphi}$. The latter equation cannot hold for generic φ and $\boldsymbol{\sigma}$, as can be seen by inserting the squeezed state $\boldsymbol{\sigma} = \mathrm{diag}(z, 1/z)$. Hence, it must be $X = xR_\vartheta$. Exploiting the rotational invariance of Y, one may therefore write the action of the whole channel on $\boldsymbol{\sigma}$ as $R_\vartheta(x^2\boldsymbol{\sigma}+y\mathbb{1})R_\vartheta^{\mathsf{T}}$, which proves the thesis (since the symplectic R_ϑ corresponds to a unitary at the Hilbert space level). The inequality (5.90) follows from applying the requirement (5.37) on phase-invariant X and Y.

Problem 5.17. (*Decomposition of phase-covariant channels*). Show that any single-mode, phase-covariant Gaussian channel is unitarily equivalent to the composition of a pure loss channel and a quantum limited amplifier $\mathcal{A}_r^1 \circ \mathcal{E}_\theta^1$.

Solution. The composition of a pure loss channel with mixing parameter θ followed by a quantum limited amplifier with parameter r results in a bosonic channel with

$$X = (\cosh r \cos\theta)\mathbb{1}_2 \quad \text{and} \quad Y = [(\cosh r)^2(\sin\theta)^2 + (\sinh r)^2]\mathbb{1}_2 \ .$$
$$(5.94)$$

Such matrices are obviously proportional to the identity and span the set of phase-covariant channels parametrised by x and y of the form $x = \cos\theta \cosh r$ and $y = [(\sin\theta)^2(\cosh r)^2 + (\sinh r)^2]$. It is left to be proven that the whole range of x and y set by the condition (5.90) may be reproduced. Given any x and y, the system above admits the following *bona fide* solution for $(\cos\theta)^2$ and $(\cosh r)^2$:

$$(\cos\theta)^2 = \frac{2x^2}{x^2+1+y} \quad \text{and} \quad (\cosh r)^2 = \frac{x^2+1+y}{2} \ , \qquad (5.95)$$

whose right-hand sides may be proven to be, respectively, smaller and greater than 1 by using the condition (5.90), thus proving the thesis. Note also that the sign of $\cos\theta$ is irrelevant to the action of the phase-covariant channel. This decomposition will come in handy in Chapter 8, when the information capacity of bosonic channels will be dealt with.

5.4   GAUSSIAN MEASUREMENTS

Our treatment of Gaussian operations continues now with a wide class of non-deterministic (non-trace-preserving) maps which encompass well-known detection schemes in optical set-ups, such as the homodyne and heterodyne protocols, with a longstanding, prominent status in engineering and signal processing.

After a cursory, preliminary remark about (non-Gaussian) photon counting detectors, to follow in this paragraph, we will build this class up from the homodyne and then heterodyne cases to a wider class of detections which have been termed "general-dyne", a terminology that we will embrace. We will then further extend this set to include noisy, imperfect detections. Along with the abstract, mathematical definition of the POVMs, we shall always explicitly construct practical schemes described in terms of optical elements and of photon counting detectors. The latter correspond to the resolution of the identity $\hat{\mathbb{1}} = \sum_{n=0}^{\infty} |n\rangle\langle n|$, where $|n\rangle$ is a number state of the observed mode, that is an eigenstate of the free Hamiltonian $\frac{1}{2}(\hat{x}^2 + \hat{p}^2)$ with associated eigenvalue $n+1/2$. This ideal projective measurement, implying the possibility of distinguishing between any two different numbers of excitations, is actually still quite challenging to realise in practice, especially at low energies. It is nevertheless rather standard, in the optical laboratory, for intense fields, which is the regime we will be mainly interested in since Gaussian measurements will always be aided by strong auxiliary laser pumps. Note that the projection on Fock states does not provide one with any information on the phase of the continuous variable system, as the characteristic functions of Fock states are invariant under phase space rotations. Although no Hermitian phase operator may be defined, phase and number are in a sense conjugate variables (an issue which will be briefly touched upon in Section 9.6, when we shall come across the quantisation of charge and flux). As we shall see, Gaussian measurements do instead bear information about the phase of the observed degrees of freedom (the optical phase, when optical fields are concerned).

All the POVMs in the general-dyne class have the property of preserving the Gaussian character of the initial state both when acting conditionally (that is, if the measurement outcome is recorded and the system is filtered according to such an outcome) and when acting unconditionally (that is, if a measurement takes place but the outcome is not recorded so that the system is not filtered according to it). Hence, they are often referred to simply as Gaussian measurements. It should perhaps be noted that the generic coarse graining of such POVMs – that is, a filtering of the system based on a set of outcomes rather than individual, granular ones – does not preserve the Gaussian character of the state, as it results in convex combinations of Gaussian states with the same second moments but different first moments, which are Gaussian themselves only if weighted by a Gaussian distribution, as in the case of classical mixing seen above (Sec. 5.3.2). This cheap way of devising non-Gaussian operations could be trivially reproduced by sampling a classical probability distribution and preparing Gaussian states based on it. Not surprisingly, this sort of non-Gaussian state is not so interesting, for reasons that will be made clear in Chapter 7, where we will also briefly point out some interesting cases of non-Gaussian post-selection.

On the other hand, there are clearly other POVMs that may send Gaussian states into Gaussian states upon the post-selection of certain outcomes. A notable specimen of this kind is photon counting, which results in a Gaussian state (the vacuum), when no photons are detected.[9] Further, let us remark

---

[9] In practice this post-selection may be achieved with very sensitive photodetectors, such

that more general, non-trivial POVMs do exist that send Gaussian states into Gaussian states upon any conditional outcome, and yet do not belong to the noisy general-dyne category. An ingenious example of such POVMs is illustrated in the following problem.[10]

Problem 5.18. (*An unusual Gaussian POVM*). Prove the following resolution of the identity:

$$\hat{\mathbb{1}} = \sum_{\lambda=0}^{1} \int_{\mathbb{R}^{2n}} \mathrm{d}\mathbf{r}\, g_\lambda(\mathbf{r})\, \hat{D}_{\mathbf{r}}^\dagger |\sigma_\lambda\rangle\langle\sigma_\lambda| \hat{D}_{\mathbf{r}} \,, \qquad (5.96)$$

where $|\sigma_\lambda\rangle$ is a pure Gaussian state with vanishing first moments and covariance matrix σ_λ (obeying the pure Gaussian state condition Det $\sigma_\lambda = 1$ for $\lambda = 0, 1$), $g_0(\mathbf{r}) = \frac{1}{(2\pi)^n}\left(1 - e^{-\mathbf{r}^T \sigma_1^{-1}\mathbf{r}}\right)$ and $g_1(\mathbf{r}) = \frac{1}{(2\pi)^n} e^{-\mathbf{r}^T \sigma_0^{-1}\mathbf{r}}$.

Solution. The characteristic representation of the identity operator is just $\mathrm{Tr}\left(\hat{D}_{-\bar{\mathbf{r}}}\right) = (2\pi)^n \delta^{2n}(\bar{\mathbf{r}})$. All we have to prove is that the characteristic function associated to the right-hand side of Eq. (5.96) is the same as this. The latter is simply the sum of integrals of the characteristic functions of pure Gaussian states, given by (4.48):

$$\sum_{\lambda=0}^{1} \int_{\mathbb{R}^{2n}} \mathrm{d}\mathbf{r}\, g_\lambda(\mathbf{r})\, e^{-\bar{\mathbf{r}}^T \Omega^T \sigma_\lambda \Omega \bar{\mathbf{r}} + i\bar{\mathbf{r}}^T \Omega^T \mathbf{r}} = (2\pi)^n \delta^{2n}(\bar{\mathbf{r}}) \,, \qquad (5.97)$$

since the $1/(2\pi)^n$ term in $g_0(\mathbf{r})$ yields the delta function once Fourier-transformed from $\mathbf{r}$ to $\bar{\mathbf{r}}$, while the other terms are designed so as to cancel each other out. The resolution of the identity above must correspond to a measurement process (with outcomes labelled by λ and $\mathbf{r}$), though which it may be in practice, the author cannot fathom.

---

as the class of avalanche photodiodes, that can reliably distinguish between zero and one or more photons (also known as "on/off detectors").

[10]Courtesy of Giacomo De Palma.

5.4.1 Homodyne detection

Let us start our discussion of Gaussian measurements with homodyne detection, which is nothing but a fancy designation, drawn from optics, for the projective measurement of a canonical operator:

Homodyne detection. A homodyne detection scheme consists in the measurement of the quadrature operator $\hat{x}_\varphi = \cos\varphi\ \hat{x} + \sin\varphi\ \hat{p}$, with outcome probabilities:

$$p(x_\varphi) = \langle x_\varphi | \varrho | x_\varphi \rangle ,\qquad (5.98)$$

where $|x_\varphi\rangle$ is an (improper) eigenvector of $\hat{x}_\varphi$.

Henceforth, we shall interchangeably refer to these projective POVMs as quadrature or homodyne measurements.

In optical set-ups, homodyne detection results from mixing the initial state ϱ with a strong coherent state $|\alpha\rangle$ (with $\alpha = |\alpha|e^{i\varphi}$ and $|\alpha| \gg 1$) at a balanced (50:50) beam splitter (given by Eq. (5.13) for $\varphi = \pi/4$), and from subtracting the detected intensities at the two outputs of the beam splitter (each measured with a photodetector). The adjective "homodyne" refers to the fact that the measured field is mixed with a probe at the same frequency.

Let us first provide the reader with a heuristic justification of why the optical scheme described above corresponds to a quadrature measurement. We will then flesh out a more detailed and rigorous argument in the following section. Notice first that the operator $\hat{x}_\varphi$ takes the following form in terms of the system ladder operators

$$\hat{x}_\varphi = \frac{e^{-i\varphi}\hat{a} + e^{i\varphi}\hat{a}^\dagger}{\sqrt{2}} .\qquad (5.99)$$

If $\hat{b}$ is the annihilation operator associated with the laser mode, prepared in the initial state $|\alpha\rangle_b$, the 50:50 beam splitter's outputs are described, in the Heisenberg picture, by the modes

$$\frac{\hat{a} + \hat{b}}{\sqrt{2}} \quad \text{and} \quad \frac{\hat{a} - \hat{b}}{\sqrt{2}} .\qquad (5.100)$$

Subtraction of the two intensities then corresponds to measuring the operator

$$\frac{1}{2}(\hat{a} + \hat{b})^\dagger(\hat{a} + \hat{b}) - \frac{1}{2}(\hat{a} - \hat{b})^\dagger(\hat{a} - \hat{b}) = \hat{a}^\dagger\hat{b} + \hat{a}\hat{b}^\dagger ,\qquad (5.101)$$

with outcome probabilities $p(x)$ which have to be evaluated on the initial state (since we described the evolution in the Heisenberg picture). If $|\alpha| \gg 1$, which we shall refer to as the "strong oscillator limit", then the state $|\alpha\rangle$, besides being by definition an eigenstate of $\hat{b}$, approximates an eigenstate of $\hat{b}^\dagger$ too, as

$$\hat{b}^\dagger|\alpha\rangle_b = \hat{b}^\dagger\hat{D}_\alpha|0\rangle_b = \hat{D}_\alpha(\hat{b}^\dagger + \alpha^*)|0\rangle_b = \alpha^*|\alpha\rangle_b + \hat{D}_\alpha|1\rangle_b \approx \alpha^*|\alpha\rangle \quad \text{for } |\alpha| \gg 1,\qquad (5.102)$$

where the subscript b emphasises belonging to the laser mode and we used $\hat{D}_\alpha^\dagger \hat{b}^\dagger \hat{D}_\alpha = (\hat{b}^\dagger + \alpha^*)$ [a re-writing of (3.16)]. Hence, the operator being measured approximates the desired quadrature:

$$_b\langle\alpha|(\hat{a}^\dagger\hat{b} + \hat{a}\hat{b}^\dagger)|\alpha\rangle_b \approx |\alpha|\sqrt{2}\,\frac{e^{-i\varphi}\hat{a} + e^{i\varphi}\hat{a}^\dagger}{\sqrt{2}} = |\alpha|\sqrt{2}\,\hat{x}_\varphi\;. \tag{5.103}$$

By varying the phase of the optical phase of the strong coherent state $|\alpha\rangle_b$, all the quadratures $\hat{x}_\varphi$ may be scanned by the same apparatus (the pre-factor $|\alpha|\sqrt{2}$ is irrelevant as it may just be handled by rescaling the measurement results).

A more sophisticated analysis of the homodyne detection scheme follows, along with an assessment of the range of validity of the approximation involved.

5.4.1.1 Homodyne generating function

A more explicit argument connecting the detection scheme described above with the homodyne POVM may be obtained by introducing the moment-generating function $f(x)$. Given a generic quantum state ϱ and self-adjoint operator $\hat{O}$, $f(x)$ is defined as $f(x) = \mathrm{Tr}\big[\varrho\, e^{x\hat{O}}\big]$, for x purely imaginary (at variance with most literature on the subject, we dispense with the i factor at the exponent, thus keeping formulae simpler). It is apparent that the function $f(x)$ contains the complete characterisation of the observable $\hat{O}$ inasmuch as the repeated differentiation of $f(x)$ yields all possible statistical moments of $\hat{O}$:

$$\mathrm{Tr}\left[\hat{O}^n\varrho\right] = \left.\frac{\partial f(x)}{\partial x^n}\right|_{x=0}. \tag{5.104}$$

We hence set out to prove that, in the strong oscillator limit, the moment-generating function of the homodyne detection scheme, obtained by considering the observable $(\hat{a}^\dagger\hat{b} + \hat{a}\hat{b}^\dagger)/(\sqrt{2}|\alpha|)$ and tracing the coherent state probe out, coincides with the homodyne generating function $\mathrm{Tr}\left[\varrho\, e^{x\hat{x}_\varphi}\right]$. This will also be an opportunity to present very standard techniques, based on group representation, to handle the ordering of non-commuting operators.

If ϱ is the state of mode a, to be measured, the moment-generating function of the homodyne scheme consisting, after the beam splitting, of the measurement of the operator determined in Eq. (5.101), reads

$$f(x) = \mathrm{Tr}_{ab}\left[e^{x\frac{\hat{a}^\dagger\hat{b}+\hat{a}\hat{b}^\dagger}{\sqrt{2}|\alpha|}}(\varrho\otimes|\alpha\rangle\langle\alpha|)\right] = \mathrm{Tr}_a\left[_b\langle\alpha|e^{x\frac{\hat{a}^\dagger\hat{b}+\hat{a}\hat{b}^\dagger}{\sqrt{2}|\alpha|}}|\alpha\rangle_b\varrho\right], \tag{5.105}$$

where the subscripts a and b stand for, respectively, the system mode, with annihilation operator $\hat{a}$, and the probe mode, with annihilation operator $\hat{b}$. In the following, we will set $\mu = x/(\sqrt{2}|\alpha|)$ to keep the notation more compact. Since mode b is in a coherent state $|\alpha\rangle_b$, it would be convenient to re-order the operators $\hat{b}$ and $\hat{b}^\dagger$ in $e^{\mu(\hat{a}^\dagger\hat{b}+\hat{a}\hat{b}^\dagger)}$ normally (i.e., with all $\hat{b}^\dagger$'s on the left and all $\hat{b}$'s on the right), so as to be able to evaluate $_b\langle\alpha|e^{\mu(\hat{a}^\dagger\hat{b}+\hat{a}\hat{b}^\dagger)}|\alpha\rangle_b$.

The key realisation in achieving such a re-ordering is that, in all cases where the Baker–Campbell–Hausdorff expansion converges, the expression of a product of exponentials of operators in terms of the operators themselves only depends on their commutation relations.[11] Conveniently, it turns out that the operators $\hat{a}\hat{b}^\dagger$, $\hat{a}^\dagger\hat{b}$ and $\hat{b}^\dagger\hat{b} - \hat{a}^\dagger\hat{a}$ form a closed algebra, in the sense that the commutators between any two such operators can be expressed in terms of the three operators alone. In our case, one has:

$$[\hat{a}\hat{b}^\dagger, \hat{a}^\dagger\hat{b}] = \hat{b}^\dagger\hat{b} - \hat{a}^\dagger\hat{a} , \tag{5.106}$$

$$[\hat{b}^\dagger\hat{b} - \hat{a}^\dagger\hat{a}, \hat{a}^\dagger\hat{b}] = -2\hat{a}^\dagger\hat{b} , \tag{5.107}$$

$$[\hat{b}^\dagger\hat{b} - \hat{a}^\dagger\hat{a}, \hat{a}\hat{b}^\dagger] = 2\hat{a}\hat{b}^\dagger . \tag{5.108}$$

The relations above are analogous to the algebra that generates $SU(2)$, upon identifying $\hat{b}^\dagger\hat{b} - \hat{a}^\dagger\hat{a}$ with σ_z, $\hat{a}^\dagger\hat{b}$ with $\sigma_- = \sigma_x - i\sigma_y$ and $\hat{a}\hat{b}^\dagger$ with $\sigma_+ = \sigma_x + i\sigma_y$ in terms of Pauli matrices. Forfeiting mathematical rigour, we will now proceed to associate our infinite dimensional operators with finite dimensional matrices, indicating such a mapping with the symbol $\leftrightarrow$, as follows

$$\hat{b}^\dagger\hat{b} - \hat{a}^\dagger\hat{a} \leftrightarrow \sigma_z = \begin{pmatrix} 1 & 0 \\ 0 & -1 \end{pmatrix} , \tag{5.109}$$

$$\hat{a}\hat{b}^\dagger \leftrightarrow \sigma_+ = \begin{pmatrix} 0 & 1 \\ 0 & 0 \end{pmatrix} , \tag{5.110}$$

$$\hat{a}^\dagger\hat{b} \leftrightarrow \sigma_- = \begin{pmatrix} 0 & 0 \\ 1 & 0 \end{pmatrix} . \tag{5.111}$$

It is straightforward to check that this mapping preserves the commutation relations (5.106)–(5.108). We can now evaluate an ordered expression for $e^{\mu(\hat{a}^\dagger\hat{b}+\hat{a}\hat{b}^\dagger)}$ by using its two-dimensional representation. Although μ is purely imaginary in the present discussion, we shall derive the re-ordering formula for the more general case of $e^{(\mu\sigma_+ - \mu^*\sigma_-)}$ with $\mu = |\mu|e^{i\phi} \in \mathbb{C}$, as such an expression will be useful later on in the book. The two-dimensional matrix exponential is easily determined as

$$e^{(\mu\sigma_+ - \mu^*\sigma_-)} = \begin{pmatrix} \cos|\mu| & e^{i\phi}\sin|\mu| \\ -e^{-i\phi}\sin|\mu| & \cos|\mu| \end{pmatrix} , \tag{5.112}$$

where we have used $(\mu\sigma_+ - \mu^*\sigma_-)^2 = -|\mu|^2\mathbb{1}_2$. We are also going to make use of the following two-dimensional matrix exponentials:

$$e^{\beta\sigma_+} = \begin{pmatrix} 1 & \beta \\ 0 & 1 \end{pmatrix} , \quad e^{\gamma\sigma_z} = \begin{pmatrix} e^\gamma & 0 \\ 0 & e^{-\gamma} \end{pmatrix} , \quad e^{\delta\sigma_-} = \begin{pmatrix} 1 & 0 \\ \delta & 1 \end{pmatrix} . \tag{5.113}$$

---

[11] Exceptions do exist, e.g., cases involving transformations that cannot be expressed as a single exponential (see page 314), for which the Baker–Campbell–Hausdorff formula breaks down (in that no solution Z to the equation $e^Z = e^X e^Y$ exists). We shall disregard this potential difficulty altogether in the present discussion.

Let us now set

$$
\begin{pmatrix} \cos|\mu| & e^{i\phi}\sin|\mu| \\ -e^{-i\phi}\sin|\mu| & \cos|\mu| \end{pmatrix} = \begin{pmatrix} e^{\gamma}+\beta\delta\,e^{-\gamma} & \beta\,e^{-\gamma} \\ \delta\,e^{-\gamma} & e^{-\gamma} \end{pmatrix} = e^{\beta\sigma_+}e^{\gamma\sigma_z}e^{\delta\sigma_-}\,,
$$

(5.114)

which is satisfied if the following conditions hold:

$$
\gamma = -\ln\cos|\mu|\,, \quad \beta = -\delta^* = e^{i\phi}\tan|\mu|\,.
$$

(5.115)

Invoking the mapping $\leftrightarrow$ defined above in Eqs. (5.109-5.111), Eqs. (5.114) and (5.115) can be brought back to the infinite dimensional representation to obtain:

Re-ordering of the beam splitter operator. Let $\mu = e^{i\phi}|\mu|$, the following relationship holds for ladder operators $\hat{a}$, $\hat{b}$ pertaining to different modes:

$$
e^{(\mu\hat{a}\hat{b}^\dagger - \mu^*\hat{a}^\dagger\hat{b})} = e^{e^{i\phi}\tan|\mu|\hat{a}\hat{b}^\dagger}\left(\cos|\mu|\right)^{(\hat{a}^\dagger\hat{a}-\hat{b}^\dagger\hat{b})}e^{-e^{-i\phi}\tan|\mu|\hat{a}^\dagger\hat{b}}\,.
$$

(5.116)

This is the re-ordering we need to determine the generating function we are after. If $\phi = \pi/2$, Eq. (5.116) immediately yields [as $\cosh(iy)=\cos(y)$ and $\tanh(iy)=i\tan(y)$]

$$
{}_b\langle\alpha|e^{\mu(\hat{a}\hat{b}^\dagger+\hat{a}^\dagger\hat{b})}|\alpha\rangle_b = e^{\tanh(\mu)\alpha^*\hat{a}}\cosh(\mu)^{\hat{a}^\dagger\hat{a}}e^{\tanh(\mu)\alpha\hat{a}^\dagger}\,{}_b\langle\alpha|\cosh(\mu)^{-\hat{b}^\dagger\hat{b}}|\alpha\rangle_b\,.
$$

(5.117)

The coherent state overlap on the right-hand side can be promptly determined by inserting the number state expansion $|\alpha\rangle_b = e^{-|\alpha|^2/2}\sum_{n=0}^{\infty}\frac{\alpha^n}{\sqrt{n!}}|n\rangle_b$, obtaining

$$
{}_b\langle\alpha|e^{\mu(\hat{a}\hat{b}^\dagger+\hat{a}^\dagger\hat{b})}|\alpha\rangle_b = e^{\tanh(\mu)\alpha^*\hat{a}}\cosh(\mu)^{\hat{a}^\dagger\hat{a}}e^{\tanh(\mu)\alpha\hat{a}^\dagger}e^{|\alpha|^2\frac{1-\cosh\mu}{\cosh\mu}}\,.
$$

(5.118)

Recalling the definition $\mu = x/(\sqrt{2}|\alpha|)$, we can now take the strong local oscillator limit ($|\alpha|\to\infty$) of the generating function $f(x)$ of Eq. (5.105):

$$
\lim_{|\alpha|\to\infty} f(x) = e^{x\frac{e^{-i\varphi}\hat{a}}{\sqrt{2}}}e^{x\frac{e^{i\varphi}\hat{a}^\dagger}{\sqrt{2}}}e^{-\frac{x^2}{4}} = e^{x\frac{e^{-i\varphi}\hat{a}+e^{i\varphi}\hat{a}^\dagger}{\sqrt{2}}} = e^{x\hat{x}_\varphi}\,,
$$

(5.119)

where we applied the Baker–Campbell–Hausdorff formula for the CCR algebra. We have thus clarified that the statistics of the homodyne detection scheme coincide with those of the quadrature operator $\hat{x}_\varphi$ in the limit of infinitely strong local oscillator.

A heuristic appreciation of the range of validity of the quadrature measurement description may be formed by considering the low-order statistical moments of the homodyne detection scheme. The first moment is just given by

$$
\mathrm{Tr}_a\left[{}_b\langle\alpha|\frac{\hat{a}\hat{b}^\dagger + \hat{a}^\dagger\hat{b}}{\sqrt{2}|\alpha|}|\alpha\rangle_b\varrho\right] = \mathrm{Tr}_a\left[\hat{x}_\varphi\varrho\right]\,,
$$

(5.120)

which is exactly the first moment of the quadrature measurement: the expectation value of the homodyne scheme coincides with that of the corresponding quadrature measurement *regardless of the energy of the local oscillator* (characterised by $|\alpha|$). Let us emphasise that, quite obviously, measurement schemes sharing the same expectation values may be dramatically different from each other. This is clearly illustrated by evaluating the second statistical moment of the homodyne scheme, which can be done by normal ordering $\hat{b}$ and $\hat{b}^\dagger$ in the operator $\hat{h} = (\hat{a}\hat{b}^\dagger + \hat{a}^\dagger\hat{b})^2$:

$$\mathrm{Tr}_a\left[{}_b\langle\alpha|\frac{\hat{h}}{2|\alpha|^2}|\alpha\rangle_b\varrho\right] = \mathrm{Tr}_a\left[{}_b\langle\alpha|\frac{\hat{a}^2\hat{b}^{\dagger 2} + \hat{a}^{\dagger 2}\hat{b}^2 + \hat{a}\hat{a}^\dagger\hat{b}^\dagger\hat{b} + \hat{a}^\dagger\hat{a}\hat{b}^\dagger\hat{b} + \hat{a}^\dagger\hat{a}}{2|\alpha|^2}|\alpha\rangle_b\varrho\right]$$

$$= \mathrm{Tr}_a\left[\hat{x}_\varphi^2\varrho\right] + \mathrm{Tr}_a\left[\frac{\hat{a}^\dagger\hat{a}}{2|\alpha|^2}\varrho\right]. \tag{5.121}$$

For the scheme to reproduce a quadrature measurement, it is hence necessary that the ratio between system and probe energy – essentially, the expectation value of $\hat{a}^\dagger\hat{a}/|\alpha|^2$ – be small. This sets the range of validity of homodyning as a quadrature measurement scheme.[12]

5.4.2 Bell measurements and heterodyne detection

The resolution of the identity (4.8) in terms of projectors on coherent states, one of the theoretical cornerstones of quantum optics, implies the existence of a corresponding physical process of measurement:

Heterodyne detection. The POVM

$$\hat{\mathbb{1}} = \int_\mathbb{C} \frac{|\alpha\rangle\langle\alpha|}{\pi}\,\mathrm{d}^2\alpha, \tag{5.122}$$

corresponds to the process of heterodyne detection, whose outcomes are labelled by α and whose probability on a quantum state of a single mode ϱ is given by $\pi^{-1}\langle\alpha|\varrho|\alpha\rangle$.

Traditionally, the optical realisation of this detection is carried out through a scheme analogous to homodyne detection, but where the mode to be measured is mixed with a laser field at a different frequency, whence the adjective "heterodyne". Here, we shall illustrate an alternative implementation that, as

---

[12] A detailed analysis of the deviation of Eq. (5.118) from the limit of infinite $|\alpha|$ might be undertaken, but turns out to be rather tedious. Let us just mention that, in taking the limit $|\alpha| \to \infty$ we assumed, somewhat boldly, that $\lim_{\mu\to 0}\cosh(\mu)^{\hat{a}^\dagger\hat{a}} = 1$. The operator $\hat{a}^\dagger\hat{a}$ is however unbounded, so one may rather consider the equivalent limit $\lim_{|\alpha|\to\infty}\exp\left(\frac{x^2\hat{a}^\dagger\hat{a}}{4|\alpha|^2}\right)$ which explicitly shows that the limit is, strictly speaking, correct if the support of the state ϱ is limited to a subspace of finite $\hat{a}^\dagger\hat{a}$. Further corrections, all vanishing if $\hat{a}^\dagger\hat{a}/|\alpha|^2$ goes to zero, crop up in the other exponential factors that occur in Eq. (5.118).

we shall see, is invested with a conceptual relevance that goes beyond its role in quantum optics.

In the previous section, we have described in detail the homodyne detection scheme that, in the limit of a strong local oscillator, allows one to approximate the measurement of any quadrature operator on a single mode $\hat{x}_\varphi$. Let us now consider a "double homodyne" set-up where a single-mode input state ϱ_s, to be measured, is mixed at a 50:50 beam splitter with a probe mode, whose state ϱ_p will be specified later on, after which the two output ports of the beam splitter undergo homodyne detections. In what follows, the labels s and p will refer to system and probe modes, respectively. The optical phases of the homodyne detections are chosen such that the operators monitored after beam splitting are the commuting operators $\hat{x}_- = (\hat{x}_s - \hat{x}_p)$ and $\hat{p}_+ = (\hat{p}_s + \hat{p}_p)$, whose common eigenvectors we will now determine.

The eigenvector equalities $\hat{x}_-|\xi\rangle = \hat{p}_+|\xi\rangle = 0$, for the unnormalised vector $|\xi\rangle = \sum_{j=0}^{\infty} |j, j\rangle$ were already established (see Problem 5.5). All other common eigenvectors are readily found by noticing that, since $\hat{x}_-$ and $\hat{p}_+$ are linear in the quadrature operators, one has, given any two-dimensional real vector $\mathbf{r} = (x, p)^\mathsf{T}$:

$$\hat{x}_- \hat{D}_\mathbf{r}^\dagger|\xi\rangle = \hat{D}_\mathbf{r}^\dagger \hat{D}_\mathbf{r}(\hat{x}_s - \hat{x}_p)\hat{D}_\mathbf{r}^\dagger|\xi\rangle = \hat{D}_\mathbf{r}^\dagger(\hat{x}_- + x)|\xi\rangle = x\hat{D}_\mathbf{r}^\dagger|\xi\rangle , \qquad (5.123)$$

$$\hat{p}_+ \hat{D}_\mathbf{r}^\dagger|\xi\rangle = \hat{D}_\mathbf{r}^\dagger \hat{D}_\mathbf{r}(\hat{p}_s + \hat{p}_p)\hat{D}_\mathbf{r}^\dagger|\xi\rangle = \hat{D}_\mathbf{r}^\dagger(\hat{p}_+ + p)|\xi\rangle = p\hat{D}_\mathbf{r}^\dagger|\xi\rangle , \qquad (5.124)$$

where $\hat{D}_\mathbf{r}$ is a displacement operator acting only on the system Hilbert space (the tensor product with the identity operator is understood).[13] Notice now that the non-normalised vectors $\hat{D}_\mathbf{r}^\dagger|\xi\rangle$ correspond, so to speak, to the infinite dimensional analogue of "maximally entangled" pure states (since the non-zero Schmidt coefficients of $|\xi\rangle$ form an infinite set and are all equal, while $\hat{D}_\mathbf{r}$ is a local unitary operator). Hence, the double homodyne detection scheme – where, instead of preparing a probe in a certain state, the whole two-mode input is measured – would allow one to project *deterministically* onto maximally entangled states, a procedure which has often been described in the literature as a "deterministic" or "unconditional" Bell measurement (i.e., a projection on a basis set of maximally entangled pure states).

This somewhat emphatic description is due to the fact that deterministic Bell measurements are difficult to implement in most other set-ups, including finite dimensional systems. For instance, in teleportation experiments with qubits, not all of the four Bell outcomes may typically be distinguished, such that the teleportation's success is conditional on the measurement outcome. Continuous variables thus enjoy a strong practical advantage in this regard which, compounded by the relatively high efficiency of homodyne detection,

---

[13]One might argue that any two-mode displacement operator whose difference in x and sum in p displacements match the required values x and p would also result in eigenvectors of $\hat{x}_-$ and $\hat{p}_+$ with the same eigenvalues. However, by exploiting the fact that $|\xi\rangle$ is the zero eigenvector of $\hat{x}_-$ and $\hat{p}_+$, it is easy to show that all such eigenvectors coincide up to an irrelevant phase factor.

has led to the first demonstration of unconditional quantum teleportation (see Chapter 8 for a description of such a protocol). It should nevertheless be noted that, because of the approximation implied in the finite local oscillator strength of the homodyne detections entering the scheme, no ideal Bell projection is feasible, even in principle, for such systems, but may only be approached in a limit.[14]

Going back to the heterodyne detection scheme, let us now see what measurement statistics are obtained by assuming a probe mode prepared in the vacuum state, i.e., by setting $\varrho_p = |0\rangle\langle 0|$, and by focusing on the system of interest alone:

$$p(x, p) = c\langle\xi|\hat{D}_{\mathbf{r}}(\varrho_s \otimes |0\rangle\langle 0|)\hat{D}_{\mathbf{r}}^{\dagger}|\xi\rangle = c\langle\xi|0\rangle_p\,\hat{D}_{\mathbf{r}}\,\varrho_s\,\hat{D}_{-\mathbf{r}}\,_p\langle 0|\xi\rangle \qquad (5.125)$$

$$= c\langle 0|\hat{D}_\alpha^{\dagger}\varrho_s\hat{D}_\alpha|0\rangle = c\langle\alpha|\varrho_s|\alpha\rangle, \qquad (5.126)$$

where $\alpha = \frac{x+ip}{\sqrt{2}}$, $\hat{D}_\alpha = \hat{D}_{-\mathbf{r}}$, in compliance with our definition in Chapter 4, and c is a generic normalisation constant (recall that we did not bother normalising $|\xi\rangle$). The constant c is then easily determined: it must equal π^{-1} since, as we know from our treatment of quantum optics (see Chapter 4), that is the quantity which normalises the projection on (non-orthogonal) coherent states, in the sense that $\hat{\mathbb{1}} = \frac{1}{\pi}\int_{\mathbb{C}}|\alpha\rangle\langle\alpha|d\alpha$. We have hence established that the measurement statistics of a double homodyne scheme are equivalent to the heterodyne POVM.

5.4.3 Ideal general-dyne detections

The heterodyne resolution of the identity (5.122) extends naturally to many modes and can be further generalised by acting on both sides with a unitary transformation, that obviously preserves the identity operator. If such a unitary is a purely quadratic unitary transformation $\hat{S}$, corresponding to the symplectic transformation S, one has:

$$\hat{\mathbb{1}} = \frac{1}{(2\pi)^n}\int_{\mathbb{R}^{2n}}d\mathbf{r}\,\hat{S}^{\dagger}\hat{D}_{-\mathbf{r}}|0\rangle\langle 0|\hat{D}_{\mathbf{r}}\hat{S} = \frac{1}{(2\pi)^n}\int_{\mathbb{R}^{2n}}d\mathbf{r}\,\hat{D}_{-\mathbf{r}}\hat{S}^{\dagger}|0\rangle\langle 0|\hat{S}\hat{D}_{\mathbf{r}}\,,$$

where we used the action of a purely quadratic operation on a displacement operator: $\hat{S}^{\dagger}\hat{D}_{\mathbf{r}}\hat{S} = \hat{D}_{S\mathbf{r}}$, changed the integration variables to $S\mathbf{r}$ and took advantage of the fact that $\text{Det}(S) = 1$ for all $S \in Sp_{2n,\mathbb{R}}$.

The measurement processes described by these POVMs correspond, if the measurement outcome is recorded, to projections on the completely generic

---

[14]On the other hand, quite trivially, all practical schemes are bound to be imperfect, and the approximation that identifies strong homodyning with a quadrature measurement is, all considered, a very reliable one.

pure Gaussian state $|\psi_G\rangle = \hat{D}_{-\mathbf{r}_m}\hat{S}^\dagger|0\rangle$. Such measurements go by the name of 'general-dyne' detections:

Ideal general-dyne detections. The POVMs

$$\hat{\mathbb{1}} = \frac{1}{(2\pi)^n}\int_{\mathbb{R}^{2n}}d\mathbf{r}_m\,\hat{D}_{-\mathbf{r}_m}\hat{S}^\dagger|0\rangle\langle 0|\hat{S}\hat{D}_{\mathbf{r}_m}\,, \tag{5.127}$$

correspond to the processes of ideal general-dyne detections, whose outcomes are labelled by $\mathbf{r}_m \in \mathbb{R}^{2n}$ and whose probability density on a quantum state ϱ is given by $(2\pi)^{-n}\langle 0|\hat{S}\hat{D}_{\mathbf{r}_m}\varrho\hat{D}_{-\mathbf{r}_m}\hat{S}^\dagger|0\rangle$.

This class of detections – that have been deemed "ideal" in contrast with their noisy counterpart, which will be the subject of the next section – is parametrised by the symplectic transformation S. The heterodyne detection scheme is recovered for $\hat{S} = \hat{\mathbb{1}}$ (projection on coherent states), while the homodyne detection schemes (projection on canonical operators' eigenstates) are attained in the limit where $\hat{S}$ is a squeezing operator with an infinite squeezing parameter: for one mode, this would correspond to $S = R_\varphi^\mathsf{T}\mathrm{diag}(z,1/z)R_\varphi$ – where the phase shifter R_φ of Eq. (5.12) determines the homodyne phase – for $z \to 0$ (this follows from what was proven in Problem 3.3). In this limit, the uncertainty on one of the canonical operators diverges while the conjugate one vanishes, and it is easy to show that the state upon which the system is projected is a canonical operator's eigenstate.

Let us mention in passing that, because of the structure of their resolution of the identity, these measurements may also be referred to as Weyl-covariant (or, simply, covariant) POVMs, following a tradition that goes back to Holevo. In practice, ideal general-dyne detections may be realised by the action on the state to be measured of the symplectic transformation S^{-1}, followed by heterodyne detection. Notice that, for consistency, we have maintained the convention whereby the action by similarity of the operator $\hat{S}^\dagger$ on the quantum state corresponds to applying the symplectic transformation S on its covariance matrix. Also note that the action of the transformation S on the general-dyne projector corresponds to acting with the inverse S^{-1} on the system state.

If the measured state is a Gaussian with covariance matrix $\boldsymbol{\sigma}$ and first moments $\bar{\mathbf{r}}$, the normalised probability density $p(\mathbf{r}_m)$ associated with the general-dyne outcome $\mathbf{r}_m$ is easily inferred from the expression for Gaussian overlaps, Eq. (4.51), as

$$p(\mathbf{r}_m) = \frac{\langle\psi_G|\varrho|\psi_G\rangle}{(2\pi)^n} = \frac{e^{-(\mathbf{r}_m-\bar{\mathbf{r}})^\mathsf{T}(\boldsymbol{\sigma}+SS^\mathsf{T})^{-1}(\mathbf{r}_m-\bar{\mathbf{r}})}}{\pi^n\sqrt{\mathrm{Det}(\boldsymbol{\sigma}+SS^\mathsf{T})}}\,. \tag{5.128}$$

The homodyne limit of infinite squeezing deserves some further comment. As mentioned above, for a single mode $(n = 1)$, the homodyne measurement of $\hat{x}$ would correspond to the limit $SS^\mathsf{T} = \lim_{z\to 0}\mathrm{diag}(z^2, 1/z^2)$ (whereas a measurement of $\hat{p}$ would correspond to $z \to \infty$, and the argument generalises to an arbitrary phase space direction by rotating this covariance matrix through R_φ). In this limit, the matrix $(\sigma + SS^\mathsf{T})^{-1}$ that appears in the expression for the general-dyne probability (5.128) tends to $\mathrm{diag}(\sigma_{11}^{-1}, 0)$ (as can be easily verified by inverting the 2×2 matrix, or by direct multiplication). Hence, only one of the two real-valued readings – x in this example, if $\mathbf{r}_m = (x, p)^\mathsf{T}$ – that ordinarily label the general-dyne outcomes of a single mode may actually be defined (and, indeed, we know that homodyne measurements give a single real outcome per mode, whilst heterodyne and other general-dyne yield two real values per mode). This also heals the divergence in the determinant that enters the normalisation factor, which is offset by the integration over the forgotten real variable. The homodyne probability density $p(x)$, in $\mathrm{d}x$, associated with a measurement of $\hat{x}$, of a Gaussian state with first moment $\langle \hat{x} \rangle = \bar{x}$ and second moment $\sigma_{11} = 2\langle(\hat{x} - \bar{x})^2\rangle$ is therefore given by

$$p(x) = \frac{\mathrm{e}^{-\frac{(x-\bar{x})^2}{\sigma_{11}}}}{\sqrt{\pi \sigma_{11}}} . \tag{5.129}$$

The variance of this distribution is $\sigma_{11}/2$, where the factor $1/2$ is just due to our definition of the covariance matrix (with the identity corresponding to the vacuum state). Notice that this result complies with the operational interpretation (4.42) of a Wigner function in Gaussian form (4.50).

Problem 5.19. (*Unrecorded general-dyne detection*). Identify the deterministic Gaussian CP-map resulting from carrying out an ideal general-dyne detection, with associated symplectic transformation S, and completely forgetting the measurement outcome.

Solution. An unrecorded general-dyne detection of the Gaussian state ϱ_G consists in the projection on the pure Gaussian state $\hat{D}_\mathbf{s}^\dagger \hat{S}^\dagger |0\rangle$, with first and second moments $\mathbf{s}$ and SS^T, integrated over all $\mathbf{s} \in \mathbb{R}^{2n}$ with probability $\langle 0|\hat{S}\hat{D}_\mathbf{s}\varrho_G\hat{D}_\mathbf{s}^\dagger\hat{S}^\dagger|0\rangle/(2\pi)^n$, that is in the overall mapping

$$\varrho_G \mapsto \int_{\mathbb{R}^{2n}} \mathrm{d}\mathbf{s} \frac{\langle 0|\hat{S}\hat{D}_\mathbf{s}\,\varrho_G\,\hat{D}_\mathbf{s}^\dagger\hat{S}^\dagger|0\rangle}{(2\pi)^n} \hat{D}_\mathbf{s}^\dagger\hat{S}^\dagger|0\rangle\langle 0|\hat{S}\hat{D}_\mathbf{s} . \tag{5.130}$$

The formula for Gaussian overlaps (4.51) implies

$$\langle 0|\hat{S}\hat{D}_\mathbf{s}\,\varrho_G\,\hat{D}_\mathbf{s}^\dagger\hat{S}^\dagger|0\rangle = \frac{2^n}{\sqrt{\mathrm{Det}\,(\sigma + SS^\mathsf{T})}} \mathrm{e}^{-(\mathbf{s}-\bar{\mathbf{r}})^\mathsf{T}(\sigma+SS^\mathsf{T})^{-1}(\mathbf{s}-\bar{\mathbf{r}})} , \tag{5.131}$$

where σ and $\bar{\mathbf{r}}$ parametrise ϱ_G. This can be inserted into Eq. (5.130), along with the Fourier-Weyl representation of $\hat{D}_\mathbf{s}^\dagger\hat{S}^\dagger|0\rangle\langle 0|\hat{S}\hat{D}_\mathbf{s}$, to get

$$\varrho_G \mapsto \int_{\mathbb{R}^{2n}} \mathrm{d}\mathbf{s} \int_{\mathbb{R}^{2n}} \mathrm{d}\tilde{\mathbf{r}} \frac{\mathrm{e}^{-(\mathbf{s}-\bar{\mathbf{r}})^\mathsf{T}(\sigma+SS^\mathsf{T})^{-1}(\mathbf{s}-\bar{\mathbf{r}})}}{\pi^n\sqrt{\mathrm{Det}\,(\sigma + SS^\mathsf{T})}} \frac{\mathrm{e}^{-\frac{1}{4}\tilde{\mathbf{r}}^\mathsf{T}SS^\mathsf{T}\tilde{\mathbf{r}}+i\tilde{\mathbf{r}}^\mathsf{T}\mathbf{s}}}{(2\pi)^n} \hat{D}_{\Omega^\mathsf{T}\tilde{\mathbf{r}}} . \tag{5.132}$$

One can then apply the change of variable $\mathbf{s} \mapsto \mathbf{s}+\tilde{\mathbf{r}}$ and the integration formula (1.18) to obtain the Gaussian mapping

$$\varrho_G \mapsto \frac{1}{(2\pi)^n} \int_{\mathbb{R}^{2n}} d\tilde{\mathbf{r}}\, e^{-\frac{1}{4}\tilde{\mathbf{r}}^\mathsf{T}(\sigma+2SS^\mathsf{T})^{-1}\tilde{\mathbf{r}}+i\tilde{\mathbf{r}}^\mathsf{T}\bar{\mathbf{r}}} \hat{D}_{\Omega^\mathsf{T}\tilde{\mathbf{r}}} . \tag{5.133}$$

The unrecorded detection thus corresponds to additive noise, that is to the deterministic Gaussian CP-map with $X = \mathbb{1}$ and $Y = 2SS^\mathsf{T}$. Heterodyne detection ($S = \mathbb{1}$) has the effect of adding two units of vacuum noise to all quadratures, whilst infinite noise is added to unmonitored quadratures in the homodyne limit.

5.4.4 Noisy measurements

The resolution of the identity (5.127) can be further generalised by applying a unital CP-map on the left- and right-hand sides of the equation, as unital maps preserve the identity operator. Since the dual of any deterministic Gaussian CP-map, characterised by Eqs. (5.64)–(5.65), is Gaussian and unital, its action on the general-dyne POVM elements will result in a larger class of detections, which we shall refer to as 'noisy' general-dyne measurement, or general-dyne measurement *tout court*, since the ideal cases are trivially encompassed.

Because of the definition of a dual map, the noise described by this class of measurements is equivalent to applying a deterministic Gaussian CP-map on the system state before carrying out a heterodyne measurement. Ideal general-dyne measurements correspond to the choice $X = S^{-1}$ and $Y = 0$, with S symplectic, with dual counterparts $X^* = S$ and $Y^* = 0$ [see Eqs. (5.64) and (5.65)]. Beyond the unitary (symplectic) case, the noise that can be modelled by this class of measurements includes the most important case of non-unit detection efficiency, described by the quantum limited attenuator where a portion of the signal is lost, as well as fuzzy measurements weighted by a Gaussian mask – by which we mean POVMs whose limited precision may be modelled as the broadening of the measured state through additive Gaussian noise, as per the classical mixing map (5.76).

In formulae, the action on the system with a Gaussian CP-map Φ followed by an ideal general-dyne measurement described as the projection on the pure Gaussian state $|\psi_G\rangle$, with covariance matrix SS^T and first moments $\mathbf{r}_m$, results in the probability density

$$p(\mathbf{r}_m) = \frac{\mathrm{Tr}\left[\Phi(\varrho_G)|\psi_G\rangle\langle\psi_G|\right]}{(2\pi)^n} = \frac{\mathrm{Tr}\left[\varrho_G\Phi^*(|\psi_G\rangle\langle\psi_G|)\right]}{(2\pi)^n} . \tag{5.134}$$

We have thus moved the effect of the evolution on the measurement process by means of the dual-map Φ^*. Let us now assume that the matrix X associated with the original CP-map Φ is invertible and analyse the action of Φ^* on the generic pure Gaussian state $|\psi_G\rangle = \hat{D}_{-\mathbf{r}_m}\hat{S}^\dagger|0\rangle$. The map Φ^* acts as the

Gaussian CP-map with matrices X^* and Y^* determined by Eqs. (5.64) and (5.65), with a further multiplicative factor $1/|\mathrm{Det}X|$. Thus, it maps $|\psi_G\rangle$ as

$$\Phi^*(|\psi_G\rangle\langle\psi_G|) = \frac{\hat{D}_{-X^{-1}\mathbf{r}_m}\varrho_m\hat{D}_{X^{-1}\mathbf{r}_m}}{|\mathrm{Det}X|} , \qquad (5.135)$$

where ϱ_m is a Gaussian state with null first moments and covariance matrix

$$\boldsymbol{\sigma}_m = X^*SS^\mathsf{T}X^{*\mathsf{T}} + Y^* . \qquad (5.136)$$

In order to express the general-dyne probability density $p(\mathbf{r}_m)$, one may safely redefine the outcome $\mathbf{r}_m$ through the mapping $\mathbf{r}_m \mapsto X\mathbf{r}_m$, which also absorbs the multiplicative factor $1/|\mathrm{Det}X|$. Besides, note that any physical covariance matrix $\boldsymbol{\sigma}_m$, such that $\boldsymbol{\sigma}_m + i\Omega \geq 0$, may be obtained as the action of a dual CP map on the covariance matrix corresponding to a pure state.[15] Therefore, we have obtained the following compact characterisation of noisy general-dyne detections:

General-dyne detections. The POVMs

$$\hat{\mathbb{1}} = \frac{1}{(2\pi)^n} \int_{\mathbb{R}^{2n}} \mathrm{d}\mathbf{r}_m\, \hat{D}_{-\mathbf{r}_m}\varrho_m\hat{D}_{\mathbf{r}_m} , \qquad (5.137)$$

where ϱ_m is a Gaussian state with zero first moments and generic covariance matrix $\boldsymbol{\sigma}_m$ such that $\boldsymbol{\sigma}_m + i\Omega \geq 0$, correspond to the processes of ideal general-dyne detections, whose outcomes are labelled by $\mathbf{r}_m$ and whose probability density on a quantum state of n modes ϱ is given by

$$p(\mathbf{r}_m) = \frac{\mathrm{Tr}[\varrho\hat{D}_{-\mathbf{r}_m}\varrho_m\hat{D}_{\mathbf{r}_m}]}{(2\pi)^n}. \qquad (5.138)$$

The covariance matrix $\boldsymbol{\sigma}_m$ parametrises the whole set of noisy as well as ideal general-dyne detections. If the state ϱ is a Gaussian with covariance matrix $\boldsymbol{\sigma}$ and vector of first moments $\bar{\mathbf{r}}$, the probability density in $\mathrm{d}\mathbf{r}_m$ is a multivariate Gaussian that can be derived from the overlap formula (4.51):

$$p(\mathbf{r}_m) = \frac{e^{(\mathbf{r}_m - \bar{\mathbf{r}})^\mathsf{T}\frac{1}{\boldsymbol{\sigma} + \boldsymbol{\sigma}_m}(\mathbf{r}_m - \bar{\mathbf{r}})}}{\pi^n\sqrt{\mathrm{Det}(\boldsymbol{\sigma} + \boldsymbol{\sigma}_m)}} . \qquad (5.139)$$

The most common and critical measurement imperfection is arguably due to non-unit efficiencies, which are modelled by assuming part of the signal is lost before entering the measuring apparatus. In our treatment, this corresponds to a quantum limited attenuator acting on the system, with $X = \cos\theta\mathbb{1}$

---

[15] This is straightforward consequence of Problem 5.10 and of the fact that any classical mixing map may be construed as a dual map by setting $X = \mathbb{1}$ (it is, in point of fact, self-dual).

and $Y = (\sin\theta)^2 \mathbb{1}$ (assuming, for simplicity, that all the modes are detected with the same efficiency). By Eqs. (5.64,5.65), this implies $X^* = (\cos\theta)^{-1}\mathbb{1}$ (note that the case $\cos\theta = 0$ is trivial, since no measurement would take place then) and $Y^* = (\tan\theta)^2 \mathbb{1}$, whence Eq. (5.136) furnishes the following notable result:

Finite detection efficiency. Let SS^T correspond to the covariance matrix of an ideal general-dyne detection (*e.g.*, $SS^\mathsf{T} = \mathbb{1}$ for heterodyne detection). Then, the corresponding detection with finite efficiency $0 < \cos\theta^2 \le 1$ is the general-dyne detection with covariance matrix $\boldsymbol{\sigma}_m = (\cos\theta)^{-2}SS^\mathsf{T} + (\tan\theta)^2 \mathbb{1}$.

Problem 5.20. (*Additive measurement noise*). Show that, for a single-mode system, the effect of additive measurement noise (with $X = \mathbb{1}$ and $Y = y\mathbb{1}$, with $y \ge 0$) on homodyne and heterodyne detections is undistinguishable from imperfect detection efficiency.

Solution. Let us first address heterodyne detection, for which $SS^\mathsf{T} = \mathbb{1}$; Eqs. (5.64) and (5.65) show that the dual of the quantum limited attenuator, which describes imperfect measurement efficiency, has $X^* = (\cos\theta)^{-1}\mathbb{1}$ and $Y^* = (\tan\theta)^2 \mathbb{1}$. The resulting general-dyne covariance matrix is hence $[1+2(\tan\theta)^2]\mathbb{1}$; the dual of the classical mixing channel (additive noise), is instead the same as the original channel, so that $X^* = \mathbb{1}$ and $Y^* = y\mathbb{1}$, whose action on the heterodyne covariance matrix leads to the covariance matrix $(1 + y)\mathbb{1}$: the two detections yield hence the very same statistics upon identifying $y = 2(\tan\theta)^2$.

For homodyne detection, with ideal covariance matrix $\lim_{z\to 0} R_\varphi^\mathsf{T}\mathrm{diag}(z, 1/z)R_\varphi$, finite efficiency changes the covariance matrix to $\lim_{z\to 0} R_\varphi^\mathsf{T}\mathrm{diag}(z, 1/z)R_\varphi + (\tan\theta)^2 \mathbb{1}$ (as the action of X^* does not affect the limit), which is the same as the additive noise case $\lim_{z\to 0} R_\varphi^\mathsf{T}\mathrm{diag}(z, 1/z)R_\varphi + y\mathbb{1}$, upon setting $y = (\tan\theta)^2$.

5.4.5 Conditional Gaussian dynamics

As we saw above, the Hilbert-Schmidt product with Gaussian states with fixed second moments and uniformly distributed first moments describes legitimate measurement processes. If the measurement outcome, labelled above by $\mathbf{r}_m$, is recorded, such measurements give rise to specific Gaussian CP-maps, which can be interpreted as the filtering of the system conditioned on recording the measurement outcome $\mathbf{r}_m$. It is instructive to understand how a Gaussian state is affected when a portion of the system modes is measured through general-dyne detections. In Chapter 6, we shall apply these formulae to derive evolution equations for continuously monitored Gaussian systems.

Given the initial Gaussian state of a system partitioned in subsystems A and B, with covariance matrix and first moments

$$\boldsymbol{\sigma} = \begin{pmatrix} \boldsymbol{\sigma}_A & \boldsymbol{\sigma}_{AB} \\ \boldsymbol{\sigma}_{AB}^\mathsf{T} & \boldsymbol{\sigma}_B \end{pmatrix}, \quad \bar{\mathbf{r}} = \begin{pmatrix} \bar{\mathbf{r}}_A \\ \bar{\mathbf{r}}_B \end{pmatrix}, \tag{5.140}$$

let us determine the final covariance matrix and first moments of the n-mode subsystem A given a general-dyne measurement outcome on the m-mode subsystem B. We need to evaluate the overlap between the initial state ϱ of subsystem B and the general-dyne Gaussian state $\varrho_G = \hat{D}_{-\mathbf{r}_m} \varrho_m \hat{D}_{\mathbf{r}_m}$, with covariance matrix $\boldsymbol{\sigma}_m$ and first moments $\mathbf{r}_m$. Note that, while $\mathbf{r}_m$ labels the outcome of the measurement, $\boldsymbol{\sigma}_m$ characterises the specific choice of (possibly noisy) general-dyne detection.

By using the Fourier-Weyl relation (4.49) in terms of the variables $\tilde{\mathbf{r}} = \Omega \mathbf{r}$, noticing that $\mathrm{Tr}[\varrho_G \hat{D}_{\mathbf{r}_B}]$ is nothing but the characteristic function of ϱ_G (up to a sign flip in the argument) and applying the multivariate Gaussian integral (1.18), one gets

$$
\begin{aligned}
\mathrm{Tr}[\varrho_G \varrho] &= \frac{1}{(2\pi)^{n+m}} \int_{\mathbb{R}^{2(n+m)}} e^{-\frac{1}{4}\tilde{\mathbf{r}}^\mathsf{T} \boldsymbol{\sigma} \tilde{\mathbf{r}} + i\tilde{\mathbf{r}}^\mathsf{T} \bar{\mathbf{r}}} \mathrm{Tr}[\varrho_G \hat{D}_{\Omega^\mathsf{T} \tilde{\mathbf{r}}}] \, d\tilde{\mathbf{r}} \\
&= \frac{1}{(2\pi)^{n+m}} \int_{\mathbb{R}^{2(n+m)}} e^{-\frac{1}{4}\tilde{\mathbf{r}}^\mathsf{T} \boldsymbol{\sigma} \tilde{\mathbf{r}} + i\tilde{\mathbf{r}}^\mathsf{T} \bar{\mathbf{r}}} \hat{D}_{\Omega^\mathsf{T} \tilde{\mathbf{r}}_A} e^{-\frac{1}{4}\tilde{\mathbf{r}}_B^\mathsf{T} \boldsymbol{\sigma}_m \tilde{\mathbf{r}}_B - i\tilde{\mathbf{r}}_B^\mathsf{T} \mathbf{r}_m} \, d\tilde{\mathbf{r}} \\
&= \frac{2^m e^{-(\mathbf{r}_m - \bar{\mathbf{r}}_B)^\mathsf{T} \frac{1}{\boldsymbol{\sigma}_B + \boldsymbol{\sigma}_m}(\mathbf{r}_m - \bar{\mathbf{r}}_B)}}{(2\pi)^{n+m}\sqrt{\mathrm{Det}(\boldsymbol{\sigma}_B + \boldsymbol{\sigma}_m)}} \times \\
&\quad \int_{\mathbb{R}^{2n}} e^{-\frac{\tilde{\mathbf{r}}_A^\mathsf{T}}{4}(\boldsymbol{\sigma}_A - \boldsymbol{\sigma}_{AB}\frac{1}{\boldsymbol{\sigma}_B + \boldsymbol{\sigma}_m}\boldsymbol{\sigma}_{AB}^\mathsf{T})\tilde{\mathbf{r}}_A} e^{i\tilde{\mathbf{r}}_A^\mathsf{T}(\bar{\mathbf{r}}_A + \boldsymbol{\sigma}_{AB}\frac{1}{\boldsymbol{\sigma}_B + \boldsymbol{\sigma}_m}(\mathbf{r}_m - \bar{\mathbf{r}}_B))} \hat{D}_{\Omega^\mathsf{T} \tilde{\mathbf{r}}_A} \, d\tilde{\mathbf{r}}_A,
\end{aligned}
\tag{5.141}
$$

which may be summarised as follows:

General-dyne filtering. The n-mode subsystem of an $(n+m)$-mode Gaussian state, with initial covariance matrix and first moments partitioned as per Eq. (5.140), undergoes the following Gaussian mapping upon the general-dyne measurement of the remaining modes:

$$\boldsymbol{\sigma}_A \mapsto \boldsymbol{\sigma}_A - \boldsymbol{\sigma}_{AB} \frac{1}{\boldsymbol{\sigma}_B + \boldsymbol{\sigma}_m} \boldsymbol{\sigma}_{AB}^\mathsf{T}, \tag{5.142}$$

$$\bar{\mathbf{r}}_A \mapsto \bar{\mathbf{r}}_A + \boldsymbol{\sigma}_{AB} \frac{1}{\boldsymbol{\sigma}_B + \boldsymbol{\sigma}_m} (\mathbf{r}_m - \bar{\mathbf{r}}_B), \tag{5.143}$$

where $\boldsymbol{\sigma}_m$, such that $\boldsymbol{\sigma}_m + i\Omega \geq 0$, characterises the general-dyne measurement, while $\mathbf{r}_m$ labels the measurement outcome.

The probability density (in $d\mathbf{r}_m$) of the general-dyne conditioning is given by Eq. (5.139) for $\boldsymbol{\sigma} = \boldsymbol{\sigma}_B$, $\bar{\mathbf{r}} = \bar{\mathbf{r}}_B$ and $n = m$. Notice that, formally, the updates of first and second moments derived above may be given in terms of Schur complements of the matrix $\sigma + (0 \oplus \sigma_m)$.

A general outstanding feature of Gaussian conditioning is immediately apparent: *the conditional evolution of the second moments – which determines all correlations and entropic quantities – does not depend on the measurement outcome*. This property is common to certain classical filters too, such as the Kalman filter, when applied to Gaussian statistics, and will have notable implications in the next chapter where continuously monitored dynamics will be addressed. We also remark that, as one should expect, if no correlations are present (i.e., if $\sigma_{AB} = 0$), the map above reduces to the identity, in that measuring subsystem B cannot have any effect on subsystem A if the two subsystems are not initially correlated. Observe also that measuring always implies a reduction in uncertainty, in the technical sense that the covariance mapping (5.142) sends σ_A into a matrix with smaller eigenvalues (as the subtracted matrix is always positive).

Let us also note here that yet another case of Gaussian conditioning is the projection on the vacuum state, realised through 'on/off' detectors that, distinguishing between the presence and absence of field excitations, implement the POVM with elements $|0\rangle\langle 0|$ and $(\hat{\mathbb{1}} - |0\rangle\langle 0|)$. If 0 is read out (on subsystem B), the original Gaussian state ϱ with statistics given above, is projected on the vacuum with probability

$$p_0 = \mathrm{Tr}_{A}[_B\langle 0|\varrho|0\rangle_B] = \frac{2^m e^{-\bar{\mathbf{r}}^{\mathsf{T}}\frac{1}{\sigma_B+1}\bar{\mathbf{r}}}}{\sqrt{\mathrm{Det}\sigma_B + \mathbb{1}}}. \tag{5.144}$$

Note that this quantity is an actual probability, and not a probability density, and differs by a factor $(2\pi)^m$ from the heterodyne probability density evaluated in $\mathbf{r}_m = 0$ (which is also proportional to a vacuum projection). The conditioning of a correlated Gaussian subsystem for vacuum projection would still be described by Eqs. (5.142) and (5.143). Notice that, with probability $(1 - p_0)$, the system would end up in a non-Gaussian state.

5.5 CHOI–JAMIOLKOWSKI DESCRIPTION OF THE MOST GENERAL GAUSSIAN CP-MAP

5.5.1 Choi isomorphism and gate teleportation

The Choi isomorphism (also known as the Choi–Jamiolkowski isomorphism) is a well-known bijective mapping between quantum states and quantum completely positive maps. Its definition is straightforward for a finite dimensional Hilbert space $\mathcal{H}$. Let us first define the maximally entangled state $|\psi\rangle = \frac{1}{\sqrt{d}}\sum_{j=1}^{d}|e_j, e_j\rangle$, where d is the dimension of the Hilbert space $\mathcal{H}$, and $\{|e_j\rangle, j = 1, \ldots, d\}$ is any orthonormal basis of $\mathcal{H}$ and $|e_j, e_j\rangle = |e_j\rangle \otimes |e_j\rangle \in \mathcal{H}^{\otimes 2}$. Then, given a CP-map Φ on $\mathcal{B}(\mathcal{H})$, representing a generic quantum operation, one can define the associated Choi state $\hat{\varphi} \in \mathcal{B}(\mathcal{H})$ as the output of the action of Φ on the first subsystem of the maximally entangled state $|\psi\rangle\langle\psi|$:

$$\hat{\varphi} = (\Phi \otimes \mathbb{1})(|\psi\rangle\langle\psi|) = \frac{1}{d^2} \sum_{j,k=1}^{d} \Phi(|e_j\rangle\langle e_k|) \otimes |e_j\rangle\langle e_k| . \tag{5.145}$$

Notice that, since Φ is completely positive, $\hat{\varphi}$ is certainly positive. If Φ is also trace-preserving, then $\hat{\varphi}$ is a proper, normalised quantum state.[16] The operator $\hat{\varphi}$ contains all the information about the CP-map.

Conversely, one may probabilistically retrieve the CP-map Φ from the normalised Choi state $\hat{\varphi}/\mathrm{Tr}(\hat{\varphi})$ by a gate teleportation procedure: given the Choi state on subsystems 1 and 2, and a generic input ϱ on subsystem 3, it is easy to show that the projection on the maximally entangled state $|\psi\rangle_{23}$ of subsystems 2 and 3 yields the quantum state $\Phi(\varrho)$ in subsystem 1:

$$_{23}\langle\psi|\frac{\hat{\varphi}}{\mathrm{Tr}[\hat{\varphi}]} \otimes \varrho|\psi\rangle_{23} = p \sum_{\substack{j,k,l,m,\\p,q=1}}^{d} \Phi(|e_j\rangle\langle e_k|) \, _{23}\langle e_l, e_l|e_j\rangle_{22}\langle e_k|\varrho_{pq}|e_p\rangle_{33}\langle e_q|e_m, e_m\rangle_{23}$$

$$= p \sum_{j,k=1}^{d} \varrho_{jk}\Phi(|e_j\rangle\langle e_k|) = p\Phi(\varrho) , \tag{5.146}$$

where $p = [d^2\mathrm{Tr}(\hat{\varphi})]^{-1}$ is the probability of the 'successful' projection (equal to $1/d^2$ for trace-preserving maps). Any CP-map on any input state ϱ can hence be obtained, with probability p, by a projective measurement in a basis containing $|\psi\rangle$ acting on the input and on half of the Choi state's subsystems. Although this retrieval is probabilistic, it has the advantage of providing one with an operational procedure to enact the CP-map. Let us mention that the quantum teleportation procedure we will describe in Section 8.1 is, up to the final feedforward which gets rid of the dependence on the measurement outcome, precisely the procedure described above for $\Phi = \mathcal{I}$ (the identity superoperator): clearly, the maximally entangled state itself is the Choi state of the identity CP-map $\mathcal{I}$.

5.5.2 Choi isomorphism in infinite dimension

In infinite dimension, the analogue of the maximally entangled state $|\psi\rangle$ is not normalisable, and hence the isomorphism should be handled with some care. To this aim, we will exploit the fact that, as we saw at the end of Section 5.1.2, a maximally entangled state can be approached by a limiting sequence of normalisable Gaussian states $|\psi_r\rangle$, defined in Eq. (5.22).

Operationally, a projection onto $|\psi_r\rangle$ in the limit $r \to \infty$ corresponds therefore to obtaining the value 0 when measuring the pair of quadratures $(\hat{x}_1 - \hat{x}_2)$ and $(\hat{p}_1 + \hat{p}_2)$. In Section 5.4.2, we saw that such a measurement can be

---

[16] In the following, we will still freely refer to $\hat{\varphi}$ as the Choi 'state', regardless of normalisation.

performed on optical systems through the double homodyne scheme that led us to heterodyne detection. As already noted, it is a distinctive practical feature of optical continuous variables that a feasible scheme (homodyne detection of combined quadratures) allows one to implement a projection on a maximally entangled state.

Given any CP-map Φ on the set of bounded operators $\mathcal{B}(L^2(\mathbb{R}))$, let us then define the Choi state for a given r as

$$\hat{\varphi}_r = (\Phi \otimes \mathbb{1})(|\psi_r\rangle\langle\psi_r|) . \tag{5.147}$$

This is still an isomorphism between CP-maps on $\mathcal{B}(L^2(\mathbb{R}))$ and states on $L^2(\mathbb{R}^2)$, even at finite r, because the Schmidt rank of the state $|\psi_r\rangle$ is infinite.[17] Besides, given any input state $\varrho \in \mathcal{B}(L^2(\mathbb{R}))$, we can reproduce Eq. (5.146) in the limit $r \to \infty$:

$$\lim_{r\to\infty} \cosh(r)^4 {}_{23}\langle\psi_r|\hat{\varphi}_r \otimes \varrho|\psi_r\rangle_{23} = \Phi(\varrho) . \tag{5.148}$$

This convergence holds element-wise in the Fock basis, and also implies the convergence of the expectation values of linear and quadratic combinations of quadrature operators on Gaussian states, which can be approximated arbitrarily well by finite truncations of the Hilbert space in the Fock basis. Such expectation values are all that we need to characterise Gaussian maps, the main point in our agenda for the present treatment.

Eq. (5.148) is useful in two respects. In the first place, it provides one with an operational recipe, based on the availability of the Choi state and on a Gaussian measurement, to approximate arbitrarily well any CP-map, albeit with arbitrarily small probability. Secondly, regardless of the probabilistic nature of the retrieval procedure, it will allow us to characterise the set of Gaussian CP-maps through the set of Gaussian states, and to infer some interesting implications concerning Gaussian operations.

Notice that the isomorphism above can be straightforwardly extended to n modes by replacing $|\psi_r\rangle$ with a tensor product of two-mode squeezed states $|\psi_r\rangle^{\otimes n}$. Such a $2n$-mode state is a Gaussian with covariance matrix $\sigma_r^{(n)}$ given by [see (Eq. 5.21)]

$$\sigma_r^{(n)} = \begin{pmatrix} \cosh(2r)\mathbb{1}_{2n} & \sinh(2r)\Sigma_n \\ \sinh(2r)\Sigma_n & \cosh(2r)\mathbb{1}_{2n} \end{pmatrix}, \quad \text{with} \quad \Sigma_n = \bigoplus_{j=1}^{n} \sigma_z . \tag{5.149}$$

5.5.3   The most general Gaussian CP-map

By virtue of the Choi isomorphism and its reversal, described by Eqs. (5.147) and (5.148), we are now in a position to characterise the class of Gaussian

---

[17] In general, this is the necessary condition for the Choi map to define an isomorphism between CP-maps and quantum states. A maximally entangled state is adopted for convenience, but is not strictly speaking necessary to obtain a bijective mapping.

CP-maps, defined as the CP-maps that send Gaussian states into Gaussian states, and that preserve the Gaussian character of the state $|\psi_r\rangle^{\otimes n}$ when acting only on its first n subsystems.[18] Note that this definition includes non-trace-preserving maps (partial or global measurements), and also encompasses all deterministic Gaussian CP-maps as special cases (dealt with in Problem 5.21 below).

Clearly, the Choi state of all such maps is a Gaussian operator and, conversely, each Gaussian state of $2n$ modes corresponds to one such map (up to normalisation). Hence, we can parametrise all Gaussian CP-maps on $\mathcal{B}(L^2(\mathbb{R}^n))$ with their Choi Gaussian state $\hat{\varphi}_r$ for given r, which in turn can be described by its $4n \times 4n$ CM σ_C and its $4n$-dimensional vector of first moments $\mathbf{d}$. By adopting the following parametrisations of σ_C in terms of $2n \times 2n$ submatrices, and of $\mathbf{d}$ in terms of $2n$-dimensional vectors

$$\sigma_C = \begin{pmatrix} \sigma_1 & \sigma_{12} \\ \sigma_{12}^\mathsf{T} & \sigma_2 \end{pmatrix} \quad , \quad \mathbf{d} = \begin{pmatrix} \mathbf{d}_1 \\ \mathbf{d}_2 \end{pmatrix}, \tag{5.150}$$

one can represent the tensor product $\hat{\varphi}_r \otimes \varrho$ entering Eq. (5.148) as

$$\sigma_C \oplus \sigma = \begin{pmatrix} \sigma_1 & \sigma_{12} & 0 \\ \sigma_{12}^\mathsf{T} & \sigma_2 & 0 \\ 0 & 0 & \sigma \end{pmatrix} \quad , \quad \mathbf{d} \oplus \bar{\mathbf{r}} = \begin{pmatrix} \mathbf{d}_1 \\ \mathbf{d}_2 \\ \bar{\mathbf{r}} \end{pmatrix}. \tag{5.151}$$

We can now apply Eqs. (5.142)–(5.143), with the substitutions $\sigma_A = \sigma_1$, $\sigma_{AB} = (\sigma_{12}\ 0)$, $\sigma_B = (\sigma_2 \oplus \sigma)$, $\sigma_m = \sigma_r^{(n)}$, $\bar{\mathbf{r}}_A = \mathbf{d}_1$, $\bar{\mathbf{r}}_B = (\mathbf{d}_2\ \bar{\mathbf{r}})^\mathsf{T}$ and $\mathbf{r}_m = 0$, to determine the effect of the (conditional) projection of system 2 and 3 onto the state $|\psi\rangle_r$ with zero first moments and second moments given by Eq. (5.149), which determines the action of the CP-map. Notice that the formalism is such that the latter maps an input in system 3 to an output in system 1, so that

$$\sigma \mapsto \lim_{r \to \infty} \left[\sigma_1 - (\sigma_{12}\ 0) \frac{1}{(\sigma_2 \oplus \sigma) + \sigma_r^{(n)}} \begin{pmatrix} \sigma_{12}^\mathsf{T} \\ 0 \end{pmatrix} \right], \tag{5.152}$$

$$\bar{\mathbf{r}} \mapsto \lim_{r \to \infty} \left[\mathbf{d}_1 - (\sigma_{12}\ 0) \frac{1}{(\sigma_2 \oplus \sigma) + \sigma_r^{(n)}} \begin{pmatrix} \mathbf{d}_2 \\ \bar{\mathbf{r}} \end{pmatrix} \right], \tag{5.153}$$

where 0 stands for a $2n \times 2n$ matrix with all zero entries. This very general characterisation will prove its worth in Chapter 7, when dealing with the distillation of Gaussian entanglement. Notice that the limit $r \to \infty$ that appears in Eqs. (5.152) and (5.153) is applied in the definition of the Choi state too, which is often defined up to such a limit, and that one should take care of handling all such limits at the end, when evaluating Eqs. (5.152) and (5.153). This issue is illustrated in the exercise that follows.

---

[18] The last proviso was added in the interest of rigour, although the author is not aware of any CP-map Φ that preserves the Gaussian characters of all Gaussian states and such that $(\Phi \otimes \mathbb{1})(|\psi_r\rangle\langle\psi_r|)$ is not Gaussian.

Problem 5.21. (*Choi state of a deterministic Gaussian CP-map*). Determine the Choi state associated with a deterministic Gaussian CP-map with generic X and Y satisfying (5.37), and verify that it acts as one would expect.

Solution. The covariance matrix of the Choi state is promptly determined by acting with the CP-map on the first subsystem of the covariance matrix $\sigma_r^{(n)}$:

$$\sigma_C = \lim_{r\to\infty} \begin{pmatrix} \cosh(2r)XX^\mathsf{T}+Y & \sinh(2r)X\Sigma_n \\ \sinh(2r)\Sigma_n X^\mathsf{T} & \cosh(2r)\mathbb{1}_{2n} \end{pmatrix}. \tag{5.154}$$

This determines the submatrices σ_1, σ_2 and σ_{12} as per Eq. (5.150). The first moments of the Choi state are obviously zero, since no displacement was applied: $\mathbf{d}_1 = \mathbf{d}_2 = 0$. In order to apply Eqs. (5.152) and (5.153) one needs to invert the matrix

$$(\sigma_2 \oplus \sigma) + \sigma_r^{(n)} = \begin{pmatrix} 2\cosh(2r)\mathbb{1}_{2n} & \sinh(2r)\Sigma_n \\ \sinh(2r)\Sigma_n & \sigma + \cosh(2r)\mathbb{1}_{2n} \end{pmatrix}. \tag{5.155}$$

This is best done by applying the Schur complement formula (1.8), yielding

$$\frac{1}{(\sigma_2 \oplus \sigma) + \sigma_r^{(n)}} = \begin{pmatrix} \frac{1}{2c\mathbb{1}-s^2\Sigma_n \frac{1}{\sigma+c\mathbb{1}}\Sigma_n} & \frac{s}{2c}\Sigma_n \frac{1}{\left(\frac{s^2}{2c}-c\right)\mathbb{1}-\sigma} \\ * & * \end{pmatrix}, \tag{5.156}$$

where we have omitted the entries that will not be relevant in the following and replaced $\cosh(2r)$ and $\sinh(2r)$ with the abridged notation c and s. Notice also that $\Sigma_n^2 = \mathbb{1}$. Now, a first major simplification will come from the fact that we will have to take the limit $r \to \infty$, whereby $\lim_{r\to\infty}\cosh(2r)/\sinh(2r) = 1$, so that s may be replaced by c. Besides, we will have to evaluate several inverses of matrices of the form $[A + c\mathbb{1}]$, where $c \to \infty$ while the matrix A is generic and constant, for which one has

$$[A + c\mathbb{1}]^{-1} = \frac{1}{c}\left[\mathbb{1} - \frac{A}{c}\right] + o(c^{-2}), \tag{5.157}$$

where $o(c^{-2})$ stands for the Landau little-o, which we will be allowed to neglect when taking the limit. Applying Eq. (5.152) then gives

$$\sigma \mapsto \lim_{c\to\infty} cXX^\mathsf{T}+Y - c^2 X\Sigma_n \frac{1}{2c\mathbb{1} - c^2\Sigma_n \frac{1}{\sigma+c\mathbb{1}}\Sigma_n}\Sigma_n X^\mathsf{T}$$

$$= \lim_{c\to\infty} cXX^\mathsf{T}+Y - c^2 X\Sigma_n \frac{1}{c\mathbb{1} + \sigma}\Sigma_n X^\mathsf{T} \tag{5.158}$$

$$= \lim_{c\to\infty} cXX^\mathsf{T}+Y - cXX^\mathsf{T}+X\sigma X^\mathsf{T} = X\sigma X^\mathsf{T}+Y,$$

as one should have expected. Similarly, the expected transformation of the first moments may be checked through Eq. (5.153):

$$\bar{\mathbf{r}} \mapsto \lim_{c \to \infty} -X \frac{c}{2} \Sigma_n^2 \frac{1}{-\frac{c}{2}\mathbb{1} - \sigma} \bar{\mathbf{r}} = \lim_{c \to \infty} X \left(\mathbb{1} - \frac{2\sigma}{c} \right) \bar{\mathbf{r}} = X\bar{\mathbf{r}}. \quad (5.159)$$

5.6 FURTHER READING

More details on the implementation of Gaussian operations may be found in quantum optics textbooks, such as [83] and [75]. Reference [26] is also a valuable resource, especially, but not only, for establishing a connection between the Gaussian formalism and experimental techniques; the standard description of heterodyne detection (involving the mixing with a field at different frequency) may be found there. The decomposition of unitary transformations invoked in the footnote on page 88 is proven in Sections 4.2 and 4.5.1 of Nielsen and Chuang's book [64]. The construction of a symplectic basis which was sketched to complete our proof of sufficiency for the Gaussian CP-map description, on page 100, is detailed in de Gosson's monograph [22].

General-dyne detections were first introduced by Wiseman and Milburn (see [90]). More general covariant measurements are dealt with in Holevo's classic book [49]. Re-ordering ('disentangling') formulae and their group theoretical backdrop are discussed in detail in Puri's book [72].

Our rendition of the infinite-dimensional Choi isomorphism followed [35]. The mathematically inclined reader may prefer Holevo's more rigorous treatment in terms of positive semidefinite forms [48]. A general study on super-operators preserving the Gaussian character of the input state is presented in [23].

CHAPTER 6

Diffusive dynamics and continuous monitoring

CONTENTS

In the last chapter, we dealt with the wide class of physical operations that preserve the Gaussian character of the initial state, treating them as discrete transformations. Here, we describe what could loosely be considered the continuous-time version of such transformations. This chapter is intended to provide the reader with the tools to describe a wide range of notable quantum dynamics, with a very broad applicative scope.

After a note on the action of linear and quadratic Hamiltonians, whose description follows directly from the theory of Gaussian states presented in Chapter 3 but was not explicitly clarified so far, we shall address systems linearly coupled to a reservoir of modes that are delta-correlated in time (a situation commonly referred to as 'white noise'), and thus give rise to Markovian dynamics which we shall refer to as "diffusive" for reasons that will be made clear in due time. Our story will commence with a derivation of the Gaussian dynamics resulting from such an environmental coupling, in the form of dynamical equations for the first and second statistical moments. Besides their interest *per se*, the latter will be instrumental in obtaining an unconventional derivation of the master equations that govern generic quantum states under the same conditions. The corresponding quantum Langevin equations for the canonical operators in the Heisenberg picture will also be derived.

Next, we will turn our attention to the study of the system's dynamics that result from the measurement of the environment. This is a topic with a longstanding tradition in quantum mechanics, grounded in the formalism of quantum stochastic calculus that emerged within the mathematical physics community as early as the seventies, and then re-introduced as the quantum trajectories approach in quantum optics almost twenty years later. The 'weak' POVMs resulting from observing a weakly coupled environment (as opposed to 'strong', projective measurements), carried out continuously in time (a prescription customarily subsumed by the term 'monitoring'), leads to a stochastic dynamics for the system, where the randomness of the measurement outcomes is compactly and most elegantly represented by infinitesimal stochastic increments, giving rise to continuous but not differentiable quantum dynamics. Existing approaches to this subject tend to be quite impervious to the uninitiated. In the quantum optics literature, this is largely due to the technicalities involved in incorporating the discontinuous update of the quantum state prescribed by the von Neumann postulate. More mathematically oriented treatments have instead typically preferred an algebraic approach where the usual difficulty of describing measurements in the Heisenberg picture is compounded by the adoption of a rather abstract hierarchy of filters, in part inspired by classical control theory, to describe different possible updates of the quantum state. This state of affairs is further complicated by the added difficulties that come with the necessary introduction of some form of quantum stochastic calculus.

Our approach to monitoring will hence take a different route. Like in our derivation of diffusion and master equations, we shall keep restricting to linear couplings with a white noise environment, forfeiting full generality.[1] This situation is well modelled by the so called input-output formalism, whose physical origin is discussed in detail in Chapter 9, which approximates reliably, among other instances, the interaction between internal resonating modes and electromagnetic radiation leaking in and out of an optical cavity. First, we shall introduce stochastic Wiener increments and the Itô rule, which will arise naturally from the identification of the continuum field's modes subject to monitoring. We shall deliberately keep our arsenal of stochastic methods to these bare essentials, sacrificing to some extent both depth and generality,[2] but still providing the reader with a powerful and seminal demonstration of the formalism. We shall then restrict to Gaussian general-dyne detections and apply the exact formalism developed in the previous chapter to the continuous filtering of the system's state conditioned by such measurements' outcomes. Next, as in our derivation of the unconditional master equations, we shall use the Gaussian dynamics as a key to unlock the full stochastic master equations

---

[1]Note that, while obviously not all-encompassing, the case of a linear coupling to the environment is very significant, since environmental couplings tend to be weak and can hence typically be linearised.

[2]For instance, no mention will be made of the equivalent Stratonovich formulation of the stochastic differential equations.

governing generic states. Finally, we will build on the quantum theory of filtering in the Gaussian regime to derive dynamical equations for linear feedback control, and also apply our theoretical tools to the enhanced production of squeezing.

6.1 LINEAR AND QUADRATIC HAMILTONIAN DYNAMICS

The action of a second-order Hamiltonian $\hat{H}$ of the form

$$\hat{H} = \frac{1}{2}\hat{\mathbf{r}}^\mathsf{T} H \hat{\mathbf{r}} + \mathbf{d}^\mathsf{T} \Omega \hat{\mathbf{r}} \,, \tag{6.1}$$

where $\hat{\mathbf{r}} = (\hat{x}_1, \hat{p}_1, \ldots, \hat{x}_n, \hat{p}_n)^\mathsf{T}$, H is a real symmetric matrix, Ω is the antisymmetric symplectic form and $\mathbf{d} \in \mathbb{R}^{2n}$, has already been analysed in Chapter 3, when we laid out the basics of the theory of Gaussian states, and taken up again in Section 5.1, when Gaussian unitary transformations were considered. Nevertheless, we shall briefly reconsider it now and explicitly state the dynamical equations such Hamiltonians engender, in both their general form and their Gaussian version.

First notice that the simultaneous action of quadratic and linear Hamiltonians in their discrete form (i.e., over a finite time interval) has already been separated in Section 5.1. In the infinitesimal time setting, such a separation is not necessary, since the evolution equations due to the two additive terms of Eq. (6.1) are obviously additive. We can hence consider such terms separately.

The general Heisenberg evolution equation for the canonical operator due to the quadratic part has already been derived in Eqs. (3.18) and (3.19) of Section 3.2.2, where we found

$$\dot{\hat{\mathbf{r}}} = \Omega H \hat{\mathbf{r}} \,. \tag{6.2}$$

This describes the evolution of any system, be it Gaussian or not, under the Hamiltonian matrix H. The differential equations governing the first and second moments $\bar{\mathbf{r}}$ and $\boldsymbol{\sigma}$ of a quantum state ϱ are promptly derived by inserting Eq. (6.2) into the definitions of $\bar{\mathbf{r}}$ and $\boldsymbol{\sigma}$:

$$\dot{\bar{\mathbf{r}}} = \frac{\mathrm{d}}{\mathrm{d}t}\langle\hat{\mathbf{r}}\rangle = \langle\dot{\hat{\mathbf{r}}}\rangle = \Omega H\langle\hat{\mathbf{r}}\rangle = \Omega H\bar{\mathbf{r}} \,, \tag{6.3}$$

$$\begin{aligned}
\dot{\boldsymbol{\sigma}} &= \frac{\mathrm{d}}{\mathrm{d}t}\langle\{(\hat{\mathbf{r}} - \bar{\mathbf{r}}), (\hat{\mathbf{r}} - \bar{\mathbf{r}})^\mathsf{T}\}\rangle = \langle\{(\dot{\hat{\mathbf{r}}} - \dot{\bar{\mathbf{r}}}), (\hat{\mathbf{r}} - \bar{\mathbf{r}})^\mathsf{T}\}\rangle + \langle\{(\hat{\mathbf{r}} - \bar{\mathbf{r}}), (\dot{\hat{\mathbf{r}}} - \dot{\bar{\mathbf{r}}})^\mathsf{T}\}\rangle \\
&= \langle\{\Omega H(\hat{\mathbf{r}} - \bar{\mathbf{r}}), (\hat{\mathbf{r}} - \bar{\mathbf{r}})^\mathsf{T}\}\rangle + \langle\{(\hat{\mathbf{r}} - \bar{\mathbf{r}}), (\hat{\mathbf{r}} - \bar{\mathbf{r}})^\mathsf{T} H\Omega^\mathsf{T}\}\rangle \\
&= \Omega H\boldsymbol{\sigma} + \boldsymbol{\sigma} H\Omega^\mathsf{T} \,, \tag{6.4}
\end{aligned}$$

where we employed our compact outer product notation (see Section 1.3.3). Clearly, the very same equations might have been derived from knowledge that the symplectic operation $\mathrm{e}^{\Omega H\mathrm{d}t} = \mathbb{1} + \Omega H\mathrm{d}t + o(\mathrm{d}t)$ acts on $\bar{\mathbf{r}}$ and $\boldsymbol{\sigma}$ as $\mathrm{e}^{\Omega H\mathrm{d}t}\bar{\mathbf{r}}$ and $\mathrm{e}^{\Omega H\mathrm{d}t}\boldsymbol{\sigma}\mathrm{e}^{H\Omega^\mathsf{T}\mathrm{d}t}$, and by keeping only the first-order terms in $\mathrm{d}t$. Such a Taylor expansion of a symplectic will come in handy in the next section.

The action of the linear Hamiltonian term $\mathbf{d}^\mathsf{T}\Omega\,\hat{\mathbf{r}}$ could easily be inferred by our analysis of the action of Weyl operators that led to Eq. (3.16), since the latter are manifestly generated by such Hamiltonians. It is however straightforward to obtain the Heisenberg equation of the canonical operators under such Hamiltonians using our compact notation:

$$\dot{\hat{\mathbf{r}}} = i\left[\mathbf{d}^\mathsf{T}\Omega\,\hat{\mathbf{r}},\hat{\mathbf{r}}\right] = i\left[\hat{\mathbf{r}},\hat{\mathbf{r}}^\mathsf{T}\Omega\,\mathbf{d}\right] = i\left[\hat{\mathbf{r}},\hat{\mathbf{r}}^\mathsf{T}\right]\Omega\,\mathbf{d} = i^2\Omega^2\,\mathbf{d} = \mathbf{d}\,. \tag{6.5}$$

The linear term in the Hamiltonian, also known as 'linear driving', simply shifts the canonical operators by a c-number (an operator proportional to the identity) at a constant rate $\mathbf{d}$. Obviously, such a change does not affect the second statistical moments of a state. This is accomplished by coupling the system quadratically to a strong classical field, whose field operators may be replaced by classical variables (see our treatment of the homodyne detection in Section 5.4.1 for a discussion of this issue). The latter could be a coupled classical current, as is common for microwave cavities, or a laser field impinging on a cavity and 'driving' it. We will encounter linear driving in the slightly different scenario of open system dynamics later on in the current chapter.

6.2 OPEN DIFFUSIVE DYNAMICS

In the following, we shall consider a system weakly coupled to a large environment, whose correlation times are much shorter than the system dynamical time scales, such that no information leaking to the environment is ever fed back into the system. A bath of this type is usually referred to as a memoryless, or Markovian, bath, while the whole set of dynamical conditions we will consider, including that of weak coupling, that basically does not alter the state of the bath, are collectively known as the Born–Markov regime.

More specifically, we will assume white quantum noise in the so-called input-output formalism, which describes a system of n modes in contact with a train of m incoming bosonic modes represented by the $2m$-dimensional vector of operators $\hat{\mathbf{r}}_{in}(t)$, which collectively interact with the system at time t for an infinitesimal interval dt and are then scattered as the output modes $\hat{\mathbf{r}}_{out}(t)$. Note that, when associated with input and output fields, the parameter t is not a dynamical variable but a label to distinguish the different interacting modes: the modes $\hat{\mathbf{r}}_{in}(t)$ are those that interact with the system at time t. The input-output formalism, which in a sense allows one to link discrete and continuous sets of modes, is derived in Section 9.1.4, where the following continuous version of the commutation relations is justified for Markovian systems in the context of field quantisation:

$$[\mathbf{r}_{in}(t),\mathbf{r}_{in}(t')] = i\Omega\delta(t - t')\,. \tag{6.6}$$

Notice that the input field operators have the dimensions of the square root of a frequency. Eq. (6.6) is known as the "white noise" condition and, as we shall see, entails a Markovian dynamics for the system.[3] We will also need to specify the second statistical moments of the input fields as

$$\langle\{\hat{\mathbf{r}}_{in}(t), \hat{\mathbf{r}}_{in}^{\mathsf{T}}(t')\}\rangle = \boldsymbol{\sigma}_{in}\, \delta(t - t')\,, \qquad \boldsymbol{\sigma}_{in} + i\boldsymbol{\Omega} \geq 0\,. \tag{6.7}$$

The white noise statistics above imply that the system interacts at each instant t with a different set of modes, completely uncorrelated with those that it encountered in the past. Otherwise, we allow for complete generality in the second-order correlations of the bath quadratures, by letting their covariance matrix $\boldsymbol{\sigma}_{in}$ be any physical CM. Note that this allows one to set a finite environmental temperature. The whiteness of the condition above hints at the fact that the profile of the noise spectrum of the input modes in the frequency domain is flat (given the delta correlation in the time domain). We are thus disregarding coloured reservoirs, which is often a good assumption in optical system but may fail in solid-state systems, where the reservoirs' frequency profiles are usually more structured.

The white noise conditions (6.6) and (6.7) may be recast as a condition on infinitesimal quantum operators, which act as counterparts of classical stochastic increments in what is known as quantum stochastic calculus. Rather than taking such a formal, axiomatic approach, let us define the operators $\delta\hat{\mathbf{r}}_{in}(t) = \int_t^{t+\delta t} \hat{\mathbf{r}}_{in}(s)ds$ for a certain arbitrary time interval δt and notice that Eqs. (6.6) and (6.7) imply:

$$[\delta\hat{\mathbf{r}}_{in}(t), \delta\hat{\mathbf{r}}_{in}^{\mathsf{T}}(t)] = i\boldsymbol{\Omega} \int_t^{t+\delta t} ds = i\boldsymbol{\Omega}\delta t\,, \tag{6.8}$$

$$\langle\{\delta\hat{\mathbf{r}}_{in}(t), \delta\hat{\mathbf{r}}_{in}^{\mathsf{T}}(t)\}\rangle = \boldsymbol{\sigma}_{in} \int_t^{t+\delta t} ds = \boldsymbol{\sigma}_{in}\delta t\,. \tag{6.9}$$

In the limit of arbitrarily small $\delta t = dt$, one has $\delta\hat{\mathbf{r}}_{in} = \hat{\mathbf{r}}_{in}\, dt$ and hence obtains

$$[\hat{\mathbf{r}}_{in}(t), \hat{\mathbf{r}}_{in}^{\mathsf{T}}(t)]dt^2 = i\boldsymbol{\Omega}\, dt\,, \tag{6.10}$$

$$\langle\{\hat{\mathbf{r}}_{in}(t), \hat{\mathbf{r}}_{in}^{\mathsf{T}}(t)\}\rangle dt^2 = \boldsymbol{\sigma}_{in}\, dt\,. \tag{6.11}$$

We now take the bold step of defining the vector of operators $\hat{\mathbf{r}}'_{in}(t)$ as canonical operators satisfying standard, "discrete" commutation relations

$$[\hat{\mathbf{r}}'_{in}(t), \hat{\mathbf{r}}'^{\mathsf{T}}_{in}(t)] = i\boldsymbol{\Omega}\,, \tag{6.12}$$

as well as the relationship $\hat{\mathbf{r}}_{in}\, dt = \hat{\mathbf{r}}'_{in}\, \sqrt{dt}$, where $\sqrt{dt}$ is an infinitesimal quantity such that $(\sqrt{dt})^2 = dt$. This ensures compliance between Eqs. (6.10) and (6.12). The operators $\hat{\mathbf{r}}'_{in}\, \sqrt{dt}$ are known as "quantum Wiener processes"

---

[3]In the sense that the solution to the dynamical equation of the system at a given time will be determined by the system state alone at any previous instant, without involving any time integral kernel.

and, when monitored, give rise to classical, stochastic, diffusive Wiener processes, such as Brownian motion or continuous random walks, whose variance grows linearly in time.[4] Therefore, the open system dynamics we are about to derive share substantial analogies with classical diffusive processes, and will hence be termed 'diffusive'. The (dimensionless) canonical operators $\hat{\mathbf{r}}'_{in}$ satisfy the CCR (6.12) and are hence associated with degrees of freedom that describe measurements in the laboratory: they are field modes whose excitations make detectors click. On the other hand, the original input operators $\hat{\mathbf{r}}_{in}$ are those that are linearly coupled to the system in the standard Hamiltonian picture: this distinction will be crucial in what follows.

Let us now introduce a completely generic quadratic coupling Hamiltonian $\hat{H}_C$ between system and input modes (the 'bath'):

$$\hat{H}_C = \hat{\mathbf{r}}^{\mathsf{T}} C \hat{\mathbf{r}}_{in} = \frac{1}{2}\hat{\mathbf{r}}^{\mathsf{T}}_{SB} H_C \hat{\mathbf{r}}_{SB} = \frac{1}{2}\hat{\mathbf{r}}^{\mathsf{T}}_{SB} \begin{pmatrix} 0 & C \\ C^{\mathsf{T}} & 0 \end{pmatrix} \hat{\mathbf{r}}_{SB}, \qquad (6.13)$$

where the real $2n \times 2m$ coupling matrix C is entirely generic (recall that any mode of the system could interact with any number of input modes) and $\hat{\mathbf{r}}^{\mathsf{T}}_{SB} = (\hat{\mathbf{r}}^{\mathsf{T}}, \hat{\mathbf{r}}^{\mathsf{T}}_{in})$.[5] Notice that the matrix C has the dimensions of the square root of a frequency (while the Hamiltonian matrices associated with discrete degrees of freedom typically have dimensions of a frequency in our formalism). We are however interested in the time evolution of the environmental quantum variables $\hat{\mathbf{r}}'_{in}$ rather than $\hat{\mathbf{r}}_{in}$, because the former are those that can be measured. In terms of the new vector of canonical operators $\hat{\mathbf{r}}'_{SB} = (\hat{\mathbf{r}}, \hat{\mathbf{r}}'_{in})^{\mathsf{T}}$, time evolution over an interval $\mathrm{d}t$ is generated by the operator $\hat{\mathbf{r}}^{\mathsf{T}}_{SB} H_C \hat{\mathbf{r}}_{SB}\, \mathrm{d}t = \hat{\mathbf{r}}'^{\mathsf{T}}_{SB} H_C \hat{\mathbf{r}}'_{SB}\sqrt{\mathrm{d}t}$. Therefore, their initial CM $\boldsymbol{\sigma} \oplus \boldsymbol{\sigma}_{in}$ evolves under the symplectic transformation

$$\mathrm{e}^{\Omega H_C \sqrt{\mathrm{d}t}} = \mathbb{1} + \Omega H_C \sqrt{\mathrm{d}t} + \frac{(\Omega H_C)^2}{2}\,\mathrm{d}t + o(\mathrm{d}t), \qquad (6.14)$$

which acts by congruence as follows

$$\mathrm{e}^{\Omega H_C \sqrt{\mathrm{d}t}}\, (\boldsymbol{\sigma} \oplus \boldsymbol{\sigma}_{in})\, \mathrm{e}^{H_C \Omega^{\mathsf{T}} \sqrt{\mathrm{d}t}} = (\boldsymbol{\sigma} \oplus \boldsymbol{\sigma}_{in}) + \boldsymbol{\sigma}_{SB}\sqrt{\mathrm{d}t} + o(\mathrm{d}t) \qquad (6.15)$$

$$+ \left(\frac{\Omega C \Omega C^{\mathsf{T}} \boldsymbol{\sigma} + \boldsymbol{\sigma} C \Omega C^{\mathsf{T}} \Omega}{2} + \Omega C \boldsymbol{\sigma}_{in} C^{\mathsf{T}} \Omega^{\mathsf{T}} \right) \oplus \tilde{\boldsymbol{\sigma}}_B\, \mathrm{d}t,$$

where $(\sqrt{\mathrm{d}t})^2 = \mathrm{d}t$ was applied, the Landau symbol $o(\mathrm{d}t)$ employed, and

$$\boldsymbol{\sigma}_{SB} = \begin{pmatrix} 0 & \Omega C \boldsymbol{\sigma}_{in} + \boldsymbol{\sigma} C \Omega^{\mathsf{T}} \\ \boldsymbol{\sigma}_{in} C^{\mathsf{T}} \Omega^{\mathsf{T}} + \Omega C^{\mathsf{T}} \boldsymbol{\sigma} & 0 \end{pmatrix}, \qquad (6.16)$$

---

[4]See Appendix C, where it is also shown that Wiener increments, which underpin the description of such processes, would satisfy deterministically a relationship analogous to $(\sqrt{\mathrm{d}t})^2 = \mathrm{d}t$, the so called Itô rule.

[5]Note that when we state that the system is "linearly coupled" to the environment, what we actually mean is that their interaction Hamiltonian is bilinear (quadratic) with each term proportional to a product of system and bath quadrature operators.

$$\tilde{\sigma}_B = \frac{\Omega C^\mathsf{T} \Omega C \sigma_{in} + \sigma_{in} C^\mathsf{T} \Omega C \Omega}{2} + \Omega^\mathsf{T} C^\mathsf{T} \sigma C \Omega . \tag{6.17}$$

Eq. (6.14) also yields the evolution of the first moment vector $\bar{\mathbf{r}}$:

$$e^{\Omega H_C \sqrt{dt}} \left(\bar{\mathbf{r}} \oplus \bar{\mathbf{r}}'_{in} \right) = (\bar{\mathbf{r}} \oplus \bar{\mathbf{r}}'_{in}) + (\Omega C \bar{\mathbf{r}}'_{in}) \oplus (\Omega C^\mathsf{T} \bar{\mathbf{r}}) \sqrt{dt}$$
$$+ \left(\frac{\Omega C \Omega C^\mathsf{T}}{2} \bar{\mathbf{r}} \right) \oplus \left(\frac{\Omega C^\mathsf{T} \Omega C}{2} \bar{\mathbf{r}}'_{in} \right) dt + o(dt) , \tag{6.18}$$

where $\bar{\mathbf{r}}'_{in}$ is the vector of first moments of the canonical operators $\hat{\mathbf{r}}'_{in}$.

In Section 6.3, we shall come back to Eqs. (6.15) and (6.18) as the starting point to study the 'conditional' dynamics resulting from monitoring the bath. Here, we shall instead analyse in detail the 'unconditional' dynamics of the system, occurring when the bath is disregarded or, equivalently, when hypothetical measurements on the bath are not recorded but rather averaged upon. The latter is obtained by taking a partial trace over the environment which, as we saw in Chapter 5, in the Gaussian formalism simply amounts to restricting to the diagonal block pertaining to the system. Doing this on the matrix equation (6.15) yields the following deterministic diffusion equation for the system covariance matrix:

$$\dot{\sigma} = A\sigma + \sigma A^\mathsf{T} + D , \tag{6.19}$$

for the so called drift and diffusion matrices A and D:

$$A = \frac{\Omega C \Omega C^\mathsf{T}}{2} , \tag{6.20}$$

$$D = \Omega C \sigma_{in} C^\mathsf{T} \Omega^\mathsf{T} . \tag{6.21}$$

The unconditional evolution of the first moments is better understood moving back to the original input operators $\hat{\mathbf{r}}_{in}$, which were defined so that their first moments $\bar{\mathbf{r}}_{in}$ satisfy $\bar{\mathbf{r}}'_{in}\sqrt{dt} = \bar{\mathbf{r}}_{in}dt$ (note that the distinction between $\hat{\mathbf{r}}'_{in}$ and $\hat{\mathbf{r}}_{in}$ is key here). The partial trace over the bath then furnishes:

$$\dot{\bar{\mathbf{r}}} = A\bar{\mathbf{r}} + \Omega C \bar{\mathbf{r}}_{in} = A\bar{\mathbf{r}} + \mathbf{d} , \tag{6.22}$$

where we defined the 'drive' $\mathbf{d} = \Omega C \bar{\mathbf{r}}_{in}$. The effect of non-vanishing input first moments $\bar{\mathbf{r}}_{in}$ is hence identical to the driving that results from a linear Hamiltonian acting directly on the system [see Eq. (6.5)]. Linear driving will be revisited with direct reference to quantum optics in Chapter 9, where our description will be rich enough to account for the interplay between the frequency of the driving field and the internal frequency of the system modes too.

We should also note that, in order to keep the derivation cleaner, we have not considered any Hamiltonian operator describing the dynamics of the system alone. If such a Hamiltonian, quadratic in the system canonical operators, is included as $\hat{H}_S = \frac{1}{2}\hat{\mathbf{r}}^\mathsf{T} H_S \hat{\mathbf{r}}$, then only the drift matrix is modified and takes the form $A = \Omega H_S + \frac{\Omega C \Omega C^\mathsf{T}}{2}$.

Our findings so far may be summarised as follows:

Diffusive Gaussian dynamics. Consider a system of n modes subject to a quadratic Hamiltonian with Hamiltonian matrix H_S, as well as to linear driving at a rate $\mathbf{d} \in \mathbb{R}^{2n}$, and linearly coupled to a white noise environment of $2m$ modes with covariance matrix $\boldsymbol{\sigma}_{in}$ through a coupling matrix C. Then, the covariance matrix $\boldsymbol{\sigma}$ and first moments $\bar{\mathbf{r}}$ of the system obey the following diffusive equations

$$\dot{\boldsymbol{\sigma}} = A\boldsymbol{\sigma} + \boldsymbol{\sigma}A^{\mathsf{T}} + D \, , \tag{6.23}$$

$$\dot{\bar{\mathbf{r}}} = A\bar{\mathbf{r}} + \mathbf{d} \, , \tag{6.24}$$

where

$$A = \Omega H_S + \frac{\Omega C \Omega C^{\mathsf{T}}}{2} \, , \tag{6.25}$$

$$D = \Omega C \boldsymbol{\sigma}_{in} C^{\mathsf{T}} \Omega^{\mathsf{T}} \, . \tag{6.26}$$

The diffusive evolutions derived above enjoy an exceptional breadth of applications, among which the most iconic and pervasive is without a doubt the description of losses in a thermal environment: this is well approximated, for each system mode, by assuming the coupling to a single-mode white noise environment in a thermal state, with $\bar{\mathbf{r}}_{in} = 0$ and $\boldsymbol{\sigma}_{in} = (2N+1)\mathbb{1} = n_{\text{th}}\mathbb{1}$ (where N is the average number of excitations in the environmental mode and n_{th} stands for the 'thermal noise', in compliance with the notation of the previous chapter), through a coupling Hamiltonian that allows for the exchange of excitations $\sqrt{\gamma} i(\hat{a}^{\dagger}\hat{a}_{in} - \hat{a}\hat{a}_{in}^{\dagger}) = \sqrt{\gamma}(\hat{p}\hat{x}_{in} - \hat{x}\hat{p}_{in})$. In our notation, this corresponds to $C = \sqrt{\gamma}\,\Omega^{\mathsf{T}}$ which, applying Eqs. (6.25) and (6.26), implies $A = \Omega C \Omega C^{\mathsf{T}}/2 = -\frac{\gamma}{2}\mathbb{1}$ (note that we are neglecting the system Hamiltonian, as would be possible in interaction picture) and $D = \gamma n_{\text{th}}\mathbb{1}$, so that Eqs. (6.19) and (6.24) take a very simple form:

Thermal diffusion. The covariance matrix $\boldsymbol{\sigma}$ and first moments $\bar{\mathbf{r}}$ of a system where each mode is coupled to an independent thermal environment with loss rate γ and number of thermal excitations $(n_{\text{th}} - 1)/2$ obey the diffusion equation

$$\dot{\boldsymbol{\sigma}} = -\gamma\boldsymbol{\sigma} + \gamma n_{\text{th}}\mathbb{1} \, , \tag{6.27}$$

$$\dot{\bar{\mathbf{r}}} = -\frac{\gamma}{2}\bar{\mathbf{r}} \, , \tag{6.28}$$

with solutions

$$\boldsymbol{\sigma}(t) = e^{-\gamma t}\boldsymbol{\sigma}(0) + (1 - e^{-\gamma t})n_{\text{th}}\mathbb{1} \, , \tag{6.29}$$

$$\bar{\mathbf{r}}(t) = e^{-\frac{\gamma}{2}t}\bar{\mathbf{r}}(0) \, . \tag{6.30}$$

At each given time, this solution taken as a discrete transformation corresponds to an attenuator channel, as we defined it in Section 5.3.3 although, at variance with the open dynamics we assumed to derive attenuator channels in the previous chapter, the exponential decay of Eq. (6.29) is the hallmark of the interaction with a large reservoir.[6] For $n_{th} = 1$, the diffusive equation above describes the case of pure losses, which models very well the leakage from a cavity or an optical fibre at high, optical frequencies.

It is interesting to consider briefly which unconditional diffusive evolutions are allowed within this framework, that is, what are the most general drift and diffusion matrices A and D. Eq. (6.23) shows that $\Omega^{\mathsf{T}} A = H_S + C\Omega C^{\mathsf{T}}/2$. Since C is completely generic (not even necessarily square),[7] $C\Omega C^{\mathsf{T}}$ is a completely generic $2n \times 2n$ real anti-symmetric matrix, while H is a completely generic symmetric matrix. Hence, A is any real square matrix, and the symmetric and antisymmetric parts of $\Omega^{\mathsf{T}} A$ furnish, respectively, H_S and $C\Omega C^{\mathsf{T}}/2$. The set of allowed D's for given A is determined as follows. Let $A_a = C\Omega C^{\mathsf{T}}$ be the anti-symmetric part of $2\Omega^{\mathsf{T}} A$. Then, D must only comply with the uncertainty relation of the bath state $\sigma_{in} + i\Omega \geq 0$, which entails $D + i\Omega C\Omega C^{\mathsf{T}}\Omega^{\mathsf{T}} = D + i\Omega A_a \Omega^{\mathsf{T}} \geq 0$. Summing up, one can characterise the most general A and D by the condition:

Bona fide diffusive dynamics. The pair of matrices A and D represent the drift and diffusion matrices of a quantum system if and only if

$$D + iA\Omega^{\mathsf{T}} - i\Omega A^{\mathsf{T}} \geq 0 . \qquad (6.31)$$

Notice that this condition is specific of quantum dynamics, since it was derived from the uncertainty principle. For a single degree of freedom, when the matrices above are all two-dimensional, the inequality (6.31) reduces to $\mathrm{Det} D \geq \mathrm{Det}(\Omega^{\mathsf{T}} A - A^{\mathsf{T}}\Omega)$, which includes all single-mode unconditional diffusive dynamics.

Let us also briefly consider the possible Gaussian steady states of the open dynamics determined above. The first and second moments at steady state

---

[6] Recall that, although only one mode is interacting with the system at any given time for a time interval dt, our formalism implies that a fresh environmental mode enters the dynamic at any time, which is analogous to assuming a very large bath.

[7] One might object that the Hamiltonian matrix H_C from which we derived our diffusion equations is not positive, which entails that the associated Hamiltonian operator is unbounded from below and hence not physical. However, once a positive system Hamiltonian matrix H_S is included, one can show that a bath Hamiltonian matrix H_B always exists that would make the global Hamiltonian matrix positive. In fact, a block matrix with $H_B > 0$ is positive if and only if the Schur complement $(H_S - CH_B^{-1}C^{\mathsf{T}})$ is positive, a condition which may be contrived for any $H_S > 0$ and any C by choosing H_B equal to a large enough constant. Hence, all the dynamics we derived are in principle possible. In practice, though, restrictions on the system and bath Hamiltonians would imply restrictions on the coupling C of a physical global Hamiltonian matrix.

$\bar{\mathbf{r}}_\infty$ and $\boldsymbol{\sigma}_\infty$ would have to satisfy

$$A\bar{\mathbf{r}}_\infty + \mathbf{d} = 0 , \tag{6.32}$$

$$A\boldsymbol{\sigma}_\infty + \boldsymbol{\sigma}_\infty A^\mathsf{T} + D = 0 . \tag{6.33}$$

It may shown that such equations admit stable solutions (in the sense that they would be resilient to small perturbation in $\mathbf{r}'_\infty$ and $\boldsymbol{\sigma}_\infty$) if and only if A is diagonalisable and the real part of all its eigenvalues is negative (a condition often referred to in engineering and control theory as A being "Hurwitz").

6.2.1 Master equations

The diffusive dynamics we have considered above, including only a linear coupling to the bath, are entirely characterised by the matrices A and D. The details of the evolution of Gaussian states, given by Eqs. (6.19) and (6.24), thus completely characterise such dynamics. Hence, the equation of motion governing the evolution of a generic quantum state ϱ, the so called master equation, can be inferred from the Gaussian dynamics.

Since, as we saw, linear driving always amounts to a Hamiltonian term which can be just added to the master equation, we can disregard it and set $\mathbf{d} = 0$. Then, it turns out that the evolution of a Gaussian state with zero first moments, which remain zero as per Eq. (6.24), is sufficient to derive the full corresponding master equation. By Eq. (4.48), a Gaussian characteristic function χ_G with null first moments may be written as $\chi_G = \mathrm{e}^{-\frac{1}{4}\tilde{\mathbf{r}}^\mathsf{T}\boldsymbol{\sigma}\tilde{\mathbf{r}}}$ (the switch to the variables $\tilde{\mathbf{r}} = \Omega\mathbf{r}$ will greatly simplify what follows) and its time evolution under the diffusive equation (6.19) is given by

$$\dot{\chi}_G = -\frac{1}{4}\tilde{\mathbf{r}}^\mathsf{T}(A\boldsymbol{\sigma} + \boldsymbol{\sigma} A^\mathsf{T} + D)\tilde{\mathbf{r}}\,\chi_G . \tag{6.34}$$

This may be re-written as a linear differential equation for a generic characteristic function χ, which will already provide one with the general dynamics of any quantum state, since any quantum state allows for a description in terms of a characteristic function. To attain such an equation, first notice that the right hand side of Eq. (6.34) above is linear in A and D, such that we can consider the effect of each of their entries separately. The contributions of each A_{jk} and D_{jk} to the time derivative of χ_G are

$$\dot{\chi}_G^{(A_{jk})} = -\frac{1}{2}A_{jk}\tilde{r}_j\left(\sum_{l=1}^{2n}\sigma_{kl}\tilde{r}_l\right)\chi_G , \tag{6.35}$$

$$\dot{\chi}_G^{(D_{jk})} = -\frac{1}{4}D_{jk}\tilde{r}_j\tilde{r}_k\,\chi_G , \tag{6.36}$$

where we adopted Einstein's summation convention for repeated indexes, which will be maintained in the remainder of this section in order to facilitate bookkeeping. The contribution of D_{jk} may hence be generalised as the multiplication by the corresponding phase space variables, be the state Gaussian or not. In order to free the A_{jk} contribution from its dependence on the Gaussian parameter σ, just notice that

$$\partial_{\tilde{r}_k}\chi_G = -\frac{1}{2}\left(\sum_{l=1}^{2n}\sigma_{kl}\tilde{r}_l\right)\chi_G, \tag{6.37}$$

so that

$$\dot{\chi}_G^{(A_{jk})} = A_{jk}\tilde{r}_j\partial_{\tilde{r}_k}\chi_G. \tag{6.38}$$

It is straightforward to show that such an action of A gives rise to an additional term for non-null first moments which is also consistent with the drift equation (6.24) (disregarding the drive which is just a Hamiltonian term anyway). One is thus led to the following general equation for a generic characteristic function χ:

$$\dot{\chi} = \left[A_{jk}\tilde{r}_j\partial_{\tilde{r}_k} - \frac{1}{4}D_{jk}\tilde{r}_j\tilde{r}_k\right]\chi. \tag{6.39}$$

Eq. (6.39) already provides one with the evolution equation for a generic quantum state. It is nevertheless interesting and instructive to translate such an equation into the Hilbert space formalism, and obtain the master equation for the density operator ϱ associated with χ.

The dictionary for such a translation is obtained by employing a notional operator $\hat{o}$ that admits a characteristic function representation $\chi_{\hat{o}}(\tilde{\mathbf{r}}) = \text{Tr}[e^{i\tilde{\mathbf{r}}^\mathsf{T}\hat{\mathbf{r}}}\hat{o}]$, where care was taken in writing down the Weyl operator in terms of the variables $\tilde{\mathbf{r}}$. The Baker-Campbell-Hausdorff relation (3.9) may be expressed, in this multivariate setting, as

$$\chi_{\hat{o}}(\tilde{\mathbf{r}}) = \text{Tr}[e^{i\tilde{\mathbf{r}}^\mathsf{T}\hat{\mathbf{r}}}\hat{o}] = \text{Tr}[e^{i\tilde{\mathbf{x}}^\mathsf{T}\hat{\mathbf{x}}}e^{i\tilde{\mathbf{p}}^\mathsf{T}\hat{\mathbf{p}}}e^{\frac{i}{2}\tilde{\mathbf{x}}^\mathsf{T}\tilde{\mathbf{p}}}\hat{o}] = \text{Tr}[e^{i\tilde{\mathbf{p}}^\mathsf{T}\hat{\mathbf{p}}}e^{i\tilde{\mathbf{x}}^\mathsf{T}\hat{\mathbf{x}}}e^{-\frac{i}{2}\tilde{\mathbf{x}}^\mathsf{T}\tilde{\mathbf{p}}}\hat{o}], \tag{6.40}$$

where we defined the n-dimensional vectors $\hat{\mathbf{x}}$ ($\hat{\mathbf{p}}$) and $\tilde{\mathbf{x}}$ ($\tilde{\mathbf{p}}$) as the vectors of odd (even) entries of the parent vectors $\hat{\mathbf{r}}$ and $\tilde{\mathbf{r}}$, such that $\hat{\mathbf{x}} \oplus \hat{\mathbf{p}} = \hat{\mathbf{r}}$ and $\tilde{\mathbf{x}} \oplus \tilde{\mathbf{p}} = \tilde{\mathbf{r}}$. Now, differentiating the last two equalities of (6.40) yields

$$\partial_{\tilde{p}_j}\chi_{\hat{o}} = i\,\text{Tr}[e^{i\tilde{\mathbf{r}}^\mathsf{T}\hat{\mathbf{r}}}\hat{o}\hat{p}_j] - \frac{i}{2}\tilde{x}_j\chi_{\hat{o}} = i\,\text{Tr}[e^{i\tilde{\mathbf{r}}^\mathsf{T}\hat{\mathbf{r}}}\hat{p}_j\hat{o}] + \frac{i}{2}\tilde{x}_j\chi_{\hat{o}}, \tag{6.41}$$

$$\partial_{\tilde{x}_j}\chi_{\hat{o}} = i\,\text{Tr}[e^{i\tilde{\mathbf{r}}^\mathsf{T}\hat{\mathbf{r}}}\hat{o}\hat{x}_j] + \frac{i}{2}\tilde{p}_j\chi_{\hat{o}} = i\,\text{Tr}[e^{i\tilde{\mathbf{r}}^\mathsf{T}\hat{\mathbf{r}}}\hat{x}_j\hat{o}] - \frac{i}{2}\tilde{p}_j\chi_{\hat{o}}, \tag{6.42}$$

whence

$$\partial_{\tilde{p}_j}\chi_{\hat{o}} = \frac{i}{2}\chi_{(\hat{p}_j\hat{o}+\hat{o}\hat{p}_j)}, \qquad \partial_{\tilde{x}_j}\chi_{\hat{o}} = \frac{i}{2}\chi_{(\hat{x}_j\hat{o}+\hat{o}\hat{x}_j)}, \tag{6.43}$$

$$\tilde{p}_j\chi_{\hat{o}} = \chi_{(\hat{x}_j\hat{o}-\hat{o}\hat{x}_j)}, \qquad \tilde{x}_j\chi_{\hat{o}} = -\chi_{(\hat{p}_j\hat{o}-\hat{o}\hat{p}_j)}. \tag{6.44}$$

Going back to the variables $\tilde{\mathbf{r}}$, the general correspondence between differential or multiplicative terms on the characteristic function and operators can be recast as

$$\partial_{\tilde{r}_j}\chi_{\hat{o}} \longleftrightarrow \frac{i}{2}(\hat{r}_j\hat{o}+\hat{o}\hat{r}_j), \qquad \tilde{r}_k\chi_{\hat{o}} \longleftrightarrow \Omega_{jk}(\hat{r}_j\hat{o}-\hat{o}\hat{r}_j). \tag{6.45}$$

Applying such a correspondence to Eq. (6.39) for the characteristic function χ and the associated density operator ϱ, yields the most general master equation obtained for white noise and linear coupling to the environment:

Diffusive master equation. The quantum state ϱ of a system of n modes, linearly coupled to a Markovian environment of n modes with covariance matrix σ_{in} through a coupling matrix C, is governed by the following master equation:

$$\dot{\varrho} = \left[\left(\frac{i}{2}\Omega A - \frac{\Omega D \Omega^{\mathsf{T}}}{4} \right)_{jk} \hat{r}_j \hat{r}_k \varrho + \text{h.c.} \right] + \left[\left(\frac{i}{2}\Omega A + \frac{\Omega D \Omega^{\mathsf{T}}}{4} \right)_{jk} \hat{r}_j \varrho \hat{r}_k + \text{h.c.} \right],$$
(6.46)

with matrices A and D given by Eqs. (6.25) and (6.26). Einstein's summation convention was adopted in the equation above.

Problem 6.1. (*Quantum optical master equation*). Show that the dynamics of a single-mode system subject to the environmental coupling $\sqrt{\gamma}i(\hat{a}^\dagger \hat{a}_{in} - \hat{a}\hat{a}_{in}^\dagger)$ with a bath in the vacuum state (pure loss) is governed by the "quantum optical master equation"

$$\dot{\varrho} = \gamma \hat{a}\varrho\hat{a}^\dagger - \frac{\gamma}{2}\left(\varrho\hat{a}^\dagger\hat{a} + \hat{a}^\dagger\hat{a}\varrho\right).$$
(6.47)

Also, modify the master equation above for an average number of excitations N in the bath.

Solution. As we saw in the previous section, pure losses correspond to $C = \sqrt{\gamma}\,\Omega^{\mathsf{T}}$ and $\sigma_{in} = \mathbb{1}$, which imply $A = -\frac{\gamma}{2}\mathbb{1}$ and $D = \gamma\mathbb{1}$. Inserting these matrices into Eq. (6.46) with $\hat{r}_1 = \hat{x}$ and $\hat{r}_2 = \hat{p}$ gives
$\dot{\varrho} = \frac{1}{4}\left[(-\hat{x}^2 - \hat{p}^2 - i\hat{x}\hat{p} + i\hat{p}\hat{x})\varrho + (\hat{x}\varrho\hat{x} + \hat{p}\varrho\hat{p} - i\hat{x}\varrho\hat{p} + i\hat{p}\varrho\hat{x}) + \text{h.c.}\right]$
and, substituting $\hat{a} = (\hat{x} + i\hat{p})/\sqrt{2}$, the desired master equation. It is customary to define the 'Liouville superoperator' $\mathcal{D}[\hat{o}] = \hat{o}\varrho\hat{o}^\dagger - \frac{1}{2}\left(\varrho\hat{o}^\dagger\hat{o} + \hat{o}^\dagger\hat{o}\varrho\right)$ for a generic operator $\hat{o}$ and write the master equation as $\dot{\varrho} = \gamma\mathcal{D}[\hat{a}]\varrho$.

Adding N average excitation in the bath just changes σ_{in}, and hence D, into $\gamma(2N+1)\mathbb{1}$. Repeating the process above for such a D yields the master equation describing losses in a thermal environment:

$$\dot{\varrho} = \gamma(N+1)\mathcal{D}[\hat{a}]\varrho + \gamma N \mathcal{D}[\hat{a}^\dagger]\varrho.$$
(6.48)

Notice that, if one assumes an environmental mode governed by a quadratic, harmonic oscillator Hamiltonian at frequency ω_B, as would be exactly the case for free electromagnetic radiation, N may be related to the environmental temperature T by $N = 1/(\exp(\beta\hbar\omega_B) - 1)$, where $\beta = 1/(kT)$ and k is the Boltzmann constant.

In the interest of rigour, one may observe that the derivation of the master equation presented above suffers from the somewhat arbitrary choice we made in bridging between the Gaussian and the general equation for the characteristic function χ. There, we tacitly admitted only combinations of terms involving two multiplications or one multiplication and one differentiation by the phase space variables. One may however convince oneself that this was legitimate by the following argument. First notice that, according to the Hilbert space translation we worked out above, this assumption is equivalent to restricting to quadratic operators (in the canonical quadratures) acting on ϱ in the master equation. Consider now the original Hamiltonian $\hat{H}_C$ of Eq. (6.13), which we handled in the Gaussian domain. In the general formalism, the master equation may be derived by acting on ϱ by similarity with the evolution operator $e^{i\hat{H}_C\,\mathrm{d}t}$ at first order in $\mathrm{d}t$, and then tracing over the input field operators. Handling the stochastic increment as we did in the Gaussian case, it may be easily seen that this procedure only leaves one with linear terms in the system canonical operators – which, as already discussed, are only Hamiltonian displacements – and quadratic terms. Hence, our assumption was a valid one. Obviously, the procedure outlined here would have led to the same master equation we derived from the Gaussian dynamics. It is mostly in the analysis of monitored, conditional dynamics, to follow, that the simplicity of the Gaussian formalism really stands out. Note that the fact that this dynamics preserves the Gaussian character of the system state hinges solely on the linearity of the coupling and on the fact that each mode only interacts for an infinitesimal time interval, *but does not depend at all on whether the state of the input field is Gaussian or not*. The only properties of the latter that affect the dynamics in the model considered are first and second statistical moments.

Let us also remark that, in the Schrödinger picture, the master equations we just derived emerge in the so called Born–Markov approximation which, in a nutshell, corresponds to stating that the system-bath coupling is weak, and hence may be treated perturbatively at first order, and that the state of the bath is not affected at all by such a coupling. Although subtleties might possibly arise in establishing a rigorous correspondence between these two treatments, it is intuitively clear that our model meets all the conditions stated above, since refreshing the bath mode after every infinitesimal time interval with another completely uncorrelated degree of freedom entails that the interacting bath mode is not at all affected by the previous interaction with the environment, while the perturbative coupling is reflected in the fact that each environmental mode couples linearly and for an infinitesimal time interval.

6.2.2 Quantum Langevin equations

The full quantum dynamics of the system may also be captured as a set of equations on its quadrature operators in the Heisenberg picture. The potential subtleties arising from the stochastic increments may be handled by remarking that, under the Hamiltonian $\hat{H}_C$, the Heisenberg evolution for a time interval dt of the vector of operators $\hat{\mathbf{r}}'_{SB} = (\hat{\mathbf{r}} \oplus \hat{\mathbf{r}}'_{in})$ (whose entries obey the CCR in their standard, discrete form) is just described by the linear symplectic transformation $e^{\Omega H_C \sqrt{dt}}$ of Eq. (6.14). Hence one has, at the lowest order:

$$d\hat{\mathbf{r}}'_{SB} = e^{\Omega H_C \sqrt{dt}} \, \hat{\mathbf{r}}'_{SB} - \hat{\mathbf{r}}'_{SB} = \Omega H_C \, \hat{\mathbf{r}}'_{SB} \, \sqrt{dt} + \frac{(\Omega H_C)^2}{2} \, \hat{\mathbf{r}}'_{SB} \, dt \ . \quad (6.49)$$

The evolution of the system operators $\hat{\mathbf{r}}$ alone is then given by $d\hat{\mathbf{r}} = A\hat{\mathbf{r}} \, dt + \Omega C \, \hat{\mathbf{r}}'_{in} \, \sqrt{dt} = A\hat{\mathbf{r}} \, dt + \Omega C \, \hat{\mathbf{r}}_{in} \, dt$. Lastly, an arbitrary system Hamiltonian $\hat{H}_S$ may be incorporated to get the "quantum Langevin equations" in their most general form (assuming white noise):

$$d\hat{\mathbf{r}} = i[\hat{H}_S, \hat{\mathbf{r}}] \, dt + A\hat{\mathbf{r}} \, dt + \Omega C \, \hat{\mathbf{r}}_{in} \, dt \ . \quad (6.50)$$

Notice that quantum or classical stochastic differential equations should be expressed in terms of infinitesimal increments, rather than time derivatives, in order to handle the stochastic increments which imply a continuous but not differentiable dynamics. The quantum Wiener process $\Omega C \hat{\mathbf{r}}_{in} dt$ adds quantum fluctuations that preserve the canonical commutation relations of the system canonical operators, and thus also ensure that the Heisenberg uncertainty principle is always satisfied by the evolving operators. We do not need to check this, since we constructed the equations from the application of a symplectic transformation on a set of canonical operators.

Problem 6.2. (*Quantum Langevin equations*). Derive the quantum Langevin equations for the ladder operator $\hat{a}$ of a single mode under pure loss. What are the quantum Langevin equations for a thermal environment?

Solution. For pure loss, we saw that $A = -\frac{\gamma}{2}\mathbb{1}$ and $C = \sqrt{\gamma}\Omega^{\mathsf{T}}$, which can be put into Eq. (6.50) along with $\hat{H}_S = 0$ to obtain, for one mode, $d\hat{x} = -\gamma\hat{x}/2 + \sqrt{\gamma}\,\hat{x}_{in} \, dt$ and $d\hat{p} = -\gamma\hat{p}/2 + \sqrt{\gamma}\,\hat{p}_{in} \, dt$, which may be combined in

$$d\hat{a} = -\frac{\gamma}{2}\hat{a} + \sqrt{\gamma}\,\hat{a}_{in} \, dt \ . \quad (6.51)$$

The Langevin equation, living in the Heisenberg picture, bears no information about the state of the bath and thus is the very same even for a thermal environment. Clearly, general observable predictions will still need the state of the bath to be specified, as is always the case in the Heisenberg picture.

The quantum Langevin equations will be re-derived, in a different form, for the specific case of pure loss in Section 9.1.4, with an emphasis on the detailed evolution of the input-output field operators and their fundamental description as travelling modes of the radiation field.

6.3 GENERAL-DYNE FILTERING OF DIFFUSIVE DYNAMICS

The finite off-diagonal block $\sigma_{SB}\sqrt{dt}$ in Eq. (6.15) is a reflection of the correlations that build up, at every instant in time, between the system and the input mode it interacted with. Hence, a measurement of the output mode corresponding to the interacting input one – that is, of the mode that scatters off the system after having interacted, see Section 9.1.4 for a more accurate definition of input and output modes – will affect the system state. Let us determine the conditional dynamics of the system when the measurement of the environmental modes is a general-dyne detection, introduced in the previous chapter. In optical set-ups, this would model the general-dyne detection of light leaking out of a cavity. These continuous, 'weak' measurements, whereby the system is not observed directly, but only through the environmental modes with which it interacted for an infinitesimal interval dt, are referred to as general-dyne "monitoring". In what follows, we shall make the crucial assumption that the reservoir is also in a Gaussian state, which in practice is the case for a bath of harmonic oscillators in a thermal Gibbs state. The latter provides one with an essentially exact description of thermal, black body electromagnetic radiation (in that the free Hamiltonian of light is quadratic) and an approximate description of oscillations in a solid (a phonon bath).

The effect of a general-dyne measurement on a subsystem of a Gaussian state was already derived in Chapter 5. By applying Eqs. (5.142) and (5.143) to the covariance matrix of Eq. (6.15), with σ_{AB} replaced by the off-diagonal block of σ_{SB}, σ taking the role of σ_A and σ_{in} in lieu of σ_B, one obtains the evolution equation of the monitored covariance matrix and of the first moments. Here, σ_m (whose only constraint is to be a physical covariance matrix: $\sigma_m + i\Omega \geq 0$) parametrises the – possibly noisy – general-dyne measurement of the bath degrees of freedom. It is thus found that the covariance matrix obeys the following deterministic Riccati (quadratic) differential equation:

$$\dot{\sigma} = A\sigma + \sigma A^{\mathsf{T}} + D - (\Omega C \sigma_{in} - \sigma C\Omega)\frac{1}{\sigma_{in} + \sigma_m}(\Omega C^{\mathsf{T}}\sigma - \sigma_{in}C^{\mathsf{T}}\Omega), \quad (6.52)$$

that can be re-written as:

General-dyne monitoring of second moments. The CM σ of a Gaussian state subject to a diffusive evolution and general-dyne monitoring of the environment satisfies

$$\dot{\sigma} = \tilde{A}\sigma + \sigma\tilde{A}^{\mathsf{T}} + \tilde{D} - \sigma BB^{\mathsf{T}}\sigma, \quad (6.53)$$

where
$$\tilde{A} = A + EB^{\mathsf{T}}, \quad \tilde{D} = D - EE^{\mathsf{T}}, \quad (6.54)$$
$$B = C\Omega\left(\sigma_{in} + \sigma_m\right)^{-1/2}, \quad E = \Omega C\sigma_{in}\left(\sigma_{in} + \sigma_m\right)^{-1/2}. \quad (6.55)$$

Here, A and D are the drift and diffusion matrices in the absence of monitoring and C is the coupling matrix to an environment with CM σ_{in}.

As already noticed when introducing general-dyne detections, the deterministic nature of the second moments' conditional evolution is a peculiar property of Gaussian POVMs, and does not depend on any specific feature of the time-continuous, noisy stochastic process the system is undergoing. The general equation (5.142), for the update of the covariance matrix of a Gaussian state under partial monitoring, was already deterministic. Note also that, as apparent from Eq. (6.52) and as one should expect, general-dyne filtering always implies a reduction of noise, in that a positive matrix is subtracted from the time derivative of σ with respect to the unconditional, unfiltered case.

The conditional evolution of the first moments is slightly more delicate to handle, as they do depend on the readout of the measurement. Eq. (5.143) yields, on our covariance matrix and first moment vector:

$$d\bar{\mathbf{r}} = (\Omega C \sigma_{in} - \sigma C \Omega) \left(\frac{1}{\sigma_{in} + \sigma_m} \right) (\mathbf{r}_m - \bar{\mathbf{r}}'_{in}) \sqrt{dt} \ , \qquad (6.56)$$

where $d\bar{\mathbf{r}}$ is the difference between the system's first moments after and before the measurement, $\bar{\mathbf{r}}'_{in}$ is the vector of first moments of the input operators $\hat{\mathbf{r}}'_{in}$ before the measurement [whose explicit expression, if need be, can be directly retrieved from the direct sum in Eq. (6.18)], and $\mathbf{r}_m$ is the outcome of the general-dyne detection of the latter, with probability density given by Eq. (5.139):

$$p(\mathbf{r}_m) = \frac{e^{-(\mathbf{r}_m - \bar{\mathbf{r}}'_{in})^{\mathsf{T}} \left(\frac{1}{\sigma_{in} + \sigma_m} \right) (\mathbf{r}_m - \bar{\mathbf{r}}'_{in})}}{\pi^n \sqrt{\mathrm{Det}(\sigma_{in} + \sigma_m)}} \ . \qquad (6.57)$$

One can now define the new vector of random variables

$$d\mathbf{w} = \left(\frac{1}{\sigma_{in} + \sigma_m} \right)^{\frac{1}{2}} (\mathbf{r}_m - \bar{\mathbf{r}}'_{in}) \sqrt{dt} \ . \qquad (6.58)$$

Inspection of Eq. (6.57) and the rule $(\sqrt{dt})^2 = dt$ reveal that the variables $d\mathbf{w}$ satisfy the following statistics:

$$\langle d\mathbf{w} \rangle = 0 \ , \quad \langle \{d\mathbf{w}, d\mathbf{w}^{\mathsf{T}}\} \rangle = \mathbb{1} dt \ . \qquad (6.59)$$

On this occasion, $\langle \cdot \rangle$ indicates a classical average with respect to the distribution of measurement outcomes. Notice that the convention we adopt is such that, as may be seen from the diagonal entries of the matrix equation in (6.59), $dw_j^2 = dt/2$. The relations (6.59), along with the Gaussian probability distribution (6.57), entail that the random variables $d\mathbf{w}$ form a vector of uncorrelated, classical multivariate Wiener increments, obeying the celebrated Itô rule, and the complete evolution of the first moments (6.56) is an example of a classical Wiener process (see Appendix C for a general definition of the latter) as per:

General-dyne monitoring of first moments. The first moments $\bar{\mathbf{r}}$ subject to a diffusive evolution with drift matrix A, coupling matrix C, and first moments of the input fields $\bar{\mathbf{r}}'_{in}$, as well as general-dyne monitoring of the environment, obey the following stochastic differential equation:

$$\mathrm{d}\bar{\mathbf{r}} = A\bar{\mathbf{r}}\,\mathrm{d}t + \mathbf{d}\,\mathrm{d}t + (E - \boldsymbol{\sigma}B)\,\mathrm{d}\mathbf{w}, \qquad (6.60)$$

where $\mathrm{d}\mathbf{w}$ is a classical stochastic process with correlations (6.59), whilst E and B are given by Eq. (6.55).

Note also that assuming a Gaussian state for the bath was instrumental in deriving a Gaussian probability distribution for the increments $\mathrm{d}\mathbf{w}$.

In this elegant formalism, the randomness of the measurement outcomes, as well as their effect on the conditional state, is captured entirely by the stochastic increments, which make the first moments diffuse in the phase space according to Eq. (6.60) above. Notice that the unconditional evolution is promptly recovered from Eq. (6.60), by noting that the average over the measurement outcomes, which is equivalent to tracing out the environment, gives $\langle \mathrm{d}\mathbf{w} \rangle = 0$, that amounts to scrapping the stochastic term altogether. The unconditional evolution of the second moments may instead be recovered from Eq. (6.52) by setting $\boldsymbol{\sigma}_m = N\mathbb{1}$ and then letting $N \to \infty$, as in that case the general-dyne detection reduces to the trivial POVM $\mathbb{1}$, where no measurement is carried out (also equivalent to tracing out the environment).

Besides their considerable applicative power, and in spite of their specificity, these solvable Gaussian cases serve as an excellent pedagogical gateway to the study of monitored dynamics. What we witnessed above is the general process whereby the *unconditional*, open dynamics, i.e., the case where the environment is not monitored, may be interpreted as an average – over the measurement outcomes – of the *conditional dynamics*, obtained by monitoring the environment. Each choice of a POVM carried out on the environment results in a specific "unravelling" of the unconditional master equation, in that the dynamics of the latter is, so to speak, unravelled into individual stochastic evolutions, which yield it back when averaged upon. We shall derive the stochastic master equations governing such conditional dynamics in the diffusive case in the next section. Let us also mention that, if the initial state of system and environment is pure, and the monitoring POVM's Kraus operators are rank one projectors, then the system state stays pure over the whole dynamics and, instead of a stochastic master equation, one may employ a stochastic Schrödinger equation for the system state vector.[8] This observation is key to the computational method known as 'quantum trajectories' whereby,

---

[8]This is a surprisingly significant consequence of the trivial fact that, for any $|\psi\rangle_{AB} \in \mathcal{H}_A \otimes \mathcal{H}_B$ and $|\varphi\rangle_B$, one has $_B\langle\varphi|\psi\rangle_{AB} \in \mathcal{H}_A$.

instead of solving the original master equation for the d^2-dimensional density operator of the system (d being the finite dimension of the Hilbert space in hand), one averages over many numerical solutions to one of the corresponding stochastic Schrödinger equations for the d-dimensional wave vector. Thus, computational time may be traded for computational memory.

As already noted, the Gaussian general-dyne unravellings determined above feature the peculiarity of deterministic second moments, whilst first moments do evolve stochastically. This means that, at every time, the unconditional evolution may be interpreted as the average over the diffusion of a hypothetical conditional Gaussian state with fixed second moments and varying first moments. The effect of the average is to enlarge the second moments of the state. We have in fact seen in Section 5.3.2 that the mixing of first moments according to a Gaussian distribution, as is the case here for the general-dyne measurement outcomes, is a Gaussian CP-map that adds a positive noise term to the covariance matrix.

It is also worth mentioning that the equations that characterise the conditional evolution of Gaussian states under general-dyne monitoring are identical to the evolution equations of a classical diffusive system under a Kalman filter, also known, in classical control theory, as linear quadratic estimation. This remarkable, and not at all trivial, analogy with classical Kalman filtering equations, which are based on the minimisation of a squared distance, may serve as a powerful pragmatic tool in achieving the real-time update of quantum systems in experiments. This correspondence with classical filtering is yet another consequence of the *apparent* classical-like nature of quantum Gaussian systems: if one restricts to Gaussian measurements and second-order interactions, our dynamics may be imitated by classical stochastic variables, with the only distinctive feature of having to obey the uncertainty principle.

One may now ask, for a given environment characterised by C and $\boldsymbol{\sigma}_{in}$ (which in turn determine A and D), what is the class of possible steady solutions to the Riccati differential equation (6.52) that may be attained by varying the general-dyne filter $\boldsymbol{\sigma}_m$. It is clear by setting $\dot{\boldsymbol{\sigma}} = 0$ in Eq. (6.52) that such states must all satisfy the necessary condition

$$A\boldsymbol{\sigma} + \boldsymbol{\sigma} A^{\mathsf{T}} + D \geq 0 , \tag{6.61}$$

since the omitted term in the equation is a positive matrix. It may be shown that such an equation, supplemented with the uncertainty principle $\boldsymbol{\sigma} + i\Omega \geq 0$, is also sufficient for a covariance matrix $\boldsymbol{\sigma}$ to be attainable at steady state. In our future studies, we will however just need the necessary condition (6.61).

The choice of the general-dyne filter employed is parametrised in our treatment in terms of the covariance matrix $\boldsymbol{\sigma}_m$ that characterises the general-dyne detection, with the only constraint that $\boldsymbol{\sigma}_m + i\Omega \geq 0$. Implicitly, we have assumed that the environment, whose state is represented by the covariance matrix $\boldsymbol{\sigma}_{in}$, has a certain number of degrees of freedom, say m, and that the continuous general-dyne measurement being performed is also parametrised by a $2m \times 2m$ covariance matrix $\boldsymbol{\sigma}_m$. However, when the state of the environment is a mixed quantum state, one could extend such a description by replacing σ_{in} with the CM corresponding to any Gaussian purification of the bath state (whose submatrix pertaining to the original degrees of freedom of the bath is still σ_{in}), and then consider any physical σ_m in the extended phase space. Not surprisingly, the access to a purified environment – unfortunately rather far-fetched in practice – turns out to be advantageous to certain tasks, such as steady-state cooling which, in ideal conditions, would always be attained perfectly by measuring a purified environment in the sense that the system state could then be purified too (see the footnote on page 153).

Notice also that, although the second moments may be stabilised, the full conditional evolution in the Hilbert space does not admit steady states, as the first moments will keep diffusing stochastically.

Finally, let us note that here we are considering a fixed general-dyne filter, not accounting for more general unravelling associated with adaptive measurements, where the choice of the measure may change in time as a function of previous measurement outcomes. These turn out to be advantageous in certain tasks where transient dynamics are relevant, such as optical phase estimation.

6.3.1 Stochastic master equations

As promised above, we shall now derive the stochastic master equation governing generic states subject to diffusive dynamics and general-dyne monitoring of the environment. Like in the unconditional case, dealt with in Section 6.2.1, we will exploit knowledge of the Gaussian dynamics to derive the full master equation.

To start with, let us notice that we expect a modification of the original unconditional master equation (6.46) by a linear term in the stochastic increments $d\mathbf{w}$. In point of fact, since the latter average to zero and we know that under such an average we should retrieve the unconditional master equation, such a linear term in $d\mathbf{w}$ is the only modification to the master equation that monitoring can give rise to. Let us then consider the effect of the first-order stochastic increment contained in Eq. (6.60) on a Gaussian characteristic function with nonzero first moments $\chi_G = e^{-\frac{1}{4}\tilde{\mathbf{r}}^T\sigma\tilde{\mathbf{r}}+i\tilde{\mathbf{r}}^T\bar{\mathbf{r}}}$:

$$d\chi_G^{(s)} = i\tilde{\mathbf{r}}^T(E - \sigma B)d\mathbf{w}\,\chi_G\,, \tag{6.62}$$

which may be turned into an equation for the stochastic increment $d\chi^{(s)}$ of a generic characteristic function χ following the same procedure as with Eq. (6.39), but noticing that, for nonzero first moments, Eq. (6.37) is modified as

$$\partial_{\tilde{r}_k}\chi_G - i\bar{r}_k\chi_G = -\frac{1}{2}\left(\sum_{l=1}^{2n}\sigma_{kl}\tilde{r}_l\right)\chi_G\,, \tag{6.63}$$

which turns (6.62) into the following general statement:

$$d\chi^{(s)} = i(E_{jk}\tilde{r}_j dw_k + 2B_{jk}\partial_{\tilde{r}_j}dw_k - 2iB_{jk}\langle\hat{r}_j\rangle dw_k)\chi\,, \tag{6.64}$$

where Einstein's convention for repeated indexes has been adopted and the Gaussian parameter $\bar{r}_j$ has been replaced with its general definition $\langle \hat{r}_j \rangle = \text{Tr}\,[\varrho\,\hat{r}_j]$. Notice that the presence of this canonical first moment renders the conditional evolution non-linear. One can now apply the correspondence (6.45) between functionals on the characteristic function and quantum operators to obtain the stochastic term entering the monitored master equation:

$$\mathrm{d}\varrho^{(s)} = i\Omega_{lj}E_{jk}[\hat{r}_l, \varrho]\mathrm{d}w_k - B_{jk}\left(\{\hat{r}_j, \varrho\} - 2\text{Tr}[\hat{r}_j\varrho]\right)\mathrm{d}w_k\,. \tag{6.65}$$

Notice that the term in E is just a combination of displacement Hamiltonians, albeit with a fluctuating coupling strength that follows the stochastic increments, whilst the anti-commutator term in B is also manifestly trace preserving but not Hamiltonian. We have thus determined the stochastic term to be added to the master equation (6.46) to account for the effect of monitoring.

Diffusive stochastic master equation. The quantum state ϱ of a system of n modes, linearly coupled to a Markovian environment of m modes with covariance matrix $\boldsymbol{\sigma}_{in}$ through a coupling matrix C, which is monitored through a general-dyne detection with covariance matrix $\boldsymbol{\sigma}_m$, is governed by the following master equation:

$$\mathrm{d}\varrho = \left[\left(\frac{i}{2}\Omega A - \frac{\Omega D\Omega^{\mathsf{T}}}{4}\right)_{jk}\hat{r}_j\hat{r}_k\varrho + \left(\frac{i}{2}\Omega A + \frac{\Omega D\Omega^{\mathsf{T}}}{4}\right)_{jk}\hat{r}_j\varrho\hat{r}_k + \text{h.c.}\right]\mathrm{d}t$$

$$+ i\Omega_{lj}E_{jk}[\hat{r}_l, \varrho]\,\mathrm{d}w_k - B_{jk}\left(\{\hat{r}_j, \varrho\} - 2\text{Tr}[\hat{r}_j\varrho]\right)\mathrm{d}w_k\,, \tag{6.66}$$

where $\mathrm{d}\mathbf{w}$ is a classical stochastic process with correlations (6.59), whilst E and B are given by Eq. (6.55).

One may wonder as to why the terms that modify the evolution of the conditional second moments (with respect to the unconditional evolution), which appear in Eq. (6.52), are not to be found in Eq. (6.66). Quite simply, in order to obtain the contribution at first order in $\mathrm{d}t$ for such a stochastic differential equation, one has to take the term in $\mathrm{d}\mathbf{w}$ at the second-order which, by the Itô rule, yields an additional term in $\mathrm{d}t$ besides the independent, deterministic one showing up in the master equation. That would give the expected correction to the conditional dynamics of the second moments.

Problem 6.3. (*Homodyne stochastic master equation*). Derive the stochastic master equation representing homodyne monitoring of the $\hat{p}$ quadrature of the environment for a single mode subject to the pure loss quantum optical master equation.

Solution. This situation is described by $C = \sqrt{\gamma}\,\Omega^{\mathsf{T}}$ and $\boldsymbol{\sigma}_{in} = \mathbb{1}$, and the corresponding unconditional master equation was already derived in Problem 6.1. The homodyne monitoring of $\hat{p}'_{in}$ is described by the covariance matrix $\boldsymbol{\sigma}_m = \lim_{z \to \infty} \mathrm{diag}(z, 1/z)$, which greatly simplifies the evaluation of $(\boldsymbol{\sigma}_{in} + \boldsymbol{\sigma}_m)^{-1/2}$ as $(\boldsymbol{\sigma}_{in} + \boldsymbol{\sigma}_m)^{-1/2} = \mathrm{diag}(0, 1)$. Note that this form of $(\boldsymbol{\sigma}_{in} + \boldsymbol{\sigma}_m)^{-1/2}$ implies that, in the homodyne limit, the first entry of the vector of stochastic increments $\mathrm{d}\mathbf{w}$ is always zero (as one should expect, since homodyning corresponds to the projective measurement of a Hermitian operator, with outcomes labelled by a single real variable): $\mathrm{d}\mathbf{w} = (0, \mathrm{d}y/\sqrt{2})^{\mathsf{T}}$, where we defined the real Wiener increment $\mathrm{d}y$, associated to the detection of $\hat{p}'_{in}$, in terms of which the master equation will be written. Notice that we have defined $\mathrm{d}y$ so that $\mathrm{d}y^2 = \mathrm{d}t$. One also has, from Eq. (6.55), $B = E = \mathrm{diag}(0, \sqrt{\gamma})$, whence the additional stochastic term to add to the master equation may be determined by using Eq. (6.66):

$$\mathrm{d}\varrho = \gamma \mathcal{D}[\hat{a}]\varrho \,\mathrm{d}t + \sqrt{\gamma}\,\mathcal{H}\,[i\hat{a}]\,\varrho \,\mathrm{d}y \,, \tag{6.67}$$

where the superoperator $\mathcal{H}[\hat{o}]$ acts on ϱ as $\mathcal{H}[\hat{o}]\varrho = \hat{o}\varrho + \varrho\hat{o}^{\dagger} - \mathrm{Tr}\left[\varrho(\hat{o} + \hat{o}^{\dagger})\right]$ for the operator $\hat{o}$, and we used the identity $\frac{i}{\sqrt{2}}[\hat{x}, \varrho] - \frac{1}{\sqrt{2}}\{\hat{p}, \varrho\} = i\hat{a}\varrho - i\varrho\hat{a}^{\dagger}$.

6.4   LINEAR FEEDBACK CONTROL

While environmental filtering is in action, one may consider the option of adjusting the parameters governing the system dynamics – typically by modifying the system Hamiltonian – on the basis of the measurement outcomes, thus enacting a form of measurement-based, closed-loop feedback control. This may be advantageous towards certain goals, as will be discussed in the following.

The simplest option in this regard is acting with a linear Hamiltonian drive that depends on the detection outcomes. In the Gaussian formalism of section 6.3, this obviously would not affect the evolution of the conditional second moments at all, and may be described by letting the parameter $\mathbf{d}$ in Eq. (6.60) depend on the record of previous measurement outcomes. Recall that the latter are contained the increments $\mathrm{d}\mathbf{w}$, which an experimentalist observing the measurement apparatus would have access to.[9] A choice that allows for an especially easy description and still leads to interesting possibilities is linear Markovian feedback, which may be defined by introducing the general-dyne 'current' $\mathbf{j} = \frac{\mathrm{d}\mathbf{w}}{\mathrm{d}t}$ and letting $\mathbf{d} = \mathbf{g} + F\mathbf{j}$, where F is a real $2n \times 2m$ matrix, while $\mathbf{g}$ is a real $2n$-dimensional vector.[10] Notice that this choice implies the idealised

---

[9] Here, we are tacitly assuming the set-up can be calibrated such that all the dynamical parameters are known.

[10] Clearly, the definition of the current $\mathbf{j}$ is not mathematically legitimate, as the Wiener process is continuous but not differentiable. Markovian feedback as modelled here may still

assumption that no temporal delay intervenes between the measurement and the feedback action.

Assume now that $\boldsymbol{\sigma}_\infty$ is a covariance matrix such that $A\boldsymbol{\sigma}_\infty + \boldsymbol{\sigma}_\infty A + D \geq 0$, which is a steady-state solution for the conditional covariance matrix under monitoring (i.e., a solution of Eq. (6.52) for $\dot{\boldsymbol{\sigma}} = 0$). It is clear from Eq. (6.60) that the choices $F = (\boldsymbol{\sigma}_\infty B - E)$ and $\mathbf{g} = -A\bar{\mathbf{r}}_\infty$, for any $\bar{\mathbf{r}}_\infty \in \mathbb{R}^{2n}$, would result in the following deterministic equation for the first moments:

$$\dot{\bar{\mathbf{r}}} = A(\bar{\mathbf{r}} - \bar{\mathbf{r}}_\infty), \qquad (6.68)$$

with solution $\bar{\mathbf{r}}(t) = e^{At}(\bar{\mathbf{r}}(0) - \bar{\mathbf{r}}_\infty) + \bar{\mathbf{r}}_\infty$. If the system is stable – that is, if A is Hurwitz, diagonalisable with all eigenvalues with negative real parts – the asymptotic solution for the first moments is $\bar{\mathbf{r}}_\infty$: linear Markovian feedback is in principle capable of cancelling the diffusion of first moments, thus stabilising the full quantum state of stable systems (since we showed that the latter, which must be still Gaussian, has steady first and second statistical moments), as well as centring it anywhere in phase space. Note also that, under general-dyne filtering and linear feedback, even the unconditional state will still be Gaussian. In the stabilised case discussed above, the action of linear feedback is such that the noise-reduced, conditional covariance matrix that stabilises Eq. (6.52) becomes the unconditional covariance matrix too, since the first moments do not diffuse and the same conditional state corresponds to any measurement outcome.

Problem 6.4. (*Feedback stabilisation*). Show that non-Markovian linear feedback would be able to stabilise the first moments even for unstable systems (with non-Hurwitz A).

Solution. It is sufficient to observe that any non-Hurwitz A can be made Hurwitz by subtracting from it an appropriate real matrix K. Notice now that an experimentalist that knows the initial state and all dynamical parameters of the system, as is reasonable to assume, will also be able to keep track of the exact value of the first moments $\bar{\mathbf{r}}$ along the evolution (which obviously depends on the whole track record of $\mathbf{dw}$, not only on the latest outcome). Then, the following non-Markovian choice for $\mathbf{d}$ may be set:

$$\mathbf{d} = -K\bar{\mathbf{r}} - (A - K)\bar{\mathbf{r}}_\infty + (\boldsymbol{\sigma}_\infty B - E)\mathbf{dj}, \qquad (6.69)$$

which leads to the same equation as (6.68) with A replaced with $(A - K)$, which is Hurwitz, and thus implies asymptotically stable first moments. Clearly, stabilisation of the whole quantum state would be possible only in the cases where monitoring alone can stabilise the second moments (on which linear feedback has no effect).

---

be understood, however, as the approximate description of a discrete process which may be carried out in practice.

Note that, if the general-dyne outcomes are recorded and the state is filtered, linear feedback cannot affect any quantity dependent on second moments such as squeezing (see the following section), entropies, nor any property associated with quantum entanglement (since linear driving corresponds to the action of local unitaries). Nonetheless, the stabilisation described above is interesting whenever one would rather deal with a stable steady state, after the transient dynamics has taken its course. In practice, this is the case, for instance, in achieving stable trapping, where one wants to localise a particle inside a deep enough potential well. In such systems, linear driving may be achieved through laser light or by shifting the centre of the trap, if possible. Moreover, there may be situations where one is better off working in a certain region of first moments – typically where they are very small such that the linear regime applies – where linear feedback control may prove very useful.

The master equation governing generic quantum states under linear feedback displays additional noise terms, due to the stochastic noise which is fed back into the system through the driving, which are not captured by the Gaussian dynamics, so that our strategy of deriving master equations from the Gaussian dynamics would fall short in such a case. We will hence refrain from reporting such a derivation and just refer the reader to the more specialised literature on the subject (see Further Reading).

Clearly, more general feedback Hamiltonian actions might also be envisaged. Let us briefly dwell on second-order Hamiltonians, which take so much space in this book: since the evolution of the second moments is deterministic and does not depend on the measurement outcome, the closed-loop application of second-order Hamiltonians would not help enhancing any properties of the Gaussian states related to the second moments and is hence not particularly interesting (although, clearly, the pre-programmed, "open-loop" application of such Hamiltonians might still have an impact). The conditional evolution associated to general-dyne filtering and subsequent second-order feedback Hamiltonians would also be Gaussian (with a first moments diffusion modified by the action of a random symplectic).

Let us also mention here that an altogether different type of feedback might also be attempted on continuous variables subject to diffusive dynamics. In the paradigm of 'coherent' feedback control, some of the output fields (essentially, the fields $\hat{r}_{in}$ after they interacted with the system) are fed back into the system as input fields, possibly after some coherent manipulation. This eliminates the measurement step, where quantum information is converted into classical information to be fed back as the (classical) conditioning of a quantum operation. It is, in this sense, an entirely coherent process. In Chapter 9, after a discussion of the input-output formalism, a coherent feedback loop will be illustrated through Problem 9.3.

Before moving on to applications, let us conclude this brief general discussion of filtering and feedback actions by acknowledging that one may well, at a first glance, find the distinction between conditional and unconditional states and dynamics somewhat disturbing. Some may find the fact that knowledge of the measurement outcomes possessed by an external agent – the experimenter – may influence the quantum state of a system rather disquieting. Steering away from even attempting a serious discussion of the subjectivist vs. objectivist interpretation of probabilities, which is in a sense the backdrop of this issue but would lead us astray, the author would like to put forward a pragmatic pedagogical approach to the matter. Much of the confusion surrounding this question may be dispelled if one regards states as preparations. The conditional quantum state is one that was prepared by letting the system undergo a certain open dynamics, measuring the environment according to a well-specified detection scheme, and then recording a certain list of measurement outcomes. This is a well-defined preparation, which 'fails' as soon as the record of outcomes is not as required, implying rejection of the system when an ensemble is considered. The fact that this preparation may only be reproduced probabilistically is altogether immaterial: once the, however likely or unlikely, specific outcomes are set, the state preparation is spelled out. It should now be apparent that this is a very different preparation from the unconditional one, where the system undergoes the same dynamics and the same measurements are carried out, but no specific record is requested (any outcomes would do, whence the average to obtain unconditional states from the conditional ones!). This dependence of the preparation on the record of outcomes is what is meant here, in operational terms, by the word "filtering".

6.5   OPTIMAL FILTERING OF QUANTUM SQUEEZING

We will now provide a demonstration of the applicative reach of our theory of general-dyne conditional dynamics. We intend in fact to optimise the production of squeezing at steady state over all continuous general-dyne detections of the environment.

We shall refer to a state as squeezed if the variance of a canonical quadrature for that state is lower than the vacuum variance (1, in our convention, where the diagonal elements of covariance matrices are defined, in terms of anti-commutators, as $2\langle \hat{r}_j \rangle$, and the vacuum has covariance matrix $\mathbb{1}$). Hence, as an indicator of squeezing, we shall take the lowest eigenvalue of the covariance matrix σ of a state (not to be confused with the symplectic eigenvalue), which we denote with $\sigma_1^\uparrow$, adopting the notation $\{\sigma_j^\uparrow\}$ for increasingly ordered eigenvalues and $\{\sigma_j^\downarrow\}$ for decreasingly ordered ones.[11] Squeezing is a useful resource in quantum metrology and sensing, where the control of continuous quantum variables with very low level of noise is paramount. In fact, historically, the quest for squeezed light, which spurred much of the quantum optics

---

[11] For multi-mode systems, the subtlety arises that an eigenvalue of the covariance matrix is not necessarily associated with the noise of a certain canonical quadrature, since the orthogonal matrix that diagonalises the covariance matrix may not be symplectic. However, this problem does not arise in single-mode systems, because all 2×2 proper orthogonal transformations are also symplectic. Since squeezing, and the optimisation that we are about to consider, is typically relevant in single-mode scenarios, we will disregard this issue.

community from the early eighties, was in part motivated by the realisation that gravitational wave detectors might have required sub-vacuum noise levels.

As a case study, let us assume a single-mode system subject to pure loss, with rate γ, and to the squeezing Hamiltonian $\chi \hat{x} \hat{p}/2$, corresponding to the Hamiltonian matrix $H_S = \chi \sigma_x/2$ in terms of the Pauli x matrix (recall also the factor $1/2$ involved in the definition of the Hamiltonian matrix). As we saw in Chapter 5, this would model a degenerate parametric oscillator, where a nonlinear crystal is pumped, and light squeezing is produced, inside a lossy optical cavity. This open system dynamics has been already discussed in the preceding sections, and is described, in our notation, by a diffusion matrix $D = \gamma \mathbb{1}$ and a drift matrix which has to be modified by the system Hamiltonian as per Eq. (6.23): $A = -\frac{\gamma}{2}\mathbb{1} + \frac{\chi}{2}\Omega \sigma_x = -\frac{\gamma}{2}\mathbb{1} + \frac{\chi}{2}\sigma_z$. Given the diagonal form of A and D it is straightforward to derive the steady-state covariance matrix $\boldsymbol{\sigma}_\infty$ for the unmonitored (unconditional) dynamics: by solving $A\boldsymbol{\sigma}_\infty + \boldsymbol{\sigma}_\infty A^\mathsf{T} + D = 0$ one obtains

$$\boldsymbol{\sigma}_\infty = \begin{pmatrix} \frac{\gamma}{\gamma-\chi} & 0 \\ 0 & \frac{\gamma}{\gamma+\chi} \end{pmatrix}. \tag{6.70}$$

Notice that the system is stable (A is Hurwitz) if and only if $|\chi| < \gamma$, which we shall assume. The asymptotic squeezing of the unconditional dynamics thus depends on the ratio between the strength of the parametric Hamiltonian and the loss rate and, in terms of the smallest asymptotic eigenvalue, is given by

$$\sigma_{1,\infty}^\uparrow = \frac{1}{1 + \frac{|\chi|}{\gamma}} > \frac{1}{2}. \tag{6.71}$$

No stable steady-state squeezing better than $1/2$ is thus possible for such open dynamics: this is often referred to as the '3 dB' squeezing limit in the literature.[12]

We intend to show here that the general-dyne detection of the optical field outside the cavity allows one to increase the asymptotic squeezing, and also determine which measurement is optimal to this aim. We shall first prove a general mathematical result which bounds the maximum attainable squeezing and then go back to our parametric oscillator and identify a detection that saturates such a bound. A nice little remark about physical covariance matrices will be instrumental in establishing our result:

Lemma. The eigenvalues of a physical covariance matrix $\boldsymbol{\sigma}$ satisfy the inequality

$$\sigma_k^\downarrow \sigma_k^\uparrow \geq 1 \tag{6.72}$$

$\sigma_k^\downarrow$ ($\sigma_k^\uparrow$) being the kth decreasingly (increasingly) ordered eigenvalue of $\boldsymbol{\sigma}$.

---

[12]Squeezing is often quantified, in decibels (dB), as $-10\log_{10}\sigma_1^\uparrow$ (note that $\sigma_1^\uparrow$ is the ratio between the variance of the squeezed canonical variable and the vacuum variance). In this instance, one has $10\log_{10}(2) \approx 3.01$.

Proof. Here, and in the remainder of this section, it will be convenient to adopt the bra-ket notation for the finite dimensional, but possibly complex, vectors on which σ acts, although they do not belong to a previously defined Hilbert space. We shall need the variational characterisation of the eigenvalues of a Hermitian matrix, which may be expressed as the so called Poincaré inequality: for all k-dimensional subspace Σ_k of a complex vector space and Hermitian L with decreasingly ordered eigenvalues $\{\lambda_j^\downarrow\}$, there exists $|v\rangle \in \Sigma_k$ such that $\frac{\langle v|L|v\rangle}{\langle v|v\rangle} \leq \lambda_k^\downarrow$. Note that, for $k = 1$, this amounts to stating $\sup_{|v\rangle} \frac{\langle v|L|v\rangle}{\langle v|v\rangle} = \lambda_1^\downarrow$. Conversely, the inequality might be cast in terms of the k-th smallest eigenvalue. For $k = 1$, that gives $\inf_{|v\rangle} \frac{\langle v|L|v\rangle}{\langle v|v\rangle} = \lambda_1^\uparrow$.

Let $|\sigma_k^\uparrow\rangle$ be the normalised eigenvector associated with the kth smallest eigenvalue $\sigma_k^\uparrow$, as well as the orthogonal vector $\Omega|\sigma_k^\uparrow\rangle$ (since σ is real and symmetric, its eigenvectors may be chosen real, which ensures $\langle \sigma_k^\uparrow|\Omega|\sigma_k^\uparrow\rangle = 0$). Multiplying the uncertainty principle $(\sigma + i\Omega) \geq 0$ on the left and right by the generic superposition $\alpha(\sigma_k^\uparrow)^{-1/2}|\sigma_k^\uparrow\rangle + \beta(\sigma_k^\uparrow)^{1/2}\Omega|\sigma_k^\uparrow\rangle$, for $\alpha \in \mathbb{C}$ and $\beta \in \mathbb{C}$ (the rescaling by $(\sigma_k^\uparrow)^{1/2}$ simplifies the expression to follow), furnishes the inequality

$$|\alpha|^2 + \langle \sigma_k^\uparrow|\Omega^\mathsf{T}\sigma\Omega|\sigma_k^\uparrow\rangle\sigma_k^\uparrow|\beta|^2 + i(\alpha\beta^* - \alpha^*\beta) \geq 0 \quad \forall \alpha, \beta \in \mathbb{C}. \quad (6.73)$$

The last term on the left hand side may be minimised by setting the complex phases of α and β to give the new inequality

$$|\alpha|^2 + \langle \sigma_k^\uparrow|\Omega^\mathsf{T}\sigma\Omega|\sigma_k^\uparrow\rangle\sigma_k^\uparrow|\beta|^2 - 2|\alpha||\beta| \geq 0 \quad \forall |\alpha|, |\beta| \in \mathbb{R}, \quad (6.74)$$

which is satisfied if and only if

$$\langle \sigma_k^\uparrow|\Omega^\mathsf{T}\sigma\Omega|\sigma_k^\uparrow\rangle \geq \frac{1}{\sigma_k^\uparrow}. \quad (6.75)$$

Consider now the k-dimensional subspace $\text{span}\{\Omega|\sigma_j^\uparrow\rangle, j \in [1, \ldots, k]\}$, spanned by the orthogonal eigenvectors $|\sigma_j^\uparrow\rangle$ transformed by Ω, which are also an orthogonal set since Ω is orthogonal. Because of the ordering of the eigenvalues (such that $1/\sigma_{k-1}^\uparrow \geq 1/\sigma_k^\uparrow$), *all* of the basis vectors of this subspace satisfy $\langle \sigma_k^\uparrow|\Omega^\mathsf{T}\sigma\Omega|\sigma_k^\uparrow\rangle \geq 1/\sigma_k^\uparrow$, which is hence satisfied by all normalised vectors in the subspace. Besides, Poincaré's inequality stated above ensures that a vector $|w\rangle$ in the k-th dimensional subspace must also satisfy $\langle w|\sigma|w\rangle \leq \sigma_k^\downarrow$. Hence, $\sigma_k^\downarrow \geq 1/\sigma_k^\uparrow$, which completes the proof. $\quad\square$

In this section, we shall only make use of this lemma in the case $k = 1$, which might have been derived through a simpler argument. However, the full statement of the lemma will turn out to be useful in Chapter 7, where we will extend our optimisation to the production of quantum entanglement. We can prove the bound:

Bound on filtered general-dyne squeezing. The smallest eigenvalue $\sigma_{1,c}^\uparrow$ of a conditional, steady covariance matrix of a diffusive system with drift and diffusion matrices A and D such that $(A + A^\mathsf{T}) < 0$, under general-dyne monitoring of the bath, satisfies

$$\sigma_{1,c}^\uparrow \geq \frac{\alpha_1^\uparrow}{\delta_1^\downarrow} , \qquad (6.76)$$

where $\{\alpha_j^\uparrow, j \in [1, \ldots, 2n]\}$ are the increasingly ordered (positive) eigenvalues of $-(A + A^\mathsf{T})$ and $\{\delta_j^\downarrow, j \in [1, \ldots, 2n]\}$ are the decreasingly ordered (positive) eigenvalues of D.

Proof. All steady conditional covariance matrices must satisfy the matrix inequality (6.61) which, if applied to the normalised eigenvector $|\sigma_{1,c}^\downarrow\rangle$ of $\boldsymbol{\sigma}$ associated with its *largest* eigenvalue $\sigma_{1,c}^\downarrow$, yields

$$\sigma_{1,c}^\downarrow \langle \sigma_{1,c}^\downarrow | (A + A^\mathsf{T}) | \sigma_{1,c}^\downarrow \rangle + \langle \sigma_{1,c}^\downarrow | D | \sigma_{1,c}^\downarrow \rangle \geq 0 . \qquad (6.77)$$

We will now just need Poincaré's standard variational characterisation of the eigenvalues, already introduced in proving the lemma above, whereby $\sup_{|v\rangle} \frac{\langle v|L|v\rangle}{\langle v|v\rangle} = \lambda_1^\downarrow$ and $\inf_{|v\rangle} \frac{\langle v|L|v\rangle}{\langle v|v\rangle} = \lambda_1^\uparrow$, where L is a Hermitian matrix with eigenvalues $\{\lambda_j\}$. Applying such relations to the previous inequality yields

$$-\sigma_{1,c}^\downarrow \alpha_1^\uparrow + \delta_1^\downarrow \geq 0 \quad \Rightarrow \quad \sigma_{1,c}^\downarrow \leq \frac{\delta_1^\downarrow}{\alpha_1^\uparrow} \qquad (6.78)$$

(recall that $-(A + A^\mathsf{T})$ is positive). We can now apply inequality (6.72), holding for the generic covariance matrix of a quantum state, to obtain $\sigma_{1,c}^\uparrow \geq 1/\sigma_{1,c}^\downarrow \geq \alpha_1^\uparrow/\delta_1^\downarrow$, thus proving the statement. $\qquad \square$

Notice, for the sake of honesty, that our bound relies on a condition, $(A + A^\mathsf{T}) < 0$, that is stronger than A being Hurwitz, which is customarily associated with stability (that is, there are examples of matrices M that are Hurwitz for which $M + M^\mathsf{T}$ is not negative). As we shall see, this still covers very useful practical cases and we will hence not worry about this technicality.

Let us now go back to our parametric oscillator and apply the bound (6.76), to infer that the optimal conditional squeezing may not beat $\sigma_{1,c}^\uparrow = \frac{\gamma}{\gamma - |\chi|}$. We are going to show that such a limit is actually achievable, assuming that the maximally squeezed steady covariance matrix reachable is one with minimum eigenvalue $\frac{\gamma}{\gamma - |\chi|}$. Since we are not restricting, at this stage, our general-dyne unravellings, we can assume them to be not noisy, which amounts to continuously projecting the environment on a pure Gaussian state. Such a procedure, as already discussed in Section 6.3, is able to keep the state of the system pure. It is thus reasonable to assume a pure steady state with a covariance matrix of the form

$$\sigma_c = \begin{pmatrix} \frac{\gamma}{\gamma-\chi} & 0 \\ 0 & \frac{\gamma-\chi}{\gamma} \end{pmatrix} , \tag{6.79}$$

where we set, to fix ideas, $0 \leq \chi \leq \gamma$ (notice that we are always assuming to work within the stable region). Let us remind the reader that purity is reflected, for a Gaussian state, by the equation $\mathrm{Det}\sigma_c = 1$. Given σ_c, it is now simple to show that inserting σ_c in lieu of σ, as well as the dynamical parameters $C = \sqrt{\gamma}\,\Omega$, $\sigma_{in} = \mathbb{1}$, $D = \gamma\mathbb{1}$ and $A = -\frac{\gamma}{2}\mathbb{1} + \frac{\chi}{2}\sigma_z$, Eq. (6.52) admits a steady solution ($\dot{\sigma} = 0$) for the choice (notice that all the matrices involved are diagonal)

$$\sigma_m = \lim_{z \to 0} \mathrm{diag}(1/z, z) . \tag{6.80}$$

This corresponds to the homodyne detection of the $\hat{p}'$ quadrature of the bath which, by virtue of the rotating wave coupling, squeezes the corresponding quadrature of the system. We have thus conclusively shown that *conditioning through environmental monitoring would be capable, for $|\chi|$ approaching the unstable value γ, to attain an arbitrarily low noise on the squeezed canonical quadrature*: a pretty impressive improvement with respect to the sharp unconditional lower bound $1/2$ of Inequality (6.71). As a side remark, let us notice that the limit of homodyne detection, while it may at first appear discouraging, actually makes the evaluation of the Schur complement involved in the update of the covariance matrix particularly simple. Notice the power of our mathematical, seemingly rather formal, approach: we have conclusively proven that this is the maximal amount of squeezing ever achievable. In the literature, this squeezing is at times referred to as 'intra-cavity', in that we have not addressed here the problem of how much squeezing it would be possible to extract from the localised continuous variable system through the input-output interface.

To complete our case study, let us also include the instance of non-unity detection efficiency η in our treatment. To do that, we have to briefly go back to our analysis of Gaussian CP-maps. A generic efficiency η, such that $0 \leq \eta \leq 1$, may be modelled by assuming that the quantum state undergoes a pure loss attenuator channel before being detected which, as discussed in Section 5.4.4, may be described by acting on the measurement covariance matrix σ_m with the dual of such a CP-map. The pure loss attenuator, in terms of the matrices X and Y that we used to parametrise deterministic Gaussian CP-maps, is given by setting $\eta = \cos^2 \theta$ and $n_{\mathrm{th}} = 1$ in Eq. (5.77)

$$X = \sqrt{\eta}\mathbb{1}_2 , \quad Y = (1 - \eta)\mathbb{1}_2 \tag{6.81}$$

($\eta = 1$ corresponding to no loss and thus to a perfect detector). The matrices corresponding to the dual map are thus, using Eqs. (5.64) and (5.65),

$$X^* = X^{-1} = \mathbb{1}_2/\sqrt{\eta}\,, \tag{6.82}$$

$$Y^* = X^{-1}YX^{-1\mathsf{T}} = \frac{1-\eta}{\eta}\mathbb{1}_2\,. \tag{6.83}$$

For the case considered above of continuous homodyne detection of the quadrature $\hat{p}'$, by working out the matrix

$$\sigma_m^* = X^*\sigma_m X^{*\mathsf{T}} + Y^* = \lim_{z\to 0}\mathrm{diag}\left(\frac{\frac{1}{z}+1-\eta}{\eta}, \frac{z+1-\eta}{\eta}\right)\,, \tag{6.84}$$

one can then obtain the conditional evolution equation (6.53) where σ_m should be substituted with σ_m^*. In particular, Eqs. (6.54) and (6.55) yield the following matrices

$$\tilde{A} = A + \mathrm{diag}(0,\eta\gamma)\,, \quad \tilde{D} = D - \mathrm{diag}(0,\eta\gamma)\,, \quad B = -\mathrm{diag}(0,\sqrt{\eta\gamma})\,. \tag{6.85}$$

The analytical solution to the Riccati equation (6.53) with $\dot{\sigma} = 0$, satisfied by the steady-state covariance matrix, can be easily obtained too (note that only one of the two solutions corresponds to a physical covariance matrix, so that no ambiguity arises here):

$$\sigma_c = \mathrm{diag}\left(\frac{\gamma}{\gamma-\chi}, \frac{\gamma(2\eta-1)-\chi+\sqrt{(\gamma+\chi)^2-4\eta\gamma\chi}}{2\eta\gamma}\right)\,. \tag{6.86}$$

The second entry in the diagonal matrix above gives the correction to the monitored squeezing due to a limited detection efficiency η. Note that the limit $\eta \to 0$ does reproduce the unconditional result (6.70), since it corresponds to performing no monitoring at all, whilst $\eta \to 1$ obviously gives the ideal conditional steady state (6.79). It is remarkable how such a comparatively simple formalism is capable of capturing quite a wide class of dynamics and monitoring processes.

Let us also mention that the question of the optimal generation of squeezing will be taken up once more in Problem 9.3, as an illustration of coherent feedback control. The latter, within the specific restrictions imposed on our set-up, will not however be able to outperform monitoring, as we shall see.

Problem 6.5. (*Scattering-induced momentum diffusion*). The scattering of a trapped particle due to interaction with background photons and other particles may be modelled through the following interaction Hamiltonian:

$$\hat{H}_C\,\mathrm{dt} = \sqrt{2\Gamma}\,\hat{x}\,\hat{x}'_{in}\,\sqrt{\mathrm{dt}} \tag{6.87}$$

(in quantum optics, it is customary to retrieve the interaction Hamiltonian considered throughout the chapter to describe a lossy environment from the Hamiltonian above via the rotating wave approximation and the assumption of weak coupling; see Chapter 9). Obtain the diffusive evolution of a system subject to the free harmonic oscillator Hamiltonian at a frequency ω, as well as to environmental scattering in a bath with average number of excitations $(n_{\text{th}} - 1)/2$. Show that such a system does not admit an unconditional steady state.

Solution. The interaction Hamiltonian above corresponds to $C = \sqrt{2\Gamma}\text{diag}(1,0)$, while the free Hamiltonian has Hamiltonian matrix $\omega\mathbb{1}$, and the thermal input covariance matrix is $\sigma_{in} = n_{\text{th}}\mathbb{1}$. This leads to the following drift and diffusion matrices governing the unconditional evolution:

$$A = \begin{pmatrix} 0 & \omega \\ -\omega & 0 \end{pmatrix}, \, , \quad D = \begin{pmatrix} 0 & 0 \\ 0 & 2\Gamma n_{\text{th}} \end{pmatrix}, \quad (6.88)$$

that is, a *purely Hamiltonian* drift matrix (corresponding to no damping for the oscillator), and a diffusion matrix where a momentum heating contribution is apparent. A steady covariance matrix σ would have to satisfy $A\sigma + \sigma A^{\mathsf{T}} + D = 0$, that reads, in terms of the entries σ_{jk} of σ:

$$\omega \begin{pmatrix} 2\sigma_{12} & \sigma_{22} - \sigma_{11} \\ \sigma_{22} - \sigma_{11} & -2\sigma_{12} + 2\frac{\Gamma n_{\text{th}}}{\omega} \end{pmatrix} = 0 , \quad (6.89)$$

which is clearly impossible to satisfy for $\Gamma \neq 0$.

Problem 6.6. (*Quantum non-demolition monitoring*). Find a general-dyne monitoring that stabilises the covariance matrix of the system of Problem 6.5 for any finite measurement efficiency η.

Solution. We will be inspired by the paradigm of 'quantum non-demolition measurements' where, in order to measure the observable $\hat{A}$ of a system non-destructively (in the sense that the measured systems survives the measurement as per a literal reading of the von Neumann postulate, and at variance with what happens, say, in a photodetector), one couples it to a probe through a Hamiltonian $\hat{A}\hat{x}_p$, and then measures $\hat{p}_p$, whose shift through the Hamiltonian $\hat{A}\hat{x}_p$ must depend on $\hat{A}$ (the probe's quadratures $\hat{x}_p$ and $\hat{p}_p$ are canonically conjugate). Since our coupling is of the form $\hat{x}\hat{x}'_{in}$, let us treat the bath as a probe and consider the measurement of $\hat{p}'_{in}$.

The continuous monitoring of the quadrature operator $\hat{p}'_{in}$ with efficiency η corresponds to the measurement covariance matrix σ_m^* of Eq. (6.84). This can be inserted into Eqs. (6.54) and (6.55), with the parameters C and σ_{in} given in the previous problem, to obtain the matrices $\tilde{A} = A$, $\tilde{D} = D$ (notice in fact that the matrix E is zero in this case) and

$$B = \begin{pmatrix} 0 & \sqrt{\omega b} \\ 0 & 0 \end{pmatrix},$$ (6.90)

where we defined $b = \frac{\Gamma \eta}{\omega[1+(n_{th}-1)\eta]}$. Let us also set $a = \frac{\Gamma}{\omega}n_{th}$ (notice that $0 \leq \eta \leq 1$ and $n_{th} \geq 1$ imply $a \geq b$). Now, Eq. (6.53) may be written, for $\dot{\sigma} = 0$, in terms of the entries of σ:

$$\omega \begin{pmatrix} 2\sigma_{12} - 2b\,\sigma_{11}^2 & \sigma_{22} - \sigma_{11} - 2b\,\sigma_{11}\sigma_{12} \\ \sigma_{22} - \sigma_{11} - 2b\,\sigma_{11}\sigma_{12} & -2\sigma_{12} + 2a - 2b\,\sigma_{12}^2 \end{pmatrix} = 0 .$$ (6.91)

The element in the lower right corner yields a positive solution

$$\sigma_{12} = \frac{\sqrt{1+4ab}-1}{2b},$$ (6.92)

which may then be used to obtain

$$\sigma_{11} = \sqrt{\frac{\sigma_{12}}{b}},$$ (6.93)

$$\sigma_{22} = \sqrt{\sigma_{12}\frac{1+4ab}{b}}.$$ (6.94)

All these entries are positive, and it is easy to verify that $\mathrm{Det}\sigma = \frac{a}{b} \geq 1$, as one should expect since the conditional covariance matrix must comply with the uncertainty principle. At variance with its corresponding unconditional diffusive equation, the conditional dynamics does hence admit a steady covariance matrix: the reduction of noise due to monitoring allows one to stabilise the second moments. Notice that this holds for any nonzero measurement efficiency.

For ideal measurements, with efficiency $\eta = 1$, and pure loss ($n_{th} = 1$), the steady state is pure. This can be inferred directly in our formalism by noting that $\mathrm{Det}\sigma = a/b = 1$ in this case, but is also a general consequence of the fact, already discussed in the context of quantum trajectories, that projective measurements of the environment keep the system state pure if the global system-environment state is pure. Hence, in principle, the monitoring of the environment would allow one to purify – essentially, cool – the state of the system. On the other hand, it may be easily seen that the steady state is not pure, even for unit efficiency, for $n_{th} \geq 1$, because such a projective purification would then require access to a purification of the environmental mode. Regardless of these fascinating musings, it should be stressed that the efficiency in the monitoring of an actual system, like a trapped particle, is severely limited by the technical difficulties involved in the collection and detection of scattered light and, even worse, particles. It is still true, however, that even imperfect monitoring helps substantially in the endeavour of trapping, cooling and squeezing such systems in thermal environments.

6.6 FURTHER READING

Monitoring is treated in great detail and generality in the text on quantum measurement and control by Wiseman and Milburn [90], a real *summa* of the field. There, the reader may find the derivation of the stochastic master equation with linear feedback at which our treatment stopped (and much else well beyond that!). Another standard reference in the area is the book by Gardiner and Zoller [31]. The seminal treatment by Carmichael is also worth noting, both concerning quantum trajectories and the derivation of master equations [16]. An introductory coverage of continuous quantum measurements, including a pedagogical discussion of some of the related stochastic methods may be found in [53], while a wider and deeper treatment of stochastic methods is contained in Gardiner's handbook [32]. The fruitful analogy between linear quantum feedback and classical linear-quadratic-Gaussian control is made in [89], where the sufficiency of Inequality (6.61) is also discussed. An independent derivation of the formalism of homodyne continuous measurements may be found in Madsen and Mølmer's contribution to [17], also available as arXiv:quant-ph/0511154. For the mathematically oriented, the vast, pioneering, often formidable and unfathomably profound body of work by Belavkin definitely deserves a mention: among the many papers therein that relate directly to the filtering of quantum continuous variables are [6] and [7].

An intriguing backstory concerning optimal general-dyne unravellings in thermal environments is told in [34]. The derivation of Poincaré's inequality may be found in Bhatia's book on matrix analysis [9]. For the optimisation of out-of-loop figures of merit under monitoring, see [65] (which focuses on entanglement but whose methods would be applicable to squeezing too).

IV

Correlations

CHAPTER 7

Entanglement of continuous variable systems

CONTENTS

The characterisation of quantum correlations, also known as entanglement, is central to quantum information theory. Operationally, entanglement gives rise to the notion of quantum non-locality, which challenges the 'reality' of quantum mechanics (adopting the terminology of Einstein, Podolski and Rosen) and arguably represents its most distinctive feature, in the sense that it calls for a departure from certain foundational tenets of classical physics. It should thus not come as a surprise that entanglement turn out to be a key resource for the implementation of a number of quantum communication protocols, and also necessary to achieve computational speed-ups with quantum hardware, in the sense that accessing regions of the Hilbert space with entangled states allows one to shorten the computational depth of certain algorithms.

Not surprisingly, generating and maintaining entanglement is generally difficult in practice, as it requires strong interactions between selected components of a system which must be at the same time isolated from the environment. Thanks to the techniques developed in quantum optical set-ups to achieve squeezed light, entanglement is however comparatively accessible with optical quantum continuous variables. As we shall see, in fact, continuous variable entanglement is closely related to the notion of squeezing. It is therefore of the utmost importance to be able to qualify and quantify the entanglement of quantum continuous variables, which is the subject matter of the present chapter.

The first part of the chapter is devoted to techniques and criteria based on second moments to detect the entanglement of Gaussian states, and also to the description of possible ways to quantify such quantum correlations, bearing in mind that the determination of entanglement monotones of clear operational significance is still an open problem, even in the restricted arena of Gaussian states. Although the criteria based on second moments are sufficient, but not necessary, for the detection of non-Gaussian entanglement too, we will push our analysis further by introducing more general criteria based on higher order statistical moments, which will be put to use to detect, on paper, the quantum correlations of an entangled cat-like state. We will also address the issues of entanglement distillation and quantum non-locality tests, in whose contexts we shall confront other non-Gaussian operations and states, such as photon subtracted states.

7.1 SEPARABILITY CRITERIA FOR GAUSSIAN STATES

Let us dive straight into the action with a very general sufficient criterion for the separability of Gaussian states (let us remind the reader that a state is entangled if and only if it is not separable, as was clarified in Section 2.6):

Sufficient condition for Gaussian separability. An n-mode Gaussian state with covariance matrix $\sigma \geq 1$ is separable across any bipartition of its modes.

Proof. As we saw in Chapter 5 (Problem 4.2), a Gaussian state ϱ_G with covariance matrix $\sigma \geq 1$ is classical, in the sense that its P-function $P(\mathbf{r})$ is a positive, proper function. Hence, it admits a P-representation (4.35) as a convex sum of multimode coherent states:

$$\varrho_G = \int_{\mathbb{R}^{2n}} d\mathbf{r} P(\mathbf{r}) \hat{D}_{\mathbf{r}}^{\dagger} |0\rangle \langle 0| \hat{D}_{\mathbf{r}} . \tag{7.1}$$

Such a state is manifestly separable, as each $\hat{D}_{\mathbf{r}}^{\dagger}|0\rangle$ is a product state since $\hat{D}_{\mathbf{r}}^{\dagger}$ is always a tensor product of unitary operators on the two local Hilbert spaces, for whatever mode bipartition one may consider. □

Besides providing an early hint as to the relationship between squeezing and entanglement,[1] such a statement will serve as a powerful lemma in what follows to establish stricter criteria for separability.

7.1.1 Separability of two-mode Gaussian states

The most basic, and arguably most relevant to experiments, separability problem in continuous variables concerns Gaussian states of two degrees of freedom. In this instance, an easily tested necessary and sufficient condition for separability can be derived, as we shall see.

To lay the groundwork of our proof, let us introduce a decomposition of the covariance matrix σ of a 2-mode state into 2×2 sub-blocks, as per

$$\sigma = \begin{pmatrix} \sigma_A & \sigma_{AB} \\ \sigma_{AB}^T & \sigma_B \end{pmatrix} , \tag{7.2}$$

where σ_A and σ_B are the local covariance matrices of the single-mode subsystems A and B, and σ_{AB} represents their correlations.

We can then move on to prove a reduction of a 2-mode covariance matrix σ to what is known as Simon normal form:

Simon normal form. Any two-mode covariance matrix σ may be reduced, by local, single-mode symplectic transformation, into the standard form σ_{sf}

$$\sigma_{sf} = \begin{pmatrix} a & 0 & c_+ & 0 \\ 0 & a & 0 & c_- \\ c_+ & 0 & b & 0 \\ 0 & c_- & 0 & b \end{pmatrix} \quad \text{with} \quad c_+ \geq c_- . \tag{7.3}$$

Proof. By the Williamson theorem, local symplectic operations may be found to bring σ_A and σ_B into their normal mode form, $a\mathbb{1}_2$ and $b\mathbb{1}_2$ respectively. Such local submatrices are obviously invariant under local rotations, which are always symplectic [they are the phase shifter transformations of Eq. (5.12)]. Through such rotations the off-diagonal block σ_{AB} may be diagonalised as per its singular value decomposition, which has also enough freedom to ensure $c_+ \geq c_-$. □

Notice that any quantity which may be affected by local unitary operations cannot play any role in determining the separability of a quantum state. Hence,

---

[1] In that we just showed that entanglement requires an eigenvalue of the covariance matrix smaller than 1 which, as discussed in Section 6.5, we refer to as "squeezing".

as far Gaussian states are concerned, first moments, which can be arbitrarily adjusted by tensor products of local and unitary displacement operators, can be safely disregarded. As for second moments of two-mode Gaussian states, only the locally invariant quantities entering Simon's normal form may have a bearing on the quantum correlations.

The normal form above may be adopted to obtain a more detailed condition for the separability of two-mode Gaussian states:

Gaussian separability lemma. Two-mode Gaussian states with $\mathrm{Det}\boldsymbol{\sigma}_{AB} \geq 0$ are separable.

Proof. Let us first take into account the case $\mathrm{Det}\boldsymbol{\sigma}_{AB} > 0$. As we just saw, the initial covariance matrix $\boldsymbol{\sigma}$ of any two-mode state can be turned by local unitary operations to the standard form $\boldsymbol{\sigma}_{sf}$ of Eq. (7.3), whose covariances may be arranged to have $a \geq b$ and $c_+ \geq c_- > 0$ without loss of generality. Let us then consider the local symplectic transformation $S_{loc} = \mathrm{diag}(\sqrt{xy}, 1/\sqrt{xy}, \sqrt{y/x}, \sqrt{x/y})$, made up by the tensor product of local squeezings. With the choices

$$x = \sqrt{\frac{c_+a + c_-b}{c_-a + c_+b}},$$

$$y = \sqrt{\frac{a/x + bx - [(a/x - bx)^2 + 4c_-^2]^{1/2}}{ax + b/x - [(ax - b/x)^2 + 4c_-^2]^{1/2}}},$$

it can be directly shown that $\boldsymbol{\sigma}' = S_{loc}^{\mathsf{T}}\boldsymbol{\sigma}_{sf}S_{loc}$ may be diagonalised by a symplectic beam splitter rotation R_θ of the form of Eq. (5.13), with a proper choice of the angle θ. *This would not be possible if c_+ and c_- had different signs.* Moreover, the smallest eigenvalue of this diagonal form is degenerate:

$$R_\theta^{\mathsf{T}}S_{loc}^{\mathsf{T}}\boldsymbol{\sigma}_{sf}S_{loc}R_\theta = \boldsymbol{\sigma}_{dg} = \mathrm{diag}\,(\kappa_1, \kappa_2, \kappa_-, \kappa_-),$$

with $\kappa_j \geq \kappa_-$ for $j = 1, 2$. But then, for such a diagonal covariance matrix, the uncertainty principle $\boldsymbol{\sigma}_{dg} + i\Omega \geq 0$ straightforwardly implies $\kappa_- \geq 1$. The Gaussian state with covariance matrix $\boldsymbol{\sigma}_{dg}$ is thus classical and therefore separable. So it is the Gaussian state with covariance matrix $\boldsymbol{\sigma}'$, related to $\boldsymbol{\sigma}_{dg}$ by a rotation, which cannot change the eigenvalues. The initial state, with covariance matrix $\boldsymbol{\sigma}$, is then separable as well, being related to the Gaussian state with covariance matrix $\boldsymbol{\sigma}'$ by local unitary operations. This completes the proof in the instance $\mathrm{Det}\boldsymbol{\sigma}_{AB} > 0$.

For $\mathrm{Det}\boldsymbol{\sigma}_{AB} = 0$, so that in $\boldsymbol{\sigma}_{sf}$ one has $c_+ \geq c_- = 0$, the proof of the lemma is easier. In this instance, a local squeezing $S_l = \mathrm{diag}\,(\sqrt{a}, 1/\sqrt{a}, \sqrt{b}, 1/\sqrt{b})$ is needed to bring $\boldsymbol{\sigma}_{sf}$ into a form $\boldsymbol{\sigma}''$ with diagonal entries $\{a^2, 1, b^2, 1\}$ and only two nonzero off-diagonal entries with value abc_+. The uncertainty principle $\sigma'' + i\Omega \geq 0$ for a matrix σ'' in such a form implies $\sigma'' \geq 1$, thus establishing the separability of the original Gaussian state with covariance matrix σ and concluding our proof. □

It is now convenient to identify the symplectic invariants introduced in Section 3.4 for the two-mode state at hand. Two independent invariants must exist (as many as the number of symplectic eigenvalues, which equals the number of modes). One of them is the determinant $\mathrm{Det}\sigma = \nu_+^2 \nu_-^2$ which, by the Binet theorem, is not affected by symplectic transformations, whose determinant is equal to 1 (ν_+ and ν_- stand for the symplectic eigenvalues of σ, with $\nu_+ \geq \nu_-$). The other independent invariant is the seralian $\Delta = \nu_+^2 + \nu_-^2 = \mathrm{Det}\sigma_A + \mathrm{Det}\sigma_B + 2\mathrm{Det}\sigma_{AB}$.

Problem 7.1. (*Seralian of two-mode states*). Show that the quantity $\Delta = \mathrm{Det}\sigma_A + \mathrm{Det}\sigma_B + 2\mathrm{Det}\sigma_{AB}$ is indeed invariant under symplectic transformations and equal to the seralian invariant $\nu_+^2 + \nu_-^2$.

Solution. Notice first that the quantity above is invariant under local symplectic transformations $S_A \oplus S_B$, because all such local transformations have determinant 1 and, acting by congruence, just multiply the 2×2 blocks without mixing them. Now, local operations may bring any initial σ in the standard form (7.3), with $\Delta = a^2 + b^2 + 2c_+c_-$. By adopting the xp-ordering (see Section 3.1) – namely, for two modes, by swapping the second and third phase space variables – the covariance matrix may be rewritten as

$$\sigma^{(xp)} = \begin{pmatrix} \sigma_x & \sigma_{xp} \\ \sigma_{xp}^T & \sigma_p \end{pmatrix}, \qquad (7.4)$$

where σ_x pertains to the variables $\hat{x}_A$ and $\hat{x}_B$, σ_p pertains to $\hat{p}_A$ and $\hat{p}_B$, while σ_{xp} contains the $\hat{x} - \hat{p}$ correlations, with associated standard form

$$\sigma_{sf}^{(xp)} = \begin{pmatrix} a & c_+ & 0 & 0 \\ c_+ & b & 0 & 0 \\ 0 & 0 & a & c_- \\ 0 & 0 & c_- & b \end{pmatrix} \quad \text{with} \quad c_+ \geq c_- . \qquad (7.5)$$

Under this different convention, the quantity Δ can be rewritten as

$$\Delta = a^2 + b^2 + 2c_+c_- = \mathrm{Tr}(\sigma_x \sigma_p) . \qquad (7.6)$$

Observe now that the beam splitter transformation of Eq. (5.13) corresponds, in xp-order, to two identical rotations acting on σ_x and σ_p: such rotations do not affect Δ, as apparent from the expression (7.6).

At the same time, such rotations, supplemented by local symplectic operations, are enough to bring $\sigma_{sf}^{(xp)}$ into normal mode form: to this aim one may in fact rotate through a beam splitter to diagonalise, say, σ_p, apply a local squeezing transformation $S_{z_1,z_2} = \mathrm{diag}(z_1, z_2, 1/z_1, 1/z_2)$ to make σ_p invariant under rotations, and then apply a second beam splitter in order to diagonalise σ_x too. Once the CM is diagonal, local squeezing transformations are sufficient to reach the normal mode form.

We have hence explicitly shown that one can bring any given two-mode covariance matrix σ to its normal form by symplectic transformations that do not affect Δ. The quantity Δ must then be completely determined by the symplectic eigenvalues ν_+ and ν_- of σ as $\Delta = \nu_+^2 + \nu_-^2$ (this relation just follows from the definition of Δ applied to the normal form $\nu_+ \mathbb{1}_2 \oplus \nu_- \mathbb{1}_2$), and is hence a symplectic invariant itself.

The invariant quantities $\mathrm{Det}\sigma$ and Δ may hence be used to determine the symplectic eigenvalues, as per Eq. (3.88), which is worth recalling here:

$$\nu_{\mp}^2 = \frac{\Delta \mp \sqrt{\Delta^2 - 4\mathrm{Det}\sigma}}{2} \, . \tag{7.7}$$

Let us also remind the reader that the positivity of a Gaussian state ϱ_G is equivalent to the condition $\sigma + i\Omega \geq 0$ on its covariance matrix σ, which can in turn be recast as $\nu_{\mp} \geq 1$ for the symplectic eigenvalues of a two-mode σ. As was shown in Section 3.4, such a condition is completely equivalent to the following set of inequalities on the covariance matrix σ of a two-mode state:

$$\mathrm{Det}\sigma - \Delta + 1 \geq 0 \, , \quad \Delta^2 \geq 4\,\mathrm{Det}\sigma \, , \quad \sigma > 0 \, . \tag{7.8}$$

The condition $\sigma > 0$ must be specified because quadratic functions of the symplectic eigenvalues such as $\mathrm{Det}\sigma$ and Δ will never be able to distinguish their sign. Also notice that $\sigma > 0$, being sufficient for the existence of a normal form, implies that the symplectic eigenvalues as determined by (7.7) must be real, and thus subsumes $\Delta^2 \geq 4\mathrm{Det}\sigma$: no positive matrix violating this inequality exists. The set of equations (7.8) forms an explicit necessary and sufficient condition for a Gaussian operator, such as ϱ_G, to be positive. This characterisation of positivity will prove useful in the following.

We are now in a position to derive a necessary and sufficient criterion for the separability of two-mode Gaussian states. In the introductory Section 2.6, we already saw that the positivity of the partial transposition $\tilde{\varrho} \geq 0$ ('PPT') is a necessary condition for the separability of any quantum state ϱ, regardless of the Hilbert space dimension (and hence its violation is a sufficient condition for entanglement). We are about to show that such a criterion is also sufficient for the separability of two-mode Gaussian states. Before proving that, let us understand how partial transposition in the Hilbert space may be represented at the phase space level. We are, in other words, interested in the effect of transposition on the canonical operators $\hat{x}$ and $\hat{p}$ of a single canonical degree of freedom. This can be determined by noticing that, in the Fock basis, the action of $\hat{a}$ and $\hat{a}^\dagger$ can be written entirely with real coefficients and, therefore, one has $\hat{a}^\mathsf{T} = \hat{a}^\dagger$ and $\hat{a}^{\dagger\mathsf{T}} = \hat{a}$ (bear in mind that the positivity of the partial transposition is invariant under a change in the transposition basis, which amounts to applying a certain local unitary to the quantum state). One is led to conclude:

$$\hat{x}^\mathsf{T} = \frac{(\hat{a} + \hat{a}^\dagger)^\mathsf{T}}{\sqrt{2}} = \hat{x} \,, \quad \hat{p}^\mathsf{T} = \frac{(\hat{a} - \hat{a}^\dagger)^\mathsf{T}}{i\sqrt{2}} = -\hat{p} \,. \tag{7.9}$$

Partial transposition can thus be represented, on the canonical operators of two degrees of freedom, by the action of the linear operator $T = \mathrm{diag}(1, 1, 1, -1)$, which mirror reflects one of the four variables. At the level of covariance matrices, this corresponds to the mapping $\boldsymbol{\sigma} \mapsto \tilde{\boldsymbol{\sigma}} = T\boldsymbol{\sigma}T$, where $\tilde{\boldsymbol{\sigma}}$ stands for the partially transposed covariance matrix. This statement may be immediately generalised to any number of degrees of freedom, to obtain:

Partial transposition in phase space. Let $\boldsymbol{\sigma}$ be the covariance matrix of a quantum state of $(m + n)$ modes. Then, the partial transposition of the state with respect to the last n modes has a covariance matrix $\tilde{\boldsymbol{\sigma}}$ given by

$$\tilde{\boldsymbol{\sigma}} = T\boldsymbol{\sigma}T \,, \tag{7.10}$$

with

$$T = \mathbb{1}_{2m} \oplus \Sigma_n \,, \quad \text{with} \quad \Sigma_n = \bigoplus_{j=1}^{n} \sigma_z \tag{7.11}$$

(T flips one of the canonical variables of each of the last n modes).

We can now prove the main result of this investigation:

PPT criterion for two-mode Gaussian states. The positivity of the partial transpose ('PPT') is necessary and sufficient for the separability of a two-mode Gaussian state.

Proof. Consider a two-mode Gaussian state ϱ_G with covariance matrix $\boldsymbol{\sigma}$ decomposed according to Eq. (7.2) and satisfying the physicality conditions (7.8). The conditions (7.8) for the positivity of a Gaussian operator only depend on $\mathrm{Det}\boldsymbol{\sigma}$, Δ and the fact that $\boldsymbol{\sigma} > 0$. Partial transposition, acting on $\boldsymbol{\sigma}$ by congruence as per $\tilde{\boldsymbol{\sigma}} = T\boldsymbol{\sigma}T$ with $\mathrm{Det}T = 1$, cannot affect the determinant: $\mathrm{Det}\boldsymbol{\sigma} = \mathrm{Det}\tilde{\boldsymbol{\sigma}}$. Nor can it affect strict positivity: $\boldsymbol{\sigma} > 0$ implies $\tilde{\boldsymbol{\sigma}} > 0$. The only relevant quantity affected by partial transposition is thus $\Delta = \mathrm{Det}\boldsymbol{\sigma}_A + \mathrm{Det}\boldsymbol{\sigma}_B + 2\mathrm{Det}\boldsymbol{\sigma}_{AB}$, which turns into $\tilde{\Delta} = \mathrm{Det}\boldsymbol{\sigma}_A + \mathrm{Det}\boldsymbol{\sigma}_B - 2\mathrm{Det}\boldsymbol{\sigma}_{AB}$, because the effect of T is flipping the signs of both a column and a row, which leaves $\mathrm{Det}\boldsymbol{\sigma}_A$ and $\mathrm{Det}\boldsymbol{\sigma}_B$ alone but flips the sign of $\text{Det}\boldsymbol{\sigma}_{AB}$. The expression of a necessary condition for separability in terms of second moments is hence

$$\text{Det}\boldsymbol{\sigma} - \tilde{\Delta} + 1 \geq 0 \ . \tag{7.12}$$

One has then only to run through three possible instances, depending on the sign of $\text{Det}\boldsymbol{\sigma}_{AB}$ and on whether (7.12) is satisfied or not.

If $\text{Det}\boldsymbol{\sigma}_{AB} \geq 0$, then the state is separable according to the separability lemma on page 174, and (7.12) is necessarily satisfied since for $\text{Det}\boldsymbol{\sigma}_{AB} \geq 0$, $\tilde{\Delta} \leq \Delta$ such that $\text{Det}\boldsymbol{\sigma} - \tilde{\Delta} + 1 \geq \text{Det}\boldsymbol{\sigma} - \Delta + 1 \geq 0$.

If $\text{Det}\boldsymbol{\sigma}_{AB} < 0$ and (7.12) is not satisfied, then the state is entangled, because a violation of (7.12) is sufficient for entanglement.

Finally, if $\text{Det}\boldsymbol{\sigma}_{AB} < 0$ and (7.12) is satisfied, then the partially transposed state $\tilde{\varrho}_G$ is a Gaussian operator that satisfies all three conditions (7.8) and for which $\text{Det}\tilde{\boldsymbol{\sigma}}_{AB} > 0$. That is, $\tilde{\varrho}_G$ is a separable physical quantum state. But then, the original state must be separable too, because partial transposition preserves separability.

The criterion (7.12) is hence necessary and sufficient for the separability of two-mode Gaussian states. □

We have thus obtained a systematic, computable procedure, based on the implications of the PPT criterion on the (experimentally accessible) second moments, to assess whether a two-mode Gaussian state is entangled. This will also lead, in Section 7.4, to a quantification of such an entanglement.

Although the criterion above is always necessary for separability, and hence its violation provides a general sufficient criterion for the entanglement of any Gaussian state, it is only sufficient in situations which end up being equivalent to the two-mode scenario, where no 'bound' entanglement is allowed, i.e., where no (undistillable) entangled states with positive partial transposition exist. Otherwise, Gaussian states may exist where the PPT criterion is not violated and yet 'bound' entanglement is present. We would hence like to address this situation and provide a more general, necessary and sufficient, mathematical criterion for the separability of a Gaussian state, that is capable of capturing bound entanglement too.

7.2   A GENERAL CRITERION FOR GAUSSIAN SEPARABILITY

Necessary and sufficient criterion for Gaussian separability. A Gaussian state of any number of modes is separable across a bipartition of the modes $A|B$ if and only if its covariance matrix $\boldsymbol{\sigma}$ satisfies

$$\boldsymbol{\sigma} \geq \boldsymbol{\sigma}_A \oplus \boldsymbol{\sigma}_B \ , \tag{7.13}$$

where $\boldsymbol{\sigma}_A$ and $\boldsymbol{\sigma}_B$ are *bona fide* local covariance matrices pertaining to the subsystems A and B ($\boldsymbol{\sigma}_j + i\Omega \geq 0$ for $j = A, B$).

Proof. If the inequality (7.13) is satisfied, then the covariance matrix of the state may be written as $\sigma = (\sigma_A \oplus \sigma_B) + Y$, where Y is a positive matrix. Hence, the Gaussian state under examination may be obtained from the separable, uncorrelated Gaussian state with covariance matrix $(\sigma_A \oplus \sigma_B)$ by the action of an additive noise CP-map, which, as we saw in Section 5.3.2, corresponds to the action of local unitaries (displacements) weighted by a Gaussian probability distribution. We have thus shown the existence of a convex decomposition into product states for the state with covariance matrix σ, and hence its separability.

Conversely, if the state ϱ_G with covariance matrix σ is separable, then one may assume that such a state is decomposable into a convex combination of (not necessarily Gaussian) product states ϱ_j with uncorrelated covariance matrices $\sigma^{(j)}$, first moments $\bar{\mathbf{r}}^{(j)}$ and positive weights λ_j. Then, the very definition of the covariance matrix yields

$$\sigma = \sum_j \lambda_j \sigma^{(j)} + \sum_j \lambda_j \{\bar{\mathbf{r}}^{(j)}, \bar{\mathbf{r}}^{(j)\mathsf{T}}\} - \{\sum_j \lambda_j \bar{\mathbf{r}}^{(j)}, \sum_k \lambda_k \bar{\mathbf{r}}^{(k)\mathsf{T}}\}. \quad (7.14)$$

The uncorrelated covariance matrix $\sigma = \sum_j \lambda_j \sigma^{(j)}$ may be identified with $(\sigma_A \oplus \sigma_B)$, and the matrix $Y = \sigma - (\sigma_A \oplus \sigma_B)$ is given, in components, by $Y_{mn} = 2\left(\sum_j \lambda_j \bar{r}_m^{(j)} \bar{r}_n^{(j)} - \sum_{j,k} \lambda_j \lambda_k \bar{r}_m^{(j)} \bar{r}_n^{(k)}\right)$, which is positive, since

$$\sum_{m,n} Y_{mn} s_m s_n = \sum_{j,k} \lambda_j \lambda_k \left(\left(\sum_m s_m \bar{r}_m^{(j)}\right) - \left(\sum_n s_n \bar{r}_n^{(k)}\right)\right)^2 \geq 0 \quad (7.15)$$

$\forall \mathbf{s} \in \mathbb{R}^{2n}$. Our proof is therefore complete. Notice that the second part of the proof did not make use of the Gaussianity of the state at any stage: the criterion (7.13) is necessary for the separability of generic states. □

The condition (7.13) is not, in practice, as accessible as the PPT criterion. However, it will allow us here to make a first attempt at the quantification of entanglement for generic Gaussian states, will be useful in studying the entanglement of multimode states, and will prove key to understanding Gaussian entanglement distillation.

Inspired by the inequality (7.13), let us put forward the following quantifier of Gaussian entanglement:

Giedke–Cirac Gaussian entanglement quantifier. Given the covariance matrix σ of a Gaussian state, let us define the "Giedke–Cirac" entanglement quantifier $\lambda(\sigma)$ as

$$\lambda(\sigma) = \sup\{p : p \leq 1, \sigma \geq p(\sigma_A \oplus \sigma_B)\}, \quad (7.16)$$

where σ_A and σ_B are physical covariance matrices of subsystems A and B, such that $\sigma_A + i\Omega \geq 0$ and $\sigma_B + i\Omega \geq 0$.

In other words, λ is the largest real number (smaller than 1) for which σ/λ is larger than the direct sum of two physical covariance matrices, and hence separable according to the condition (7.13). Clearly, $\lambda = 1$ for all separable states. Also, λ approaches zero in the limit of maximally entangled states, such as two-mode squeezed states at infinite squeezing. In Section 7.5, we will show that λ is monotone under Gaussian LOCC (although we will not study its behaviour under general LOCC): this will be instrumental to the analysis of Gaussian entanglement distillation.

In Section 7.4, we will see that the Giedke–Cirac quantifier is closely related, though not identical, to the logarithmic negativity.

7.3 MULTI-MODE GAUSSIAN STATES

We are now going to extend the sufficiency of the PPT separability condition beyond the two-mode case that was analysed in the previous section. We will in fact show that such a condition is sufficient for all 1 vs. n mode bipartitions of a Gaussian state. Further, we shall see that the PPT criterion also holds for 'isotropic' states (i.e., states with a fully degenerate symplectic spectrum, including all pure states), as well as for all locally symmetric states (bipartite states where one subsystem is invariant under the exchange of any two modes).

7.3.1 1 vs. n mode Gaussian states

The approach we will take relies heavily on the properties of Schur complements, and allows one to reproduce the proof for the two-mode result as well, in a few swift strokes.[2]

We intend to prove the following:

PPT is necessary and sufficient for Gaussian states of 1 vs. n modes. Let σ be the global covariance matrix of a multi-mode Gaussian state ϱ_G with system A comprised of a single mode and system B comprising n modes. The state ϱ_G is separable across the bipartition $A|B$ if and only if

$$\sigma \geq \begin{pmatrix} i\Omega_1 & 0 \\ 0 & -i\Omega_n \end{pmatrix}, \tag{7.17}$$

which corresponds to the positivity of the partial transposition of ϱ_G.

Proof. Let us first clarify the equivalence between the inequality (7.17) and the PPT criterion. The PPT inequality $T\sigma T + i\Omega \geq 0$, where T is the mirror reflection of all p variables of one subsystem (say B here) that enacts partial transposition in phase space, is equivalent to $T(T\sigma T +$

---

[2]The inclusion of this proof is courtesy of Ludovico Lami.

$i\Omega)T = \sigma + iT\Omega T = \sigma + i(\Omega_1 \oplus (-\Omega_n)) \geq 0$ (since $\sigma_z \Omega_1 \sigma_z = -\Omega_1$), which is the inequality above.

Also, let us define the local covariance blocks as per

$$\sigma = \begin{pmatrix} \sigma_A & \sigma_{AB} \\ \sigma_{AB}^{\mathsf{T}} & \sigma_B \end{pmatrix} . \tag{7.18}$$

Notice also that, if $(\sigma_B + i\Omega_n)$ has a zero eigenvalue, then it is easy to show that one of the symplectic eigenvalues of σ_B must equal 1, which implies that a local symplectic transformation on B exists that puts one of its modes (the normal mode associated with the unit symplectic eigenvalue) into a pure state: such a mode cannot be correlated with any other system and trivially satisfies (7.17), and can hence be disregarded. We can therefore assume an invertible $(\sigma_B + i\Omega_n) > 0$, without loss of generality. Besides, since the sets defined by the inequalities (7.13) and (7.17) are topologically closed, it will be sufficient to show that their interiors coincide,[3] so we can assume strict inequalities for both of them in what follows.

The general separability criterion (7.13) reveals that σ belongs to the interior of the set of separable states if and only if $\sigma - (\gamma_A \oplus 0_B) > (0_A \oplus \gamma_B)$, for some γ_A and γ_B that satisfy the uncertainty principle. We will now invoke the variational characterisation of the Schur complement (1.7) which says that, for fixed $\gamma_A < \sigma_A$, the supremum of the valid γ_B is given by the Schur complement $(\sigma - (\gamma_A \oplus 0_B))/(\sigma_A - \gamma_A)$. Separability is therefore equivalent to the existence of γ_A such that $\gamma_A \geq i\Omega_1$ and such that $(\sigma - (\gamma_A \oplus 0_B))/(\sigma_A - \gamma_A) > (-i\Omega_n)$. Because of the positivity characterisation through the Schur complement (1.6), the latter condition is equivalent to $(\sigma - (\gamma_A \oplus 0_B)) > 0_A \oplus (-i\Omega_n)$, that is

$$\sigma \geq \gamma_A \oplus (-i\Omega_n) . \tag{7.19}$$

Inequality (7.19), where the semi-definite sign has been restored, yields another general separability criterion, equivalent to (7.13). By taking the other Schur complement of (7.19), one obtains yet another necessary and sufficient separability criterion, expressed as the existence of a 2×2 real matrix γ_A such that

$$\sigma_A - \sigma_{AB}(\sigma_B + i\Omega_n)^{-1}\sigma_{AB}^{\mathsf{T}} \geq \gamma_A \geq i\Omega_1 . \tag{7.20}$$

But, if both (7.17) and the uncertainty principle in the form $\sigma - i\Omega > 0$ are strictly satisfied, as we can assume, one can apply the Schur positivity condition (1.6) to them and obtain

$$\sigma_A - \sigma_{AB}(\sigma_B \mp i\Omega_n)^{-1}\sigma_{AB}^{\mathsf{T}} \geq i\Omega_1 . \tag{7.21}$$

Then, the 2×2 real matrix γ_A of (7.20) always exists since, given two 2×2 Hermitian matrices H and J such that $H \leq J$ and $H \leq J^*$, a real matrix exists R such that $H \geq R \geq J$.[4] Therefore, the condition (7.17), always necessary, is also sufficient for separability. □

The PPT criterion may not be further extended to multi-mode Gaussian states unless, as we shall see, additional specific conditions are imposed on them. In fact, it may be shown that bound entangled Gaussian states, i.e., entangled states with positive partial transposition (which cannot be distilled since LOCC preserve PPT), exist for partitions as small as 2 vs. 2 modes.

7.3.2 Locally symmetric states

PPT is necessary and sufficient for locally symmetric Gaussian states. The positivity of the partial transposition is necessary and sufficient for the separability of bipartite Gaussian states which are invariant under mode permutations within one of the two local subsystems.

Proof. We will prove this statement by showing that any locally symmetric covariance matrix may be reduced, through a local symplectic transformation, to the direct sum of a 1 vs. n mode state and uncorrelated single-mode states. The results from the previous section will hence imply sufficiency of PPT for such states.

With reference to the local blocks of Eq. (7.18), assuming party A comprises now m modes, local exchange symmetry imposes the following forms for σ_A and σ_{AB}:

$$\sigma_A = \begin{pmatrix} \alpha & \delta & \cdots & \delta \\ \delta & \alpha & & \vdots \\ \vdots & & \ddots & \delta \\ \delta & \cdots & \delta & \alpha \end{pmatrix}, \quad \sigma_{AB} = \begin{pmatrix} \varepsilon_1 & \varepsilon_2 & \cdots & \varepsilon_n \\ \varepsilon_1 & \varepsilon_2 & \cdots & \varepsilon_n \\ \vdots & \vdots & & \vdots \\ \varepsilon_1 & \varepsilon_2 & \cdots & \varepsilon_n \end{pmatrix},$$
$$(7.22)$$

where α, δ and ε_j, for $j = 1, \ldots, n$ are 2×2 matrices.

It is now convenient to decompose the real phase space of system A as $\mathbb{R}^{2m} = \mathbb{R}^m \otimes \mathbb{R}^2$ and adopt the bra-ket notation on such a space to

---

[3] The two sets will then be proven to be the same, since their closures will also coincide.

[4] Assume in fact $a = \mathrm{Im}(H_{12}) > 0$ and $b = -\mathrm{Im}(J_{12}) > 0$ (without loss of generality, since the hypothesis is invariant under complex conjugation of H and J). Let $p = b/(a+b) < 1$, then the matrix $R = pH + (1-p)J$ is real and such that $H \leq R \leq J$. Also note that the cases $a = 0$ and $b = 0$ are trivial, since then $R = H$ or $R = J$ would be valid choices.

re-write the equations above as

$$\boldsymbol{\sigma}_A = \mathbb{1} \otimes (\boldsymbol{\alpha} - \boldsymbol{\delta}) + m|+\rangle\langle+| \otimes \boldsymbol{\delta} , \quad \boldsymbol{\sigma}_{AB} = \sqrt{m} \sum_{j=1}^{n} |+\rangle\langle j| \otimes \boldsymbol{\varepsilon}_j , \quad (7.23)$$

where $|+\rangle = \frac{1}{\sqrt{m}} \sum_{j=1}^{m} |j\rangle$, where $\{|j\rangle, j \in [1,\dots,m]\}$ is the canonical basis of the subspace $\mathbb{R}^m$. Observe now that, under the tensor product decomposition we adopted, one has $\Omega_m = \mathbb{1}_m \otimes \Omega_1$, so that the orthogonal transformation $O \times \mathbb{1}_2$, with $O \in O(m)$ is symplectic. Let now O be the orthogonal such that $O|+\rangle = |1\rangle$, then

$$(O \otimes \mathbb{1}_2)\boldsymbol{\sigma}_A(O \otimes \mathbb{1}_2)^T = |1\rangle\langle 1| \otimes (\boldsymbol{\alpha} + (m-1)\boldsymbol{\delta}) + \sum_{j=2}^{m} |j\rangle\langle j| \otimes (\boldsymbol{\alpha} - \boldsymbol{\delta})$$

$$= \begin{pmatrix} \boldsymbol{\alpha} + (m-1)\boldsymbol{\delta} & 0 & \cdots & 0 \\ 0 & \boldsymbol{\alpha} - \boldsymbol{\delta} & & \vdots \\ \vdots & & \ddots & 0 \\ 0 & \cdots & 0 & \boldsymbol{\alpha} - \boldsymbol{\delta} \end{pmatrix} , \quad (7.24)$$

$$(O \otimes \mathbb{1}_2)\boldsymbol{\sigma}_{AB} = \sqrt{m} \sum_{j=1}^{n} |1\rangle\langle j| \otimes \boldsymbol{\varepsilon}_j$$

$$= \begin{pmatrix} \sqrt{m}\boldsymbol{\varepsilon}_1 & \sqrt{m}\boldsymbol{\varepsilon}_2 & \cdots & \sqrt{m}\boldsymbol{\varepsilon}_n \\ 0 & 0 & \cdots & 0 \\ \vdots & \vdots & & \vdots \\ 0 & 0 & \cdots & 0 \end{pmatrix} . \quad (7.25)$$

$$(7.26)$$

All the correlations, quantum and classical, between A and B are therefore accounted for entirely by the first mode of A and the PPT criterion (7.17) for 1 vs. n mode applies. □

Notice that our proof also offers a systematic recipe to identify the correlated mode, given the covariance matrix of any locally symmetric state.

7.3.3 Pure and isotropic Gaussian states

It is well known that the PPT criterion is in general sufficient for pure states to be separable.[5] Here, we will re-derive this result for pure Gaussian states by a direct symplectic argument, in keeping with the spirit of our monograph.

---

[5]This may be seen by a direct inspection of the Schmidt decomposition of a state. In actual fact, one can also show the stronger statement that a bound entangled state must have at least rank 4.

As a byproduct we will obtain an interesting decomposition of multi-mode pure Gaussian states under local symplectic transformations.

Mode-wise entanglement of pure Gaussian states. Each pure Gaussian state of $m + n$ modes is equivalent, up to local unitary operations, to a tensor product of two-mode squeezed states and single-mode vacua, with global covariance matrix:

$$\boldsymbol{\sigma} = \begin{pmatrix} c_1 \mathbb{1}_2 & s_1 \sigma_z & 0 & \cdots & \cdots & \cdots & \cdots & 0 \\ s_1 \sigma_z & c_1 \mathbb{1}_2 & & & & & & \vdots \\ 0 & & \ddots & & & & & \vdots \\ \vdots & & & c_j \mathbb{1}_2 & s_j \sigma_z & & & \vdots \\ \vdots & & & s_j \sigma_z & c_j \mathbb{1}_2 & & & \vdots \\ \vdots & & & & & \mathbb{1}_2 & & \vdots \\ \vdots & & & & & & \ddots & 0 \\ 0 & \cdots & \cdots & \cdots & \cdots & \cdots & 0 & \mathbb{1}_2 \end{pmatrix}, \qquad (7.27)$$

where $c_j = \cosh(2r_j)$ and $s = \sinh(2r_j)$ for a set of two-mode squeezing parameters $\{r_j\}$.

Proof. Let $\boldsymbol{\sigma}$ be the covariance matrix of a pure Gaussian state, partitioned in blocks according to Eq. (7.18). Through local symplectic operations, one can turn $\boldsymbol{\sigma}_A$ and $\boldsymbol{\sigma}_B$ into their normal forms $\boldsymbol{\nu}_A = \bigoplus_{j=1}^{m} \nu_{A,j} \mathbb{1}_2$ and $\boldsymbol{\nu}_B = \bigoplus_{j=1}^{n} \nu_{B,j} \mathbb{1}_2$, where m and n are the number of modes pertaining to A and B, respectively, whilst a generic, possibly rectangular, $\boldsymbol{\sigma}_{AB}$ may be expressed in terms of 2×2 blocks, each containing the correlations between a pair of normal modes of A and B, as

$$\boldsymbol{\sigma}_{AB} = \begin{pmatrix} \boldsymbol{\varepsilon}_{11} & \cdots & \boldsymbol{\varepsilon}_{1n} \\ \vdots & \boldsymbol{\varepsilon}_{jk} & \vdots \\ \boldsymbol{\varepsilon}_{m1} & \cdots & \boldsymbol{\varepsilon}_{mn} \end{pmatrix}. \qquad (7.28)$$

A pure Gaussian state has all symplectic eigenvalues equal to 1, which is in turn equivalent to the matrix equality $\boldsymbol{\sigma}\Omega\boldsymbol{\sigma}\Omega = -\mathbb{1}$. In terms of the blocks $\boldsymbol{\nu}_A$, $\boldsymbol{\nu}_B$ and $\boldsymbol{\sigma}_{AB}$, the latter equation may be written as (making use of the fact that the normal mode covariance matrices $\boldsymbol{\nu}_A$ and $\boldsymbol{\nu}_B$

commute with their local symplectic forms Ω_m and Ω_n):

$$\Omega_m \sigma_{AB} \Omega_n \sigma_{AB}^\mathsf{T} = \nu_A^2 - \mathbb{1}_{2m} , \tag{7.29}$$

$$\Omega_n \sigma_{AB}^\mathsf{T} \Omega_m \sigma_{AB} = \nu_B^2 - \mathbb{1}_{2n} , \tag{7.30}$$

$$\nu_A^{-1} \Omega_m \sigma_{AB} \Omega_n \nu_B = \sigma_{AB} , \tag{7.31}$$

$$\nu_A \Omega_m \sigma_{AB} \Omega_n \nu_B^{-1} = \sigma_{AB} . \tag{7.32}$$

In turn, Eq. (7.31) leads to the following equation for the 2×2 blocks $\{\varepsilon\}$:

$$\nu_{A,j} \varepsilon_{jk} = \nu_{B,k} \Omega_1 \varepsilon_{jk} \Omega_1 , \tag{7.33}$$

which has two main implications. First, it tells us that $\varepsilon_{jk} = 0$ unless $\nu_{A,j} = \nu_{B,k}$ (this can be seen, for instance, by observing that the matrices on the left- and right-hand sides must have the same singular values): for correlations between A and B to exist, some of the local symplectic eigenvalues of A must be associated with equal symplectic eigenvalues of B, as only off-diagonal blocks of covariances that relate normal modes with the same local symplectic eigenvalue may be nonzero. Furthermore, the latest equation also implies

$$\varepsilon_{jk} = \Omega_1 \varepsilon_{jk} \Omega_1 , \tag{7.34}$$

which bounds ε_{jk} to be of the form $\varepsilon_{jk} = a_{jk} \sigma_z + b_{jk} \sigma_x$. Observe that this implies $\sigma_{AB} \Omega_n = \Omega_m \sigma_{AB}$.

Note also that, because of Eqs. (7.29) and (7.30), all blocks ε_{jk} pertaining to symplectic eigenvalues $\nu_{A,j}$ or $\nu_{B,k}$ equal to 1 must be zero (this should not come as a surprise: it is just saying that a mode in a locally pure state cannot be correlated with any other modes).

All locally non-degenerate symplectic eigenvalues different from 1 must instead come in pairs, for A and B, each pair corresponding to a pure two-mode state. What is left to show is that, on subspaces with degenerate symplectic eigenvalues different from 1, a local symplectic transformation exists that allows one to bring the whole state into a tensor product of pure two- and single-mode states.

To this aim, let us focus on a degenerate subspace comprising s modes of A and t modes of B, all with local symplectic eigenvalue $\nu > 1$, assuming $s \leq t$. If the canonical variables of the degenerate subspace are rearranged in xp-order as in Eq. (3.4), the transformation $(O \oplus O)$ with O orthogonal is symplectic, while the off-diagonal block σ_{AB} reads

$$\sigma_{AB} = \begin{pmatrix} A & B \\ B^\mathsf{T} & -A \end{pmatrix} \tag{7.35}$$

with $A_{jk} = a_{jk}$ and $B_{jk} = b_{jk}$. One can then pick two local orthogonal symplectic transformations $(O_A \oplus O_A)$ and $(O_B \oplus O_B)$ to put A in singular value form: $D = O_A A O_B^\mathsf{T}$, where D is an $s \times t$ matrix with $D_{jk} = a_j \delta_{jk}$ and the $\{a_j, j \in [1 \ldots, s]\}$ are the (positive semidefinite) singular values of A. This transformation is orthogonal and hence does not affect the local, degenerate normal form covariance matrices, which are proportional to the identity. Putting A in singular value form corresponds to setting the diagonal elements of all ε_{jk} with $j \neq k$ to zero. Then Eqs. (7.29) and (7.30), which read, in terms of the submatrices ε_{jk} within the degenerate subspace,

$$\sum_{l=1}^{t} \varepsilon_{jl} \varepsilon_{kl} = \delta_{jk}(\nu^2 - 1)\mathbb{1}_2 \quad \text{for} \quad j, k = 1, \ldots, s, \qquad (7.36)$$

$$\sum_{l=1}^{s} \varepsilon_{lj} \varepsilon_{lk} = \delta_{jk}(\nu^2 - 1)\mathbb{1}_2 \quad \text{for} \quad j, k = 1, \ldots, t, \qquad (7.37)$$

can be written explicitly in components as

$$a_j b_{kj} - a_k b_{jk} = 0, \quad a_k b_{kj} - a_j b_{jk} = 0, \quad \text{for} \quad j, k = 1 \ldots, s, \qquad (7.38)$$

$$a_j^2 + \sum_{l=1}^{s} b_{jl}^2 = a_j^2 + \sum_{l=1}^{s} b_{lj}^2 = (\nu^2 - 1), \quad \text{for} \quad j = 1 \ldots, s. \qquad (7.39)$$

The off-diagonal element of Eq. (7.37) for $j \leq s$ and $k > s$ implies that either $b_{jk} = 0$ for $k > s$, or $a_j = 0$ for all j. In the latter case, $A = 0$ and B can be put in diagonal form by a singular value decomposition. In both cases, this ensures that the $t - s$ modes of B in excess are uncorrelated and hence pertain to a symplectic eigenvalue equal to 1. In other words, B may be taken square without loss of generality (i.e., degenerate symplectic eigenvalues must also come in pairs shared by A and B), and we have in fact assumed this already in limiting the indexes of (7.38) and (7.39).

Let us now split the singular values $\{a_j\}$ into zero and nonzero ones. Eq. (7.38) shows that all the $\{b_{jk}\}$ that are related to a pair of zero and nonzero singular values must vanish. Hence, the B matrix must be block diagonal with respect to the subspaces associated with zero and nonzero singular values of A. The block corresponding to the zero subspace can be put in diagonal form by the local symplectic transformations effecting its singular value decomposition, since the zero diagonal block of A would obviously be unaffected in the process (see Eq. (7.35) to picture how a local symplectic would act on σ_{AB}). As for the block corresponding to nonzero singular values, Eq. (7.38) shows that it must be such that either $b_{jk} = 0$ or $a_j = a_k$ and $b_{jk} = b_{kj}$: all such nonzero symmetric blocks may be diagonalised by acting with the same local orthogonal symplectic on both subsystems. Notice that in this case we would not be allowed to apply different orthogonal transformations, since the entries of A are not zero, but the rotational invariance of the block of A involved (due to $a_j = a_k$), as well as the symmetry $b_{jk} = b_{kj}$, come to the rescue here.

We have thus shown that the A and B associated to degenerate subspaces must be square and may always be diagonalised with local symplectic transformations which, together with what we already established about the non-degenerate symplectic eigenvalues, corresponds precisely to reducing the covariance matrix to that of a direct sum of pure single-mode states (always locally equivalent to the vacuum state) and two-mode states. The covariance matrix of the latter is determined by Eq. (7.39) with the additional prescription that, once all the off-diagonal entries of B have been set to zero, the diagonal b_{jj} can also be by local phase shifters. Then one has, setting $\nu = \nu_{A,j} = \cosh(2r_j)^2$, $a_j^2 = -\sinh(2r_j)^2$, whose sign may also be flipped by a local phase shifter with symplectic matrix Ω_1, finally yielding the covariance matrix (7.27) upon an appropriate rearrangement of the modes. □

Although rather involved, due to the technical necessity of addressing potential degeneracies, this proof is in a sense nothing but the symplectic counterpart of the Schmidt decomposition, which is reflected in the pairs of identical local symplectic eigenvalues proper to two-mode squeezed states.

Clearly, the very same argument would apply to mixed isotropic states, with all symplectic eigenvalues equal to ν, whose covariance matrix is merely the covariance matrix of a pure state times ν. Interestingly, *this implies that PPT is necessary and sufficient for such states too.*

7.4   LOGARITHMIC NEGATIVITY OF GAUSSIAN STATES

The logarithmic negativity, introduced in Section 2.6.2, is an entanglement monotone that, while not amenable to a clear-cut operational interpretation, can often be evaluated exactly. This is the case for Gaussian states too, as we will show in the present section.

The logarithmic negativity $E_\mathcal{N}$ of a bipartite quantum state ϱ is defined, in terms of its partial transpose $\tilde{\varrho}$, as

$$E_\mathcal{N} = \log_2 \|\tilde{\varrho}\|_1 , \tag{7.40}$$

where $\|\hat{o}\|_1$ denotes the trace norm of operator $\hat{o}$, i.e., the sum of the absolute value of its eigenvalues of $\hat{o}$, if diagonalisable. Note that partial transposition cannot change the trace of an operator, so that the sum of the eigenvalues of a partially transposed state $\tilde{\varrho}$ is still 1. The quantity $\|\tilde{\varrho}\|_1$ may hence be different from, and in particular larger than 1, if and only if $\tilde{\varrho}$ has negative eigenvalues. It follows that $E_\mathcal{N}$ is equal to 0 for all states with positive partial transpose, and larger than 0 for all states which violate the PPT criterion. The adoption of the logarithmic negativity is therefore somewhat unsatisfactory, in that as already mentioned there exist entangled states, known as bound entangled states, that have positive partial transpose, for which the logarithmic negativity is zero. Even so, the logarithmic negativity is still a consistent entanglement monotone – a quantity that does not increase under local operations and classical communication – and is related to entanglement distillation in the sense that it provides an upper bound to the asymptotic conversion rate between the state in question and entangled Bell pairs through LOCC. The $\log_2$ occurring in Eq. (7.40) is needed for the logarithmic negativity to provide one with such an upper bound.

The quantity $\|\tilde{\varrho}_G\|_1 = \mathrm{Tr}|\tilde{\varrho}_G|$ is promptly evaluated for Gaussian states of $m + n$ modes with covariance matrix $\boldsymbol{\sigma}$. Let $\tilde{\boldsymbol{\sigma}} = T\boldsymbol{\sigma}T$ be the partially transposed covariance matrix, where partial transposition with respect to the last n modes is described by $T = \bigoplus_{j=1}^{m} \mathbb{1}_2 \oplus \bigoplus_{j=1}^{n} \sigma_z$, as per Eq. (7.11). Now, let $\{\tilde{\nu}_j, j \in [1, \ldots, m+n]\}$ be the symplectic eigenvalues of the partially transposed covariance matrix $\tilde{\boldsymbol{\sigma}}$. The constructive characterisation of Gaussian states we went through in Chapter 3, and in particular Eq. (3.60), allows us to cast the Gaussian operator $\tilde{\varrho}_G$ in the form

$$\tilde{\varrho}_G = \hat{D}_{\mathbf{r}}^\dagger \hat{S}^\dagger \left(\bigotimes_{j=1}^{m+n} \left(\sum_{k=0}^{\infty} 2\frac{(\tilde{\nu}_j - 1)^k}{(\tilde{\nu}_j + 1)^{k+1}} |k\rangle_{jj}\langle k| \right) \right) \hat{S}\hat{D}_{\mathbf{r}} \qquad (7.41)$$

(where we just replaced ν_j in Eq. (3.60) with $\tilde{\nu}_j$) for some unitary operators $\hat{S}$ and $\hat{D}_{\mathbf{r}}$. As one should expect, only partially transposed normal modes with $\tilde{\nu}_j < 1$ contribute negative eigenvalues to the tensor product spectrum. The operator $|\tilde{\varrho}_G|$ is simply

$$|\tilde{\varrho}_G| = \hat{D}_{\mathbf{r}}^\dagger \hat{S}^\dagger \left(\bigotimes_{j=1}^{m+n} \left(\sum_{k=0}^{\infty} 2\frac{|\tilde{\nu}_j - 1|^k}{(\tilde{\nu}_j + 1)^{k+1}} |k\rangle_{jj}\langle k| \right) \right) \hat{S}\hat{D}_{\mathbf{r}} . \qquad (7.42)$$

Clearly, the unitary operations $\hat{D}_{\mathbf{r}}$ and $\hat{S}$ do not affect the trace of $|\tilde{\varrho}_G|$. Also, the trace is multiplicative under tensor products $[\mathrm{Tr}(\varrho_A \otimes \varrho_B) = \mathrm{Tr}(\varrho_A)\mathrm{Tr}(\varrho_B)$ for trace class ϱ_A and $\varrho_B]$, so that each term in the central tensor product contributes multiplicatively to $\mathrm{Tr}(|\tilde{\varrho}_G|)$. The factors with $\tilde{\nu}_j \geq 1$ are the same as those of a physical Gaussian state, and hence normalised. It will therefore suffice to consider the contribution of operators with $\tilde{\nu}_j < 1$, which are straightforwardly evaluated as geometric sums:

$$\frac{2}{\tilde{\nu}_j + 1} \sum_{k=0}^{\infty} \left(\frac{1 - \tilde{\nu}_j}{\tilde{\nu}_j + 1} \right)^k = \frac{2}{\tilde{\nu}_j + 1} \frac{\tilde{\nu}_j + 1}{2\tilde{\nu}_j} = \frac{1}{\tilde{\nu}_j} . \qquad (7.43)$$

Hence, one obtains

$$\|\tilde{\varrho}_G\|_1 = \prod_{j=1}^{m+n} \max\left\{ 1, \frac{1}{\tilde{\nu}_j} \right\} \qquad (7.44)$$

and

$$E_{\mathcal{N}}(\varrho_G) = \sum_{j=1}^{m+n} \max\{0, -\log_2(\tilde{\nu}_j)\} . \tag{7.45}$$

The situation is particularly simple for two-mode systems where, when partially transposing a physical state (i.e., one with $\sigma + i\Omega \geq 0$) with respect to one mode, only one of the two partially transposed symplectic eigenvalues can be smaller than 1. The two partially transposed eigenvalues are obtained by transposing the invariant Δ in Eq. (7.7):

$$\tilde{\nu}_{\mp}^2 = \frac{\tilde{\Delta} \mp \sqrt{\tilde{\Delta}^2 - 4\mathrm{Det}\sigma}}{2} , \tag{7.46}$$

and the logarithmic negativity of a two-mode state $\varrho_{G,1+1}$ reads

$$E_{\mathcal{N}}(\varrho_{G,1+1}) = \max\{0, -\log_2 \tilde{\nu}_-\} . \tag{7.47}$$

Given their relevance to applications, it is worthwhile to recap our findings in the specific case of two-mode states:

Entanglement of a two-mode Gaussian state. Summing up, a systematic recipe to verify and quantify the entanglement of a two-mode Gaussian state ϱ_G with covariance matrix

$$\sigma = \begin{pmatrix} \sigma_A & \sigma_{AB} \\ \sigma_{AB}^{\mathsf{T}} & \sigma_B \end{pmatrix} \tag{7.48}$$

comprises the following steps:

1. Determine the two partially transposed symplectic invariants $\mathrm{Det}\sigma$ and $\tilde{\Delta} = \mathrm{Det}\sigma_A + \mathrm{Det}\sigma_B - 2\mathrm{Det}\sigma_{AB}$ (notice the minus sign that distinguishes $\tilde{\Delta}$ from Δ).

2. Determine $\tilde{\nu}_-$ according to Eq. (7.46). The state is entangled if and only if $\tilde{\nu}_- < 1$.

3. Determine the logarithmic negativity according to Eq. (7.47).

The agreeable property of having at most one partially transposed symplectic eigenvalue smaller than 1, which simplifies greatly the evaluation of the logarithmic negativity as we have seen in the two-mode case above, is common to all 1 vs. n mode bipartitions. In fact, one may show the following:

Little lemma. Let σ be the *bona fide* covariance matrix of a state of an $(m+n)$-mode system, with $m \leq n$. Let $\tilde{\sigma} = T\sigma T$ be the partial transposition of σ with respect to any of the two subsystems. Then, at most m of the symplectic eigenvalues $\{\tilde{\nu}_j\}$ of $\tilde{\sigma}$ can violate the inequality $\tilde{\nu}_j \geq 1$.

Proof. Suppose the transposition is performed in the m-mode subsystem. Let $\mathcal{D}(M)$ be the dimension of the subspace upon which the generic matrix M is negative definite. Since T reduces to the identity on a $(2n + m)$-dimensional subspace, the inequality $\tilde{\sigma} + i\Omega \geq 0$ reduces to the (definitely satisfied) inequality (3.19) on such a subspace, thus implying $\mathcal{D}(\tilde{\sigma} + i\Omega) \leq m$. One has then $\mathcal{D}(\tilde{\nu} + i\Omega) = \mathcal{D}(\tilde{\sigma} + i\Omega) \leq m$, because $\tilde{\nu} + i\Omega = S^T(\tilde{\sigma} + i\Omega)S$ for $S \in Sp_{2(m+n),\mathbb{R}}$ and the signature is preserved under congruence. The eigenvalues of $\tilde{\nu} + i\Omega$ are given by $\{\tilde{\nu}_j \mp 1\}$, thus proving the lemma as the $\{\tilde{\nu}_j\}$ have to be positive ($T\sigma T > 0$ because $\sigma > 0$). Notice that the choice of the transposed subsystem is not relevant. In the symplectic formalism, this can be seen directly since the action by congruence of the matrix $\oplus_1^{m+n}\sigma_z$ turns $\tilde{\sigma}$ into the partial transpose under the n-mode subsystem, $\tilde{\sigma}'$, and Ω into $-\Omega$, and $\tilde{\sigma}' - i\Omega \geq 0 \Leftrightarrow \tilde{\sigma}' + i\Omega \geq 0$. $\qquad\square$

Notice that, as a consequence of their equivalence with 1 vs. n mode systems, shown in Section 7.3.2, locally symmetric states also admit at most one partially transposed symplectic eigenvalue smaller than 1, on which their logarithmic negativity entirely depends.

Before proceeding to apply our findings to specific practical cases, let us dwell on a last piece of theoretical inquiry, by relating the logarithmic negativity, determined above, to the Giedke–Cirac quantifier from Section 7.2:

Giedke–Cirac quantifier and negativity. For states where the PPT criterion is necessary and sufficient for entanglement, one has $\lambda = \tilde{\nu}_-$: the Giedke–Cirac quantifier coincides with the smallest partially transposed symplectic eigenvalue.

Proof. Because of the definition of λ, the following matrix inequality holds for some physical σ_A and σ_B: $\sigma \geq \lambda(\sigma_A \oplus \sigma_B)$, which can be partially transposed acting by congruence with the operator T to obtain $\tilde{\sigma} \geq \lambda(\sigma'_A \oplus \sigma_B)$, where σ'_A is also a physical CM. Now, the squared, smallest, partially transposed symplectic eigenvalue may be determined as the smallest eigenvalue of the matrix $\Omega^T\tilde{\sigma}\Omega\tilde{\sigma} \geq \lambda^2\Omega^T(\sigma'_A \oplus \sigma_B)\Omega(\sigma'_A \oplus \sigma_B)$ (where we just used the fact that, if $A \geq B \geq 0$ and $C \geq 0$, then $AC \geq BC$). But, up to the factor λ^2, the smallest eigenvalue of the latter matrix is just the smallest symplectic eigenvalue of a physical (non partially-transposed) covariance matrix, which is larger than one. Since $A \geq B \geq 0$ implies that the smallest eigenvalue of A is larger than the smallest eigenvalue of B, one obtains

$$\tilde{\nu}_- \geq \lambda\,. \tag{7.49}$$

Notice now that all the symplectic eigenvalues of the matrix $\tilde{\sigma}/\tilde{\nu}_-$ must be larger than one. Hence, for a system where the PPT criterion is sufficient for separability, the matrix $\sigma/\tilde{\nu}_-$ must be separable. Because of the definition of λ as a supremum, this entails

$$\lambda \geq \tilde{\nu}_- \qquad (7.50)$$

which, combined with (7.49), proves the statement. □

The quantity λ is thus monotonic with the logarithmic negativity for systems where PPT is necessary and sufficient for separability. Clearly, the two quantifiers cannot coincide in situations where bound entanglement is allowed, since λ would be able to spot such entanglement ($\lambda < 1$ for bound entangled states) whereas $\tilde{\nu}_-$ would not ($\tilde{\nu}_- \geq 1$ for bound entangled states).

Let us now apply our methods to determine the logarithmic negativity of the pure two mode squeezed state with covariance matrix σ_r of Eq. (5.21), certainly the most iconic of all Gaussian entangled states. For such a state, one has the invariants $\tilde{\Delta} = 2\cosh(2r)^2 + 2\sinh(2r)^2 = 2\cosh(4r)$ and $\text{Det}\sigma_r = 1$ (since the state is pure), whence $\tilde{\nu}_- = \cosh(4r) - |\sinh(4r)| = e^{-4|r|}$. The associated logarithmic negativity is $E_{\mathcal{N}} = 4|r|/\log_2(e)$: interestingly, it grows linearly with the two-mode squeezing parameter $|r|$. Notice that the local von Neumann entropy, also a measure of entanglement for such pure states, may also be easily determined as $s_V(\cosh(2r))$, in terms of the function s_V entering Eq. (3.92).

Needless to say, the quantification of entanglement finds a wide variety of applications. Some of them will be covered in the following thread of related exercises.

Problem 7.2. (*Thermally seeded two-mode squeezed states*). A more realistic sort of entangled Gaussian resource than pure two-mode squeezed states is obtained by letting a thermal state with covariance matrix $n_{\text{th}}\mathbb{1}_4 \geq \mathbb{1}_4$, rather than the vacuum, through a parametric amplifier, that transforms it with the two-mode squeezing operation S_r of Eq. (5.18). Determine a condition on n_{th} and r for the output state to be entangled, as well as the logarithmic negativity of such a state.

Solution. This is just the same as the pure state case just analysed above where the covariance matrix is multiplied by the thermal noise factor $n_{\text{th}} \geq 1$. Hence, the partially transposed symplectic eigenvalue must also change linearly: $\tilde{\nu}_- = n_{\text{th}}e^{-4|r|}$. The state at issue is hence entangled if and only if

$$e^{4|r|} > n_{\text{th}} , \qquad (7.51)$$

that is, if two-mode squeezing "prevails" over the thermal noise. The associated logarithmic negativity is $E_{\mathcal{N}} = -\log_2(n_{\text{th}}) + 4|r|/\log_2(e)$.

Problem 7.3. (*Entanglement from single-mode squeezing*). Another standard way of obtaining two-mode entanglement in practice is mixing two single-mode squeezed states, with covariance matrices $\mathrm{diag}(z_1, 1/z_1)$ and $\mathrm{diag}(z_2, 1/z_2)$, with $z_{1,2} > 0$, at a 50:50 beam splitter, described by the symplectic transformation $R_{\pi/4}$ of Eq. (5.13) for $\cos\theta = \sin\theta = 1/\sqrt{2}$. Show that the two-mode output of the beam splitter is not entangled for $z_1 = z_2$. Next, determine the entanglement of the output state for generic z_1 and z_2, and show that, if an upper bound to the squeezing is assumed, $1 \leq z_{1,2} \leq \bar{z}$, then the maximal entanglement is achieved for $z_1 = 1/z_2 = \bar{z}$.

Solution. The effect of the beam splitter transformation is rotating equally two principal submatrices of the covariance matrix. For the initial state $\mathrm{diag}(z_1, z_1^{-1}, z_1, z_1^{-1})$, such principal submatrices are both proportional to the identity and hence the covariance matrix is left unchanged, and still completely uncorrelated, let alone entangled.

As for the more general instance, since all global states involved are pure we can quantify entanglement through the local symplectic eigenvalue (an increasing function of the local von Neumann entropy) or, equivalently, the determinant of the local covariance matrix, equal to the squared symplectic eigenvalue (an analysis in terms of the logarithmic negativity would yield analogous results). The output of the beam splitter has covariance matrix

$$
S_{\pi/4}\mathrm{diag}(z_1, z_1^{-1}, z_2, z_2^{-1})S_{\pi/4}^{\mathsf{T}} =
$$
$$
\frac{1}{2}\begin{pmatrix} \mathrm{diag}(z_1 + z_2, z_1^{-1} + z_2^{-1}) & \mathrm{diag}(z_1 - z_2, z_1^{-1} - z_2^{-1}) \\ \mathrm{diag}(z_1 - z_2, z_1^{-1} - z_2^{-1}) & \mathrm{diag}(z_1 + z_2, z_1^{-1} + z_2^{-1}) \end{pmatrix},
$$
$$
(7.52)
$$

with local determinant $\frac{1}{4}(z_1 + z_2)(z_1^{-1} + z_2^{-1}) = \frac{1}{2} + \frac{z_1}{4z_2} + \frac{z_2}{4z_1}$: this state is hence entangled for all values of z_1 and z_2 such that $z_1 \neq z_2$ (as the determinant above is strictly greater than 1 for these values). The local von Neumann entropy can be evaluated as per Eq. (3.92), and reads $s_V\left(\sqrt{\frac{1}{2} + \frac{z_1}{4z_2} + \frac{z_2}{4z_1}}\right)$.

If z_1 and z_2 are capped at $\bar{z}$, the entanglement is maximum for $z_1 = z_2^{-1} = \bar{z}$ (notice also that inverting both z_1 and z_2 does not affect the entanglement), and reads $s_V\left(\sqrt{\frac{1}{2} + \frac{\bar{z}^2}{2} + \frac{1}{2\bar{z}^2}}\right)$ in terms of entropy of entanglement. In this case, it may be shown that the smallest partially transposed symplectic eigenvalue takes the value $1/\bar{z}$, which is the minimum possible within the constraints given, as will be proven in the next problem.

Problem 7.4. (*Entangling power of passive transformations*). Show that the smallest partially transposed symplectic eigenvalue $\tilde{\nu}_-$ satisfies

$$\tilde{\nu}_-^2 \geq \sigma_1^{\uparrow} \sigma_2^{\uparrow}, \tag{7.53}$$

where $\{\sigma_j^{\uparrow}\}$ are the eigenvalues of σ in increasing order.

Solution. Let $\tilde{\Omega} = T\Omega T$ be the (anti-symmetric) partially transposed symplectic form. Given a generic real, normalised vector $\mathbf{v}$, one may define the related normalised vector $\mathbf{w} = \tilde{\Omega}\sigma^{1/2}\mathbf{v}/\sqrt{\mathbf{v}^T\sigma\mathbf{v}}$ such that $\mathbf{v}^T\sigma^{1/2}\mathbf{w} = 0$ (due to the skew symmetry of $\tilde{\Omega}$ and symmetry of $\sigma^{1/2}$) and satisfying $(\mathbf{v}^T\sigma\mathbf{v})(\mathbf{w}^T\sigma\mathbf{w}) = \mathbf{v}^T\sigma^{1/2}\tilde{\Omega}\sigma\tilde{\Omega}\sigma^{1/2}\mathbf{v}$. On the other hand, the variational characterisation of the smallest eigenvalue is such that

$$\tilde{\nu}_-^2 = \min_{|\mathbf{v}|=1} \mathbf{v}^T(\sigma^{1/2}\tilde{\Omega}^T\sigma\tilde{\Omega}\sigma^{1/2})\mathbf{v} = \min_{|\mathbf{v}|=1} \mathbf{v}^T\sigma\mathbf{v}\ \mathbf{w}^T\sigma\mathbf{w}, \tag{7.54}$$

where we used $\tilde{\Omega}^T = -\tilde{\Omega}$ (note also that, because of the cyclicity of the characteristic polynomial's coefficients, the eigenvalues of $\sigma^{1/2}\tilde{\Omega}^T\sigma\tilde{\Omega}\sigma^{1/2}$ are the same as the eigenvalues of $\tilde{\Omega}^T\sigma\tilde{\Omega}\sigma$). The right hand side of Eq. (7.54) can be lower bounded by relaxing the minimisation to all vectors $\mathbf{v}$ and $\mathbf{w}$ such that $\mathbf{v}^T\sigma^{1/2}\mathbf{w} = 0$, which are clearly a wider set than the pairs $\mathbf{v}$ and $\mathbf{w}$ related as above. Hence, one gets:

$$\tilde{\nu}_-^2 \geq \min_{\mathbf{v}^T\sigma^{1/2}\mathbf{w}=0} \mathbf{v}^T\sigma\mathbf{v}\ \mathbf{w}^T\sigma\mathbf{w} = \sigma_1^{\uparrow}\sigma_2^{\uparrow}, \tag{7.55}$$

which demonstrates the statement. In order to prove the last identity, notice that the minimum must be achieved in the subspace associated with the two smallest eigenvectors and that, in such a subspace, all pairs of vectors fulfilling $\mathbf{v}^T\sigma^{1/2}\mathbf{w} = 0$ yield the same value for the product to be minimised, that is $\sigma_1^{\uparrow}\sigma_2^{\uparrow}$.

Note that the inequality (7.53) poses a fundamental limit on how much entanglement may be generated through passive (energy preserving) operations, that are represented by orthogonal symplectic transformations and as such cannot alter the eigenvalues. In fact, it may be shown in general that a passive transformation whose output saturates the bound, as shown explicitly in the specific 50:50 beam splitter case of Problem 7.3, always exists. This statement reaffirms the connection between squeezing (in the sense of the covariance matrix's minimum eigenvalue being smaller than 1) and entanglement, that was already pointed out in the sufficient separability criterion on page 172, and makes it quantitative.

Problem 7.5. (*Decay of quantum correlations*). An initial two-mode squeezed state with squeezing parameter r interacts with a thermal environment, described by the same quantum attenuator channel (5.77), with loss factor $(\cos\theta)^2 = \Gamma$ and thermal noise $n_{th} \geq 1$, acting on both modes. Determine for what values of Γ, n_{th} and r the final state is entangled, and evaluate its logarithmic negativity as a function of such parameters.

Solution. The action of the attenuator CP-maps on the input state is

$$\Gamma \begin{pmatrix} c\mathbb{1} & s\sigma_z \\ s\sigma_z & c\mathbb{1} \end{pmatrix} + (1-\Gamma)n_{th}\mathbb{1}_4 = \begin{pmatrix} [\Gamma c + (1-\Gamma)n_{th}]\mathbb{1} & \Gamma s\sigma_z \\ \Gamma s\sigma_z & [\Gamma c + (1-\Gamma)n_{th}]\mathbb{1} \end{pmatrix},$$
(7.56)

with $c = \cosh(2r)$ and $s = \sinh(2r)$. Now, let $a = [\Gamma c + (1-\Gamma)n_{th}]$ and $b = \Gamma s$, then the partially transposed seralian is $\tilde{\Delta} = 2(a^2 + b^2)$ and the determinant is $\text{Det}\boldsymbol{\sigma} = (a^2 - b^2)^2$ which, when inserted in Eq. (7.46), yield $\tilde{\nu}_-^2 = (a^2 + b^2) - 2ab = (a - b)^2$. Hence, one has $\tilde{\nu}_- = |a-b| = \Gamma e^{-2|r|} + (1-\Gamma)n_{th}$ (squeezing parameters of both signs are thus encompassed). The state is entangled if and only if

$$\tilde{\nu}_- < 1 \quad \Leftrightarrow \quad e^{-2|r|} < \frac{1 - (1-\Gamma)n_{th}}{\Gamma}, \tag{7.57}$$

while the logarithmic negativity may be calculated via Eq. (7.47). Eq. (7.57) deserves a few words of commentary: if $n_{th} = 1$, then the entanglement of the initial state does decrease but never vanishes, for any value of Γ. In a dynamical situation (where, under typical Markovian conditions, $\Gamma = e^{-\gamma t}$ for some coupling strength γ, see Chapter 6) the entanglement of an initial pure state experiencing decoherence in a zero temperature (pure loss) environment never disappears completely, as the state is entangled at all finite times. If, instead, thermal noise is present ($n_{th} > 1$), then there exists a value of Γ such that the state becomes separable, a circumstance which has been spectacularised in the literature under the name of "entanglement sudden death".

Problem 7.6. (*Steady-state entanglement in parametric down-conversion*). Parametric down-conversion is described, in the interaction picture, where the effect of local Hamiltonians is absorbed in the rotation of the local bases, by the Hamiltonian $\chi(\hat{x}_1\hat{p}_2 + \hat{p}_1\hat{x}_2)/2$ for a two-mode system. Determine the logarithmic negativity, at steady-state, for a system of two modes that interact via the Hamiltonian above and are each subject to pure loss in an independent zero-temperature environment, with loss factor γ (the same for both modes).

Solution. The parametric Hamiltonian under examination corresponds to the Hamiltonian matrix $H_s = \frac{\chi}{2} \begin{pmatrix} 0 & \sigma_x \\ \sigma_x & 0 \end{pmatrix}$. In the notation of Chapter 6, the system is governed by a diffusive dynamics with coupling matrix $C = \sqrt{\gamma}(\Omega_1^\mathsf{T} \oplus \Omega_1^\mathsf{T})$ and $\sigma_{in} = \mathbb{1}_4$, for which Eqs. (6.26) and (6.23) yield the diffusion matrix $D = \gamma \mathbb{1}_4$ and the drift matrix $A = \frac{1}{2} \begin{pmatrix} -\gamma \mathbb{1}_2 & \chi \sigma_z \\ \chi \sigma_z & -\gamma \mathbb{1}_2 \end{pmatrix}$. The steady-state covariance matrix σ_∞ solves Eq. (6.33), and may be determined with elementary algebra since A and D are the direct sum of 2×2 principal submatrices. One thus gets

$$\sigma_\infty = \frac{1}{1 - \frac{\chi^2}{\gamma^2}} \begin{pmatrix} \mathbb{1}_2 & \frac{\chi}{\gamma}\sigma_z \\ \frac{\chi}{\gamma}\sigma_z & \mathbb{1}_2 \end{pmatrix} . \tag{7.58}$$

Clearly, the system admits a steady-state only if $|\chi| < \gamma$ (for which the matrix A is Hurwitz), as we shall assume in what follows. It is convenient to carry out the evaluation of the transposed symplectic eigenvalue $\tilde{\nu}_-$ disregarding the multiplicative factor in front of σ_∞, which can be simply reinstated at the end. One has then the rescaled transposed seralian $\tilde{\Delta}' = 2(1 + \frac{\chi^2}{\gamma^2})$ and determinant $\mathrm{Det}\sigma'_\infty = (1 - \frac{\chi^2}{\gamma^2})^2$, which give a rescaled symplectic eigenvalue $\tilde{\nu}'_- = (1 - \frac{|\chi|}{\gamma})$. The actual smallest partially transposed symplectic eigenvalue at steady-state reads

$$\tilde{\nu}_{-,\infty} = \frac{1 - \frac{|\chi|}{\gamma}}{1 - \frac{\chi^2}{\gamma^2}} = \frac{1}{1 + \frac{|\chi|}{\gamma}} . \tag{7.59}$$

Note that, in the stable region (i.e., for the values of the dynamical parameters that admit a steady-state), one has $\tilde{\nu}_{-,\infty} > 1/2$, such that the associated asymptotic logarithmic negativity $E_{\mathcal{N},\infty}$, given by Eq. (7.47), is constrained by a most simple upper bound:

$$E_{\mathcal{N},\infty} < 1 . \tag{7.60}$$

The steady-state is nevertheless entangled for all $\chi \neq 0$. Notice that the bound (7.59) is analogous to the 3 dB squeezing limit (6.71) we determined when assessing steady-state squeezing in analogous conditions.

Problem 7.7. (*Optimal steady-state entanglement under monitoring*). An experimentalist operates a non-degenerate parametric oscillator, described exactly as in the previous problem, set up in a cavity, and is determined to improve the modest steady-state entanglement calculated above by observing the light leaking out of the cavity through continuous general-dyne detections. Assume a stable region ($|\chi| < \gamma$) and determine the maximum logarithmic negativity the experimentalist may ever obtain by general-dyne monitoring.

Solution. In Problem 7.4 we proved the inequality (7.53), showing that the smallest partially transposed symplectic eigenvalue $\tilde{\nu}_-$ of a covariance matrix $\boldsymbol{\sigma}$ satisfies $\tilde{\nu}_-^2 \geq \sigma_1^\uparrow \sigma_2^\uparrow$, where $\sigma_{1,2}^\uparrow$ are the two smallest eigenvalues of $\boldsymbol{\sigma}$. On the other hand, in Chapter 6, as a preliminary step in the optimisation of squeezing under monitoring, we adopted Poincaré's variational principle to prove the inequality (6.72) which, if combined with the inequality above, leads to $\tilde{\nu}_-^2 \geq 1/(\sigma_1^\downarrow \sigma_2^\downarrow)$. In order to maximise the entanglement at steady-state (i.e., minimise $\tilde{\nu}_-$), we would hence like to bound the two largest eigenvalues $\sigma_1^\downarrow$ and $\sigma_2^\downarrow$ from above. This can be done by recalling the necessary condition (6.61); optimality will then be shown by identifying an explicit general-dyne monitoring that achieves the bound. Let $\mathbf{v}_j$ be the normalised eigenvector of $\boldsymbol{\sigma}$ associated with the eigenvalue $\sigma_j^\downarrow$, for $j = 1, 2$. The inequality (6.61) implies

$$\mathbf{v}_j^\mathsf{T}\left(\sigma_j^\downarrow(A + A^\mathsf{T}) + D\right)\mathbf{v}_j = \sigma_j^\downarrow \mathbf{v}_j^\mathsf{T}(A + A^\mathsf{T})\mathbf{v}_j + \gamma \geq 0 \quad \text{for} \quad j = 1, 2 \tag{7.61}$$

(where we used $D = \gamma \mathbb{1}_4$, determined in the previous problem). But $\mathbf{v}_j^\mathsf{T}(A + A^\mathsf{T})\mathbf{v}_j \leq -(\gamma - |\chi|)$ (which is the largest eigenvalue of the negative definite matrix $A + A^\mathsf{T}$, also determined in the previous problem, in the stable region). The previous equation then yields a bound on the largest symplectic eigenvalues of $\boldsymbol{\sigma}$:

$$\sigma_j^\downarrow \leq \frac{\gamma}{\gamma - |\chi|} \,. \tag{7.62}$$

Combining the inequalities obtained above yields the absolute bound on the entanglement

$$\tilde{\nu}_- \geq \left(\sigma_1^\uparrow \sigma_2^\uparrow\right)^{1/2} \geq \left(\sigma_1^\downarrow \sigma_2^\downarrow\right)^{-1/2} \geq 1 - \frac{|\chi|}{\gamma} \,. \tag{7.63}$$

This is much better than the unmonitored bound (7.59) in that, for one thing, the entanglement would now be allowed to approach infinity (which is the case if $\tilde{\nu}_-$ approaches 0) if the coupling strength approaches instability ($|\chi| \lesssim \gamma$). More generally, it is straightforward to verify that the bound to $\tilde{\nu}_-$ in Eq. (7.63) is, in the stable region, always lower than $\tilde{\nu}_{-,\infty}$ of Eq. (7.59).

Hopes are hence on the rise now. In order to become really ecstatic, the experimentalist would have to determine which, if any, general-dyne detection of the environment saturates the bound (7.63). The optimal strategy can in this instance be easily determined: notice in fact that both A and D can be diagonalised by a 50:50 beam splitter (the transformation (5.13) with $\theta = \pi/4$). In the transformed, decoupled coordinates, we can use the analysis carried out in Section 6.5 to optimise the monitored squeezing. It can thus be seen that two homodyne detections of the form of Eq. (6.80), lead to two single-mode pure squeezed states with covariance matrices $\boldsymbol{\sigma}_c$ of (6.79), that saturate the inequalities $\sigma_{c,j}^{\uparrow} \geq (\sigma_{c,j}^{\downarrow})^{-1} \geq 1 - |\chi|/\gamma$, for $j = 1, 2$. If the detections are chosen with opposite phases (inverting z for one of the two detections), then the two-mode steady-state has covariance matrix $\mathrm{diag}(1/(1-|\chi|/\gamma), 1-|\chi|/\gamma, 1-|\chi|/\gamma, 1/(1-|\chi|/\gamma))$. Hence, the final inversion of the 50:50 beam splitter leads to a pure two-mode squeezed state with $\tilde{\nu}_- = 1 - \frac{|\chi|}{\gamma}$ (see Problem 7.3), that does saturate all the bounds of (7.63). The bound is thus achievable (and we have provided the reader with yet another explicit illustration of the intimate connection between continuous variable squeezing and entanglement).

It is not all champagne and roses for our experimentalist though, as the optimal strategy we just determined corresponds to the double homodyne detection of the environmental quadratures: it is therefore a global measurement of commuting quadratures such as $\hat{x}'_{in,1} - \hat{x}'_{in,2}$ and $\hat{p}'_{in,1} + \hat{p}'_{in,2}$, that would involve the mixing of the environmental modes at a beam splitter, a very difficult deed to pull off.

Problem 7.8. (*Gaussian quantum discord*). The 'quantum discord' is an alternative way to quantify quantum correlations based on a discrepancy between two quantities that are identical in classical information, but differ for quantum states. In general, given a bipartite state ϱ_{AB}, the quantum discord D_A is defined as

$$D_A(\varrho_{AB}) = S_V(\varrho_A) - S_V(\varrho_{AB}) + \min_{\hat{K}_\mu^{(B)}} \sum_\mu S_V\left(\mathrm{Tr}_B(\hat{K}_{\mu,B}\varrho_{AB}\hat{K}_{\mu,B}^\dagger)\right),$$

(7.64)

where S_V stands for the von Neumann entropy and the minimisation is carried out over all possible POVMs $\{\hat{K}_{\mu,B}\}$ acting on the local system B. The 'Gaussian' quantum discord $G_A(\varrho_{AB})$ is then defined by restricting the POVMs $\{\hat{K}_{\mu,B}\}$ to general-dyne detections, defined by Eq. (5.137). Determine the Gaussian discord of a thermally seeded two-mode squeezed state with covariance matrix $n_{th}\boldsymbol{\sigma}_r$, with $\boldsymbol{\sigma}_r$ given by Eq. (5.21) and $n_{th} \geq 1$.

Solution. The global entropy $S_V(\varrho_{AB})$ is given by $2s_V(n_{th})$, in terms of the entropic function s_V that relates the von Neumann entropy to the symplectic eigenvalues as per Eq. (3.92) (by construction, the global covariance matrix has two degenerate symplectic eigenvalues equal to n_{th}). The local entropy $S_V(\varrho_A)$ is instead $s_V(n_{th}\cosh(2r))$ (the local covariance matrix is already in normal form).

Let us now turn to the last term in Eq. (7.64). It is convenient to set $a = n_{th}\cosh(2r)$ and $b = n_{th}\sinh(2r)$. A first simplification of the minimisation over general-dyne POVMs of the entropy of the conditional state of A comes from the fact, discussed at length in Chapters 5 and 6, that the covariance matrix, and hence the entropy, of a such state does not depend on the measurement outcome, so that one can disregard the average over μ altogether. The local conditional covariance matrix $\sigma_{A,c}$ is found by applying Eq. (5.142) and is given by the Schur complement

$$\sigma_{A,c} = a\mathbb{1} - b^2\sigma_z\frac{1}{a\mathbb{1} + \sigma_m}\sigma_z , \tag{7.65}$$

where $\sigma_m \geq i\Omega$ is a single-mode covariance matrix that parametrises the general-dyne measurement. As we saw in Problem 5.7, $\sigma_m = \nu R_\varphi\mathrm{diag}(z, 1/z)R_\varphi^\mathsf{T}$, for a phase shifter transformation R_φ of Eq. (5.12), $z > 0$ and $\nu \geq 1$. Our aim now is minimising the determinant $\mathrm{Det}\sigma_{A,c}$, which determines the von Neumann entropy of the single-mode state (see Section 3.5), over all possible choices of σ_m, parametrised as above. One has

$$\sigma_{A,c} = R_\varphi^\mathsf{T}(a\mathbb{1} - b^2\sigma_z\frac{1}{a\mathbb{1} + \nu\,\mathrm{diag}(z, 1/z)}\sigma_z)R_\varphi , \tag{7.66}$$

where we used the property $\sigma_z R_\varphi = R_\varphi^\mathsf{T}\sigma_z$. The determinant is thus independent from φ and the rotation may be disregarded. The determinant of the diagonal matrix above is handy to evaluate:

$$\mathrm{Det}\sigma_{A,c} = \left(a - \frac{b^2}{a + \nu z}\right)\left(a - \frac{b^2}{a + \nu/z}\right) , \tag{7.67}$$

which is obviously always an increasing function of ν, so that $\nu = 1$, corresponding to ideal general-dyne measurements, is optimal (the optimality of pure state projections could also have been inferred on more general theoretical grounds). For $\nu = 1$, one gets

$$\mathrm{Det}\sigma_{A,c} = a^2 - b^2\left(\frac{a^2 + n_{th}^2 + af(z)}{a^2 + 1 + af(z)}\right) , \tag{7.68}$$

with $f(z) = z + 1/z$. Differentiation with respect to f gives $\partial_f\mathrm{Det}\sigma_{A,c} = ab^2(n_{th}^2 - 1)/(a^2 + 1 + af(z))^2 \geq 0$, so that the minimal value of $f(z)$ is also optimal, and is given by $f(z) = 2$ for $z = 1$, which corresponds to heterodyne detection. The symplectic eigenvalue

τ that minimises the last term in the Gaussian discord is, therefore, the square root of the determinant above for $f(z) = 1$:

$$\tau = n_{th} \sqrt{\cosh(2r)^2 - \sinh(2r)^2 \left(\frac{n_{th}^2 \cosh(2r)^2 + 2n_{th} \cosh(2r) + n_{th}^2}{n_{th}^2 \cosh(2r)^2 + 2n_{th} \cosh(2r) + 1} \right)},$$

$$(7.69)$$

and the Gaussian discord reads

$$G_A(\varrho_{AB}) = s_V(n_{th} \cosh(2r)) - 2s_V(n_{th}) + s_V(\tau). \qquad (7.70)$$

Through a seminal result on the additivity of the minimal output entropy of certain Gaussian CP-maps, which will be derived in the following chapter, one can show that in this case the Gaussian discord $G_A(\varrho_{AB})$ gives the actual quantum discord $D_A(\varrho_{AB})$. For $n_{th} = 1$, when the global state is pure, $\tau = 1$, and the discord reduces to the entanglement entropy $s_V(\cosh(2r))$ (a general property). For $r = 0$, the state is completely uncorrelated, $\tau = n_{th}$ and the discord is zero, as it should be.

7.5 ENTANGLEMENT DISTILLATION

Entanglement distillation is the process whereby two parties, A and B, sharing a certain number of copies of an entangled state, apply local operations and classical communication (LOCC) in order to obtain fewer copies of a state that is more entangled than the original one. In the customary formulation of this question, one aims at obtaining a certain number of maximally entangled target states. Here, we will show that the entanglement distillation of Gaussian states through Gaussian LOCC is impossible, by establishing the following, stronger, result:

No-go to Gaussian entanglement distillation. The Giedke–Cirac quantifier λ, defined by Eq. (7.16), may only increase for a Gaussian state under Gaussian LOCC operations.

Proof. First, some characterisation of Gaussian LOCC is in order. We will achieve it very generally through the Choi-Jamiolkowski map of Section 5.5.3, by exploiting the simple observation that if a CP-map belongs to the LOCC class, then its Choi state is separable (across the bipartition $A|B$, the ancillae to A and B, that double the Hilbert space in the Choi description of CP-maps, will still be entangled with A and B, respectively). In turn, this just means that, by virtue of Inequality (7.13), the Choi state has covariance matrix $\boldsymbol{\sigma}_C$ such that $\boldsymbol{\sigma}_C \geq \boldsymbol{\sigma}_{C,A} \oplus \boldsymbol{\sigma}_{C,B}$ for some physical, local $\boldsymbol{\sigma}_{C,A}$ and $\boldsymbol{\sigma}_{C,B}$. We will characterise the class of operations with respect to which we intend to prove λ's monotonicity through this condition. In doing so, we are actually considering the larger class of 'entanglement breaking' – or 'separable' – operations, which include the set of LOCC, and hence proving an even stronger statement (see Section D.2 of Appendix D for more details on entanglement breaking channels). Note also that $\boldsymbol{\sigma}_C = (\boldsymbol{\sigma}_{C,A} \oplus \boldsymbol{\sigma}_{C,B}) + Y$ for a positive $Y \geq 0$ which, as discussed in Section 5.3.2, may always be added through classical mixing by Gaussian distributed displacements, and can only increase λ. We can hence set $\boldsymbol{\sigma}_C = (\boldsymbol{\sigma}_{C,A} \oplus \boldsymbol{\sigma}_{C,B})$. Notice also that we have, as a byproduct, established the result that Gaussian separable operations may always be decomposed into completely local ones and classical mixing by random displacements. Let

$$\boldsymbol{\sigma}_{C,A} = \begin{pmatrix} A_1 & A_{12} \\ A_{12}^\mathsf{T} & A_2 \end{pmatrix}, \quad \boldsymbol{\sigma}_{C,B} = \begin{pmatrix} B_1 & B_{12} \\ B_{12}^\mathsf{T} & B_2 \end{pmatrix}. \tag{7.71}$$

By the definition of λ, it is also known that $\boldsymbol{\sigma} \geq \lambda(\boldsymbol{\sigma}_A \oplus \boldsymbol{\sigma}_B)$. Then, applying the mapping (5.152) yields an output covariance matrix $\boldsymbol{\sigma}_{out}$ (the first moments are irrelevant since λ does not depend on them), that fulfils

$$\begin{aligned}
\boldsymbol{\sigma}_{out} &= \lim_{r \to \infty} A_1 \oplus B_1 - G \frac{1}{A_2 \oplus B_2 \oplus \boldsymbol{\sigma} + \boldsymbol{\sigma}_r} G^\mathsf{T} \\
&\geq \lim_{r \to \infty} A_1 \oplus B_1 - G \frac{1}{A_2 \oplus B_2 \oplus \lambda \boldsymbol{\sigma}_A \oplus \lambda \boldsymbol{\sigma}_B + \boldsymbol{\sigma}_r} G^\mathsf{T} \\
&= \lambda \boldsymbol{\gamma}_A \oplus \lambda \boldsymbol{\gamma}_B,
\end{aligned} \tag{7.72}$$

where $\boldsymbol{\sigma}_r$ is the covariance matrix $\boldsymbol{\sigma}_r^{(n)}$ of Eq. (5.149) for the appropriate number of modes, and $G = \begin{pmatrix} A_{12} \oplus B_{12} & 0 \end{pmatrix}$, where 0 is a null matrix filling up the necessary number of entries. The right hand side in the inequality is a direct sum of local matrices, which we have denoted with $\lambda \boldsymbol{\gamma}_A \oplus \lambda \boldsymbol{\gamma}_B$ (since only $\boldsymbol{\sigma}$ could contain correlations between A and B). Hence, we can evaluate the submatrices $\boldsymbol{\gamma}_A$ and $\boldsymbol{\gamma}_B$ through the inversion formula (1.8), and take the limit in r through Eq. (5.157):

$$\begin{aligned}
\boldsymbol{\gamma}_A &= A_1 - A_{12} \frac{1}{A_2 + \lambda \Sigma \boldsymbol{\sigma}_A \Sigma} A_{12}^\mathsf{T} = A_1 - A_{12} \Sigma \frac{1}{\Sigma A_2 \Sigma + \lambda \boldsymbol{\sigma}_A} \Sigma A_{12}^\mathsf{T} \\
&\geq A_1 - A_{12} \Sigma \frac{1}{\Sigma A_2 \Sigma - i \lambda \Omega} \Sigma A_{12}^\mathsf{T},
\end{aligned} \tag{7.73}$$

where we used $\Sigma^2 = \mathbb{1}$ (the matrix Σ is defined as Σ_n in Eq. (5.149), where n is the appropriate number of modes, which is left unspecified here) and $\boldsymbol{\sigma}_A + i\Omega \geq 0$.[6] The same expression holds for $\boldsymbol{\gamma}_B$. Our strategy now is showing that $\boldsymbol{\gamma}_A$ and $\boldsymbol{\gamma}_B$ are physical covariance matrices.

We still haven't used the fact that $\sigma_{C,A}$ is a physical covariance matrix: $\sigma_{C,A} + i\Omega \geq 0$, which implies $\sigma_{C,A} + i\lambda\Omega \geq 0$, as $\lambda \leq 1$. Applying the transformation $(\mathbb{1} \oplus \Sigma)$ by congruence on this inequality yields

$$(\mathbb{1}\oplus\Sigma)(\sigma_{C,A}+i\lambda\Omega)(\mathbb{1}\oplus\Sigma) = \begin{pmatrix} A_1 + i\lambda\Omega & A_{12}\Sigma \\ \Sigma A_{12}^\mathsf{T} & \Sigma A_2 \Sigma - i\lambda\Omega \end{pmatrix} \geq 0 \,, \quad (7.74)$$

where we used $\Sigma\Omega\Sigma = -\Omega$. Let us now recast the inequality (7.74) in terms of the Schur complement, as per Eq. (1.6), to obtain

$$A_1 - A_{12}\Sigma\frac{1}{\Sigma A_2 \Sigma - i\lambda\Omega}\Sigma A_{12}^\mathsf{T} + i\lambda\Omega \geq 0 \,. \quad (7.75)$$

But this inequality is just telling us that the right hand side of inequality (7.73) times $1/\lambda$ is a *bona fide* covariance matrix. Hence, γ_A is *bona fide* too, and the same goes for γ_B. Therefore, inequality (7.72) just states that the output covariance matrix σ_{out} is greater than λ times a direct sum of covariance matrices, where λ is the input entanglement quantifier, whence $\lambda(\sigma_{out}) \geq \lambda(\sigma)$. Hence, such a quantifier may only grow or stay constant through Gaussian separable operations, encompassing all Gaussian LOCC. □

Notice that the statement above applies to any number of degrees of freedom owned by A and B (this is the case since the dimensions of A and B are generic) as well as, crucially, to any number of input and output modes, in the sense that discarding modes cannot do the two parties any good. This is simply because λ cannot decrease under partial traces, which can be shown by noting that the inequality $\sigma \geq \lambda(\gamma_A \oplus \gamma_B)$ must apply to all principal submatrices, which means that the partial trace's covariance matrix σ^* will also have to satisfy $\sigma^* \geq \lambda(\gamma_A^* \oplus \gamma_B^*)$ (where the shorthand notation $*$ indicates the discarding of a certain set of modes), whence $\lambda(\sigma^*) \geq \lambda(\sigma)$. Moreover, probabilistic CP-maps are implicitly included in the Choi–Jamiolkowski description, and hence also subject to the no-go. Hence, any possibility of achieving entanglement distillation by Gaussian means is conclusively ruled out by the statement above.

Bear in mind that, because of the correspondence we established in the previous section with the smallest partially transposed symplectic eigenvalue, this result does apply to the logarithmic negativity itself for all states where the PPT criterion is necessary and sufficient for separability (1 vs. n mode, pure, isotropic and bisymmetric states, as we saw).

---

[6] $A_2 + i\Omega$ (and hence $A_2 + i\lambda\Omega$) may be assumed to be invertible. If not, it would mean a normal mode of A_2 is in a pure state, and thus uncorrelated, so that the corresponding A_{12} would not contribute to the Schur complement. A similar argument is spelled out on page 181, in proving of the multimode separability criterion (7.17).

7.5.1 Distilling Gaussian entanglement from non-Gaussian states

The no-go established in the previous section leaves open the possibility of achieving the distillation of continuous variable entanglement by Gaussian LOCC starting from weakly entangled non-Gaussian states. This would be desirable, given the importance of entanglement distillation in the context of quantum communication – whenever the necessity arises to establish long-distance entanglement, always hampered by noisy transmission lines – and the comparative accessibility of Gaussian operations. Here, we will dip our toes into this issue by showcasing a classic example of a probabilistic distillation protocol, due to Browne, Eisert, Scheel and Plenio, that concentrates the entanglement of non-Gaussian states into fewer copies of approximately Gaussian states with higher entanglement.

Imagine A and B share two copies of the state $\varrho_{AB} \otimes \varrho_{AB}$. For clarity, the two pairs of local systems in which the two copies of the state are embodied will be referred to as A_1, A_2, on A's side, and B_1, B_2, on B's side. The protocol runs as follows:

1. A and B mix subsystems $A_1 - A_2$ and $B_1 - B_2$ at a 50:50 beam splitter.

2. A and B measure subsystems A_1 and B_1 through an on/off detector that just distinguishes between the presence and absence of photons.

3. If the outcome is zero ('no click') on both sides, the systems A_1 and B_1 are projected on the vacuum, and the state of systems A_2 and B_2 is kept, possibly for another iteration of the protocol (with another pair of successfully generated distilled states). Otherwise, the systems A_2 and B_2 are discarded.

Notice that, since the projection on the vacuum state is a Gaussian operation, and this vanilla version of the protocol simply aborts the run if that does not take place, the whole operation is a Gaussian LOCC, and hence subject to the no-go theorem of the previous section. In this regard, it is an interesting exercise to establish the preliminary result below.

Problem 7.9. (*Fixed points of the distillation map*). Show that if ϱ_{AB} is Gaussian, it is left unchanged by a run of the protocol above.

Solution. The action of the two beam splitters, each given by Eq. (5.13) for $\theta = \pi/4$, on the most general Gaussian state of the whole system reads:

$$\frac{1}{2}\begin{pmatrix} R_{\frac{\pi}{4}} & 0 \\ 0 & R_{\frac{\pi}{4}} \end{pmatrix}\begin{pmatrix} \boldsymbol{\sigma}_A & 0 & \boldsymbol{\sigma}_{AB} & 0 \\ 0 & \boldsymbol{\sigma}_A & 0 & \boldsymbol{\sigma}_{AB} \\ \boldsymbol{\sigma}_{AB}^{\mathsf{T}} & 0 & \boldsymbol{\sigma}_B & 0 \\ 0 & \boldsymbol{\sigma}_{AB}^{\mathsf{T}} & 0 & \boldsymbol{\sigma}_B \end{pmatrix}\begin{pmatrix} R_{\frac{\pi}{4}}^{\mathsf{T}} & 0 \\ 0 & R_{\frac{\pi}{4}}^{\mathsf{T}} \end{pmatrix}$$

$$= \begin{pmatrix} \boldsymbol{\sigma}_A & 0 & \boldsymbol{\sigma}_{AB} & 0 \\ 0 & \boldsymbol{\sigma}_A & 0 & \boldsymbol{\sigma}_{AB} \\ \boldsymbol{\sigma}_{AB}^{\mathsf{T}} & 0 & \boldsymbol{\sigma}_B & 0 \\ 0 & \boldsymbol{\sigma}_{AB}^{\mathsf{T}} & 0 & \boldsymbol{\sigma}_B \end{pmatrix}. \tag{7.76}$$

The covariance matrix does not change at all. In particular, no correlations are built between the two copies of the system, across the bipartition $1|2$, such that whatever measurement outcome in step 2 of the protocol will not have any effect on A_2 and B_2, which are still to be found, in this case deterministically, in the state they started from.

Gaussian states are hence fixed points of the protocol. In order to understand how the protocol acts on non-Gaussian states, let us specialise to pure input states of the form $|\psi\rangle_{AB} = \sum_n \alpha_n |n,n\rangle$, where $|n,n\rangle$ is a tensor product of Fock states. This will also be a good opportunity to practice with the manipulation of non-Gaussian states through optical elements. As we saw in Section 5.1.2, the 50:50 beam splitter is generated by the Hamiltonian $(i\hat{a}_1^\dagger \hat{a}_2 - i\hat{a}_1 \hat{a}_2^\dagger)$, determined in Eq. (5.16) in terms of annihilation and creation operators of the two modes involved, acting for a time $t = \pi/4$: $\hat{R}_{\frac{\pi}{4}} = e^{\frac{\pi}{4}(\hat{a}_1 \hat{a}_2^\dagger - \hat{a}_1^\dagger \hat{a}_2)}$. The effect of this operator on Fock states is better appreciated by applying the reordering (5.116) to $\hat{R}_{\frac{\pi}{4}}^\dagger$ (in compliance with our convention thus far, the action of the operation on states will be conjugated), which gives

$$\hat{R}_{\frac{\pi}{4}}^\dagger = e^{\frac{\pi}{4}(\hat{a}_1^\dagger \hat{a}_2 - \hat{a}_1 \hat{a}_2^\dagger)} = e^{-\hat{a}_1 \hat{a}_2^\dagger} \left(\frac{1}{\sqrt{2}}\right)^{\hat{a}_1^\dagger \hat{a}_1 - \hat{a}_2^\dagger \hat{a}_2} e^{\hat{a}_1^\dagger \hat{a}_2}. \tag{7.77}$$

Using this equation, one can determine the action of $\hat{R}_{\frac{\pi}{4}}^\dagger$ on the generic Fock state $|m,n\rangle$:

$$\hat{R}_{\frac{\pi}{4}}^\dagger |m,n\rangle = \sum_{j=0}^{n} \sum_{k=0}^{m+j} (-1)^k \left(\frac{1}{\sqrt{2}}\right)^{m-n+2j} \frac{(m+j)!n!}{j!k!(n-j)!} \times$$

$$\times \sqrt{\frac{(n-j+k)!}{m!n!(m+j-k)!}} |m+j-k, n-j+k\rangle. \tag{7.78}$$

The projection on the vacuum state of the first subsystem greatly simplifies the state of the remaining subsystem (for the moment being, let us disregard normalisation):

$$\langle 0|\hat{R}_{\frac{\pi}{4}}|m,n\rangle = \sum_{j=0}^{n}(-1)^{m+j}\left(\frac{1}{\sqrt{2}}\right)^{m-n+2j}\binom{n}{j}\sqrt{\binom{m+n}{m}}|m+n\rangle$$

$$= (-1)^{m}\left(\frac{1}{\sqrt{2}}\right)^{m+n}\sqrt{\binom{m+n}{m}}|m+n\rangle. \tag{7.79}$$

Let us see what becomes of the original state $|\psi_{AB}\rangle$, up to normalisation, when the two parties act on their two local subsystems with the transformation above:

$$\sum_{m,n=0}^{\infty}\alpha_m\alpha_n\langle 0|\hat{R}_{\frac{\pi}{4}}|m,n\rangle \otimes \langle 0|\hat{R}_{\frac{\pi}{4}}|m,n\rangle = \sum_{n=0}^{\infty}\left[2^{-n}\sum_{m=0}^{n}\alpha_m\alpha_{n-m}\binom{n}{m}\right]|n,n\rangle, \tag{7.80}$$

where we changed one of the summation variables as per $(m+n)\mapsto n$ (and therefore also changed the summation extrema of m). The probabilistic protocol applied maintains the input state's purity as well as its Fock Schmidt basis (in that the state is still diagonal in the basis $\{|n,n\rangle, n \in [0,\ldots,\infty]\}$). Up to normalisation (which may always be recovered and is irrelevant in assessing the entanglement of a pure state), one iteration of the protocol maps the original coefficients $\{\alpha_n^{(i)}, n \in [0\ldots,\infty]\}$ into the $\{\alpha_n^{(i+1)}, n \in [0\ldots,\infty]\}\}$, given by

$$\alpha_n^{(i+1)} = \left[2^{-n}\sum_{m=0}^{n}\alpha_m^{(i)}\alpha_{n-m}^{(i)}\binom{n}{m}\right]. \tag{7.81}$$

The only fixed points of this map are of the form $\alpha_n = \beta^n$, as can be seen by direct substitution and by noting that setting $\alpha_0^{(i+1)} = \alpha_0^{(i)}$ implies $\alpha_0 = 1$ and that hence the (free) choice of α_1 fixes all the higher coefficients through the condition $\alpha_n^{(i+1)} = \alpha_n^{(i)}$.

Now, normalised quantum states of the form $(1-\beta)\sum_{n=0}^{\infty}\beta^n|n,n\rangle$ are nothing but Gaussian two-mode squeezed states, as can be seen by setting $\beta = \tanh(r)$ in Eq. (5.22).[7] We have thus shown that two-mode squeezed states are the only fixed points of our iterative protocol.

It is left to be shown that starting states exist that do converge to such Gaussian entangled fixed points, and that the entanglement of the protocol's output is higher than the initial one (for each individual copy of the input

---

[7] In point of fact, states of this form are the only Gaussian states with a Fock Schmidt basis: the eigenvalues of the partial trace of a state of the form $\sum_{n=0}^{\infty}\alpha_n|n,n\rangle$ (which would still be a Gaussian state if the original state were Gaussian) are given by $|\alpha_n|^2$. But inspection of the general formula (3.60) reveals that a Gaussian state must have eigenvalues scaling like a geometric progression: β^n is then the only possibility for them. Notice that in setting $\tanh(r) = \beta$ we are disregarding the case of a complex β, that could be included but which we are not inclined to discuss here and now.

and output states). We will prove this by assuming $\alpha_0^{(0)} \neq 0$, which hence may be set as $\alpha_0^{(0)} = 1$, and we will also assume $\alpha_1^{(0)} > 0$ (we don't want to commit to a full analysis of the distillation protocols, but rather provide the reader with a proof of principle of its effectiveness). Then, Eq. (7.81) gives $\alpha_0^{(i+1)} = \alpha_0^{(i)2} = 1$, $\forall i$, as well as $\alpha_1^{(i)} = \alpha_1^{(0)}$, $\forall i$. Notice also that the hierarchy of equations (7.81) is such that the increment of each element $(\alpha_n^{(i+1)} - \alpha_n^{(i)})$ is a sum of the lower elements with positive coefficients. Hence, for $\alpha_0^{(0)}, \alpha_1^{(0)} > 0$, which is the case we are considering, the repeated application of the protocol will at some point make all elements positive. In studying the limit $i \to \infty$, we can hence set $\alpha_n^{(i)} > 0$ $\forall n$. We can now show the convergence to the sequence β^n by induction. First notice that, since $\alpha_1^{(i)} = \alpha_1^{(0)}$,

$$\alpha_2^{(i+1)} = \frac{1}{2}\left(\alpha_2^{(i)} + \alpha_1^{(i)2}\right) \quad \Rightarrow \quad \lim_{i \to \infty} \frac{\alpha_2^{(i)}}{\alpha_1^{(i)}} = \alpha_1^{(0)}. \tag{7.82}$$

We can hence make the inductive assumption $\lim_{i \to \infty} \frac{\alpha_m^{(i)}}{\alpha_{m-1}^{(i)}} = \alpha_1^{(0)}$ $\forall m \in [1, \ldots, n]$, which implies $\lim_{i \to \infty} \alpha_m^{(i)} = \alpha_1^{(1)m}$ in the same range. This can be inserted into Eq. (7.81) to show that

$$\lim_{i \to \infty} \left(\alpha_{n+1}^{(i+1)} - \frac{\alpha_{n+1}^{(i)}}{2^n}\right) = \left(1 - \frac{1}{2^n}\right) \alpha_1^{(0)\,(n+1)}, \tag{7.83}$$

whence one finally has

$$\lim_{i \to \infty} \alpha_n^{(i)} = \alpha_1^{(0)\,n} \quad \forall n \in [0, \ldots, \infty]. \tag{7.84}$$

It is thus shown that, under the initial conditions $\alpha_0^{(0)}, \alpha_1^{(0)} > 0$ for initial states with a Fock Schmidt basis, the iteration of the Gaussian LOCC distillation protocol above approaches a two-mode squeezed state. In particular, it converges to a two-mode squeezed state with $\tanh r = \alpha_1^{(0)}$.

We still have to gauge the actual advantage the protocol gives in terms of entanglement, if any. Besides, nothing has been said about the success probability of the protocol. We shall elaborate at some length on these two issues through a specific example in the two problems that follow.

Problem 7.10. (*Proof of entanglement distillation*). Show that an initial quantum state exists whose entanglement will grow under successful iterations of the protocol above.

Solution. It is wise to consider an initial state of the simplest possible form matching the conditions assumed above: $\frac{1}{\sqrt{1-\beta^2}}(|0,0\rangle + \beta|1,1\rangle)$. Notice that, in our notation, where $\alpha_0^{(0)}$ was fixed, this corresponds to $\alpha_0^{(0)} = 1$ and $\alpha_1^{(0)} = \beta > 0$. Eq. (7.84) shows that this state will converge towards the normalised two-mode squeezed state $\sqrt{1-\beta^2}\sum_{n=0}^{\infty}\beta^n|n,n\rangle$ (normalisation may always be worked out at the end). Since these are pure states, their entanglement may be quantified through the von Neumann entropy of the local states, obtained by tracing out one of the two systems. The local states $\varrho^{(0)}$ and $\varrho^{(\infty)}$ for initial and asymptotic states read, respectively,

$$\varrho^{(0)} = \frac{1}{1+\beta^2}\left(|0\rangle\langle 0| + \beta^2|1\rangle\langle 1|\right), \quad \varrho^{(\infty)} = (1-\beta^2)\sum_{n=0}^{\infty}\beta^{2n}|n\rangle\langle n| .$$
(7.85)

At this point, one could be content with computing the von Neumann entropies numerically (in terms of the eigenvalues λ_n that were derived above, these are given by $S_V = -\sum_n \lambda_n \log_2(\lambda_n)$ and would show that, say, for $\beta = 1/\sqrt{2}$ the entanglement goes from about 0.92 ebits to exactly 2 ebits. An even more convincing argument is that, for $\beta \to 1$, one would have $\tanh(r) \to 1$, and hence $r \to \infty$: in such a case the fixed point of the protocol is the maximally entangled continuous variable state that came to the fore in Section 5.1.2.2, whose entanglement in ebits is infinite!

In fact, there is an elegant and general way to show that the entanglement increases for all $\beta < 1$, that does not require any numerics: it consists in showing that the state $\varrho^{(0)}$ *majorises* the state $\varrho^{(\infty)}$. We shall say that a trace class, normalised, positive operator C, with decreasingly ordered eigenvalues $\gamma_n^{\downarrow}$, 'strictly majorises' D, with decreasingly ordered eigenvalues $\delta_n^{\downarrow}$, in symbols $C \succ D$, if $\sum_{n=0}^{j}\gamma_n^{\downarrow} > \sum_{n=0}^{j}\delta_n^{\downarrow}$ $\forall j \in [0, \ldots, \infty]$. If $C \succ D$, then $S_V(C) < S_V(D)$.[a]

It is simple to show that $\varrho^{(0)} \succ \varrho^{(\infty)}$ for all values of $0 < \beta < 1$, since

$$\frac{1}{1+\beta^2} > (1-\beta^2) \quad \text{and} \quad 1 > (1-\beta^2)\sum_{n=0}^{j}\beta^{2n}$$
(7.86)

for all finite j. Hence, we have also proven that $S_V(\varrho^{(0)}) < S_V(\varrho^{(\infty)})$ for all $0 < \beta < 1$.

The hardcore sceptic might still be dissatisfied with knowing that the entanglement only asymptotically converges to a higher value than the initial one. Hence, let us also assess the effect of a single iteration of the protocol on the initial state $\frac{1}{\sqrt{1+\beta^2}}(|0,0\rangle + \beta|1,1\rangle)$. Inspection of Eq. (7.81) shows that the only coefficient which is nonzero after a single iteration, besides α_0 and α_1, is $\alpha_2 = \beta^2/2$. The normalised output state is hence $\frac{1}{\sqrt{1+\beta^2+\beta^4/4}}(|0,0\rangle + \beta|1,1\rangle + \frac{\beta^2}{2}|2,2\rangle)$, with local state $\varrho^{(1)} = \frac{1}{1+\beta^2+\beta^4/4}(|0\rangle\langle 0| + \beta^2|1\rangle\langle 1| + \frac{\beta^4}{4}|2\rangle\langle 2|)$. Again, it is easy to show that $\varrho^{(0)} \succ \varrho^{(1)}$, for all β: the entanglement does grow, for all β even after a single iteration. The entanglement distillation protocol presented, ineffective within the Gaussian regime, does work when fed appropriate non-Gaussian entangled states.

---

[a]The customary definition of the majorisation relation does not employ a strict inequality. We slightly twisted the standard definition into the "strict" version above, because we are interested in claiming that the local von Neumann entropy of the distilled state is *strictly* larger than the original one. This follows from strict majorisation because the von Neumann entropy S_V is strictly concave: $S_V(p\varrho_1 + (1-p)\varrho_2) \geq pS_V(\varrho_1) + (1-p)S_V(\varrho_2)$ with equality if and only if $\varrho_1 = \varrho_2$.

Problem 7.11. (*Success probability of entanglement distillation*). Determine the probability of success of the protocol at the $(i+1)^{\text{th}}$ iteration, given the coefficients $\{\alpha_n^{(i)}\}$ after the ith iteration. Further, write down the probability of success at the first iteration for the initial state $\frac{1}{\sqrt{1+\beta^2}}(|0,0\rangle + \beta|1,1\rangle)$.

Solution. We will need to take care of the state's normalisation in order to predict probabilities: for a generic set of coefficients, each of the pair of states is normalised by $N = \sqrt{\sum_{n=0} \alpha_n^{(i)2}}$ (we keep considering only real coefficients, for simplicity). The success probability $p^{(i+1)}$ may then be extracted from Eq. (7.80), taking the norm of the vector on the right hand side divided by N^4 (not just N^2, since the original state was the tensor product of two states):

$$p^{(i+1)} = \frac{\sum_{n=0}^{\infty} \left[2^{-n} \sum_{m=0}^{n} \alpha_m^{(i)} \alpha_{n-m}^{(i)} \binom{n}{m} \right]^2}{\left(\sum_{n=0} \alpha_n^{(i)2} \right)^2}. \tag{7.87}$$

For the initial state $\frac{1}{\sqrt{1+\beta^2}}(|0,0\rangle + \beta|1,1\rangle)$, the only nonzero coefficients are $\alpha_0^{(0)} = 1$ and $\alpha_1^{(0)} = \beta$, and the formula above yields

$$p^{(1)} = \frac{1 + \beta^2 + \frac{1}{4}\beta^4}{1 + 2\beta^2 + \beta^4}. \tag{7.88}$$

This is a decreasing function of β. Now, the higher the initial $\beta \in [0,1]$, the higher the entanglement gained through distillation. However, the probability of success goes down. This trade-off, illustrated here only in a very specific case, is typical of distillation processes.

7.6 HIGHER-ORDER SEPARABILITY CRITERIA

All the entanglement criteria reviewed so far are based on second moments. All such criteria may be applied outside the Gaussian domain, where they provide general sufficient conditions for entanglement. However, such conditions are often too weak to detect the entanglement of non-Gaussian states. It is expedient to illustrate this situation with a specific instance: let us consider the normalised, entangled cat-like state of two modes

$$|\psi_\alpha\rangle = \frac{|\alpha, \alpha\rangle - |-\alpha, -\alpha\rangle}{\sqrt{2 - 2e^{-|2\alpha|^2}}} \,, \tag{7.89}$$

where $|\alpha, \alpha\rangle$ stands for the tensor product of coherent states $|\alpha\rangle \otimes |\alpha\rangle$, and the normalisation reflects the fact that coherent states are not orthogonal, see Eq. (4.7). We will assume $|\alpha| > 0$, so that the normalisation factor is well defined (disregarding the trivial case where the state above reduces to the obviously separable vacuum). Such a state is obviously entangled, since it is globally pure but with mixed local states. However, the PPT criterion at the level of second moments, that is the inequality (7.12), that would be necessary and sufficient for the separability of a two-mode Gaussian states, fails to detect this entangled state.

Problem 7.12. (*A stealthy cat*). Demonstrate the last statement.

Solution. To start with, bear in mind that any property or criterion associated with entanglement alone must be invariant under local unitaries. One is therefore free to apply local phase rotations – whose symplectic matrix is given by Eq. (5.12) – that act on the coherent state by just multiplying α by a phase, and to set $\alpha = |\alpha|$ (real) without loss of generality. The covariance matrix is then more conveniently reconstructed working in terms of the vector of ladder operators $\hat{a}$, defined on page 36, as $\sigma_a = \langle (\hat{a} - \alpha), (\hat{a} - \alpha)^\dagger \rangle$ in our outer product notation, where α is the (complex) vector of first moments. Given the symmetries of this two-mode case, it is promptly seen that all the entries of σ_a will be determined by its first row. Besides, the first moments α vanish:

$$\alpha = \langle\psi_\alpha|\hat{a}|\psi_\alpha\rangle \propto \langle\alpha, \alpha|\hat{a}|\alpha, \alpha\rangle + \langle-\alpha, -\alpha|\hat{a}|-\alpha, -\alpha\rangle$$
$$- \langle\alpha, \alpha|\hat{a}|-\alpha, -\alpha\rangle - \langle-\alpha, -\alpha|\hat{a}|\alpha, \alpha\rangle = 0 \quad (7.90)$$

(which might have been inferred from the fact that the state is invariant under the point reflection about the phase space origin). Let us then determine the entries in the first row of σ_a:

$$\sigma_{a,11} = \langle 2\hat{a}_1^\dagger \hat{a}_1 + 1 \rangle = 1 + 2|\alpha|^2 \frac{1 + e^{-4|\alpha|^2}}{1 - e^{-4|\alpha|^2}} , \tag{7.91}$$

$$\sigma_{a,12} = \langle 2\hat{a}_1^2 \rangle = 2|\alpha|^2 , \tag{7.92}$$

$$\sigma_{a,13} = \langle 2\hat{a}_1 \hat{a}_2^\dagger \rangle = 2|\alpha|^2 \frac{1 + e^{-4|\alpha|^2}}{1 - e^{-4|\alpha|^2}} , \tag{7.93}$$

$$\sigma_{a,14} = \langle 2\hat{a}_1 \hat{a}_2 \rangle = 2|\alpha|^2 , \tag{7.94}$$

where we just used eigenvalue equations like $\hat{a}_j |\alpha, \alpha\rangle = \alpha |\alpha, \alpha\rangle$ and $\langle -\alpha, -\alpha | \alpha, \alpha \rangle = e^{-4|\alpha|^2}$ [from Eq. (4.7)]. From the four values above, the whole covariance matrix may be reconstructed as:

$$\sigma_a = \begin{pmatrix} \mathbb{1}_2 + 2|\alpha|^2 \Upsilon & 2|\alpha|^2 \Upsilon \\ 2|\alpha|^2 \Upsilon & \mathbb{1}_2 + 2|\alpha|^2 \Upsilon \end{pmatrix} , \tag{7.95}$$

with

$$\Upsilon = \begin{pmatrix} \eta & 1 \\ 1 & \eta \end{pmatrix} , \quad \text{and} \quad \eta = \frac{1 + e^{-4|\alpha|^2}}{1 - e^{-4|\alpha|^2}} > 1 . \tag{7.96}$$

We could now go back to x and p quadratures through the block-diagonal unitary operator $\bar{U}$ of Eq. (3.6), and then try to apply the criteria for two-mode covariance matrices as we wrote them earlier. However, we do not need to: the determinant of the off-diagonal block of σ_a is $4|\alpha|^2(\eta^2 - 1) > 0$ for all $|\alpha| > 0$. By the lemma on page 174, this means that the corresponding two-mode Gaussian state would be separable (notice that the action of the unitary, block-diagonal $\bar{U}$ would not change the determinant at issue), and hence that the partially transposed symplectic eigenvalues of the associated covariance matrix σ_a are all greater than one. In other words, the criterion (7.12) is satisfied and fails to detect the cat-like entanglement of $|\psi_\alpha\rangle$.

Fortunately, more sophisticated criteria than those based on the second moments may be obtained to qualify continuous variable entanglement. We will present here a constructive method, due to Shchukin and Vogel, to obtain inequalities involving higher-order statistical moments of the canonical operator and thus identify entangled non-Gaussian states that may be elusive to criteria based on second moments. It should be stressed that these inequalities will be based on the positivity of partial transposition, and hence will not be capable of detecting bound entangled states, which do exist, even for two modes, when generic states are considered.[8] However, distillable, 'NPT' entanglement (of states with non-positive partial transpose) is in a sense the most valuable, so that these criteria are of notable practical interest. Also,

---

[8] Bound entangled states exist in composite Hilbert spaces of dimension 8 and above.

let us remark that the PPT criterion on second moments is conclusive to establish whether a Gaussian state is PPT or not, so that the criteria we are about to derive are redundant as far as Gaussian states are concerned *assuming the whole covariance matrix may be reconstructed* (which may not always be practical).

The construction begins by noticing that a positive, self-adjoint operator $\hat{A}$ must be such that $\mathrm{Tr}\left(\hat{A}\hat{f}^{\dagger}\hat{f}\right) \geq 0$ for any operator $\hat{f}$, since $\hat{f}^{\dagger}\hat{f}$ is positive. In fact, one may restrict to operators $\hat{f}$ that can be normal-ordered (normal ordering was defined in Section 4.3).[9] This simple argument may be applied to the partial transposition of the density matrix $\tilde{\varrho}$. For n modes, the most general $\hat{f}$ is given by $\hat{f} = \sum_{j_1,...j_{2n}=0}^{\infty} c_{j_1,...j_{2n}}(\hat{a}_1^{\dagger j_1}\hat{a}_1^{j_2}\ldots\hat{a}_n^{\dagger j_{2n-1}}\hat{a}_n^{j_{2n}})$ for $c_{j_1,...j_{2n}} \in \mathbb{C}$. The positivity of the partial transpose then may be expressed as the positivity of a quadratic form:

$$\mathrm{Tr}\left(\tilde{\varrho}\hat{f}^{\dagger}\hat{f}\right) = \sum_{\substack{j_1,...j_{2n}=0 \\ k_1,...k_{2n}=0}}^{\infty} c_{j_1,...j_{2n}}^{*}\, c_{k_1,...k_{2n}} M_{j_1,...j_{2n},k_1,...k_{2n}} \,, \qquad (7.97)$$

with

$$\begin{aligned}
M_{j_1,...j_{2n},k_1,...k_{2n}} &= \mathrm{Tr}\left(\tilde{\varrho}\,\hat{a}_1^{\dagger j_2}\hat{a}_1^{j_1}\hat{a}_1^{\dagger k_1}\hat{a}_1^{k_2}\ldots\hat{a}_n^{\dagger j_{2n}}\hat{a}_n^{j_{2n-1}}\hat{a}_n^{\dagger k_{2n-1}}\hat{a}_n^{k_{2n}}\right) \\
&= \mathrm{Tr}\left(\tilde{\varrho}\prod_{s=1}^{n}\hat{a}_s^{(j_{2s-1},j_{2s},k_{2s-1},k_{2s})}\right) \,, \qquad (7.98)
\end{aligned}$$

where we have defined the operators

$$\hat{a}_s^{(j,k,p,q)} = \hat{a}_s^{\dagger k}\hat{a}_s^{j}\hat{a}_s^{\dagger p}\hat{a}_s^{q} \,. \qquad (7.99)$$

Crucially, the inequality (7.97) must hold for all choices of $c_{j_1,...j_{2n}}$: by virtue of Sylvester's criterion, this is the case if and only if all the principal minors of the

---

[9]This is because $\hat{A}$ is positive if and only if $\langle\psi|\hat{A}|\psi\rangle = \mathrm{Tr}\left(|\psi\rangle\langle\psi|\hat{A}\right) \geq 0$ for all $|\psi\rangle \in \mathcal{H}$, and the projector on any quantum state $|\psi\rangle\langle\psi|$ may be written as $|\psi\rangle\langle\psi| = \hat{f}^{\dagger}\hat{f}$, where $\hat{f}$ may be chosen in normal order as follows. First note that any vector can be expanded in the Fock basis: $|\psi\rangle = \sum_{n=0}^{\infty}\psi_n|n\rangle = \sum_{n=0}^{\infty}\frac{\psi_n}{\sqrt{n!}}\hat{a}^{\dagger n}|0\rangle$. Then, write $|\psi\rangle\langle\psi| = \sum_{m,n=0}^{\infty}\frac{\psi_n}{\sqrt{n!}}\frac{\psi_m^{*}}{\sqrt{m!}}\hat{a}^{\dagger n}|0\rangle\langle 0|0\rangle\langle 0|\hat{a}^m$ and identify $\hat{f} = \sum_{m=0}^{\infty}\frac{\psi_m^{*}}{\sqrt{m!}}|0\rangle\langle 0|\hat{a}^m$. The operator $\hat{f}$ may be normally ordered since the vacuum projector $|0\rangle\langle 0|$ may be expressed as $|0\rangle\langle 0| = \sum_{j=0}^{\infty}\frac{(-1)^j}{j!}\hat{a}^{\dagger j}\hat{a}^j =: e^{-\hat{a}^{\dagger}\hat{a}}:$, where the colons stand for the normal-ordered form of the operator enclosed. In turn, this expression for the vacuum projector may be derived by normal ordering the displacement operator in Eq. (4.14) through Baker-Campbell-Hausdorff:

$$\pi|0\rangle\langle 0| = \int_{\mathbb{C}}\mathrm{d}^2\gamma\, e^{-|\gamma|^2}e^{\gamma\hat{a}^{\dagger}}e^{-\gamma^{*}\hat{a}} = \int_{\mathbb{C}}\mathrm{d}^2\gamma\, e^{-|\gamma|^2}\sum_{j,k=0}^{\infty}(-1)^j\frac{\gamma^{*j}\gamma^k}{j!k!}\hat{a}^{\dagger k}\hat{a}^j = \pi\sum_{j=0}^{\infty}\frac{(-1)^j}{j!}\hat{a}^{\dagger j}\hat{a}^j,$$

where the integral formula (4.16) was applied.

quadratic form M (that is, all the determinants of its principal submatrices) are positive.

The quadratic form M is infinite dimensional (since each of the labels j_n's and k_n's goes from 0 to ∞). However, identifying any negative, finite-dimensional principal minor is enough to certify the negativity of the partial transpose, and hence entanglement: this provides one with an infinite hierarchy of sufficient conditions for entanglement, that can be pushed well beyond the second moments. Let us elaborate a little more on the manipulation of M needed to obtain such inequalities, and then illustrate the method in the case of the entangled cat-like state defined above.

First, let us express M in terms of observable moments of the state ϱ, rather than as a function of its partial transpose as above. To this aim, let us split the system into modes with ladder operators $\{\hat{a}_1, \ldots, \hat{a}_m\}$ and $\{\hat{b}_1, \ldots, \hat{b}_n\}$, and suppose we intend to check for bipartite entanglement across such a mode partition. Then, we would have to partially transpose ϱ for one of the two sets of modes, say the second set. We can now use the fact that the partial transposition preserves the trace and that applying it twice to an operator leaves it unchanged, as well as the cyclicity of the trace, to turn Eq. (7.104) into

$$
\begin{aligned}
&M_{j_1,\ldots j_{2(m+n)},k_1,\ldots k_{2(m+n)}} = \\
&= \mathrm{Tr}\left(\varrho \prod_{s=1}^{m} \hat{a}_s^{(j_{2s-1},j_{2s},k_{2s-1},k_{2s})} \prod_{t=1}^{n} \hat{b}_t^{(j_{2(t+n)-1},j_{2(t+n)},k_{2(t+n)-1},k_{2(t+n)})\mathsf{T}} \right),
\end{aligned}
$$
$$(7.100)$$

where the partial transposition results in the complete transposition of the operators $\hat{b}_t^{(j_{2(t+n)-1},j_{2(t+n)},k_{2(t+n)-1},k_{2(t+n)})\mathsf{T}}$. We already know how to handle partial transposition on the canonical operators: in deriving Eq. (7.9), it was shown that transposition in the Fock basis is nothing but Hermitian conjugation on the ladder operators: $\hat{b}^{\mathsf{T}} = \hat{b}^{\dagger}$. Therefore, one is left with

$$
\begin{aligned}
&M_{j_1,\ldots j_{2(m+n)},k_1,\ldots k_{2(m+n)}} = \\
&= \mathrm{Tr}\left(\varrho \prod_{s=1}^{m} \hat{a}_s^{(j_{2s-1},j_{2s},k_{2s-1},k_{2s})} \prod_{t=1}^{n} \hat{b}_t'^{(j_{2(t+n)-1},j_{2(t+n)},k_{2(t+n)-1},k_{2(t+n)})} \right),
\end{aligned}
$$
$$(7.101)$$

where

$$
\hat{b}_t'^{(j,k,p,q)} = \hat{b}_t^{\dagger q} \hat{b}_t^p \hat{b}_t^{\dagger j} \hat{b}_t^k .
$$
$$(7.102)$$

The discrete multi-indexes of M can be reorganised to obtain an infinite-dimensional, two-indexed matrix M_{jk}, with $j, k \in \mathbb{N}$. Then, one may evaluate any principal minor of M: its positivity is necessary for separability, so that its negativity would be sufficient for entanglement. The rearrangement of the multi-indexes is a tedious process, which is not worth showing in generality.

Let us instead specialise to the two-mode case, where M would be defined as

$$M_{j_1,,j_2,j_3,j_4,k_1,k_2,k_3,k_4} = \text{Tr}\left(\varrho\, \hat{a}^{\dagger j_2} \hat{a}^{j_1} \hat{a}^{\dagger k_1} \hat{a}^{k_2} \hat{b}^{\dagger k_4} \hat{b}^{k_3} \hat{b}^{\dagger j_3} \hat{b}^{j_4}\right), \qquad (7.103)$$

and show what the first entries of the rearranged matrix might look like:

$$M = \begin{pmatrix}
1 & \langle \hat{a} \rangle & \langle \hat{a}^\dagger \rangle & \langle \hat{b}^\dagger \rangle & \langle \hat{b} \rangle & \langle \hat{a}\hat{a}^\dagger \rangle & \langle \hat{a}\hat{b}^\dagger \rangle & \cdots \\
\langle \hat{a}^\dagger \rangle & \langle \hat{a}^\dagger \hat{a} \rangle & \langle \hat{a}^{\dagger 2} \rangle & \langle \hat{a}^\dagger \hat{b}^\dagger \rangle & \langle \hat{a}^\dagger \hat{b} \rangle & \langle \hat{a}^\dagger \hat{a}\hat{a}^\dagger \rangle & \langle \hat{a}^\dagger \hat{a}\hat{b}^\dagger \rangle & \cdots \\
\langle \hat{a} \rangle & \langle \hat{a}^2 \rangle & \langle \hat{a}\hat{a}^\dagger \rangle & \langle \hat{a}\hat{b}^\dagger \rangle & \langle \hat{a}\hat{b} \rangle & \langle \hat{a}^2 \hat{a}^\dagger \rangle & \langle \hat{a}^2 \hat{b}^\dagger \rangle & \cdots \\
\langle \hat{b} \rangle & \langle \hat{a}\hat{b} \rangle & \langle \hat{a}^\dagger \hat{b} \rangle & \langle \hat{b}^\dagger \hat{b} \rangle & \langle \hat{b}^2 \rangle & \langle \hat{a}\hat{a}^\dagger \hat{b} \rangle & \langle \hat{a}\hat{b}^\dagger \hat{b} \rangle & \cdots \\
\langle \hat{b}^\dagger \rangle & \langle \hat{a}\hat{b}^\dagger \rangle & \langle \hat{a}^\dagger \hat{b}^\dagger \rangle & \langle \hat{b}^{\dagger 2} \rangle & \langle \hat{b}\hat{b}^\dagger \rangle & \langle \hat{a}\hat{a}^\dagger \hat{b}^\dagger \rangle & \langle \hat{a}\hat{b}^{\dagger 2} \rangle & \cdots \\
\langle \hat{a}\hat{a}^\dagger \rangle & \langle \hat{a}\hat{a}^\dagger \hat{a} \rangle & \langle \hat{a}\hat{a}^{\dagger 2} \rangle & \langle \hat{a}\hat{a}^\dagger \hat{b}^\dagger \rangle & \langle \hat{a}\hat{a}^\dagger \hat{b} \rangle & \langle \hat{a}\hat{a}^\dagger \hat{a}\hat{a}^\dagger \rangle & \langle \hat{a}\hat{a}^\dagger \hat{a}\hat{b}^\dagger \rangle & \cdots \\
\langle \hat{a}^\dagger \hat{b} \rangle & \langle \hat{a}^\dagger \hat{a}\hat{b} \rangle & \langle \hat{a}^{\dagger 2}\hat{b} \rangle & \langle \hat{a}^\dagger \hat{b}^\dagger \hat{b} \rangle & \langle \hat{a}^\dagger \hat{b}^2 \rangle & \langle \hat{a}^\dagger \hat{a}\hat{a}^\dagger \hat{b} \rangle & \langle \hat{a}^\dagger \hat{a}\hat{b}^\dagger \hat{b} \rangle & \cdots \\
\vdots & \vdots & \vdots & \vdots & \vdots & \vdots & \vdots & \ddots
\end{pmatrix}$$

$$(7.104)$$

Note that the first line and the first row determine the choices for the inner expectation values, with the rule of thumb that, for $\hat{a}$ and $\hat{a}^\dagger$ operators, the one appearing in the first column stays on the left of the product whereas, for $\hat{b}$ operators, the one showing up in the first row stays on the left (by the effect of partial transposition) This is illustrated in the entries M_{22} and M_{44}.

It is worthwhile to summarise this approach by setting out an explicit sufficient condition issued from it:

> **Higher-order NPT test.** The negativity of the determinant of any of the principal submatrices of the matrix M of Eq. (7.104) is a sufficient condition for the non-positivity of the partial transposition of a bipartite quantum state of two modes, and hence for its entanglement.

Now, we are finally in a position to go back to our stealthy cat-like state, that avoided detection through second moments. Let us pick rows and columns 1, 4 and 7 in the matrix above, and obtain the following general, sufficient condition for entanglement:

$$\text{Det}\begin{pmatrix}
1 & \langle \hat{b}^\dagger \rangle & \langle \hat{a}\hat{b}^\dagger \rangle \\
\langle \hat{b} \rangle & \langle \hat{b}^\dagger \hat{b} \rangle & \langle \hat{a}\hat{b}^\dagger \hat{b} \rangle \\
\langle \hat{a}^\dagger \hat{b} \rangle & \langle \hat{a}^\dagger \hat{b}^\dagger \hat{b} \rangle & \langle \hat{a}^\dagger \hat{a}\hat{b}^\dagger \hat{b} \rangle
\end{pmatrix} < 0. \qquad (7.105)$$

This will catch the cat state.

> **Problem 7.13.** (*The cat is spotted*). Prove that condition (7.105) is able to detect the entanglement of the cat-like state $|\psi_\alpha\rangle$ for all $\alpha \in \mathbb{C}$.
>
> *Solution.* Let us go back to the state $|\psi_\alpha\rangle$ and evaluate the necessary expectation values. As in the previous problem, we can set $\alpha = |\alpha|$ through local phase space rotations, that correspond to local unitaries.

In solving that problem, we already calculated (notice that, there, we referred to $\hat{a}$ as $\hat{a}_1$ and $\hat{b}$ as $\hat{a}_2$) $\langle \hat{b} \rangle = \langle \hat{b}^\dagger \rangle = 0$ and $\langle \hat{b}^\dagger \hat{b} \rangle = \langle \hat{a}^\dagger \hat{b} \rangle = \langle \hat{a} \hat{b}^\dagger \rangle = |\alpha|^2 \eta$ with

$$\eta = \frac{1 + e^{-4|\alpha|^2}}{1 - e^{-4|\alpha|^2}} > 1 . \tag{7.106}$$

By applying Eq. (4.7) and the coherent state eigenvalue equation, we just need to determine the remaining entries of the matrix in (7.105):

$$\langle \hat{a} \hat{b}^\dagger \hat{b} \rangle = \langle \hat{a}^\dagger \hat{b}^\dagger \hat{b} \rangle = 0 , \tag{7.107}$$

$$\langle \hat{a}^\dagger \hat{a} \hat{b}^\dagger \hat{b} \rangle = |\alpha|^2 . \tag{7.108}$$

Let us now apply the Sarrus rule to the 3×3 determinant

$$\mathrm{Det} \begin{pmatrix} 1 & 0 & |\alpha|^2\eta \\ 0 & |\alpha|^2\eta & 0 \\ |\alpha|^2\eta & 0 & |\alpha|^4 \end{pmatrix} = |\alpha|^6 \eta(1 - \eta^2) < 0 \quad \forall\, |\alpha| > 0 . \tag{7.109}$$

The entanglement of the cat-like state is hence always detected by inequality (7.105).

In general, it should be clear how such conditions may be tailored to the situation in hand, even considering experimental limitations on the available measurements that always arise in practical situations, so that tests constructed through the general procedure above may be helpful for Gaussian states too, when the whole covariance matrix cannot be reconstructed.

7.7 PROBABILISTIC ENTANGLEMENT ENHANCEMENT: PHOTON SUBTRACTED STATES

In view of the, repeatedly recalled, classical-like nature of Gaussian states (a notion that will be put on firmer ground in the next section), non-Gaussian entangled states are certainly very valuable resources.

They are, however, difficult to generate and control in practice. Probably, the simplest avenue to obtain a non-Gaussian entangled state is through the process of photon subtraction, which may be carried out as follows:

- Two modes prepared in a Gaussian entangled state (typically a two-mode squeezed state, or a noisy version thereof) are both mixed with the vacuum state at a beam splitter with transmittivity $(\cos\theta)^2$.

- The reflected outputs of the beam splitters which, so to speak, carry a portion $(\sin\theta)^2$ of the original covariances, are then detected through on/off detectors, which click only if any positive number of excitations are detected.

- When both detectors click, the protocol is successful: the initial state has been projected into an "inconclusively" photon subtracted, non-Gaussian entangled state (where inconclusively refers to the fact that the number of excitations subtracted is unknown).

In a sense, photon subtraction is the converse of the "Gaussianification" protocol that was described in Section 7.5.1, in that it runs through the same series of operations but its final success is conditional on obtaining the opposite readings (projections on the excited subspaces rather than vacuum projections).

Here, let us assume that the input state is the pure two-mode squeezed state $|\psi_r\rangle = \cosh(r)^{-1} \sum_n \tanh(r)^n |n,n\rangle$, and determine the corresponding output photon subtracted state. The total initial state, including the modes in the vacuum is $|\psi_r\rangle \otimes |00\rangle = \sum_n \frac{\tanh(r)^n}{\cosh(r)} |n,n,0,0\rangle$, and we shall denote with $\hat{a}$, $\hat{b}$, $\hat{c}$ and $\hat{d}$ the ladder operators of the four modes ordered as in the state just given, which we shall refer to with the corresponding letters. The beam splitter operator that mixes the modes ac $\hat{R}_\theta^{(ac)\dagger} = e^{\theta(\hat{a}^\dagger\hat{c}-\hat{a}\hat{c}^\dagger)}$ (and likewise for b and d) may be re-ordered through Eq. (5.116) to obtain

$$\hat{R}_\theta^{(ac)\dagger} = e^{-\tan|\theta|\hat{a}\hat{c}^\dagger}(\cos\theta)^{\hat{a}^\dagger\hat{a}-\hat{c}^\dagger\hat{c}}e^{\tan|\theta|\hat{a}^\dagger\hat{c}}, \qquad (7.110)$$

which allows one to evaluate the state $|\eta_r\rangle = \hat{R}_\theta^{(ac)\dagger}\hat{R}_\theta^{(bd)\dagger}(|\psi_r\rangle \otimes |00\rangle)$ after all the beam splitting:

$$|\eta_r\rangle = \sum_{n=0}^{\infty}\sum_{j=0}^{n}\sum_{k=0}^{n} \frac{[(\cos\theta)^2\tanh(r)]^n}{\cosh(r)}(-\tan|\theta|)^{j+k}\frac{\hat{a}^j\hat{b}^k}{j!k!}|n,n,j,k\rangle. \quad (7.111)$$

It would be possible now to write down the projection on the excited subspace of the two auxiliary modes exactly. This would yield a mixed entangled state for the two modes a and b, whose properties are in principle computable. However, such an approach would involve rather cumbersome evaluations without granting much general insight. It is perhaps more interesting here to illustrate photon subtraction in the specific regime $0 < \theta \ll 1$, where a reasonable approximation renders the analytical treatment much cleaner.

We will in fact consider the Taylor expansion in θ of the state above, and truncate its projection on the excited subspace at the first non-vanishing order. One has $(\cos\theta)^2 = 1 - \theta^2 + o(\theta^2)$ and $\tan\theta = \theta + o(\theta^2)$, so that

$$|\eta_r\rangle = (|\psi_r\rangle - \theta^2|\varphi_r\rangle)\otimes|00\rangle - \theta\hat{a}|\psi_r\rangle\otimes|10\rangle - \theta\hat{b}|\psi_r\rangle\otimes|01\rangle + \theta^2\hat{a}\hat{b}|\psi_r\rangle\otimes|11\rangle + o(\theta^2),$$
$$(7.112)$$

with $|\varphi_r\rangle = \sum_{n=0}^{\infty}\frac{n\tanh(r)^n}{\cosh(r)}|n,n\rangle$. Projecting on the excited subspaces of modes c and d then yields

$$(\hat{1} - |0\rangle\langle0|)_c(\hat{1} - |0\rangle\langle0|)_d|\eta_r\rangle = \theta^2\hat{a}\hat{b}|\psi_r\rangle + o(\theta^2). \qquad (7.113)$$

At leading order (θ^4 for the density operator), the state of the system is updated to a pure, entangled non-Gaussian state where one photon has been subtracted from each mode, with success probability

$$p_s = \sinh(r)^2 (2\sinh(r)^2 + 1)\theta^4 + o(\theta^4) \,. \tag{7.114}$$

Notice that, even if $\theta \ll 1$, the success probability may still in principle be boosted, to some extent, by increasing the two-mode squeezing parameter r. In this regard, let us clarify that, if one increases r, the higher-order terms in θ become relevant and heal the apparent divergence in the leading order term (so that $p_s \leq 1$).

Problem 7.14. (*Probability of photon subtraction*). Prove Eq. (7.114).

Solution. By virtue of Eq. (7.113), the success probability at leading order is given by $\theta^4 \langle \psi_r | \hat{a}^\dagger \hat{a} \hat{b}^\dagger \hat{b} | \psi_r \rangle = \theta^4 \sum_{n=0}^{\infty} \frac{n^2 \tanh(r)^{2n}}{\cosh(r)^2}$, which can be evaluated by using $\sum_{n=0}^{\infty} n^2 z^n = \frac{z(z+1)}{(1-z)^3}$ for $|z| < 1$ (which is obtained by deriving the geometric series twice). One thus obtains $p_s = \frac{\tanh(r)^2(\tanh(r)^2+1)}{(1-\tanh(r)^2)^2} \theta^4 + o(\theta^4)$, which may be re-arranged as in Eq. (7.114).

Eq. (7.114) also provides us with the normalisation of the state. We have shown that a non-Gaussian state may be obtained:

Photon subtracted twin beam. The probabilistic photon subtraction procedure described above with $\theta \ll 1$, applied on a two-mode squeezed state $|\psi_r\rangle = \sum_{n=0}^{\infty} \frac{\tanh(r)^n}{\cosh(r)} |n, n\rangle$, approximately yields the photon subtracted state

$$|\zeta_r\rangle = \frac{\hat{a}\hat{b}|\psi_r\rangle}{\sinh(r)\sqrt{2\sinh(r)^2 + 1}} = \sum_{n=0}^{\infty} \frac{2\tanh(r)^{n+1}(n+1)|n, n\rangle}{\sinh(2r)\sqrt{2\sinh(r)^2 + 1}} \,. \tag{7.115}$$

Let us mention in passing that an analogous procedure, where the input state interacts with the vacuum through degenerate parametric downconversion rather than beam splitting, and the preparation is conditioned by the detection of one photon in the spare output port, may be employed to generate 'photon added' states, such as $\hat{a}^\dagger|\psi_r\rangle$ (up to normalisation). If the input state is the vacuum state, this procedure yields a single-photon state $\hat{a}^\dagger|0\rangle = |1\rangle$.[10]

Clearly, the protocol above can be adapted to subtract a single photon from a single mode. It is, however, when acting jointly on correlated states

---

[10]Another probabilistic scheme to obtain single-photon states consists of simply letting a weak two-mode squeezed state (the state $|\psi_r\rangle$ above with small r) through a beam splitter and then performing on/off photodetection at one output port: if photons are detected, the other outgoing mode is highly likely to have collapsed into the state $|1\rangle$.

that photon subtraction bears substantial advantages. To start with, it may be shown that the photon subtracted state $|\zeta_r\rangle$ is more entangled than the parent two-mode squeezed state $|\psi_r\rangle$ for all values of r.[11] As well as an enhanced entanglement, photon subtracted states also enjoy a more subtle and consequential property: their Wigner function, obviously non-Gaussian, is not everywhere positive. It is in fact a classic result by Hudson (see Further Reading) that the only pure states with positive Wigner function are Gaussian pure states. This will play a major role in the next section.

Problem 7.15. (*Characteristic function of photon subtracted twin beams*). Show that the characteristic function of the photon subtracted state $|\zeta_r\rangle$ is given by

$$\chi_{|\zeta_r\rangle\langle\zeta_r|} = C\left[\partial^2_{x_a x_a} + \partial^2_{p_a p_a} + \frac{p_a^2}{4} + \frac{x_a^2}{4} + p_a\partial_{p_a} + x_a\partial_{x_a} + 1\right] \times$$
$$\times \left[\partial^2_{x_b x_b} + \partial^2_{p_b p_b} + \frac{p_b^2}{4} + \frac{x_b^2}{4} + p_b\partial_{p_b} + x_b\partial_{x_b} + 1\right] e^{-\frac{\mathbf{r}^\mathsf{T}\sigma_{-r}\mathbf{r}}{4}}, \tag{7.116}$$

where $C = [4\sinh(r)^2(2\sinh(r)^2 + 1)]^{-1}$, σ_r is the covariance matrix of the two-mode squeezed state (5.21), and $\mathbf{r} = (x_a, p_a, x_b, p_b)^\mathsf{T}$.

Solution. We would like to understand the representation of the photon subtraction map in the characteristic function formalism. To this aim, we can use the correspondence (6.45). Let us define, as usual, $\sqrt{2}\hat{a} = \hat{x} + i\hat{p}$, and work for convenience in terms of the quadrature operators. In order to understand how the representation works, it will be sufficient to consider only one mode. From (6.45), one obtains the following correspondences for the characteristic function χ of a generic state ϱ:

$$\hat{x}\varrho \leftrightarrow \left(-i\partial_x + \frac{p}{2}\right)\chi, \quad \varrho\hat{x} \leftrightarrow \left(-i\partial_x - \frac{p}{2}\right)\chi, \tag{7.117}$$

$$\hat{p}\varrho \leftrightarrow \left(-i\partial_p - \frac{x}{2}\right)\chi, \quad \varrho\hat{p} \leftrightarrow \left(-i\partial_p + \frac{x}{2}\right)\chi. \tag{7.118}$$

The photon subtraction map on ϱ may be written as $\hat{a}\varrho\hat{a}^\dagger = \frac{1}{2}(\hat{x}\varrho\hat{x} + \hat{p}\varrho\hat{p} + i\hat{p}\varrho\hat{x} - i\hat{x}\varrho\hat{p})$ which, combined with the previous relations, yields the representation

$$\hat{a}\varrho\hat{a}^\dagger \leftrightarrow -\frac{1}{2}\left[\partial^2_{xx} + \partial^2_{pp} + \frac{p^2}{4} + \frac{x^2}{4} + p\partial_p + x\partial_x + 1\right]\chi. \tag{7.119}$$

---

[11]This may be explicitly proven by showing that the squared Schmidt coefficients of $|\psi_r\rangle$ strictly majorise the squared Schmidt coefficients of $|\zeta_r\rangle$ which, as discussed in solving Problem 7.10, implies that the entanglement entropy (i.e., the local von Neumann entropy) of $|\zeta_r\rangle$ is higher than that of $|\psi_r\rangle$.

Defining the canonical operators $\sqrt{2}\hat{a}_j = \hat{x}_j + i\hat{p}_j$ for $j = a, b$ and the associated phase space variables, collected in $\mathbf{r}$, one may then just let both photon subtraction mappings act on the characteristic function of the two-mode squeezed state $e^{-\frac{\mathbf{r}^T \Omega^T \sigma_r \Omega \mathbf{r}}{4}} = e^{-\frac{\mathbf{r}^T \sigma_{-r} \mathbf{r}}{4}}$ to obtain the expression above. Note that the factors $1/2$ have been incorporated, together with the state normalisation, already determined, in the constant C.

We have thus explicitly shown that the characteristic function of photon subtracted states is not Gaussian. Its Fourier transform, yielding the Wigner function, will not be positive.

7.8 QUANTUM NONLOCALITY WITH CONTINUOUS VARIABLES

We shall conclude our survey of quantum correlations with a concise discussion of their utmost operational consequence: the notion of quantum nonlocality. Before moving on to specific considerations concerning continuous variable systems, let us include a brief reminder on how quantum nonlocality may be discussed on quantitative, operational grounds through nonlocality tests.

Assume parties A and B are each allowed to measure two local observables, say $\hat{A}_1$ and $\hat{A}_2$ for A and $\hat{B}_1$ and $\hat{B}_2$ for B (we adopt the notation with a hat for future convenience, though this argument does not imply any quantum mechanical Hilbert space structure). Say all such observables are probabilistic and dichotomic, each admitting two possible outcomes, denoted in what follows with a_1, a_2, b_1 and b_2 conventionally set at ∓ 1. Then, one has $a_1(b_1 + b_2) + a_2(b_1 - b_2) \leq 2$, because either $b_1 = b_2$ or $b_1 = -b_2$. If one assumes that 'real', pre-assigned values of such outcomes actually exist, and that the probabilistic character of the observables is just due to the observers' ignorance about the actual state of a hypothetical field of 'hidden variables', then the inequality we just wrote on the outcomes must hold under the average over such hidden variables. One thus gets the celebrated CHSH (Clauser, Horne, Shimony, Holt) inequality:

$$\langle \hat{A}_1 \hat{B}_1 \rangle + \langle \hat{A}_1 \hat{B}_2 \rangle + \langle \hat{A}_2 \hat{B}_1 \rangle - \langle \hat{A}_2 \hat{B}_2 \rangle \leq 2 \,, \qquad (7.120)$$

which is probably the most common formulation of a Bell test. Quantum mechanics violate this inequality: under a proper choice of observables, strongly entangled quantum mechanical states allow for correlations that may reach the value $2\sqrt{2}$ for the sum on the left hand side (the so called 'Tsirelson bound'). The well-known solution to this apparent incongruity is that quantum mechanics does not admit a realistic description in terms of hidden variables. We do not intend here to engage in a foundational analysis, but just to emphasise that nonlocality tests are the most distinctive feature of quantum mechanics, the one that calls for a dramatic departure from classical determinism, in that they reveal that mere ignorance – about hypothetical, hidden degrees of freedom – is incapable of accounting for the strength of quantum correlations. This strongly suggests that any technological exploitation of quantum correlations that aims to grant a true quantum advantage over competing classical schemes must be related to the possibility of the system in use to violate Bell inequalities like (7.120).

This leads us back to our continuous variable systems. As already stated more than once, Gaussian states cannot violate Bell inequalities if one restricts to general-dyne measurements. This may be understood through a reasoning that goes back to Bell himself. In Chapter 5 we saw that all such measurements are equivalent, up to the addition of ancillary modes, to homodyne measurements.[12] But, Eq. (4.42) shows that homodyne statistics are described by marginals of the Wigner function (i.e., by its integration along the phase space variables that are not being measured). Notice now that a positive Wigner function is an actual classical probability distribution (since all Wigner functions are normalised). Thus, any general-dyne Bell test with states with a positive Wigner function (like Gaussian states, but not only) admits a natural, realistic hidden variable description, the hidden variables being just the phase space variables that are not measured. *No such state will ever be able to violate any general-dyne Bell test.*

This situation can be addressed in two ways: by employing non-Gaussian measurements on Gaussian states, or by employing non-Gaussian states. In theoretical terms, the validity of both approaches is somewhat trivial: in fact, the Hilbert space does not care about the Gaussian or non-Gaussian character of a state. If one is allowed any measurement, the strongly entangled coherent superposition $|\psi_r\rangle = \sqrt{1 - \beta^2} \sum_{n=0}^{\infty} \beta^n |n, n\rangle$ (a Gaussian two-mode squeezed state with $\tanh(r) = \beta$) is bound to exhibit strong non-locality. A classic choice of operators to test entanglement with such Gaussian states is the 'displaced parity' test, where A and B can choose between the two dichotomic observables $(-1)^{\hat{a}^\dagger \hat{a}}$ and $\hat{D}_{\mathbf{r}}(-1)^{\hat{a}^\dagger \hat{a}} \hat{D}_{\mathbf{r}}^\dagger$ for a displacement operator $\hat{D}_{\mathbf{r}}^\dagger$, with outcomes ∓ 1 (note that $\mathbf{r}$ must be chosen large enough for the two observables to be distinguishable). Unfortunately, observables like the parity one are very difficult to implement in practice. Nonetheless, this simple argument should dispel any doubt, if there ever was one, concerning the genuine quantum nature of Gaussian states. The case for pursuing Gaussian quantum technologies will be further strengthened in the next chapter, where we will review a few convincing technological applications of such states in the quantum domain (most notably quantum key distribution, whose security hinges on the uncertainty principle, a genuine quantum effect applying to the Gaussian domain).

While the theoretical question concerning the, so to speak, proof of principle possibility of violating Bell inequalities was deemed trivial above, the

---

[12]This is the case since all such measurements are equivalent to a combination of deterministic Gaussian operations and heterodyne detection, which in turn corresponds to the homodyne detection of two combined quadratures after mixing with the (Gaussian) vacuum state; see Sections 5.4.2 and 5.4.4.

practical question of identifying achievable resources to that aim *when measurement schemes are fixed* is everything but. In order to demonstrate nonlocality with accessible technology, it would be desirable to do it with homodyne tests, given the reliability and high efficiency of such detection schemes. As was just stated, to this aim one needs to employ states with a negative Wigner function. The photon subtracted two-mode squeezed states $|\zeta_r\rangle$, defined in the previous section, are an example of such states within experimental reach. It can be shown that, if the homodyne measurements are dichotomised by simply assigning the value $+1$ to all positive quadrature readings and -1 to all negative values (note that infinite variations are possible in this regard), then the state $|\zeta_r\rangle$ does allow for the violation of inequality (7.120).

As a final remark, let us mention that this application of photon subtracted states exemplifies very well the value of non-Gaussian resources, which are clearly much sought after in the laboratory. Their dynamical generation requires strong nonlinearities or anharmonicities, which can be realised, among other ways, by strongly coupling the continuous variables to finite dimensional systems. In Chapter 9 we will discuss, as an example, the creation of a non-Gaussian cat-like state (which already made appearances in Problem 4.3 and Section 7.6). Alternately, one can rely on non-Gaussian detection events, as in the photon added or subtracted states. As we mentioned in the previous section, such probabilistic techniques also allow one to generate single-photon states too: in Section 9.1.6, we will study a scheme, also probabilistic, to entangle two such states to form another type of non-Gaussian entangled resource.

7.9 FURTHER READING

The separability problem for two-mode Gaussian states was solved by Simon [80], in a classic paper whose approach we followed rather closely. The necessity of PPT for the separability of 1 vs. n mode Gaussian states was first established in [86] (where the separability criterion (7.13) is also derived and bound entangled states of 2 vs. 2 modes are explicitly constructed), albeit here we followed the treatment of [58], which also contains an account of locally symmetric (termed "mono-symmetric" there) and isotropic states. The equivalence of the latter to the tensor product of two-mode squeezed states was first pointed out by Botero and Reznik [11], who proved it directly at the Hilbert space level through the Schmidt decomposition, while the phase space approach reported here is analogous to the content of [36].

The notion of negativity and of logarithmic negativity emerged in several slight variations in the late nineties. The form adopted here was established in [82], where it is also evaluated for Gaussian states. Such a measure was proven to be monotone under LOCC in [70]. Optimal passive entangling transformations are discussed in [91]. The entanglement of formation, an entanglement monotone not included in our coverage, was determined for symmetric two-mode Gaussian states in [37].

The Gaussian version of quantum discord was introduced in [38] and [1], evaluated for all Gaussian states in the latter, and shown to be the actual quantum discord of a family of states including thermal two-mode squeezed states in [69].

The impossibility of Gaussian distillation with Gaussian devices was simultaneously revealed by the papers [35, 25, 27]. Here, we followed the treatment given in [35], from which what we referred to as the 'Giedke–Cirac' quantifier was also taken. The entanglement distillation protocol we presented was introduced in [15].

The separability criteria based on higher moments that we discussed were derived in [78].

A systematic analysis of entanglement enhancement via photon addition and subtraction is contained in [62]. More details on photon subtracted states may also be found in [26], which also contains a quantitative survey on the violation of Bell inequalities with continuous variables. The violation of Bell inequalities with photon subtracted states and the dichotomised homodyne scheme was first suggested and analysed in [63, 30]. Bell's original argument on positive Wigner functions is reproduced as Chapter 21 of his classic volume of collected papers [8].

V

Technologies

CHAPTER 8

Quantum information protocols with continuous variables

CONTENTS

The theoretical toolbox we have developed as the book unfolded will now be put to use to describe a series of tasks, related to the transmission or extraction of information, which may be implemented with quantum continuous variable systems. We will open with quantum teleportation, the transmission of a quantum state through shared quantum entanglement and classical communication, and determine certain thresholds it must attain to be considered genuinely quantum; we shall then proceed to establish the maximum classical communication capacity of a class of Gaussian channels; next, the optimal estimation of a dynamical parameter using Gaussian states as probes will be considered; finally, our enquiry will close with an illustration of the fact that Gaussian states may be used for the secure distribution of a cryptographic key. It is worth mentioning that, in discussing teleportation thresholds and communication capacities, although the optimal schemes we shall identify will turn out to be Gaussian, we will confront full optimisations at the Hilbert space level, and will thus introduce methods that go beyond the Gaussian description.

Gaussian quantum key distribution has already been commercialised, and all the other techniques we will touch on in this chapter are gradually approaching the status of full-blown engineering technologies, which would arguably form the most convincing empirical evidence one may have in support of the underlying physical theory.

8.1 QUANTUM TELEPORTATION

Quantum teleportation is a protocol where a quantum state is transmitted between two distant parties, say A and B, through classical communication and previously established quantum entanglement. If A were sending the state ϱ_0, embodied in system C, via the shared entangled state ϱ_{AB}, an ideal quantum teleportation protocol would go as follows:

1. A projectively measures the input system C as well as its share of the entangled state ϱ_{AB} in a maximally entangled basis $\{|\lambda\rangle_{AC}, \lambda\}$, labelled by λ.

2. A communicates the classical measurement outcome λ to B.

3. B performs a unitary operation $\hat{U}_\lambda$ depending on λ on its share of the original entangled state.

It may be shown that, under ideal conditions, the unitary $\hat{U}_\lambda$ may be chosen so that the final state of B is ϱ_0 with probability 1. Notice that the measurement performed by A, besides, in a loose sense, 'comparing' systems A and C, also disentangles the state of system B from the rest of the system. Also, let us remark that the notion of the final unitary implemented by B being informed by a previous measurement on a different system is sometimes referred to as 'feedforward'.

Quantum teleportation is a fundamental primitive of quantum communication, and might end up playing a direct role in quantum computing too, where coherent quantum information (i.e., arbitrary quantum states) will need to be moved around a processor in much the same way as classical information is transferred along the wired buses of a classical computer. In these contexts, quantum teleportation, even in the presence of noise, might arguably offer a cleaner transfer than quantum wires – chains of interacting quantum systems, such as spin chains or, for continuous variables, harmonic chains.[1]

In quantum communication scenarios, quantum teleportation requires the establishment of high-quality, long-distance entanglement. The latter may be achieved, in realistic noisy environments, by the procedure of entanglement swapping in networks of 'quantum repeaters' where, in order to establish long-distance entanglement between two extremal nodes, short-distance entanglement is first generated among nearest-neighbouring nodes in a network, and then the entanglement is swapped to the extremal nodes by a set of measurements in entangled bases at all the intervening nodes. When supplemented with entanglement distillation which, as we saw in Section 7.5, just requires a classical communication channel among the nodes, this procedure allows one in principle to establish highly entangled states across long distances and hence to attain the transmission of quantum information via teleportation.

It is hence clear that the establishment of valuable long-distance entanglement in quantum repeaters must rely on redundancy and probabilistic protocols (this was also highlighted in our discussion of entanglement distillation in the previous chapter). In turn, the latter are only viable if multiple copies of the quantum states may be stored and accessed on demand: in other words, efficient quantum communication begs for efficient quantum memories. Incidentally, certain aspects of the theoretical treatment of continuous variable quantum teleportation that we will develop below may be carried over to assess the efficiency of quantum memories.

Without further ado, let us now go back to the quantum teleportation scenario. We are interested in the case where the three systems involved are continuous variable modes. In Chapter 5, we have identified a class of Gaussian maximally entangled states, with covariance matrix given by Eq. (5.149) in the limit $r \to \infty$ and, in describing the heterodyne detection scheme, we have also learned how to realise the projection on such maximally entangled states: party A has to mix systems A and C at a 50:50 beam splitter and then perform the homodyne detection of complementary quadratures, thus realising the measurement of the quadratures $\hat{x}_C - \hat{x}_A$ and $\hat{p}_C + \hat{p}_A$. As can be seen from Eqs. (5.123) and (5.124), this procedure realises a projection on the orthogonal set $\hat{D}_{\mathbf{r}}^{\dagger}|\delta\rangle$; here, $|\delta\rangle$ is the two-mode Gaussian state with covariance matrix given by Eq. (5.21) for $r \to \infty$, $\hat{D}_{\mathbf{r}}$ is a displacement that acts only on system C, and the two-dimensional real vector $\mathbf{r}$ is the measurement outcome. Let us remind the reader that, in the Fock basis, one has $|\delta\rangle = \lim_{r \to \infty} \cosh(r)^{-1} \sum_{j=0}^{\infty} \tanh(r)^j |j, j\rangle$. Note that, up to the deviations of actual homodyne detections from ideal ones (mainly due to the non-infinite

---

[1]See Problem 3.4 for the theoretical description of a harmonic chain. In practice, an example of a harmonic chain would be an array of trapped ions, where the canonical variables describing the motion of the ions interact through linearised Coulomb couplings (see Section 9.4); coupled nanomechanical oscillators are another possible candidate system to this aim. Notice that anharmonic (higher-order) couplings may well be of interest too in this context.

laser power available in practice), which we shall disregard in the present analysis, this projection can be realised in the limit $r \to \infty$.

Hypothetically, if the shared entangled state ϱ_{AB} were the infinitely entangled state $|\delta\rangle$, perfect teleportation of any state could be achieved by having B perform a simple displacement operation. It is interesting to look into this aspect: we will just need to prove it holds for a generic pure state expanded in the Fock basis as $|\varphi\rangle = \sum_{j=0}^{\infty} \varphi_j |j\rangle$, and the same will follow for mixed states by the linearity of the process and convexity of the state space. Let us in the following disregard normalisation, and just determine the final Hilbert space vector associated with system B, assuming projection on the eigenstate $\hat{D}_{\mathbf{r}}^{\dagger}|\delta\rangle$. Let $|B\rangle$ be the conditional, unnormalised vector of B after the measurement:

$$
\begin{aligned}
|B\rangle\langle B| &= \sum_{\substack{h,j,k,l, \\ m,n=0}}^{\infty} {}_{CA}\langle m,m|\hat{D}_{\mathbf{r}}(\varphi_h \varphi_j^*|h\rangle\langle j|_C \otimes |k,k\rangle\langle l,l|_{AB})\hat{D}_{\mathbf{r}}^{\dagger}|n,n\rangle_{CA} \\
&= \sum_{h,j,k,l=0}^{\infty} \langle k|\hat{D}_{\mathbf{r}}|h\rangle\langle j|\hat{D}_{\mathbf{r}}^{\dagger}|l\rangle\varphi_h \varphi_j^*|k\rangle\langle l|,
\end{aligned}
\tag{8.1}
$$

where, in the first line, we labelled all operators with the system they act upon (A, B, C, or combinations thereof), with the exception of $\hat{D}_{\mathbf{r}}$, which was defined as acting on C. The last equation above yields

$$
|B\rangle = \sum_{h,k=0}^{\infty} \langle k|\hat{D}_{\mathbf{r}}|h\rangle\varphi_h|k\rangle = \sum_{h,k=0}^{\infty} |k\rangle\langle k|\hat{D}_{\mathbf{r}}|h\rangle\varphi_h = \sum_{h=0}^{\infty} \hat{D}_{\mathbf{r}}|h\rangle\varphi_h = \hat{D}_{\mathbf{r}}|\varphi\rangle.
$$
$$\tag{8.2}$$

Hence, as we stated above, all B would have to do to wrap up the teleportation and recover the initial state perfectly is displace its state with $\hat{D}^{\dagger}\mathbf{r}$.

8.1.1 Quantum teleportation of Gaussian states

However, infinite entanglement cannot be achieved in practice. Therefore, we shall study quantum teleportation in the more realistic case where A and B share a two-mode squeezed state with finite squeezing parameter r, which certainly represents the more stringent practical limitation to the protocol. Also, we shall restrict to input Gaussian states. To be clear, let us explicitly state the continuous variable version of the teleportation protocol: A intends to send the Gaussian state ϱ_G, with covariance matrix $\boldsymbol{\sigma}_0$ and first moments $\mathbf{r}_0$, to B. A and B share a Gaussian entangled state with covariance matrix $\boldsymbol{\sigma}_r$ and null first moments.

1. By double homodyning, A projects the state ϱ_G and its share of the entangled state on the state $\hat{D}_{\mathbf{r}}^{\dagger}|\delta\rangle$.

2. A sends to B the measurement outcome $\mathbf{r}$.

3. B applies a unitary displacement with parameter $\mathbf{f}(\mathbf{r})$ – function of $\mathbf{r}$ – to its share of the original entangled state.

The initial Gaussian state of the system has the following second and first moments:

$$\begin{pmatrix} c\mathbb{1}_2 & s\sigma_z & 0 \\ s\sigma_z & c\mathbb{1}_2 & 0 \\ 0 & 0 & \sigma_0 \end{pmatrix}, \quad \begin{pmatrix} 0 \\ 0 \\ \mathbf{r}_0 \end{pmatrix}, \tag{8.3}$$

where $c = \cosh(2r)$, $s = \sinh(2r)$ and we have placed the systems in the order B, A, C (this will make the application of the formulae from Chapter 5 easier). We now need to determine the conditional state of B after systems AC are projected on the Gaussian state $\hat{D}_{\mathbf{r}}^{\dagger}|\delta\rangle$, with second and first moments given by

$$\sigma_m = \lim_{c' \to \infty} \begin{pmatrix} c'\mathbb{1}_2 & c'\sigma_z \\ c'\sigma_z & c'\mathbb{1}_2 \end{pmatrix}, \quad \mathbf{r}_m = \begin{pmatrix} 0 \\ \mathbf{r} \end{pmatrix} \tag{8.4}$$

(where the fact that $\lim_{r' \to \infty} \cosh(2r') = \lim_{r \to \infty} \sinh(2r') = \lim_{c' \to \infty} c'$ was incorporated). We can now straightforwardly apply the formulae (5.142) and (5.143) to get the conditional first and second moments σ_B and $\mathbf{r}_B$ of B. The inversion of the matrix $(c\mathbb{1} \oplus \sigma_0) + \sigma_m$ may be handled through the Schur complement formula (1.8), and simplifies considerably, in the limit $c' \to \infty$, because of Eq. (5.157):

$$[(c\mathbb{1} \oplus \sigma_0) + \sigma_m]^{-1} = \begin{pmatrix} (c\mathbb{1}_2 + \sigma_z\sigma_0\sigma_z)^{-1} & \sigma_z(c\mathbb{1}_2 + \sigma_0)^{-1} \\ (c\mathbb{1}_2 + \sigma_0)^{-1}\sigma_z & (c\mathbb{1}_2 + \sigma_0)^{-1} \end{pmatrix}. \tag{8.5}$$

One thus obtains:

$$\sigma_B = c\mathbb{1}_2 - s^2(c\mathbb{1}_2 + \sigma_0)^{-1}, \quad \mathbf{r}_B = -s(c\mathbb{1}_2 + \sigma_0)^{-1}(\mathbf{r} - \mathbf{r}_0). \tag{8.6}$$

As one should expect, letting $r \to \infty$ (infinite shared entanglement, with $c \to \infty$ and $c/s \to 1$), the insertion of Eq. (5.157) yields the limits $\sigma_B \to \sigma_0$ and $\mathbf{r}_B \to (\mathbf{r}_0 - \mathbf{r})$. In this limit, as was already shown for general states above, the implementation of a displacement $\hat{D}_{\mathbf{r}}^{\dagger}$ on B's part realises a perfect teleportation of any Gaussian input state (note that the action of $\hat{D}_{\mathbf{r}}^{\dagger}$ on a state from the left shifts its first moments by $+\mathbf{r}$).

Inserting the bottom right block of the matrix (8.5) into Eq. (5.139) determines the probability density of the measurement outcomes $\mathbf{r}$ (in $d\mathbf{r}$)

$$p(\mathbf{r}) = \frac{e^{-(\mathbf{r}-\mathbf{r}_0)^{\mathsf{T}}(c\mathbb{1}_2 + \sigma_0)^{-1}(\mathbf{r}-\mathbf{r}_0)}}{\pi\sqrt{\mathrm{Det}(c\mathbb{1}_2 + \sigma_0)}} = \frac{e^{-(\mathbf{r}-\mathbf{r}_0)^{\mathsf{T}}\gamma^{-1}(\mathbf{r}-\mathbf{r}_0)}}{\pi\sqrt{\mathrm{Det}\gamma}}, \tag{8.7}$$

which will be useful presently. Given its central role in the following, we also defined the matrix $\gamma = (c\mathbb{1}_2 + \sigma_0)$.

Even for finite r, the inversion of the 2×2 matrix $\boldsymbol{\gamma} = (c\mathbb{1}_2 + \boldsymbol{\sigma}_0)$ can be worked out explicitly:

$$(c\mathbb{1}_2 + \boldsymbol{\sigma}_0)^{-1} = \frac{c\mathbb{1}_2 + \Omega \boldsymbol{\sigma}_0 \Omega^{\mathsf{T}}}{c^2 + \mathrm{Tr}(\boldsymbol{\sigma}_0)c + \mathrm{Det}(\boldsymbol{\sigma}_0)} \, , \tag{8.8}$$

where Ω is the anti-symmetric form of Eq. (3.2) in dimension 2. Assuming now that B enacts a displacement operator with parameter $\mathbf{f}(\mathbf{r})$ which depends on the feedforward measurement outcome $\mathbf{r}$, one obtains the statistics of the teleported state as

$$\boldsymbol{\sigma}_B = c\mathbb{1}_2 - s^2 \frac{c\mathbb{1}_2 + \Omega \boldsymbol{\sigma}_0 \Omega^{\mathsf{T}}}{c^2 + \mathrm{Tr}(\boldsymbol{\sigma}_0)c + \mathrm{Det}(\boldsymbol{\sigma}_0)} \, , \tag{8.9}$$

$$\mathbf{r}_B = -s \frac{c\mathbb{1}_2 + \Omega \boldsymbol{\sigma}_0 \Omega^{\mathsf{T}}}{c^2 + \mathrm{Tr}(\boldsymbol{\sigma}_0)c + \mathrm{Det}(\boldsymbol{\sigma}_0)} (\mathbf{r} - \mathbf{r}_0) + \mathbf{f}(\mathbf{r}) \, . \tag{8.10}$$

Clearly, the teleportation is not perfect in this case. In general, B is aware of the measurement outcome $\mathbf{r}$, but does not know anything about $\boldsymbol{\sigma}_0$ or $\mathbf{r}_0$. If B is restricted to displacement operators, it is not even immediately obvious how to identify an optimal displacement $\mathbf{f}(\mathbf{r})$, as a function of $\mathbf{r}$, to best approximate the input state.

We need a figure of merit to provide B with some guidance here. We shall adopt the overlap $F = \mathrm{Tr}(\varrho_0 \varrho_{tel})$ between initial and final states ϱ_0 and ϱ_{tel} as a benchmark to assess the quality of the teleportation protocol which, with a marginal abuse of terminology, we shall term the 'teleportation fidelity' F. Notice that this figure of merit only corresponds to the – generally better behaved – Uhlmann fidelity $\mathrm{Tr}[\sqrt{\sqrt{\varrho_0} \varrho_{tel} \sqrt{\varrho_0}}]$ between quantum states only if the input state is pure. This is not a major concern to us. For one thing, the Uhlmann fidelity is also not endowed with a direct, general operational interpretation.[2] Besides, and more importantly, even if the overlap $\mathrm{Tr}(\varrho_1 \varrho_2)$ is not per se a good measure of similarity for two mixed states (which should be clear, since $\mathrm{Tr}(\varrho_1^2) \neq 1$ for a mixed state ϱ_1), which is what one deals with in any realistic, noisy situation, it can still be reconstructed from raw data in several notable teleportation experiments, and thus serves as a perfectly adequate benchmark, as will be made apparent in the next section. The easily computable overlap is thus an ideal compromise in this situation, as it allows for a detailed theoretical analysis and is also applicable to experiments. We will hence make use of the expression for the overlap between two Gaussian states, which has been evaluated in Eq. (4.51). As mentioned above, if one of the two states is pure the overlap yields the actual quantum fidelity between the states.

Eq. (4.51) may be applied to Eqs. (8.9) and (8.10) to assess the overlap

---

[2] In this regard, one may rather want to consider the trace norm distance $\|\varrho_0 - \varrho_{tel}\|_1$, that is operationally meaningful in terms of distinguishability of the states involved. Like the Uhlmann fidelity, though, this quantifier would make our analysis considerably heavier, if at all possible.

between the teleported state and the initial state ϱ_0, with covariance matrix $\boldsymbol{\sigma}_0$ and first moments $\mathbf{r}_0$, thus obtaining the teleportation fidelity F. Let us assess what the fidelity F would be if B chose $\mathbf{f}(\mathbf{r}) = \mathbf{r}$, which we know to be optimal when infinite entanglement is shared, getting

$$F = \frac{2\,e^{-(\mathbf{r}-\mathbf{r}_0)^{\mathsf{T}}\frac{(\mathbb{1}_2 - s\boldsymbol{\gamma}^{-1})^2}{\boldsymbol{\gamma} - s^2\boldsymbol{\gamma}^{-1}}(\mathbf{r}-\mathbf{r}_0)}}{\sqrt{\mathrm{Det}(\boldsymbol{\gamma} - s^2\boldsymbol{\gamma}^{-1})}}, \tag{8.11}$$

where $\boldsymbol{\gamma} = (c\mathbb{1}_2 + \boldsymbol{\sigma}_0)$ (in simplifying a seemingly convoluted expression, we just made use of the fact that functions of the same matrix commute with each other). This choice of feedforward is particularly favourable since, as may be seen from Eq. (8.7), the probability density $p(\mathbf{r})$ for outcome $\mathbf{r}$ also depends only on $(\mathbf{r} - \mathbf{r}_0)$. This means that the teleportation fidelity averaged over the measurement outcomes, $\langle F \rangle$, a very relevant figure of merit in experiments, is independent from the first moments of the teleported state $\mathbf{r}_0$:

$$
\begin{aligned}
\langle F \rangle &= \int_{\mathbb{R}^2} d\mathbf{r} \, \frac{2\,e^{-(\mathbf{r}-\mathbf{r}_0)^{\mathsf{T}}\left(\frac{(\mathbb{1}_2 - s\boldsymbol{\gamma}^{-1})^2}{\boldsymbol{\gamma} - s^2\boldsymbol{\gamma}^{-1}} + \boldsymbol{\gamma}^{-1}\right)(\mathbf{r}-\mathbf{r}_0)}}{\pi\sqrt{\mathrm{Det}\boldsymbol{\gamma}}\sqrt{\mathrm{Det}(\boldsymbol{\gamma} - s^2\boldsymbol{\gamma}^{-1})}} \\
&= \int_{\mathbb{R}^2} d\mathbf{r} \, \frac{2\,e^{-\mathbf{r}^{\mathsf{T}}\frac{2(\boldsymbol{\gamma} - s\mathbb{1}_2)}{\boldsymbol{\gamma}^2 - s^2\mathbb{1}_2}\mathbf{r}}}{\pi\sqrt{\mathrm{Det}(\boldsymbol{\gamma}^2 - s^2\mathbb{1}_2)}} = \frac{1}{\sqrt{\mathrm{Det}(\boldsymbol{\gamma} - s\mathbb{1}_2)}} = \frac{1}{\sqrt{\mathrm{Det}(\boldsymbol{\sigma}_0 + e^{-2r}\mathbb{1}_2)}}.
\end{aligned}
\tag{8.12}
$$

These findings may be summarised as follows:

Average teleportation fidelity of a Gaussian state. The average overlap $\langle F \rangle$ between an input Gaussian state with statistical moments $\mathbf{r}_0$ and $\boldsymbol{\sigma}_0$ and its teleportation output, obtained through the protocol described on page 226 with $\mathbf{f}(\mathbf{r}) = \mathbf{r}$, is given by

$$\langle F \rangle = \frac{1}{\sqrt{\mathrm{Det}(\boldsymbol{\sigma}_0 + e^{-2r}\mathbb{1}_2)}}, \tag{8.13}$$

where r is the two-mode squeezing parameter of the shared entangled state.

If the limit of infinite shared entanglement ($r \to \infty$), and pure initial Gaussian state ($\mathrm{Det}\boldsymbol{\sigma}_0 = 1$), the ideal average teleportation fidelity of 1 is indeed approached, as one should expect. Notice also that, as we saw in Section 7.4, the quantity e^{-2r} coincides with the smallest symplectic eigenvalue $\tilde{\nu}_-$ of the entangled state $|\psi\rangle_r$, so that in this case the average teleportation fidelity $\langle F \rangle = \mathrm{Det}(\boldsymbol{\sigma}_0 + \tilde{\nu}_-\mathbb{1}_2)^{-1/2}$, being monotone with the logarithmic negativity, provides one with a direct operational characterisation of the quantum entanglement of the shared resource.

Problem 8.1. (*Modified teleportation*). In the slightly more lenient scenario where $\boldsymbol{\sigma}_0$ is supposed to be known (and the only teleported coherent information is encoded in the unknown $\mathbf{r}_0$), one may choose the different feedforward $\mathbf{f}(\mathbf{r}) = s\frac{c\mathbb{1}_2+\Omega\boldsymbol{\sigma}_0\Omega^\mathsf{T}}{c^2+\mathrm{Tr}(\boldsymbol{\sigma}_0)c+\mathrm{Det}(\boldsymbol{\sigma}_0)}\mathbf{r}$, which gets rid of the contribution of $\mathbf{r}$ to the first moments, and thus makes the teleported state independent from the measurement outcome. Assume $\boldsymbol{\sigma}_0 = \mathbb{1}_2$ and derive the teleportation fidelity F and average teleportation fidelity $\bar{F}$ in this instance.

Solution. Replacing $\boldsymbol{\sigma}_0$ with $\mathbb{1}_2$ in Eq. (8.9) yields the output covariance matrix $\mathbb{1}_2$: the covariance matrix of coherent states is teleported faithfully. Cancelling the contribution proportional to $\mathbf{r}$ in Eq. (8.10) gives first moments that can be written as $s\boldsymbol{\gamma}^{-1}\mathbf{r}_0 = s(c\mathbb{1}_2+\boldsymbol{\sigma}_0)^{-1}\mathbf{r}_0 = \mathbf{r}_0 s/(c + 1)$. This should be compared with the initial state with covariance matrix $\boldsymbol{\sigma}_0$ and first moments $\mathbf{r}_0$ through the formula (4.51):

$$F = \mathrm{e}^{-\frac{1}{2}\left(\frac{c+1-s}{c+1}\right)^2|\mathbf{r}_0|^2} . \tag{8.14}$$

Since the final teleported state, as well as the teleportation fidelity, do not depend on the measurement outcome $\mathbf{r}$ at all, it must be $F = \langle F \rangle$.

8.1.2 Classical threshold for coherent states

In practice, it is customary to assess the effectiveness of a quantum teleportation protocol by comparing its performance, in terms of the average overlap $\bar{F}$, against the so called "classical" threshold. The latter is evaluated by optimising the output fidelity, over a given input alphabet, of a "measure-and-prepare" channel (or, equivalently, of an *entanglement breaking* channel, as explained in Section D.2 of Appendix D). This corresponds to a situation where no entanglement is shared between the two parties and all the sender can do is measure its copy of the input state through a POVM and send the outcome to the receiver through the classical channel; the receiver will then prepare a state dependent on such an outcome. In formulae, a generic measure-and-prepare channel Φ may be defined as

$$\Phi(\varrho) = \sum_\mu \varrho_\mu \, \mathrm{Tr}(\hat{K}_\mu^\dagger \hat{K}_\mu \varrho) , \tag{8.15}$$

for quantum states ϱ_μ and a POVM such that $\sum_\mu \hat{K}_\mu^\dagger \hat{K}_\mu = \hat{\mathbb{1}}$.[3]

---

[3] Notice that these strategies are not classical in the sense that they are described by the classical limit of quantum mechanics, as their optimisation does require the full formalism of quantum mechanics. The term refers here to the absence of entanglement and hence quantum non-locality as an available resource, a facet that represents the fundamental distinction between classical and quantum mechanics, in the sense that these procedures may

The classical threshold does depend on the 'alphabet' (i.e., the distribution) of input quantum states which, in any practical proof of principle demonstration of quantum teleportation, would have to be specified. Although more general scenarios might be considered, we shall specialise our treatment to the case of coherent states, following an argument by Hammerer, Wolf, Polzik and Cirac. This will allow us to illustrate some of the techniques involved in this line of inquiry in a simpler context, which played a prominent role historically, because of the ready availability of coherent states.

By setting $\sigma_0 = \mathbb{1}_2$ – recall that coherent states may be defined as the class of Gaussian states with covariance matrix equal to that of the vacuum state – Eq. (8.13) yields

$$\langle F \rangle = \frac{1}{1 + e^{-2r}} \ . \tag{8.16}$$

The quantum teleportation overlap averaged over the measurement outcomes, $\langle F \rangle$, does not depend on the input coherent state at all, and hence will not depend on how such states are distributed. It is hence the quantity that should be compared with the classical threshold averaged over the distribution of initial states, which we shall term $\bar{F}_{th}$.

Let us then assume a normal distribution of single-mode input coherent states $\{|\alpha\rangle\}$, each prepared with probability $p(\alpha) = \frac{\lambda}{\pi} e^{-\lambda|\alpha|^2}$, and define

$$\bar{F}_{th} = \sup_{\Phi} \int_{\mathbb{C}} d^2\alpha \, p(\alpha) \langle \alpha | \Phi(|\alpha\rangle\langle\alpha|) | \alpha \rangle \ , \tag{8.17}$$

where the supremum is taken over all measure-and-prepare channels of the form of Eq. (8.15).

Let us also recall that, according to our notation, $|\alpha\rangle = \hat{D}_{-\mathbf{r}}|0\rangle$ for $\hat{D}_{-\mathbf{r}} = e^{i\mathbf{r}^\mathsf{T}\Omega^\mathsf{T}\hat{\mathbf{r}}}$, with $\alpha = \frac{x+ip}{\sqrt{2}}$ and $\mathbf{r} = (x,p)^\mathsf{T}$, and proceed to prove the following:

Classical threshold for coherent states. The optimal average fidelity $\bar{F}_{th}$ achieved by a measure-and-prepare strategy over a normal distribution $p(\alpha) = \frac{\lambda}{\pi} e^{-\lambda|\alpha|^2}$ of coherent states $\{|\alpha\rangle\}$ is

$$\bar{F}_{th} = \frac{1+\lambda}{2+\lambda} \ , \tag{8.18}$$

and is attained by the measure-and-prepare heterodyne protocol

$$\Phi_{het}(\varrho) = \frac{1}{\pi} \int_{\mathbb{C}} d^2\mu \, \langle \mu | \varrho | \mu \rangle | \frac{\mu}{\lambda+1} \rangle \langle \frac{\mu}{\lambda+1} | , \tag{8.19}$$

consisting of the heterodyne detection of the input state followed by the preparation of the coherent state $|\frac{\mu}{\lambda+1}\rangle$ upon the heterodyne outcome μ.

---

be mocked up with classical means. While such classical protocols can in principle teleport classical probability distributions perfectly – since classical states may be distinguished deterministically, their effectiveness is limited when quantum states are to be transmitted.

Proof. In order to prove Eq. (8.18) we will first upper bound the classical threshold by the right hand side, and then show that such a bound is attained by the heterodyne measure-and-prepare channel Φ_{het}.

First, let us note that, by diagonalising the positive operators ϱ_μ and $\hat{K}_\mu^\dagger \hat{K}_\mu$ and properly redefining indexes, one can express the most general measure-and-prepare channel of Eq. (8.15) as

$$\Phi(\varrho) = \sum_j |v_j\rangle\langle v_j| \langle w_j|\varrho|w_j\rangle \,, \qquad (8.20)$$

where the vectors $\{|v_j\rangle\}$ are normalised; the vectors $\{|w_j\rangle\}$ are not normalised but satisfy the POVM condition

$$\sum_j |w_j\rangle\langle w_j| = \mathbb{1} \,. \qquad (8.21)$$

Notice that the sum over j is generic and generalisable to an integral.

It will be convenient, in order to make ready use of previous findings from the book, to switch to the variables $\mathbf{r} = (x, p)^\mathsf{T}$ such that $\alpha = \frac{x+ip}{\sqrt{2}}$, with a probability distribution of the input states given by $p(\mathbf{r})d\mathbf{r} = \frac{\lambda}{2\pi}e^{-\frac{\lambda}{2}|\mathbf{r}|^2}d\mathbf{r}$.

The classical threshold which we intend to determine reads

$$\bar{F}_{th} = \sup_{\substack{\{|v_j\rangle\} \\ \{|w_j\rangle\}}} \sum_j \int_{\mathbb{R}^2} d\mathbf{r} \, p(\mathbf{r}) |\langle v_j|\hat{D}_\mathbf{r}|0\rangle|^2 |\langle w_j|\hat{D}_\mathbf{r}|0\rangle|^2 \,, \qquad (8.22)$$

where the supremum is taken over all possible sets of normalised Hilbert space vectors $\{|v_j\rangle\}$ and non-normalised $\{|w_j\rangle\}$ that fulfil Eq. (8.21).

Given a vector $|w_j\rangle$, define now the associated operator

$$\hat{A}_{|w_j\rangle} = \int_{\mathbb{R}^2} d\mathbf{r} \, p(\mathbf{r}) |\langle w_j|\hat{D}_\mathbf{r}|0\rangle|^2 \hat{D}_\mathbf{r}|0\rangle\langle 0|\hat{D}_\mathbf{r}^\dagger \,, \qquad (8.23)$$

which allows us to recast the threshold as

$$\bar{F}_{th} = \sup_{\{|w_j\rangle\}} \sum_j \|\hat{A}_{|w_j\rangle}\|_\infty \qquad (8.24)$$

(let us remind the reader that the operator norm $\|\hat{A}_{|w_j\rangle}\|_\infty$ equals the largest eigenvalue of the positive operator $\hat{A}_{|w_j\rangle}$). This implicitly takes care of the optimisation over the $\{|v\rangle_j\}$: the optimal choice for each j is given by the normalised eigenvector of $\hat{A}_{|w_j\rangle}$ corresponding to its largest eigenvalue. The desired upper bound to the classical threshold may then be obtained by proving the inequality

$$\|\hat{A}_{|w_j\rangle}\|_\infty \le \frac{1+\lambda}{2+\lambda}\|\hat{A}_{|w_j\rangle}\|_1 \tag{8.25}$$

which, if applied to Eq. (8.24), leads to

$$\bar{F}_{th} \le \frac{1+\lambda}{2+\lambda} \sup_{\{|w_j\rangle\}} \sum_j \|\hat{A}_{|w_j\rangle}\|_1 = \frac{1+\lambda}{2+\lambda}, \tag{8.26}$$

since (recall that the operator norm $\|\cdot\|_1$ of a positive operator is just its trace)

$$\sum_j \|\hat{A}_{|w_j\rangle}\|_1 = \int_{\mathbb{R}^2} d\mathbf{r}\, p(\mathbf{r}) \sum_j |\langle w_j|\hat{D}_\mathbf{r}|0\rangle|^2 \mathrm{Tr}\left[\hat{D}_\mathbf{r}|0\rangle\langle 0|\hat{D}_\mathbf{r}^\dagger\right]$$

$$= \int_{\mathbb{R}^2} d\mathbf{r}\, p(\mathbf{r})\langle 0|\hat{D}_\mathbf{r}^\dagger \left(\sum_j |w_j\rangle\langle w_j|\right)\hat{D}_\mathbf{r}|0\rangle = 1, \tag{8.27}$$

where the property (8.21) was exploited.

Hence, the inequality (8.25) is equivalent to the upper bound we are after. In order to prove it, we will need to resort to a trick which finds wide application in quantum information theory, which consists essentially in expressing the trace of a product of operators as the trace of a tensor product of operators:

$$\|\hat{A}_{|w_j\rangle}\|_p^p = \mathrm{Tr}\left(\hat{A}_{|w_j\rangle}^p\right) = \int_{\mathbb{R}^{2p}} d\mathbf{r}_1 \ldots d\mathbf{r}_p\, p(\mathbf{r}_1)\ldots p(\mathbf{r}_p)$$

$$\times |\langle w_j|\hat{D}_{\mathbf{r}_1}|0\rangle|^2 \ldots |\langle w_j|\hat{D}_{\mathbf{r}_p}|0\rangle|^2$$

$$\times \mathrm{Tr}\left(\hat{D}_{\mathbf{r}_1}|0\rangle\langle 0|\hat{D}_{\mathbf{r}_1}^\dagger \hat{D}_{\mathbf{r}_2}|0\rangle \ldots \langle 0|\hat{D}_{\mathbf{r}_{p-1}}^\dagger \hat{D}_{\mathbf{r}_p}|0\rangle\langle 0|\hat{D}_{\mathbf{r}_p}^\dagger\right)$$

$$= \mathrm{Tr}\left(|w_j\rangle\langle w_j|^{\otimes p}\hat{B}\right), \tag{8.28}$$

$$\|\hat{A}_{|w_j\rangle}\|_1^p = \mathrm{Tr}\left(\hat{A}_{|w_j\rangle}\right)^p = \int_{\mathbb{R}^{2p}} d\mathbf{r}_1 \ldots d\mathbf{r}_p\, p(\mathbf{r}_1)\ldots p(\mathbf{r}_p)$$

$$\times |\langle w_j|\hat{D}_{\mathbf{r}_1}|0\rangle|^2 \ldots |\langle w_j|\hat{D}_{\mathbf{r}_p}|0\rangle|^2 = \mathrm{Tr}\left(|w_j\rangle\langle w_j|^{\otimes p}\hat{C}\right), \tag{8.29}$$

where the operators $\hat{B}$ and $\hat{C}$ are given by

$$\hat{B} = \int_{\mathbb{R}^{2p}} d\mathbf{r}_1 \ldots d\mathbf{r}_p \, p(\mathbf{r}_1) \ldots p(\mathbf{r}_p) \langle 0|\hat{D}_{\mathbf{r}_1}^\dagger \hat{D}_{\mathbf{r}_2}|0\rangle \ldots \langle 0|\hat{D}_{\mathbf{r}_{p-1}}^\dagger \hat{D}_{\mathbf{r}_p}|0\rangle \langle 0|\hat{D}_{\mathbf{r}_p}^\dagger \hat{D}_{\mathbf{r}_1}|0\rangle$$

$$\bigotimes_{k=1}^{p} \hat{D}_{\mathbf{r}_k}|0\rangle\langle 0|\hat{D}_{\mathbf{r}_k}^\dagger , \tag{8.30}$$

$$\hat{C} = \bigotimes_{k=1}^{p} \int_{\mathbb{R}^2} d\mathbf{r}_k \, p(\mathbf{r}_k) \hat{D}_{\mathbf{r}_k}|0\rangle\langle 0|\hat{D}_{\mathbf{r}_k}^\dagger . \tag{8.31}$$

A close inspection reveals that the operators $\hat{B}$ and $\hat{C}$ are Gaussian, as they are just the output of random, multivariate Gaussian distributed displacements acting on the vacuum state, which follow the same description as the classical mixing map given in Section 5.3.2. We will hence be able to analyse them with the Gaussian toolbox developed in Chapters 3 and 5. The Gaussian character is easily recognised in operator $\hat{C}$, which may be re-written as

$$\hat{C} = \int_{\mathbb{R}^{2p}} d\mathbf{s} \, \frac{e^{-\mathbf{s}^\mathsf{T} Y^{-1}\mathbf{s}}}{\pi^p \sqrt{\text{Det} Y}} \hat{D}_{\mathbf{s}}|0\rangle\langle 0|\hat{D}_{\mathbf{s}}^\dagger , \quad \text{with} \quad Y = \frac{2}{\lambda}\mathbb{1}_{2p} , \tag{8.32}$$

where $\mathbf{s} = (x_1, \ldots, x_p, p_1, \ldots, p_p)^\mathsf{T}$ and $|0\rangle$ stands for the multi-mode vacuum state. The advantage of moving to xp-ordering, with symplectic form J as in Eq. (3.4), will soon become clear. The operator $\hat{C}$ is hence a Gaussian state with covariance matrix $(\mathbb{1} + Y) = \frac{\lambda+2}{\lambda}\mathbb{1}$ (resulting from the addition of Y to the vacuum covariance matrix). The p symplectic eigenvalues of $\hat{C}$ are $\frac{\lambda+2}{\lambda}$ and its diagonal form in the Fock basis is given by Eq. (3.60), as

$$\hat{C} = \left(\frac{\lambda}{\lambda+1}\right)^p \bigotimes_{j=1}^{p} \sum_{m_j=0}^{\infty} \left(\frac{1}{\lambda+1}\right)^{m_j} |m_j\rangle\langle m_j| . \tag{8.33}$$

Handling $\hat{B}$ requires a little more work and savvy: the coherent state overlaps that enter its expression, which are worked out according to Eq. (4.7), may be incorporated along with the probability distributions in the exponential, resulting in

$$\hat{B} = \int_{\mathbb{C}^p} d\boldsymbol{\alpha} \, \frac{\lambda^p \, e^{-\boldsymbol{\alpha}^\dagger W^{-1}\boldsymbol{\alpha}}}{\pi^p} \hat{D}_{\boldsymbol{\alpha}}^\dagger|0\rangle\langle 0|\hat{D}_{\boldsymbol{\alpha}} , \tag{8.34}$$

where we switched to the complex vectors $\boldsymbol{\alpha} = (\alpha_1, \ldots, \alpha_p, \alpha_1^*, \ldots, \alpha_p^*)^\mathsf{T}$, with $\hat{D}_{\boldsymbol{\alpha}} = \otimes_{j=1}^p \hat{D}_{\alpha_j}$ (for consistency, we maintained the convention

$\hat{D}_{\alpha_j} = \hat{D}^{\dagger}_{\mathbf{r}_j}$, although this does not really matter here, as our distribution is invariant under a sign flip in $\boldsymbol{\alpha}$), and the matrix W has been defined by setting

$$W^{-1} = \frac{\lambda + 1}{2}\mathbb{1} - \frac{1}{2}\left(T \oplus T^{\mathsf{T}}\right) \tag{8.35}$$

with

$$T = \begin{pmatrix} 0 & 0 & 0 & \cdots & 0 & 1 \\ 1 & 0 & 0 & \ddots & \ddots & 0 \\ 0 & 1 & 0 & \ddots & \ddots & \vdots \\ \vdots & 0 & \ddots & \ddots & \ddots & 0 \\ 0 & \vdots & 0 & 1 & 0 & 0 \\ 0 & 0 & \cdots & 0 & 1 & 0 \end{pmatrix}. \tag{8.36}$$

The matrix T is nothing but a translation operator that shifts the position along a closed ring of p sites by one site. It is diagonalised by a unitary matrix U with entries $U_{jk} = \mathrm{e}^{ijk\frac{2\pi}{p}}$, for $j, k \in [0, \ldots, p-1]$, which represents a change of variables to Fourier modes (whose real linear combinations allowed us to decouple the potential of the closed, translationally invariant harmonic chain encountered in Problem 3.4). The eigenvalues $\{t_j\}$ of T are hence also straightforward to determine, and are given by $t_j = \mathrm{e}^{-ij\frac{2\pi}{p}}$, for $j \in [0, \ldots, p-1]$.

Notice now that any 'passive', orthogonal symplectic transformation may be written, for the complex variables $\boldsymbol{\alpha}$, quite simply as the direct sum of a unitary matrix and its complex conjugate [this is shown explicitly in Appendix B, Eq. (B.19)]. The symplectic transformation $(U \oplus U^*)$ hence diagonalises W^{-1} acting by congruence, yielding eigenvalues $\{e_j, \ j \in [0, \ldots, p-1]\}$ given by

$$e_j = \frac{\lambda + 1 - t_j}{2} = \frac{\lambda + 1 - \mathrm{e}^{-ij\frac{2\pi}{p}}}{2} \tag{8.37}$$

(since the part proportional to the identity is invariant under unitary transformations), and must correspond to a unitary $\hat{U}$ at the Hilbert space level (through the mapping (3.24), which obviously carries over to complex variables). Let us clarify that, because this symplectic transformation is also orthogonal, it does not affect the Gaussian state $\hat{C}$, whose covariance matrix is proportional to the identity and whose first moments are null, so that the state is invariant under phase space rotations. This implies that the operators $\hat{C}$ and $\hat{B}$ commute, and the Fock states that diagonalise $\hat{C}$ may be chosen as the ones corresponding to the Fourier modes. The operator $\hat{B}$ in the Fourier modes takes the decoupled form

$$\hat{B} = \bigotimes_{j=1}^{p} \int_{\mathbb{C}} d^2\alpha_j \, \frac{\lambda \, e^{-2e_j|\alpha_j|^2}}{\pi} \, \hat{D}_{\alpha_j}^\dagger |0\rangle\langle 0| \hat{D}_{\alpha_j}$$

$$= \bigotimes_{j=1}^{p} \int_{\mathbb{C}} d\mathbf{r}_j \, \frac{\lambda \, e^{-e_j|\mathbf{r}_j|^2}}{2\pi} \, \hat{D}_{\mathbf{r}_j} |0\rangle\langle 0| \hat{D}_{\mathbf{r}_j}^\dagger . \qquad (8.38)$$

Notice now that our analysis of classical mixing in Section 5.3.2 was based on the Gaussian integral (1.18), which may still be applied here since, even if the matrix $e_j \mathbb{1}$ is not positive (its eigenvalues are not real), it is clearly Hurwitz (i.e., $\mathrm{Re}(e_j) > 0 \; \forall \; j$), so that the Gaussian integral still converges. Taking into account the integration factor, the operator $\hat{B}$ admits then the following Gaussian Fourier-Weyl expansion:

$$\hat{B} = \frac{\lambda^p}{2^p} \left(\prod_{j=1}^{p} \frac{1}{e_j} \right) \bigotimes_{k=1}^{p} \left[\frac{1}{2\pi} \int_{\mathbb{R}^2} d\mathbf{r}_k \, e^{-\frac{1}{4}\frac{e_k+1}{e_k}\mathbf{r}_k^\mathsf{T} \Omega^\mathsf{T} \mathbb{1}\Omega\mathbf{r}_k} \, \hat{D}_{\mathbf{r}_k} \right] . \qquad (8.39)$$

The operators in square brackets are all normalised Gaussian operators, with symplectic eigenvalues

$$\tau_j = \frac{e_j + 1}{e_j} = \frac{\lambda + 3 - t_j}{\lambda + 1 - t_j} . \qquad (8.40)$$

Although the τ_j's are complex, their real parts are greater than or equal to 1, so that the formula (3.45) that provides one with the diagonal form of a normalised Gaussian operator can still be applied. Hence, in the very same Fock basis $\{|m_1, \ldots, m_p\rangle\}$ that diagonalises $\hat{C}$, one has, taking into account all factors:[4]

$$\hat{B} = \frac{\lambda^p}{2^p} \left(\prod_{j=1}^{p} \frac{1}{e_j} \right) \left(\prod_{j=1}^{p} \frac{2}{\tau_j + 1} \right) \bigotimes_{j=1}^{p} \sum_{m_j=0}^{\infty} \left(\frac{\tau_j - 1}{\tau_j + 1} \right)^{m_j} |m_j\rangle\langle m_j|$$

$$= \lambda^p \left(\prod_{j=1}^{p} \frac{1}{2e_j + 1} \right) \sum_{m_j=0}^{\infty} \left(\frac{1}{2e_j + 1} \right)^{m_j} |m_j\rangle\langle m_j|$$

$$= \frac{\lambda^p}{(\lambda + 2)^p - 1} \sum_{m_j=0}^{\infty} \left(\frac{1}{\lambda + 2 - t_j} \right)^{m_j} |m_j\rangle\langle m_j| . \qquad (8.41)$$

We are at last in a position to go back to Eqs. (8.28) and (8.29) and, denoting with $c_{m_1,\ldots,m_p}$ the generic coefficients of the vector $|w\rangle_j^{\otimes p}$ in the

Fock basis, and recalling Eq. (8.33), write

$$\mathrm{Tr}\left(|w_j\rangle\langle w_j|^{\otimes p}\hat{B}\right) = \frac{\lambda^p}{(\lambda+2)^p-1}\left|\sum_{\substack{m_1 \\ \vdots \\ m_p=0}}^{\infty}\prod_{j=0}^{p}\left(\frac{1}{\lambda+2-t_j}\right)^{m_j}|c_{m_1,\ldots,m_p}|^2\right|$$

$$\leq \frac{\lambda^p}{(\lambda+2)^p-1}\sum_{\substack{m_1 \\ \vdots \\ m_p=0}}^{\infty}\prod_{j=0}^{p}\left(\frac{1}{\lambda+1}\right)^{m_j}|c_{m_1,\ldots,m_p}|^2$$

$$= \frac{(\lambda+1)^p}{(\lambda+2)^p-1}\mathrm{Tr}\left(|w_j\rangle\langle w_j|^{\otimes p}\hat{C}\right) \qquad (8.42)$$

which, taking the limit $p \to \infty$ and recalling the identities Eqs. (8.28) and (8.29), yields the upper bound (8.25).

All that is left to show now is that the heterodyne strategy attains the upper bound $(\lambda+1)/(\lambda+2)$, which may be done directly by inserting the map Φ_{het} in the expression for the average heterodyne fidelity $\bar{F}_{het}$:

$$\bar{F}_{het} = \int_{\mathbb{C}}\mathrm{d}^2\alpha\int_{\mathbb{C}}\mathrm{d}^2\mu\frac{\lambda e^{-\lambda|\alpha|^2}}{\pi^2}|\langle\mu|\alpha\rangle|^2|\langle\frac{\mu}{\lambda+1}|\alpha\rangle|^2$$

$$= \int_{\mathbb{C}}\mathrm{d}^2\alpha\int_{\mathbb{C}}\mathrm{d}^2\mu\frac{\lambda e^{-\lambda|\alpha|^2}}{\pi^2}e^{-|\alpha-\mu|^2-|\alpha-\frac{\mu}{\lambda+1}|^2} = \frac{\lambda+1}{\lambda+2}, \qquad (8.43)$$

which may be computed by turning to real variables and utilising the Gaussian integral (1.18). The classical threshold for normally distributed coherent states is thus determined. □

An experimental demonstration of the quantum teleportation of an alphabet of coherent states must hence beat the threshold (8.18) to be successful. In this regard, notice that the benchmark obtained above applies not only to the assessment of quantum teleportation experiments, but also to the storage of the same alphabet of states in a quantum memory (that may be abstractly interpreted as "teleportation in time", rather than space). A quantum memory – like the ones that, for light, may be built with the atomic ensembles described in Section 9.5, which were not able to surpass the threshold (8.18), would be less effective than the 'classical' storage strategy consisting of the following steps: heterodyne detection of the input state; classical storage of the outcome μ; eventual preparation of the coherent state $|\frac{\mu}{\lambda+1}\rangle$ (when the memory needs to be accessed).

---

[4]We will use the relationship $\prod_{j=0}^{p-1}(x - e^{-ij\frac{2\pi}{p}}) = (x^p - 1)$. In order to prove it, just notice that the left hand side is a polynomial of order p in x whose roots are the p p^{th} roots of the identity $e^{-ij\frac{2\pi}{p}}$, for $p = 0,\ldots,p-1$.

Problem 8.2. (*Modified vs. standard teleportation*). Reconsider the modified teleportation protocol of Problem 8.1: assume an input alphabet of single-mode coherent states $\{\hat{D}_{\mathbf{r_0}}|0\rangle\}$ with weight $\frac{\lambda e^{-\frac{\lambda}{2}|\mathbf{r_0}|^2}}{2\pi}d^2r_0$, and determine for what values of λ such a protocol outperforms the standard teleportation protocol.

Solution. The outcome-independent fidelity of the modified protocol is given by Eq. (8.14): one just needs to average it over the input alphabet to obtain the average teleportation fidelity:

$$\bar{F}_{mod} = \int_{\mathbb{R}^2} e^{-\frac{1}{2}\left(\frac{c+1-s}{c+1}\right)^2|\mathbf{r_0}|^2} \frac{\lambda e^{-\frac{\lambda}{2}|\mathbf{r_0}|^2}}{2\pi} d\mathbf{r_0} = \frac{\lambda(\cosh(2r)+1)}{\lambda(\cosh(2r)+1)+2e^{-2r}},$$
(8.44)

which should be contrasted with the standard average fidelity $\bar{F} = 1/(1+e^{-2r})$ (independent from λ). For any finite λ, both strategies yield ideal teleportation in the limit of infinite shared entanglement $r \to \infty$. For given, finite r, however, one has $\lim_{\lambda \to \infty} \bar{F}_{mod} = 1 \geq \bar{F}$ (not surprisingly, since in the modified protocol previous knowledge of σ_0 is assumed): there must hence be a region of values of λ, corresponding to peaked-enough input distributions, where the modified protocol is better than the standard one. This region is bounded, since $\lim_{\lambda \to 0} \bar{F}_{mod} = 0$. Setting $\bar{F}_{mod} = \bar{F}$ furnishes the minimum value λ_{mod} for which the modified protocol outperforms the standard one, just given by $\lambda_{mod} = \frac{2}{\cosh(2r)+1}$.

8.1.3 Classical thresholds for squeezed states

In order to extend the results above to more general sets of input states, we will restrict now to the idealised case of states which are uniformly displaced in phase space (i.e., to the limit $\lambda \to 0$ in the notation of the previous section). Under such a proviso, and certain additional assumptions, we will be able to determine the classical threshold for input ensembles of single-mode squeezed states with random optical squeezing phase and uniform displacements. Besides its inherent interest, this study will allow us to showcase methods for the solution of symmetric optimisation problems, among which is a characterisation of Weyl-covariant channels that may well be useful in other contexts.

We are considering the following set of input states:

$$\varrho_{in} = \lim_{\lambda \to 0} \int_{\mathbb{R}^2} d\mathbf{r} \frac{e^{-\frac{\lambda}{2}|\mathbf{r}|^2}}{2\pi} \int_0^{2\pi} \frac{d\varphi}{2\pi} \Phi\left(\hat{D}_{\mathbf{r}}\hat{R}_{\varphi}|z\rangle\langle z|\hat{R}_{\varphi}^{\dagger}\hat{D}_{\mathbf{r}}^{\dagger}\right),$$
(8.45)

where $\hat{R}_{\varphi}$ is the unitary associated with the phase shifter transformation (5.12) and $|z\rangle$ is the pure Gaussian state with null first moments and covariance matrix $\mathrm{diag}(z, 1/z)$, with $z > 0$. The ensemble of states above clearly depends on z. Let us then denote with $\bar{F}_z(\Phi)$ the average overlap one would obtain by applying the measure-and-prepare strategy Φ:

$$\bar{F}_z(\Phi) = \lim_{\lambda \to 0} \int_{\mathbb{R}^2} d\mathbf{r} \frac{e^{-\frac{\lambda}{2}|\mathbf{r}|^2}}{2\pi} \int_0^{2\pi} \frac{d\varphi}{2\pi} \langle z|\hat{R}_\varphi^\dagger \hat{D}_\mathbf{r}^\dagger \Phi \left(\hat{D}_\mathbf{r} \hat{R}_\varphi |z\rangle \langle z|\hat{R}_\varphi^\dagger \hat{D}_\mathbf{r}^\dagger \right) \hat{D}_\mathbf{r} \hat{R}_\varphi |z\rangle , \tag{8.46}$$

which we intend to maximise:

$$\bar{F}_{th,z} = \sup_\Phi \bar{F}_z(\Phi) , \tag{8.47}$$

over all measure-and-prepare channels Φ.

Since the measure $\lim_{\lambda \to 0} \frac{e^{-\frac{\lambda}{2}|\mathbf{r}|^2}}{2\pi} d\mathbf{r}$ is invariant under phase space translations, enacted by Weyl operators, we will now assume the optimal value $F_{th,z}$ to be attained by a *Weyl-covariant* measure-and-prepare channel Φ, that is, by a channel such that $\Phi(\hat{D}_\mathbf{r} \varrho \hat{D}_\mathbf{r}^\dagger) = \hat{D}_\mathbf{r} \Phi(\varrho) \hat{D}_\mathbf{r}^\dagger$ for all density operators ϱ and $\mathbf{r} \in \mathbb{R}^2$.

If the covariant group G were compact, as would be the case for, e.g., rotations in any finite dimension, then any channel Φ could be associated to a covariant one $\tilde{\Phi}$, defined as

$$\tilde{\Phi}(\varrho) = \int_G dg\, \hat{U}_g^\dagger \Phi(\hat{U}_g \varrho \hat{U}_g^\dagger) \hat{U}_g , \tag{8.48}$$

where dg is the invariant Haar measure over the group G, and $\hat{U}_g$ is a unitary representation of G acting on the Hilbert space. The covariant CP-map $\tilde{\Phi}$ then performs at least as well as Φ: $\bar{F}_z(\tilde{\Phi}) \geq \bar{F}_z(\Phi)$ (since the functional F is concave), which would justify the restriction of the optimisation to covariant CP-maps.

In the case of the non-compact group of translations over $\mathbb{R}^2$ which, as we saw, are projectively represented by single-mode Weyl displacement operators $\hat{D}_\mathbf{r}$, the covariant map $\tilde{\Phi}$ may not be defined through an integral as above, as its output would not in general be trace-class. Nonetheless, an analogous argument for a figure of merit like the teleportation overlap carries over, and one is still allowed to restrict to Weyl-covariant CP-maps. The rigorous justification of this restriction is rather technical and would require the introduction of formal tools that lie outside the scope of our treatment. It will hence not be expounded here, but just assumed.

In fact, we will extend the optimisation to the larger set of Weyl-covariant 'NPT-breaking' channels which, when acting on a subsystem of a composite Hilbert space, always turn the original state into one with positive partial transpose (a "PPT" state).[5] As shown in Section D.3.1 of Appendix D,

---

[5] Measure-and-prepare channels, as made clear in Appendix D, are the same as entanglement breaking channels, and thus send any state into a separable one when acting on a subsystem. Since a separable state is necessarily PPT while the converse is not true, all entanglement breaking channels are NPT-breaking, but the converse is not true. We will thus be evaluating the supremum on a wider set of CP-maps.

NPT-breaking Weyl-covariant channels are those whose dual Φ^* acts on Weyl operators as per

$$\Phi^*(\hat{D}_\mathbf{r}) = f(\mathbf{r})\hat{D}_\mathbf{r} , \qquad (8.49)$$

where $f(\mathbf{r})$ is a quantum characteristic function, so that $f(\sqrt{2}\mathbf{r}) = \mathrm{Tr}[\tau \hat{D}_{-\sqrt{2}\mathbf{r}}]$ for some density operator τ. The assumption of Weyl-covariance, as well as the associated characterisation of Weyl-covariant channels, will allow us to prove what follows.

Classical threshold for squeezed states. The optimal average fidelity $\bar{F}_{th,z}$ achieved by a measure-and-prepare strategy over a distribution of uniformly rotated and displaced squeezed states with fixed squeezing z is

$$\bar{F}_{th,z} = \frac{\sqrt{z}}{z+1} , \qquad (8.50)$$

and is attained by the measure-and-prepare heterodyne protocol

$$\Phi_{het}(\varrho) = \frac{1}{\pi} \int_\mathbb{C} \mathrm{d}^2\mu \, \langle\mu|\varrho|\mu\rangle |\mu\rangle\langle\mu| , \qquad (8.51)$$

which consists of the heterodyne detection of the input state followed by the preparation of the coherent state $|\mu\rangle$ upon the heterodyne outcome μ.

Proof. Like for coherent states, we shall prove Eq. (8.50) by first proving that $\bar{F}_{th,z}$ is upper bounded by the quantity on the left hand side, and then by showing that the latter is attained by the heterodyne strategy.

Weyl-covariance ensures that the optimal channel produces the same overlap if the input state is displaced. We are thus dispensed from averaging over displacements, and the threshold satisfies

$$\bar{F}_{th,z} \leq \sup_\Phi \int_0^{2\pi} \frac{\mathrm{d}\varphi}{2\pi} \, \langle z|\hat{R}_\varphi^\dagger \Phi\left(\hat{R}_\varphi|z\rangle\langle z|\hat{R}_\varphi^\dagger\right) \hat{R}_\varphi|z\rangle . \qquad (8.52)$$

Where the inequality is due to the fact that we are maximising over the broader set of NPT-breaking, rather than entanglement breaking, channels. The action of the dual Φ^*, specified by Eq. (8.49), implies that Φ maps the characteristic function $\chi(\mathbf{r}) = e^{-\frac{1}{4}\mathbf{r}^\mathsf{T}\Omega^\mathsf{T}\sigma\Omega\mathbf{r}}$ of the centred Gaussian state $\hat{R}_\varphi|z\rangle$ (with covariance matrix $\sigma = R_\varphi^\mathsf{T}\mathrm{diag}(z, 1/z)R_\varphi$) into $f(-\sqrt{2}\mathbf{r})\chi(\mathbf{r}) = \mathrm{Tr}(\tau\hat{D}_{\sqrt{2}\mathbf{r}})\chi(\mathbf{r})$. The Fourier-Weyl expansion (4.18) then allows one to write

$$\bar{F}_{th,z} \leq \sup_\tau \int_0^{2\pi} \frac{\mathrm{d}\varphi}{2\pi} \frac{1}{2\pi} \int_{\mathbb{R}^2} \mathrm{d}\mathbf{r} \, \mathrm{Tr}(\tau\hat{D}_{\sqrt{2}\mathbf{r}}) e^{-\frac{1}{4}\mathbf{r}^\mathsf{T}\Omega^\mathsf{T}\sigma\Omega\mathbf{r}} \langle z|\hat{R}_\varphi^\dagger \hat{D}_\mathbf{r} \hat{R}_\varphi|z\rangle$$

$$= \sup_\tau \int_0^{2\pi} \frac{\mathrm{d}\varphi}{2\pi} \int_{\mathbb{R}^2} \frac{\mathrm{d}\mathbf{r}}{2\pi} \, \mathrm{Tr}(\tau\hat{D}_{\sqrt{2}\mathbf{r}}) e^{-\frac{1}{2}\mathbf{r}^\mathsf{T}\Omega^\mathsf{T}\sigma\Omega\mathbf{r}} , \qquad (8.53)$$

where we turned the optimisation over NPT-channels into one over density operators τ and inserted $\chi(\mathbf{r}) = \langle z|\hat{R}_\varphi^\dagger \hat{D}_\mathbf{r} \hat{R}_\varphi|z\rangle$ (notice that $\chi(\mathbf{r}) = \chi(-\mathbf{r})$ for a centred Gaussian state). Observe also that $e^{-\frac{1}{2}\mathbf{r}^\mathsf{T}\Omega^\mathsf{T}\sigma\Omega\mathbf{r}} = \chi(\sqrt{2}\mathbf{r})$. The orthonormality of the set of Weyl operators with respect to the Hilbert–Schmidt norm [Eq. (4.21)] may then be exploited backward to obtain a relationship on density operators alone:

$$\bar{F}_{th,z} \le \sup_\tau \int_0^{2\pi} \frac{\mathrm{d}\varphi}{2\pi} \int_{\mathbb{R}^2} \frac{\mathrm{d}\mathbf{r}}{2\pi} \langle z|\hat{R}_\varphi^\dagger \hat{D}_{-\sqrt{2}\mathbf{r}} \hat{R}_\varphi|z\rangle \mathrm{Tr}(\tau \hat{D}_{\sqrt{2}\mathbf{r}})$$

$$= \frac{1}{2}\sup_\tau \int_0^{2\pi} \frac{\mathrm{d}\varphi}{2\pi} \langle z|\hat{R}_\varphi^\dagger \tau \hat{R}_\varphi|z\rangle = \left\|\int_0^{2\pi} \frac{\mathrm{d}\varphi}{4\pi} \hat{R}_\varphi|z\rangle\langle z|\hat{R}_\varphi^\dagger\right\|_\infty , \quad (8.54)$$

where the factor $1/2$ is a consequence of the change of variables $\sqrt{2}\mathbf{r} \mapsto \mathbf{r}$. The operator norm now subsumes the maximisation over all possible density operators τ. The flat average over phase space rotations just sets the off diagonal elements in the Fock basis to zero,[6] so that one just has to optimise the Fock number overlap of $|z\rangle$, which was worked out in Problem 3.2 and yields the desired result:

$$\bar{F}_{th,z} \le \frac{1}{2}\sup_{|m\rangle} |\langle m|z\rangle|^2 = \frac{1}{2}|\langle 0|z\rangle|^2 = \frac{\sqrt{z}}{z+1} . \quad (8.55)$$

Let us now show that the entanglement breaking, heterodyne protocol Φ_{het} achieves the bound, and thus complete the proof. The Gaussian CP-map Φ_{het} acts on a Gaussian input state ϱ_G with covariance matrix σ and first moments $\bar{\mathbf{r}}$ as

$$\Phi_{het}(\varrho_G) = \int_{\mathbb{R}^2} \frac{\mathrm{d}\mathbf{r}}{2\pi} \langle 0|\hat{D}_\mathbf{r}^\dagger \varrho \hat{D}_\mathbf{r}|0\rangle \hat{D}_\mathbf{r}|0\rangle\langle 0|\hat{D}_\mathbf{r}^\dagger$$

$$= \int_{\mathbb{R}^2} \frac{\mathrm{d}\mathbf{r}}{2\pi} \frac{2}{\sqrt{\mathrm{Det}(\sigma+\mathbb{1})}} e^{-(\mathbf{r}-\bar{\mathbf{r}})^\mathsf{T}\frac{1}{\sigma+\mathbb{1}}(\mathbf{r}-\bar{\mathbf{r}})} \hat{D}_\mathbf{r}|0\rangle\langle 0|\hat{D}_\mathbf{r}^\dagger$$

$$= \int_{\mathbb{R}^2} \frac{\mathrm{d}\mathbf{r}}{2\pi} \frac{2}{\sqrt{\mathrm{Det}(\sigma+\mathbb{1})}} e^{-\mathbf{r}^\mathsf{T}\frac{1}{\sigma+\mathbb{1}}\mathbf{r}} \hat{D}_{\bar{\mathbf{r}}}\hat{D}_\mathbf{r}|0\rangle\langle 0|\hat{D}_\mathbf{r}^\dagger\hat{D}_{\bar{\mathbf{r}}}^\dagger , \quad (8.56)$$

where the formula for the Gaussian overlap (4.51) was applied. The output of Eq. (8.56) is, up to the action of the displacement $\hat{D}_{\bar{\mathbf{r}}}$ that fixes its first moments to $\bar{\mathbf{r}}$, nothing but the application of classical mixing (normally distributed displacements) on the vacuum state: it is hence a Gaussian state with first moments $\bar{\mathbf{r}}$ and covariance matrix $\mathbb{1}+\sigma+\mathbb{1} = \sigma+2\mathbb{1}$. In the language of Chapter 5, it is the Gaussian CP-map with $X = \mathbb{1}$ and $Y = 2\mathbb{1}$. It is a Weyl- and phase-covariant (invariant under displacements and phase rotations) channel whose output overlap does not depend on either φ or $\mathbf{r}$, and may simply evaluated on the centred squeezed Gaussian state with covariance matrix $\mathrm{diag}(z, 1/z)$. Yet another application of the formula (4.51) then gives

$$\bar{F}_{th,z} = \frac{2}{\sqrt{\mathrm{Det}(2\mathrm{diag}(z, 1/z) + 2\mathbb{1})}} = \frac{\sqrt{z}}{z+1} . \tag{8.57}$$

□

Notice that, setting $z = 1$, one retrieves the limit of uniformly distributed coherent states $\lim_{\lambda \to 0} \frac{1+\lambda}{2+\lambda} = \frac{1}{2}$, consistent with the previous section.

8.2   CLASSICAL COMMUNICATION OVER BOSONIC CHANNELS

Let parties A and B share a quantum communication link that alters the input state like a CP-map Φ. The quantum channel Φ can be exploited as a medium to transfer *classical* information as follows: the sender A could prepare states ϱ_x depending on a classical variable x, taken from an alphabet with probability $p(x)$, and transmit it to B through the channel. The receiver B can then measure the received state $\Phi(\varrho_x)$ by some POVM with elements $\hat{K}_y$ and outcomes labelled with y. Thus, B obtains a random classical variable y whose conditional probability distribution $p(y|x) = \mathrm{Tr}(\hat{K}_y \Phi(\varrho_x) \hat{K}_y^\dagger)$ depends on the encoded x. As customary, the statistical ensembles associated with x and y will be referred to with X and Y below.

Classical information theory tells us that the information that B can extract about x is given by the mutual information $I(Y;X) = S(X) + S(Y) - S(X,Y)$, where the quantities $S(X)$, $S(Y)$ and $S(X,Y)$ are the classical Shannon entropies of the marginal ensembles X and Y, and of the global ensemble (X,Y).[7] This quantity, maximised over all the possible POVMs, corresponds to the number of classical bits per use of the channel that can be transmitted asymptotically (i.e., in the limit of an infinitely long string) *if only uncorrelated inputs over different uses of the channel are allowed*. In general, the maximisation of the classical capacity over all POVMs is a difficult endeavour. However, a seminal general result due to Holevo allows one to establish the following upper bound on the mutual information $I(X;Y)$:

$$I(X;Y) \le C_\chi(\Phi) = \sup_\varrho \chi(\Phi(\varrho)) = \sup_\varrho \left[S_V(\Phi(\varrho)) - \inf_{p_x, \varrho_x} \sum_x p_x S_V(\Phi(\varrho_x)) \right],$$
$$\tag{8.58}$$

---

[6]To appreciate this, note that $\hat{R}_\varphi = e^{i\varphi \hat{a}^\dagger \hat{a}}$, whence $\int_0^{2\pi} \frac{\mathrm{d}\varphi}{2\pi} \hat{R}_\varphi |m\rangle\langle n| \hat{R}_\varphi^\dagger = \int_0^{2\pi} \frac{\mathrm{d}\varphi}{2\pi} e^{i(m-n)\varphi} |m\rangle\langle n| = \delta_{mn} |n\rangle\langle n|.$

[7]Technically, the classical entropy $S(X) = -\sum_x p(x) \log_2(p(x))$ represents the number of bits per letter, in the asymptotic limit of infinitely long sequences, to which the alphabet X may be compressed. The mutual information $I(X;Y)$ is instead the number of bits that can be spared in such a compression *given knowledge of the correlated variable Y* (less bits will be needed, since something about X has been learned through Y).

where the last infimum is taken with respect to all convex decomposition of the quantum state ϱ, such that $\sum_x p_x \varrho_x = \varrho$, while S_V is the von Neumann entropy. A derivation of the bound is reported in Section D.1 of Appendix D.

The quantity $\chi(\varrho)$, often referred to as the "Holevo χ", provides one with an upper bound to the 'accessible information' contained in a quantum state ϱ. Its supremum over all possible outputs of a quantum channel then yields the maximal mutual information over uncorrelated inputs $C_\chi(\Phi)$, defined above. Heuristically, the form of the Holevo χ may be understood by considering that a higher overall entropy corresponds to more information, as it allows for more possibility in the alphabet of symbols, while a low entropy of the individual states in the decomposition corresponds to the capability of distinguishing the different symbols (bear in mind that the Holevo χ is a property of the 'alphabet' of quantum symbols employed, and not of the individual states sent through the channel).

The classical channel capacity $C(\Phi)$ subsumes the possibility of entangled inputs over different uses of the channel, and is hence defined as the regularisation

$$C(\Phi) = \lim_{n \to \infty} \frac{C_\chi(\Phi^{\otimes n})}{n} . \tag{8.59}$$

This regularised capacity is asymptotically achievable (yet another seminal contribution by Holevo, building on work by Hausladen, Jozsa, Schumacher and Westmoreland), and is hence the actual ultimate classical communication limit of a quantum channel.

Problem 8.3. (*Capacity of the identity superoperator*). Show that the classical communication capacity of the identity channel in a Hilbert space of dimension d is $\log_2(d)$.

Solution. If A and B share n links that act as ideal identity superoperators, A can encode an alphabet X in the following decomposition of the maximally mixed state $\varrho = \frac{1}{d^n} \sum_x |x\rangle\langle x|$, where $\{|x\rangle, x \in [1 \ldots, d^n]\}$ form an orthonormal basis of the Hilbert space $\mathcal{H}^{\otimes n}$. This decomposition clearly saturates the Holevo bound for any n, since the first term in the expression for χ attains the maximum entropy $n \log_2 d$, while the second term is zero (and hence maximum too). Therefore, the capacity is additive in this case and the regularised capacity per use of the channel is $\log_2 d$.

One still has to show that the bound can actually be attained for finite n: since the channel leaves the state unscathed, B can measure in the same basis, now labelled with y to comply with the terminology above, obtaining a conditional probability distribution for y given x that reads $p(y|x) = \langle y|x\rangle\langle x|y\rangle = \delta_{xy}$: x and y are perfectly correlated. The distribution of each variable is flat, given by $p(x) = p(y) = 1/d$, while the global probability distribution $p(x, y)$ is $p(x, y) = p(y|x)p(x) = \delta_{xy}/d$ (where we utilised the Bayes rule for conditional probabilities). Hence, the mutual information is $S(X) + S(Y) - S(X, Y) = \log_2 d + \log_2 d - \log_2 d = \log_2 d$. This value saturates the Holevo bound and must be the ultimate classical capacity of the channel. Notice that this innocent example implies, not surprisingly, that an infinite amount of information can in principle be transmitted, per use of the channel, through an infinite dimensional Hilbert space. In practice, as we shall see, a cap on the available energy prevents such idealistic exploitations.

It is clear from the expression (8.58) for the Holevo χ that the channel capacity is intimately related to the maximum and minimum output entropy of the channel. The latter, in particular, is typically more challenging to determine and key to obtaining an analytical expression for the capacity. A property that simplifies the issue considerably is the additivity of the minimum output purity over several uses of the channel, which implies on general grounds that the classical capacity is itself additive. Additivity is crucial, as it implies that all regularised quantities such as $C(\Phi)$ coincide with their single-shot version: $C(\Phi) = C_\chi(\Phi)$ in our notation. There exist, however, examples of superadditive quantum channels: additivity does not hold in general, although it may for specific channels.

Over the next three sections, we will derive analytical expressions for the classical capacity of phase-insensitive bosonic Gaussian channels, dealing first with the maximum and then with the minimum output purity (we will have to bound the input energy allowed and obtain a corresponding finite maximum output entropy). We will show that, for such channels, both these quantities are additive and attained by Gaussian inputs. We refer, as customary in the literature, to phase-insensitive channels as 'phase-covariant' and 'contravariant' channels. As we saw in Problems 5.16 and 5.17, single-mode phase-covariant channels are, up to a unitary phase shifter operation, deterministic Gaussian CP-maps with X and Y (see Section 5.3 for the characterisation of Gaussian channels in terms of these real matrices) of the form:

$$X = x\mathbb{1}_2 , \quad Y = y\mathbb{1}_2 , \quad y \geq |x^2 - 1| \tag{8.60}$$

which admit the general decomposition $\mathcal{A}_r^0 \circ \mathcal{E}_\theta^0$, in terms of a quantum limited amplifier $\mathcal{A}_r^0$ and pure loss $\mathcal{E}_\theta^0$. In Section D.4 of Appendix D, single-mode phase-contravariant channels are shown to be characterised, up to a phase space rotation, by

$$X = x\sigma_z , \quad Y = y\mathbb{1}_2 , \quad y \geq (x^2 + 1) \tag{8.61}$$

and shown to admit the general decomposition $\bar{\mathcal{A}}_r^0 \circ \mathcal{E}_\theta^0$, in terms of the conjugate to a quantum limited amplifier $\bar{\mathcal{A}}_r^0$ and a pure loss channel $\mathcal{E}_\theta^0$. Conjugate channels are defined and discussed in detail in Section D.4.1. Since a final unitary phase-space rotation is irrelevant to output entropies and capacities, we can adopt the simple forms above in our derivation.

Also note that, since entropies do not depend on the first moments, we will be allowed to set them to zero and disregard them in determining extremal entropies.

8.2.1 Maximum output entropy of phase-insensitive channels: Gaussian extremality

As anticipated, in order to study the maximal output entropy of a channel, one has to set a cap, that would certainly be there in practice, on the average input energy. It is convenient to define the set of allowed input states ϱ_{in} as those with a maximum expectation value nE for the 'free', rescaled n-mode Hamiltonian $2\sum_{j=1}^{n}(\hat{x}_j^2 + \hat{p}_j^2)$: $\mathcal{E} = \{\varrho_{in} \; : \; \text{Tr}[\varrho_{in} 2\sum_{j=1}^{n}(\hat{x}_j^2 + \hat{p}_j^2)] \leq nE\}$. Notice that such an expectation value is readily expressed in terms of the covariance matrix $\boldsymbol{\sigma}_{in}$ of the (Gaussian or less) input state ϱ_{in}: $\text{Tr}[\varrho_{in} 2(\hat{x}^2 + \hat{p}^2)] = \text{Tr}\boldsymbol{\sigma}$ (recall that we are setting input first moments to zero).

The maximum entropy over n uses of the quantum channel Φ is then defined as

$$S_{max}^{(n)}(E) = \sup_{\varrho_{in} \in \mathcal{E}} S_V(\Phi^{\otimes n}(\varrho_{in})) \,. \tag{8.62}$$

For a phase-insensitive bosonic Gaussian channel with X and Y given by Eqs. (8.60,8.61), the transformation of the input second moments, be they associated with a Gaussian state or not, is fixed, and so is the output energy E_{out} which, given an input energy per mode E_{in}, is just $E_{out} = nx^2 E_{in} + 2ny^2$. Now, from statistical mechanics, we know that the state with maximum entropy for a given expectation value of the Hamiltonian is just the Gibbs state associated with that Hamiltonian (an application of Lagrange multipliers). But that is precisely how we defined Gaussian states, as thermal states of quadratic Hamiltonians! In this case, therefore, the state of maximum entropy for given *maximum* input energy will be the output Gaussian state corresponding to the Hamiltonian matrix $4\mathbb{1}$ for the maximum possible value of the output energy (obviously, the entropy of the Gibbs state grows monotonically with its energy). For n uses of the channel, this is achieved for the uncorrelated input with covariance matrix $\boldsymbol{\sigma}_{in} = \frac{E}{2}\mathbb{1}_{2n}$, and corresponding output $\boldsymbol{\sigma}_{out} = (x^2\frac{E}{2} + y)\mathbb{1}_{2n}$. The maximum output entropy $S_{max}^{(n)}(E)$ is therefore additive, and its regularisation may be determined as

$$S_{max}(E) = \lim_{n\to\infty} \frac{S_{max}^{(n)}(E)}{n} = S_{max}^{(1)}(E) = s_V\left(x^2\frac{E}{2} + y\right), \tag{8.63}$$

where we have utilised the function s_V, defined in Eq. (3.93) as

$$s_V(x) = \frac{x+1}{2}\log_2\left(\frac{x+1}{2}\right) - \frac{x-1}{2}\log_2\left(\frac{x-1}{2}\right), \tag{8.64}$$

that relates the von Neumann entropy to the symplectic eigenvalue of the single-mode Gaussian state. The latter was immediately apparent from the expression for σ_{out} above, which is already in normal form.

In general, Gaussian states may be proven to be quantum states of maximal entropy for given second moments, and this also extends to any other strongly subadditive and unitarily invariant functional.

8.2.2 Minimum output entropy of phase-insensitive channels

We shall now determine the minimum output entropy of phase-insensitive bosonic channels. Although these findings first appeared in a complete form in work by Giovannetti, García-Patrón, Cerf and Holevo, we will hereafter follow very closely a revisited treatment by Mari, Giovannetti and Holevo:

Minimum output entropy of phase-insensitive channels. The minimum output entropy S_{min} of all phase-insensitive bosonic channels is attained by coherent state inputs (each coherent state yields the same output entropy for such channels). For a channel with $X = x\mathbb{1}$ and $Y = y\mathbb{1}$ (or for the phase-contravariant channel with $X = x\sigma_z$ and the same Y) it is given by:

$$S_{min} = s_V(x^2 + y) \qquad (8.65)$$

for s_V defined in Eq. (8.64).

Proof. To inquire into their minimum output entropy, we shall make use of the general reduction of single-mode phase-insensitive channels as compositions of pure loss channels and quantum limited amplifiers: $\mathcal{A}_r^0 \circ \mathcal{E}_\theta^0$, or $\bar{\mathcal{A}}_r^0 \circ \mathcal{E}_\theta^0$ for phase contravariant channels (where $\bar{\mathcal{A}}_r^0$ if the conjugate to a quantum limited amplifier). The reader is referred to Problem 5.17 and Appendix D for their proof.

Now, the minimum output purity of the pure loss channel is additive and trivial to determine, as the purity of any coherent input state is preserved through the channel: this is immediately obvious since its covariance matrix (initially equal to $\mathbb{1}_2$) is transformed into $\cos^2\theta\mathbb{1}_2 + \sin^2\theta\mathbb{1}_2 = \mathbb{1}_2$ (still corresponding to a pure coherent state). The first moments are dampened by a factor $\cos\theta$, but the output still corresponds to a coherent state (or a product thereof, for many uses of the channel).

Hence, if we could prove that the minimum output entropy of the amplifier channel is also additive and attained for input coherent states, the problem would be solved for all phase-insensitive channels. Using the fact, highlighted in Appendix D, that a channel $\mathcal{A}_r^0$ and its conjugate $\bar{\mathcal{A}}_r^0$ must have the same minimum output entropy, we can further restrict to the phase-contravariant conjugates of quantum limited amplifier channels, with

$$X = \sinh(r)\sigma_z \ , \quad Y = \cosh(r)^2 \ . \qquad (8.66)$$

These channels have the, as we shall see very agreeable, property of being entanglement breaking: their action on one subsystem of a bipartite entangled state yields a separable state. This can be promptly seen by applying one such channel on one of the modes of the covariance matrix σ_r of Eq. (5.21) in the limit $r \to \infty$ (which we denote with r' below to distinguish it from the finite channel parameter r), which corresponds to the maximally entangled Gaussian state $|\delta\rangle$. This gives the output covariance matrix

$$\lim_{r' \to \infty} \begin{pmatrix} \sinh(r)^2 \cosh(2r')\mathbb{1} + \cosh(r)^2\mathbb{1}_2 & \sinh(r)\sinh(2r')\mathbb{1}_2 \\ \sinh(r)\sinh(2r')\mathbb{1}_2 & \cosh(2r')\mathbb{1}_2 \end{pmatrix} . \quad (8.67)$$

Regardless of all other parameters, and of the limiting procedure, this is a two-mode covariance matrix with a positive determinant of the off-diagonal block, and hence corresponds to a separable state by the separability lemma on page 174. In Appendix D, it is shown that all channels that send the maximally entangled state $|\delta\rangle = \lim_{r' \to \infty} \frac{1}{\cosh(r')} \sum_{j=0}^{\infty} \tanh(r')^j |j, j\rangle$ into a separable one by acting on one subsystem are in fact entanglement breaking (i.e., they break the entanglement of any bipartite state by acting locally). A very general argument, reviewed in Appendix D, implies that *the minimum output entropy of entanglement breaking channels is additive*. We have thus obtained another major simplification of the problem: we can in fact focus on the minimisation of the *single-shot* output entropy of the contravariant amplifier channel (8.66) or, equivalently, of its conjugate.

In Section D.4.1 of Appendix D it is also shown that the channel $\bar{\mathcal{A}}_r^0$ of Eq. (8.66) admits the following decomposition:

$$\bar{\mathcal{A}}_r^0 = T \circ \mathcal{A}_r^0 \circ \mathcal{E}_\theta^0, \quad \text{with} \quad \theta = \arccos(\tanh(r)) . \quad (8.68)$$

The final transposition T does not alter the spectrum of the output state, and can therefore be ignored. For any pure input state $|\psi\rangle\langle\psi|$, one has

$$S_V(\mathcal{A}_r^0(|\psi\rangle\langle\psi|)) = S_V(\bar{\mathcal{A}}_r^0(|\psi\rangle\langle\psi|)) = S_V(\mathcal{A}_r^0 \circ \mathcal{E}_\theta^0(|\psi\rangle\langle\psi|))$$
$$= S_V\left[\sum_j p_j \mathcal{A}_r^0(|\psi_j\rangle\langle\psi_j|)\right], \quad (8.69)$$

where $\sum_j p_j |\psi_j\rangle\langle\psi_j|$ is an ensemble decomposition of the output of the attenuator channel: $\mathcal{E}_\theta^0(|\psi\rangle\langle\psi|) = \sum_j p_j |\psi_j\rangle\langle\psi_j|$ (which, let us remind the reader, consists in mixing the input state with the vacuum at a beam splitter).

Assume now that $|\psi\rangle$ is an optimal input, resulting in the minimum entropy. In general, concavity of the entropy and Eq. (8.69) yield

$$S_V(\mathcal{A}_r^0(|\psi\rangle\langle\psi|)) \geq \sum_j p_j S_V(\mathcal{A}_r^0(|\psi_j\rangle\langle\psi_j|)) \,. \tag{8.70}$$

But if $|\psi\rangle$ is optimal, by hypothesis, then it must also be $S_V(\mathcal{A}_r^0(|\psi\rangle\langle\psi|)) = \sum_j p_j S_V(\mathcal{A}_r^0(|\psi_j\rangle\langle\psi_j|))$. From the strict concavity of the entropy, this is only possible if and only if all the output states $\mathcal{A}_r^0(|\psi_j\rangle\langle\psi_j|)$ coincide: $\mathcal{A}_r^0(|\psi_j\rangle\langle\psi_j|) = \varrho_{out}$ for all j. The amplifier channel maps a generic characteristic function $\chi(\mathbf{r})$ into $e^{-\frac{\sinh(r)^2}{4}\mathbf{r}^\mathsf{T}\mathbf{r}}\chi(\cosh(r)\mathbf{r})$: clearly, any two different characteristic functions would be mapped into different characteristic functions, and the same holds for states. Hence, the states $|\psi_j\rangle\langle\psi_j|$ in the ensemble $\sum_j p_j|\psi_j\rangle\langle\psi_j|$ corresponding to the optimal input must all be the same. In other words, we have proven that the optimal input must be such that the output of the quantum limited attenuator $\mathcal{E}_\theta^0$ is pure. As we saw in Problem 5.12, among all quantum states, the only class with this property is the class of coherent states, which are as a consequence the only optimal inputs.

The final evaluation of Eq. (8.65) is straightforward, since the output covariance matrix corresponding to the coherent state input ($\boldsymbol{\sigma}_{in} = \mathbb{1}$) is already in normal form, with the symplectic eigenvalue in full display. The proof is therefore complete. □

Besides the determination of the communication capacity, which we shall achieve in the next section, the additivity of the minimal output entropies of phase-insensitive channels has other interesting consequences, in situations where full optimisations at the Hilbert space level are called for. To name one, let us mention that the additivity proven above implies that the Gaussian quantum discord evaluated in Problem 7.8 is the actual quantum discord, where the minimisation is carried out over all possible POVMs.

Problem 8.4. (*Entanglement breaking phase insensitive Gaussian channels*). Prove that a single-mode, phase-insensitive Gaussian channel with $X = x\mathbb{1}$ and $Y = y\mathbb{1}$ is entanglement breaking if and only if

$$y \geq x^2 + 1 \,. \tag{8.71}$$

Solution. As already mentioned, and clarified in Appendix D, a Gaussian channel is entanglement breaking if and only if its action on a subsystem of the 'maximally entangled' Gaussian state, with covariance matrix $\boldsymbol{\sigma}_r$ of Eq. (5.21) in the limit $r \to \infty$, results in a separable state. The covariance matrix $\boldsymbol{\sigma}_{out}$ corresponding to the partial action of the CP-map is given by Eq. (5.53) as

$$\sigma_{out} = \lim_{c \to \infty} \begin{pmatrix} (x^2 c + y)\mathbb{1}_2 & xc\sigma_z \\ xc\sigma_z & c \end{pmatrix} , \qquad (8.72)$$

where we adjusted the notation to include the limit $\lim_{r \to \infty} \cosh(2r) = \lim_{r \to \infty} \sinh(2r) = \lim_{c \to \infty} c$. We can now apply the inequality (7.12) to establish whether such an output is entangled or not. The limits of the partially transposed symplectic invariants, $\mathrm{Det}\,\sigma_{out}$ and $\tilde{\Delta}_{out}$, are extremely simple to express:

$$\lim_{c \to \infty} \mathrm{Det}\,\sigma_{out} = \lim_{c \to \infty} y^2 c^2 , \quad \lim_{c \to \infty} \tilde{\Delta}_{out} = \lim_{c \to \infty} (x^2 + 1)^2 c^2 , \quad (8.73)$$

so that the channel is entanglement breaking if and only if

$$\lim_{c \to \infty} \left(\mathrm{Det}\,\sigma_{out} - \tilde{\Delta} + 1 \right) = \lim_{c \to \infty} (y^2 - (x^2 + 1))c^2 \geq 0 , \quad (8.74)$$

which is indeed equivalent to Eq. (8.71). Notice that all channels that satisfy such an inequality are physical since $x^2 + 1 \geq |x^2 - 1|$, so that they surely satisfy the uncertainty relation for phase insensitive channels, (5.90). Hence, physical, non-entanglement-breaking Gaussian CP-maps populate the region of parameters $y \in [|x^2 - 1|, (x^2 + 1)]$.

8.2.3 Classical capacity of phase-insensitive channels

The previous two sections show that both the minimum and maximum (at given energy) output entropies of phase-insensitive Gaussian channels are additive. Hence, the regularisation of Eq. (8.59) is identical to the single-shot version of Eq. (8.58): $C(\Phi) = C_\chi(\Phi)$. Further, an input alphabet ϱ_{in} that attains the Holevo bound of Eq. (8.58) is promptly envisaged:

$$\varrho_{in} = \int_{\mathbb{R}^2} \mathrm{d}\mathbf{r} \, \frac{e^{-\left(\frac{E}{2} - 1\right)^{-1}\mathbf{r}^\mathsf{T}\mathbf{r}}}{\pi \left(\frac{E}{2} - 1\right)} \hat{D}_\mathbf{r}|0\rangle\langle 0|\hat{D}_\mathbf{r}^\dagger , \qquad (8.75)$$

which corresponds to acting on the vacuum state with the classical mixing channel studied in Section 5.3.2, with $Y = (E/2 - 1)\mathbb{1}$.[8]

The state ϱ_{in} is the Gaussian state with covariance matrix $(E/2)\mathbb{1}$ that maximises the output entropy, and hence the first term in the quantity C_χ, while each state of the decomposition above is a coherent state $\hat{D}_\mathbf{r}|0\rangle$, and

---

[8]Note that $E/2 \geq 1$ since the energy was defined as $\mathrm{Tr}\sigma$ and equals 2 for the ground state (the vacuum). Besides, the case $E = 2$, where the classical mixing channel would not be defined as Y must be positive definite, is trivial and can be disregarded, since it implies that the vacuum would be the only state that can be sent, resulting in a null capacity.

hence maximises the second term. Putting together Eqs. (8.65) and (8.63) then provides us with an explicit expression for the classical capacity:

Classical capacity of phase-insensitive channels. The classical capacity C at maximum input energy E of a phase-insensitive bosonic channel with $X = x\mathbb{1}$ and $Y = y\mathbb{1}$ (or of the phase-contravariant channel with $X = x\sigma_z$ and the same Y) is:

$$C = s_V\left(x^2\frac{E}{2} + y\right) - s_V(x^2 + y) \qquad (8.76)$$

for s_V defined in Eq. (8.64). Here, the energy is defined, up to a constant, as the expectation value of the free Hamiltonian: $E = \sup_{\sigma_{in}} \text{Tr}\,\sigma_{in}$ over all possible input covariance matrices σ_{in} (not necessarily associated with a Gaussian state).

This may be contrasted (see the exercise below) with the formula for the capacity of a classical channel with additive Gaussian white noise, the celebrated Shannon–Hartley formula, which reads

$$C_{cl} = \frac{1}{2}\log_2\left(1 + \frac{E}{N}\right), \qquad (8.77)$$

where E/N is the signal-to-noise ratio, that characterises the allowed inputs as well as the Gaussian classical noise model, described as a conditional probability on a single real random variable.

Problem 8.5. (*Shannon–Hartley formula*). Show that the phase-insensitive classical mixing channel, with $X = \mathbb{1}$ and $Y = (N/2)\mathbb{1}$ has, in the limit $E, N \to \infty$ whilst keeping a finite E/N, a capacity equal to $2C_{cl}$ of Eq. (8.77). What is the reason behind the factor 2?

Solution. The function $s_V(z)$ of Eq. (8.64) may be recast as

$$s_V(z) = \frac{z}{2}\log_2\left(\frac{z+1}{z-1}\right) + \frac{1}{2}\log_2\left(\frac{z^2-1}{4}\right). \qquad (8.78)$$

Utilising L'Hôpital's rule shows that the first term tends to 0 in the limit $z \to \infty$, hence $\lim_{z\to\infty} s_V(z) = \lim_{z\to\infty}\log_2(z/2)$. Inserting this limit into the expression (8.76) for C, after the replacements $x = 1$ and $y = N/2$ – since the energy was defined as $\text{Tr}\,\sigma$, this definition of y would correspond to an energy N associated with the noise – yields

$$\lim_{N\to\infty} C = \lim_{N\to\infty}\left[\log_2\left(\frac{E+N}{4}\right) - \log_2\left(\frac{N}{4}\right)\right] = \log_2\left(1 + \frac{E}{N}\right). \qquad (8.79)$$

The 1/2 discrepancy is due to the fact that, in the classical limit, the two quadratures of the field turn into two independent real classical variables (subject to independent Gaussian noise sources), while the formula (8.77) applies to a single real classical variable.

8.3 QUANTUM METROLOGY

Imagine one has access to quantum means and wants to determine the value of an unknown real parameter, say ϑ, that characterises an operation Φ_ϑ, whose form is otherwise known. In practice, ϑ would typically be a temperature, or a parameter that characterises a Hamiltonian operator (such as an interaction time, or strength, or the product of the two). This is the vanilla version of the problem of 'parameter estimation' in quantum metrology. Occasionally, a variation of this issue when one attempts to establish whether a certain parameter is different from zero is referred to as 'quantum sensing'.

A very natural approach to determining ϑ is to prepare a certain quantum state ϱ, let it undergo the unknown operation Φ_ϑ, and finally measure the state through some POVM $\sum_\mu \hat{K}_\mu^\dagger \hat{K}_\mu = \hat{\mathbb{1}}$. After several measurements, one would like to reconstruct the value of ϑ as well as its standard deviation $\Delta\vartheta$, which gives a reliable quantitative estimate of the determination's precision. If the POVM is fixed, this boils down to a classical problem, where the standard deviation $\Delta\vartheta$ needs to be determined from sampling the conditional distribution $p(\mu|\vartheta) = \mathrm{Tr}(\hat{K}_\mu \varrho_\vartheta \hat{K}_\mu^\dagger)$ of the correlated classical variable μ (the measurement outcome). The solution to this question is given by the classical Fisher information $I_{\hat{K}_\mu, \vartheta}$ (where we emphasise the dependence on the chosen POVM in the context pictured above), given by

$$I_{\hat{K}_\mu, \vartheta} = \sum_\mu p(\mu|\vartheta) \left[\partial_\vartheta \ln \left(p(\mu|\vartheta)\right)\right]^2 = \sum_\mu \frac{(p'(\mu|\vartheta))^2}{p(\mu|\vartheta)} \tag{8.80}$$

(where the prime denotes the partial derivative with respect to the parameter ϑ and the summation might be replaced with an integral over a suitable measurable set), and the associated Cramér–Rao bound is

$$\Delta\vartheta \geq \frac{1}{\sqrt{N I_{\hat{K}_\mu, \vartheta}}}, \tag{8.81}$$

where N is the number of measurements carried out (see Appendix E for the derivation of this inequality). Notice that the lower bound above will in general depend on the value of the parameter ϑ.

The optimisation of the classical Fisher information over all possible POVMs gives rise to the quantum Fisher information I_ϑ:

$$I_\vartheta = \sup_{\hat{K}_\mu} I_{\hat{K}_\mu, \vartheta}, \tag{8.82}$$

where, as above, the symbol $\hat{K}_\mu$ stands for the whole POVM, and the associated quantum Cramér–Rao bound:

$$\Delta\vartheta \geq \frac{1}{\sqrt{N I_\vartheta}} \, . \tag{8.83}$$

Since this inequality may be shown to be achievable, it represents the ultimate bound to quantum parameter estimation.

Here, and henceforth, let $\{\varrho_\vartheta\}$ denote the set of states $\Phi_\vartheta(\varrho)$, parameterised by ϑ, in terms of which the estimation problem could have been cast (without reference to the operation Φ_ϑ). Remarkably, the quantum Fisher information of Eq. (8.82) is amenable to the following general characterisation:

$$I_\vartheta = \mathrm{Tr}\left(\varrho_\vartheta \hat{\mathcal{L}}_\vartheta^2\right) \, , \tag{8.84}$$

where the 'symmetric logarithmic derivative' operator is defined implicitly as the self-adjoint operator that satisfies the following equation:

$$2\varrho_\vartheta' = \hat{\mathcal{L}}_\vartheta \varrho_\vartheta + \varrho_\vartheta \hat{\mathcal{L}}_\vartheta \, , \tag{8.85}$$

and thus characterises the sensitivity of the set ϱ_ϑ to variations in the parameter ϑ. These general results are collected and derived in Appendix E. Let us note in passing that an equivalent characterisation of the quantum Fisher information may also be given in terms of the 'Bures distance' between quantum states. We will not concern ourselves here with such a connection, but just take the symmetric logarithmic derivative path.

Because of their ready availability, the pervasive nature of Gaussian operations, and their ease of manipulation and description, Gaussian states are obvious, major candidates as metrological probes. This provides a compelling reason for investigating the quantum Fisher information of a set of Gaussian states $\{\varrho_\vartheta\}$, which is the subject of the section to follow.

8.3.1   Gaussian quantum Fisher information

Technically, in order to determine the Fisher information, one has to obtain an expression for the symmetric logarithmic derivative operator $\hat{\mathcal{L}}_\vartheta$, defined in Eq. (8.85). For a set of n-mode Gaussian states ϱ_ϑ, we shall put forward the *ansatz* that the symmetric logarithmic derivative must be at most quadratic, and write (we adopt here and in what follows Einstein's convention of summation over repeated indexes):

$$\hat{\mathcal{L}}_\vartheta = L^{(0)} + L_l^{(1)} \hat{r}_l + L_{jk}^{(2)} \hat{r}_j \hat{r}_k \, , \tag{8.86}$$

where the vector $\hat{\mathbf{r}} = (\hat{x}_1, \hat{p}_1, \ldots, \hat{x}_n, \hat{p}_n)^{\mathsf{T}}$ is our standard vector of canonical operators, $L^{(0)} \in \mathbb{R}$, $\mathbf{L}^{(1)} \in \mathbb{R}^{2n}$ and $L^{(2)}$ is a symmetric, real $2n \times 2n$ matrix (whose symmetry ensures the overall Hermiticity of the operator).

We have already developed, in a different context, the formal tools to determine the operator $\hat{\mathcal{L}}_{\vartheta}$. In Chapter 6, when general master equations for the density operator where derived from Gaussian dynamics, the bridging between the phase space and the Hilbert space description was provided by the two relationships (6.45), which allow one to establish the characteristic function correspondences

$$-2i\partial_{\tilde{r}_j}\chi \leftrightarrow \hat{r}_j\varrho + \varrho\hat{r}_j , \tag{8.87}$$

$$\frac{1}{2}(\Omega_{jj'}\tilde{r}_{j'}\Omega_{kk'}\tilde{r}_{k'} - 4\partial_{\tilde{r}_k}\partial_{\tilde{r}_j})\chi \leftrightarrow \hat{r}_k\hat{r}_j\varrho + \varrho\hat{r}_j\hat{r}_k . \tag{8.88}$$

Recall that the variables $\tilde{r}_j$ are defined as $\Omega_{jj'}r_{j'}$, so that the characteristic function χ_G of a generic Gaussian state with covariance matrix σ and first moments $\mathbf{d}$, that both depend on ϑ, can be written as

$$\chi_G = e^{-\frac{1}{4}\sigma_{jk}\tilde{r}_j\tilde{r}_k + i\tilde{r}_l d_l} , \tag{8.89}$$

whence

$$\partial_{\tilde{r}_j}\chi_G = \left(id_j - \frac{1}{2}\sigma_{jj'}\tilde{r}_{j'}\right)\chi_G , \tag{8.90}$$

$$\partial_{\tilde{r}_k}\partial_{\tilde{r}_j}\chi_G = \left[\left(id_k - \frac{1}{2}\sigma_{kk'}\tilde{r}_{k'}\right)\left(id_j - \frac{1}{2}\sigma_{jj'}\tilde{r}_{j'}\right) - \frac{1}{2}\sigma_{jk}\right]\chi_G , \tag{8.91}$$

$$\chi_G' = \left(i\tilde{r}_p d_p' - \frac{1}{4}\sigma_{lm}'\tilde{r}_l\tilde{r}_m\right)\chi_G , \tag{8.92}$$

where a prime $'$ stands for the derivative with respect to the estimation parameter ϑ. Eqs. (8.86), (8.87) and (8.88) allow one to rephrase the symmetric logarithmic derivative equation (8.85) as the following condition:

$$2i\tilde{r}_p d_p' - \frac{\sigma_{lm}'}{2}\tilde{r}_l\tilde{r}_m = 2L^{(0)} + L_p^{(1)}(2d_p + i\sigma_{pp'}\tilde{r}_{p'}) + L_{jk}^{(2)}\left(\frac{\Omega_{jj'}\tilde{r}_{j'}\Omega_{kk'}\tilde{r}_{k'}}{2}\right.$$
$$\left. - \frac{\sigma_{jj'}\tilde{r}_{j'}\sigma_{kk'}\tilde{r}_{k'}}{2} + \sigma_{jk} + 2d_j d_k + id_j\sigma_{kk'}\tilde{r}_{k'} + id_k\sigma_{jj'}\tilde{r}_{j'}\right) \tag{8.93}$$

(we have divided by χ_G, which is allowed since it is never zero). This must hold for all $\tilde{r}$, so that we can equate the different orders of the last equation independently. It is convenient to switch back to a geometric representation of the matrices involved, without indexes, to find

$$\sigma' = \sigma L^{(2)}\sigma + \Omega L^{(2)}\Omega , \tag{8.94}$$

$$\mathbf{L}^{(1)} = 2\sigma^{-1}\mathbf{d}' - 2L^{(2)}\mathbf{d} , \tag{8.95}$$

$$L^{(0)} = -\frac{1}{2}\text{Tr}[\sigma L^{(2)}] - \mathbf{L}^{(1)\mathsf{T}}\mathbf{d} - \mathbf{d}^\mathsf{T} L^{(2)}\mathbf{d} . \tag{8.96}$$

Note that $A_{jk}B_{jk} = \text{Tr}[AB]$, which is just the Hilbert–Schmidt inner product if A and B are real and symmetric. Notice also that, in keeping with our general notation, $\mathbf{L}^{(1)}$ stands for the vector with components $L_j^{(1)}$. Once $L^{(2)}$ is determined by Eq. (8.94), $\mathbf{L}^{(1)}$ and $L^{(0)}$ are given by Eqs. (8.95) and (8.96).

In order to determine $L^{(2)}$, we need to analyse the linear functional $\mathcal{A}_\sigma$ that maps real matrices into matrices according to $\mathcal{A}_\sigma(M) = \sigma M\sigma + \Omega M\Omega$. Clearly, if $\mathcal{A}_\sigma$ were invertible, one would have $\mathcal{A}_\sigma^{-1}(\sigma')$. The inverse $\mathcal{A}_\sigma^{-1}$ may be obtained fairly easily by the symplectic diagonalisation of σ. In fact, let S^{-1} be the symplectic that turns σ in normal form such that, without loss of generality, $\sigma = S\nu S^\mathsf{T}$ with $\nu = \bigoplus_{j=1}^n \nu_j \mathbb{1}_2$. Then, one has

$$\mathcal{A}_\sigma(M) = S\left(\nu S^\mathsf{T} M S\nu + \Omega S^\mathsf{T} M S\Omega\right) S^\mathsf{T} = S\mathcal{A}_\nu(S^\mathsf{T} M S)S^\mathsf{T}, \qquad (8.97)$$

where we used the symplectic conditions $S^{-1}\Omega = \Omega S^\mathsf{T}$ and $\Omega S^{\mathsf{T}-1} = S\Omega$. The inverse of the map $\mathcal{A}_\sigma$, which we are aiming for, may hence be written as

$$\mathcal{A}_\sigma^{-1}(M) = S^{\mathsf{T}-1}\mathcal{A}_\nu^{-1}(S^{-1}MS^{\mathsf{T}-1})S^{-1}. \qquad (8.98)$$

Now, the maps $\Omega \cdot \Omega = -\Omega \cdot \Omega^\mathsf{T}$ and $\nu \cdot \nu$, given by the congruence action of two block matrices, one of which has all blocks proportional to the identity, clearly commute: $\nu\Omega M\Omega\nu = \Omega\nu M\nu\Omega$. A set $\mathfrak{A}$ of common 'eigenmatrices' for such maps is promptly seen to be the following:

$$\mathfrak{A} = \frac{1}{\sqrt{2}}\{\Omega_1^{(jk)}, \ \sigma_z^{(jk)}, \ \mathbb{1}_2^{(jk)}, \ \sigma_x^{(jk)}, \ j,k \in [1,\dots,n]\}$$
$$= \{A_l^{(jk)}, \ l \in [0,\dots,3], \ j,k \in [1,\dots,n]\}, \qquad (8.99)$$

where the superscript (jk) indicates that the matrix is zero everywhere except for the 2×2 block in position jk, whose entries equal the indicated matrix (recall that Ω_1 stands the single-mode symplectic form Ω, while σ_z and σ_x are standard Pauli matrices). It is easy to verify that $\Omega \cdot \Omega$ has eigenvalue -1 for $\mathbb{1}_2^{(jk)}$ and $\Omega_1^{(jk)}$ and $+1$ for $\sigma_z^{(jk)}$ and $\sigma_x^{(jk)}$, whilst ν has eigenvalue $\nu_j\nu_k$ for each eigenmatrix with superscript (jk). Besides, the eigenmatrices above have been chosen so that they are orthonormal with respect to the Hilbert–Schmidt inner product.

The eigenvalues of the map $\mathcal{A}_\nu$ are hence $\nu_j\nu_k \mp 1$ (each of the eigenmatrices $M_l^{(jk)}$, ordered as per Eq. (8.99), has associated eigenvalue $(\nu_j\nu_k - (-1)^l)$). Such a map is therefore invertible if and only if $\nu_j \neq 1$ for all j, that is, if all the local states of the normal modes of the global state with covariance matrix σ are mixed. If any such state is pure, with $\nu_j = 1$, then the full inversion of the map $\mathcal{A}_\nu$, and hence $\mathcal{A}_\sigma$, is generally not possible (notice that the inversions of symplectic matrices that occur in the relationship between the two maps are always possible since $\mathrm{Det}\,S = 1$: no issue may arise there). Even then, the matrix $L^{(2)}$ may still be determined, according to Eq. (8.94), if the argument σ' is orthogonal to the singular eigenvalues. As we will see in detail in the next section in the case of a single mode, this is equivalent to stating that a change in the parameter ϑ is not able to turn the pure state with covariance matrix σ into a mixed state: if that is the case, the symmetric logarithmic derivative, and hence the quantum Fisher information, are still well defined. At a more fundamental, mathematical level, this is a reflection of the fact that turning a pure state into a mixed one means suddenly changing the rank of the state (incidentally, in the Gaussian case, the rank jumps from 1 to ∞, as Gaussian states admit only such two values for the rank): when that happens, the symmetric logarithmic derivative operator may not exist. Although this clarification was in order, we will now proceed assuming the inverse $\mathcal{A}_\sigma^{-1}$ exists: pathological cases related to the singularity of $\mathcal{A}_\sigma$ will emerge at the end of our discussion as divergences in the quantum Fisher information. Formally, our treatment will be equivalent to perturbing slightly, by a quantity ϵ, the incriminated symplectic eigenvalues, and then determining the limit $\epsilon \to 0$ at the end of the evaluation. Notice also that the case $\nu_j + \epsilon$ is indistinguishable from ν_j for all practical purposes.

We can then proceed to invert Eq. (8.94) and write

$$L^{(2)} = \mathcal{A}_\sigma^{-1}(\sigma') . \tag{8.100}$$

In a practical calculation, given σ' and $\sigma = S\nu S^\mathsf{T}$, one would simply parametrise the matrix $S^{-1}\sigma' S^{\mathsf{T}-1}$ as a superposition of eigenmatrices:

$$S^{-1}\sigma' S^{\mathsf{T}-1} = a_l^{(jk)} A_l^{(jk)} , \tag{8.101}$$

and then explicitly have

$$L^{(2)} = \mathcal{A}_\sigma^{-1}(\sigma') = \frac{a_l^{(jk)}}{\nu_j \nu_k - (-1)^l} S^{\mathsf{T}-1} A_l^{(jk)} S^{-1} . \tag{8.102}$$

Now that we have obtained a formula for the Gaussian symmetric logarithmic derivative $\hat{\mathcal{L}}_\vartheta$, we can proceed to insert it into the expression for the quantum Fisher information (8.84). This task is simplified by noticing that Eqs. (8.85) and (8.84) may be combined to obtain the following expression for the quantum Fisher information

$$I_\vartheta = \mathrm{Tr}[\varrho'_\vartheta \hat{\mathcal{L}}_\vartheta] , \tag{8.103}$$

and that we already know the characteristic function associated with ϱ'_ϑ, which is nothing but χ'_G of Eq. (8.92). From Eqs. (6.45) we know that the characteristic function $\chi_{\hat{o}\hat{r}_j}$ associated with operator $\hat{o}\hat{r}_j$ may be derived from the characteristic function $\chi_{\hat{o}}$ associated with operator $\hat{o}$ as $\chi_{\hat{o}\hat{r}_j} = (-i\partial_{\tilde{r}_j} - \frac{1}{2}\Omega_{jj'}\tilde{r}_{j'})\chi_{\hat{o}}$. We can then apply the powerful property of the characteristic function whereby the expectation value of the associated operator is just the characteristic function evaluated in 0, to obtain, recalling the quadratic form (8.86):

$$
\begin{aligned}
I_\vartheta = \mathrm{Tr}[\varrho_\vartheta' \hat{\mathcal{L}}_\vartheta] &= \left[L^{(0)} + L_j^{(1)}(-i\partial_{\tilde{r}_j} - \tfrac{1}{2}\Omega_{jj'}\tilde{r}_{j'}) \right. \\
&\quad \left. + L_{jk}^{(2)}(-i\partial_{\tilde{r}_k} - \tfrac{1}{2}\Omega_{kk'}\tilde{r}_{k'})(-i\partial_{\tilde{r}_j} - \tfrac{1}{2}\Omega_{jj'}\tilde{r}_{j'}) \right] \left(i\tilde{r}_p d_p' - \tfrac{1}{4}\sigma_{lm}'\tilde{r}_l\tilde{r}_m \right) \chi_G \Big|_{\tilde{\mathbf{r}}=0} \\
&= L_j^{(1)} d_j' + \tfrac{1}{2}L_{jk}^{(2)}\sigma_{jk}' + 2L_{jk}^{(2)} d_j' d_k \ ,
\end{aligned}
\tag{8.104}
$$

which, going back to a geometric notation, without indexes, and replacing $\mathbf{L}^{(1)}$ with its expression (8.95), becomes

$$
I_\vartheta = \frac{1}{2}\mathrm{Tr}[L^{(2)}\boldsymbol{\sigma}'] + 2\mathbf{d}'^{\mathsf{T}}\boldsymbol{\sigma}^{-1}\mathbf{d}' \ .
\tag{8.105}
$$

It is certainly worthwhile to summarise our findings as follows:

Gaussian quantum Fisher information. Given a set of Gaussian states with covariance matrix $\boldsymbol{\sigma}_\vartheta$ and first moments $\mathbf{d}_\vartheta$ depending on one real parameter ϑ, the quantum Fisher information I_ϑ associated with the optimal estimation of ϑ may be determined as follows:

1. Determine the normal mode decomposition $\boldsymbol{\sigma}_\vartheta = S\left(\bigoplus_{j=1}^n \nu_j \mathbb{1}_2\right)S^{\mathsf{T}}$.

2. Evaluate the matrix $S^{-1}\boldsymbol{\sigma}'S^{\mathsf{T}-1}$, where $\boldsymbol{\sigma}' = \partial_\vartheta\boldsymbol{\sigma}_\vartheta$, and determine the coefficients $a_l^{(jk)} = \mathrm{Tr}[A_l^{(jk)}S^{-1}\boldsymbol{\sigma}'S^{\mathsf{T}-1}]$, where $A_l^{(jk)}$ are the basis matrices defined in Eq. (8.99), such that $S^{-1}\boldsymbol{\sigma}'S^{\mathsf{T}-1} = a_l^{(jk)}A_j^{(jk)}$.

3. Evaluate the matrix $L^{(2)} = \dfrac{a_l^{(jk)}}{\nu_j\nu_k - (-1)^l}S^{\mathsf{T}-1}A_l^{(jk)}S^{-1}$.

4. Evaluate $I_\vartheta = \frac{1}{2}\mathrm{Tr}[L^{(2)}\boldsymbol{\sigma}'] + 2\mathbf{d}'^{\mathsf{T}}\boldsymbol{\sigma}_\vartheta^{-1}\mathbf{d}'$, where $\mathbf{d}' = \partial_\vartheta\mathbf{d}_\vartheta$.

The optimal measurement to be carried out in order to attain the quantum Cramér–Rao bound is the projective von Neumann measurement in the basis which diagonalises the symmetric logarithmic derivative (see Appendix E) and may also be worked out, with the techniques developed in Chapter 3, essentially by determining the normal mode decomposition of the matrix $L^{(2)}$, associated with the quadratic operator $\hat{\mathcal{L}}_\vartheta$ through Eq. (8.86). It will hence correspond to a number measurement in the Fock basis that diagonalises $\hat{\mathcal{L}}_\vartheta$.

Note that we have not dealt with the possibility of multi-parameter estimation, where the Cramér–Rao bound turns into a matrix inequality on the error covariance matrix, nor have we discussed the problem of choosing input probe states, on which the estimation of a partially unknown quantum operation will in general strongly depend. In the following section, we will specialise our treatment to the single-mode case and show some examples that should provide the reader with some intuition as to how the findings above apply in practical cases.

8.3.2 Quantum estimation with single-mode Gaussian states

The reduction of the general recipe above to single-mode states is both very instructive and extremely relevant to applications. Single-mode states enjoy the specificity of possessing a single symplectic eigenvalue ν_1, so that $\boldsymbol{\sigma} = SS^{\mathsf{T}}\nu_1$.[9] Let us now expand the generic matrix $S^{-1}\boldsymbol{\sigma}'S^{\mathsf{T}-1}$ in the basis of Eq. (8.99), comprising the three matrices $\mathbb{1}_2$, σ_z and σ_x (Ω_1 is redundant here, since $S^{-1}\boldsymbol{\sigma}'S^{\mathsf{T}-1}$ is symmetric):

$$S^{-1}\boldsymbol{\sigma}'S^{\mathsf{T}-1} = a_1 \frac{\sigma_z}{\sqrt{2}} + a_2 \frac{\mathbb{1}_2}{\sqrt{2}} + a_3 \frac{\sigma_x}{\sqrt{2}} , \qquad (8.106)$$

in terms of the three generic real coefficients a_1, a_2 and a_3. Notice that this implies

$$\boldsymbol{\sigma}' = S \left(a_1 \frac{\sigma_z}{\sqrt{2}} + a_2 \frac{\mathbb{1}_2}{\sqrt{2}} + a_3 \frac{\sigma_x}{\sqrt{2}} \right) S^{\mathsf{T}} . \qquad (8.107)$$

The following expression for $L^{(2)}$ immediately ensues:

$$L^{(2)} = S^{\mathsf{T}-1} \left(\frac{a_1}{\nu_1^2 + 1} \frac{\sigma_z}{\sqrt{2}} + \frac{a_2}{\nu_1^2 - 1} \frac{\mathbb{1}_2}{\sqrt{2}} + \frac{a_3}{\nu_1^2 + 1} \frac{\sigma_x}{\sqrt{2}} \right) S^{-1} . \qquad (8.108)$$

The evaluation of Eq. (8.105) is now made straightforward by the fact that the eigenmatrices we have chosen are orthonormal with respect to the Hilbert–Schmidt scalar product, which is precisely what needs to be computed between $L^{(2)}$ and $\boldsymbol{\sigma}'$. One then gets (note that similarity transformations preserve the trace)

$$I_\vartheta = \frac{1}{2} \left(\frac{a_1^2}{\nu_1^2 + 1} + \frac{a_2^2}{\nu_1^2 - 1} + \frac{a_3^2}{\nu_1^2 + 1} \right) + 2\mathbf{d}'^{\mathsf{T}} \boldsymbol{\sigma}^{-1} \mathbf{d}' . \qquad (8.109)$$

This equation may be recast in a more direct and appealing form, by singling out the effect of the change in parameter on the purity $\mu = \mathrm{Tr}(\varrho_\vartheta^2)$ of the Gaussian states under scrutiny. For Gaussian states of one mode, we know that $\mu = 1/\sqrt{\mathrm{Det}\boldsymbol{\sigma}} = \nu_1^{-1}$. Furthermore, $\boldsymbol{\sigma}^{-1} = S^{\mathsf{T}-1}S^{-1}\nu_1^{-1}$. By evaluating $\mathrm{Det}(\boldsymbol{\sigma} + \boldsymbol{\sigma}'d\vartheta)$ at first order in $d\vartheta$, it is easy to verify that $(\mathrm{Det}\boldsymbol{\sigma})' = \mathrm{Tr}[\boldsymbol{\sigma}^{-1}\boldsymbol{\sigma}']\mathrm{Det}\boldsymbol{\sigma} = \sqrt{2}\nu_1 a_2$, where we used Eq. (8.107) for $\boldsymbol{\sigma}'$. In terms of the purity μ, one has $\mu' = -\frac{1}{2}\frac{(\mathrm{Det}\boldsymbol{\sigma})'}{(\mathrm{Det}\boldsymbol{\sigma})^{3/2}} = -\frac{a_2\mu^2}{\sqrt{2}}$.

Notice now that

$$\mathrm{Tr}[(\boldsymbol{\sigma}^{-1}\boldsymbol{\sigma}')^2] = \mu^2(a_1^2 + a_2^2 + a_3^2)$$

$$= \mu^2(1 + \mu^2) \left(\frac{a_1^2}{1 + \mu^2} + \frac{a_2^2}{1 - \mu^2} + \frac{a_3^2}{1 + \mu^2} \right) - \frac{2\mu^4 a_2^2}{1 - \mu^2}$$

$$= \mu^2(1 + \mu^2) \left(\frac{a_1^2}{1 + \mu^2} + \frac{a_2^2}{1 - \mu^2} + \frac{a_3^2}{1 + \mu^2} \right) - \frac{4\mu'^2}{1 - \mu^2} , \qquad (8.110)$$

---

[9] In fact, the argument developed in this section could be generalised to 'isotropic' multimode states – with fully degenerate symplectic spectrum – for which the same identity holds, since the symplectic transformation can be carried through the normal mode form (which is proportional to the identity for such states).

which can be inserted into Eq. (8.109), recalling that $\nu_1 = \mu^{-1}$, to obtain the following notable result:

Single-mode Gaussian quantum Fisher information. The quantum Fisher information I_ϑ of a set of single-mode Gaussian states with covariance matrices σ_ϑ and first moments d_ϑ is given by

$$I_\vartheta = \frac{1}{2} \frac{\mathrm{Tr}[(\sigma^{-1}\sigma')^2]}{1+\mu^2} + \frac{2\mu'^2}{1-\mu^4} + 2d'^\mathsf{T}\sigma^{-1}d', \qquad (8.111)$$

where $\mu = 1/\sqrt{\mathrm{Det}\,\sigma_\vartheta}$ is the purity of the quantum states and the prime $'$ denotes differentiation with respect to the parameter ϑ.

Notice that this parametrisation clearly isolates the term that may lead to diverging Fisher information as the one dependent on the derivative of the purity. As discussed in the previous section, this is related to the impossibility of defining the symmetric logarithmic derivative and is due to the fact that such a term would be responsible for a sudden change in the rank of the Gaussian state, jumping from 1 to ∞. The formula above may be applied to a number of diverse circumstances. We shall briefly examine a few of them in the problems that follow.

Problem 8.6. (*Optical phase estimation*). An experimentalist, capable of generating coherent states $\hat{D}_\mathbf{r}|0\rangle$ with respect to a stable phase reference, is given a phase plate which acts as a unitary phase shift operation, with associated symplectic R_φ of Eq. (5.12), but is not sure about the value of φ. Determine the quantum Fisher information associated with the estimation of φ that would need to be carried out in order to calibrate the phase shifter.

Solution. The covariance matrix of the coherent states exiting the phase plate is just the identity: $\sigma = \mathbb{1}_2$ (a constant), so that the first two terms in Eq. (8.111), depending on the derivative of the covariance, do not contribute to the quantum Fisher information in this case. Without loss of generality, given the rotational symmetry of the problem, one can pick the first moment vector $\mathbf{r}$ of the coherent state along x: $\mathbf{r} = (|\mathbf{r}|, 0)^\mathsf{T}$, so that the action of the phase plate rotates it to $\mathbf{r}_\varphi = |\mathbf{r}|(\cos\varphi, -\sin\varphi)^\mathsf{T}$. Hence, its derivative is $\mathbf{r}'_\varphi = |\mathbf{r}|(-\sin\varphi, -\cos\varphi)^\mathsf{T}$, which may be inserted as d' into (8.111) to get the simple result

$$I_\varphi = 2|\mathbf{r}|^2 = N, \qquad (8.112)$$

where N is the average number of excitations in the probe coherent state with displacement $\mathbf{r}$. Clearly, a stronger laser, with a larger amplitude with respect to the vacuum noise, allows for a better discrimination of φ. In this very basic instance, the Fisher information is independent from the value of the parameter φ.

Problem 8.7. (*Squeezing for optical phase estimation*). After a major overhaul of the lab, our experimentalist is capable of generating displaced squeezed states, with covariance matrix $Z = \mathrm{diag}(z, 1/z)$ and first moments $\mathbf{r}$, and is still keen on estimating the unknown phase of an R_φ of Eq. (5.12), as in the previous problem. Determine the quantum Fisher information associated with such an estimation with the new squeezed resource.

Solution. It is crucial to note that $R'_\varphi = R_{\varphi+\frac{\pi}{2}} = \Omega R_\varphi$ and hence $R'_{-\varphi} = -R_{-\varphi+\frac{\pi}{2}} = -R_{-\varphi}\Omega$, where $\Omega = R_{\frac{\pi}{2}}$ and the abelian nature of $SO(2)$ were employed. Thus, since $\boldsymbol{\sigma} = R_\varphi Z R_{-\varphi}$, $\boldsymbol{\sigma}' = \Omega\boldsymbol{\sigma} - \boldsymbol{\sigma}\Omega$, whence $(\boldsymbol{\sigma}^{-1}\boldsymbol{\sigma}')^2 = 2(\boldsymbol{\sigma}^{-1}\Omega\boldsymbol{\sigma}\Omega^{\mathsf{T}} - \mathbb{1}) = 2(\boldsymbol{\sigma}^{-2} - \mathbb{1})$ (in the last step, we used the identity $\Omega Z \Omega^{\mathsf{T}} = Z^{-1}$). The first term in (8.111) is hence determined by $\mathrm{Tr}[(\boldsymbol{\sigma}^{-1}\boldsymbol{\sigma}')^2] = 2\mathrm{Tr}[Z^{-2}] - 4 = 2(z^2 + 1/z^2 - 2)$, which is independent of the parameter φ. The second term is instead always zero, as the purity does not change under unitary phase shifters.

The term depending on the first moments can be dealt with as in the previous problem, with $\mathbf{d}_\varphi = R_\varphi \mathbf{r}$, for some arbitrary initial $\mathbf{r}$, and $\mathbf{d}' = R_\varphi \Omega \mathbf{r}$, so that $\mathbf{d}'^{\mathsf{T}}\boldsymbol{\sigma}\mathbf{d}' = \mathbf{r}^{\mathsf{T}}\Omega^{\mathsf{T}}Z\Omega\mathbf{r}$. Notice that the unknown yet identical rotations acting on squeezing and first moments cancel out, to yield a value independent of φ. Let us then set, without loss of generality, $z > 1$ and assume $\mathbf{r}$ to be prepared along the optimal direction such that $\mathbf{r}^{\mathsf{T}}\Omega^{\mathsf{T}}Z\Omega\mathbf{r} = z|\mathbf{r}|^2$.

The total quantum Fisher information is therefore

$$I_\varphi = \frac{(z^2 + \frac{1}{z^2} - 2)}{2} + 2z|\mathbf{r}|^2 = \frac{(z - \frac{1}{z})^2}{2} + 2z|\mathbf{r}|^2 . \qquad (8.113)$$

Not surprisingly, for given $|\mathbf{r}|^2$, squeezing grants an improvement over the coherent state case (8.112). Note that, in terms of the average number of excitations in the probe states, N, one has $4N + 2 = \mathrm{Tr}\,\boldsymbol{\sigma} + 2|\mathbf{r}|^2 = (z + 1/z) + 2|\mathbf{r}|^2$. If one sets $|\mathbf{r}| = 0$ and puts all the energy into squeezing, the formula above becomes

$$I_\varphi = \frac{(z - \frac{1}{z})^2}{2} = \frac{(z + \frac{1}{z})^2 - 4}{2} = 8(N^2 + N) , \qquad (8.114)$$

which yields a better scaling with N than the linear one obtained above in Eq. (8.112). This argument has however to be taken *cum grano salis*, as squeezing and displacement are hardly two interchangeable resources.

Problem 8.8. (*A Gaussian thermometer*). Our experimentalist, steeped in Gaussian lore, decides to build a thermometer with Gaussian quantum states. The vacuum state is hence allowed to interact with a thermal environment for a certain time, which is formally equivalent to letting it through a lossy attenuator channel, defined by Eq. (5.77). The value of θ, reflecting interaction time and coupling to the environment, is known, but n_{th} is not – the thermal noise n_{th} is related to the inverse temperature β through the Bose law, as $n_{th} = \frac{e^{\beta\omega}+1}{e^{\beta\omega}-1}$, where ω is the (known) system frequency (natural units are assumed). Determine the Fisher information associated with the estimation of n_{th}. Would the use of coherent states with finite amplitude as probes improve the thermometre?

Solution. The input covariance matrix is mapped into $\sigma_{n_{th}} = (\cos^2\theta + n_{th}\sin^2\theta)\mathbb{1}$, with $\sigma' = \sin^2\theta\mathbb{1}$. Because all covariance matrices involved are already in normal form, with symplectic eigenvalue $\nu_1 = (\cos^2\theta + n_{th}\sin^2\theta)$, it is handier in this case to employ the formula (8.109). One has simply $a_2 = \sqrt{2}\sin^2\theta$, while $a_1 = a_3 = 0$. Then

$$I_{n_{th}} = \frac{\sin^4\theta}{(\cos^2\theta + n_{th}\sin^2\theta)^2 - 1} = \frac{1}{(n_{th}-1)}\frac{\sin^2\theta}{(n_{th}-1)\sin^2\theta + 2}. \tag{8.115}$$

This thermometre might in principle attain an asymptotic precision of $1/(n_{th}^2 - 1)$ in the limit $\sin\theta \to 1$ (which corresponds to an infinite interaction time). Note that the scaling of the signal-to-noise ratio achieved by this thermometre by attaining the quantum Cramér-Rao bound would be

$$\frac{n_{th}}{\Delta n_{th}} = \sqrt{N}\frac{n_{th}}{|\sin\theta|}\sqrt{(n_{th}-1)[(n_{th}-1)\sin^2\theta + 2]}, \tag{8.116}$$

where N is the number of measurements taken. Higher temperatures allow for higher relative precisions at the rate $\sim n_{th}^2$.
Coherent states of finite amplitude would not help, as the parameter n_{th} does not affect the first moments at all, and hence no contribution would come from the last term of Eq. (8.109) in this case. Squeezing would help ... or would it?

8.4   QUANTUM KEY DISTRIBUTION

Quantum cryptography, or, more specifically, 'quantum key distribution', is primarily concerned with the distribution of a secret classical key between two distant parties linked by a quantum channel. Like in most other areas of quantum communication, the implementation of cryptographic protocols using Gaussian quantum states of light is particularly appealing in practice, so that Gaussian quantum cryptography has grown to become a rather bulky subfield, with several developments and ramifications in terms of both security proofs and protocols. Here, we will not make any attempt to cover such developments: we will limit ourselves to demonstrating the possibility of continuous variable quantum cryptography by outlining a seminal proposal in this direction, put forward by Grosshans and Grangier.

Before proceeding with the continuous variable case, let us briefly review the classic BB84 qubit protocol (named after Charles Bennet and Gilles Brassard), which will allow us to highlight some of the basic features and methods common to most quantum cryptographic protocols.

The aim of the game is for two distant parties A and B to share a common, secret "one-time pad", a string of classical bits that they can add to their messages to encrypt and decrypt them. They can proceed as follows:

1. A has a one-time pad to send, say, for the sake of example, $\{100101\}$.

2. A uses a random number generator that yields a sequence of preparation bases out of a choice of two, say in our example $\{\sigma_x, \sigma_z, \sigma_z, \sigma_z, \sigma_x, \sigma_x\}$.

3. Alice encodes the secret key into certain quantum states depending on the value of the bit and the associated preparation basis, according to $\{0, \sigma_z\} \mapsto |0\rangle$, $\{1, \sigma_z\} \mapsto |1\rangle$, $\{0, \sigma_x\} \mapsto |+\rangle$, $\{1, \sigma_x\} \mapsto |-\rangle$, where $|0\rangle$ and $|1\rangle$ are eigenstates of the standard Pauli z matrix σ_z with eigenvalues -1 and $+1$ in the qubit Hilbert space, while $|\pm\rangle = (|0\rangle \pm |1\rangle)/\sqrt{2}$. In the example, A would prepare $\{|-\rangle, |0\rangle, |0\rangle, |1\rangle|+\rangle, |-\rangle\}$.

4. A sends the states to B. A third party E might be lurking and dropping eaves along the transmission line. This possibility will be discussed below.

5. B has a random number generator that dictates the choice of measurement basis, say $\{\sigma_x, \sigma_x, \sigma_z, \sigma_z, \sigma_x, \sigma_z\}$. Measuring in such a basis, and assuming that no noise affected the states in the intervening transmission, B would get the outcomes $\{1, ?, 0, 1, 1, ?\}$, where the question marks denote a 0 or 1 with probability $1/2$ each.

6. A and B publicly announce their respective sequences of preparation and measurement bases: only digits where they used the same basis are kept. The others (the question marks above) are discarded.

7. A and B must still do something about the presence of noise and of a possible third party E, against which "secrecy" should be established, during step 4: they hence sacrifice a portion of their key by publicly announcing it. Thus, they estimate an error rate ε. If ε is small enough, the presence of E may be ruled out (see later).

8. The rate ε might still be too large for practical purposes, so A and B may perform "information reconciliation", a classical error correction protocol based on substrings' parity checks, in order to reduce it.

9. Finally, A and B add a further layer of security by performing the so called "privacy amplification" whereby, rather than using the 'bare' keys, they take the parity of strings with a certain length.

The last two steps are classical: some suitably modified versions of them will be employed in the continuous variable protocol that shall be described in the next section.

The crucial aspect that distinguishes quantum key distribution protocols like BB84 from a classical transmission is the analysis of possible information 'attacks' by E during step 4, which must be carried out by the tenets of quantum mechanics. The formal classifications of possible types of attack is potentially rather convoluted and debatable. Here, we will merely distinguish the attacks as *individual*, where E intercepts and manipulates one transmitted qubit at a time, or *collective*, where E accesses and manipulates multiple transmitted qubits at a time. At some stage, E will want to perform a measurement by some POVM, with the ultimate objective of extracting the classical information encoded in the transmitted qubit. Afterwards, E will send a state along to B, which could be any state depending on the measurement outcome (such as the state obtained via the von Neumann postulate after the measurement). Clearly, E has, in general, a wide set of options available both in terms of choice of measurement and of state preparation. For instance, a 'weak' (non-projective) POVM would grant less information but also disturb the state less, and might be more advantageous than a projective von Neumann measurement on the transmitted qubit. We will not venture to face the complications of an actual security proof here, but just present some general heuristic comments on the security of the BB84 protocol against individual attacks.

If the transmission were a mere classical bit, E could just determine it by measuring and then send a copy to B. The same is not allowed in the quantum domain by the celebrated no-cloning theorem, a simple consequence of linearity, by whose virtue a single specimen of a quantum state cannot be copied perfectly. Still, party E could intercept the qubit and measure it, in the σ_x or σ_z basis, since knowledge of the protocol is public, thus getting the right classical bit in 50% of the cases. Notice that E even knows which bits to keep, as in step 6 of the protocol A and B announce the bases used for encoding. However, when E gets the basis wrong, the qubit state will collapse in a linear superposition of the original plus an orthogonal state (assuming E sends a state corresponding to the collapsed one: since nothing is known about the transmitted state, that turns out to be as good a choice as any). Thus, in such cases, B has a final probability of 50% of getting a wrong bit. Overall, this accounts for a probability of error for A in B, calculated in step 7, of at least 25%! Such a huge error rate flags the presence of the eavesdropper and makes the exchange secure. A slightly more sophisticated argument to justify security, taking into account a noisy transmission line, will be presented below in a continuous variable scenario.

8.4.1 Quantum key distribution with coherent states

If A and B are linked by a lossy bosonic channel modelled by Eq. (5.77) with loss factor $\cos^2 \theta = \eta$ and thermal noise n_{th} – such as an optical fibre at telecommunication frequencies – they can exchange a secret key as follows:

1. A samples two normally distributed random real variables x and p, with mean zero and variance $V/2$ (the same on both variables), and prepares the coherent state $|x + ip\rangle$.

2. A sends the state to B over the noisy channel.

3. B performs homodyne detections of either $\hat{x}$ or $\hat{p}$, according to a random sequence of measurement bases, obtaining the real variables y_B.

4. B announces publicly what measurement bases were used. Somewhat similar to the original BB84 protocol, A will accordingly discard half of the original x and p values that were used in preparing the state, keeping only those corresponding to the measured bases. This defines the random variable y_A.

Let us now pause the protocol for a moment of discussion. At this stage, the random variables y_A and y_B – arising, respectively, from state preparations and measurements, the two boundaries between the classical and the quantum worlds – are a pair of classical, correlated Gaussian variables. As we shall see, the variance of y_B can be evaluated as $\eta(V + 1)/2 + (1 - \eta)n_{th}/2$, while the variance of y_A is $V/2$. The Shannon–Hartley formula (8.77), encountered in the study of bosonic channels and re-derived in Problem 8.5 as a limiting instance of a quantum capacity, can hence be applied for energy modulation $E = V/2$ and additive noise $N = 1/2 + n_{th}(1 - \eta)/(2\eta)$, to determine the maximum classical mutual information (in bits) that A and B can attain:[10]

$$I_{AB} = \frac{1}{2} \log_2 \left(1 + \frac{\eta V}{\eta + n_{th}(1 - \eta)} \right) . \tag{8.118}$$

---

[10] It is easy to see from the probability distribution of homodyne detection (5.129) that the variable y_B is governed by the following conditional probability density $dp(y_B|y_A)$:

$$dp(y_B|y_A) = \frac{e^{-\frac{(y_B - y_A)^2}{\eta V + \eta + n_{th}(1 - \eta)}}}{\sqrt{\pi(\eta V + \eta + n_{th}(1 - \eta))}} \, dy_B . \tag{8.117}$$

Given the knowledge of η, the effect of the attenuation η on A's energy modulation can be reabsorbed by rescaling the variables, which leaves one with a conditional variance equal to $V/2 + 1/2 + n_{th}(1 - \eta)/(2\eta)$ (recall that the variance of a Gaussian distribution is half the denominator at the exponential), on which the Shannon–Hartley formula applies with signal-to-noise ratio $E/N = V/(1 + n_{th}(1 - \eta)/\eta)$, as in the main text.

The Shannon limit (8.118) may be reached in practice by a classical 'sliced reconciliation' protocol, with which the protocol would now resume, leaving A and B with a shared key. After that, standard privacy amplification would follow, as in the BB84 case. We will not dwell on any such classical techniques, but rather discuss the issue of security during the transmission of the quantum signal, which represents the distinctive quantum cryptographic facet of the protocol.

In order to provide a convincing security argument, let us assume an eavesdropper party E capable of ideal operations within the standard restrictions of quantum mechanics. In order to be completely undetectable, E will want to hide in the noise. We shall hence constrain E's manipulation of the transmitted state to produce a state, to be sent to B, which is the same as the state that would result from the noisy transmission line[11] (which E knows inside out). At the same time, E will aim at preparing and then measuring another state which is as close as possible to the one that will be sent to B. The relationship of this task to the no-cloning theorem is thus apparent. In particular, E is confronted with the problem of achieving optimal cloning within the restrictions imposed by the theorem. The limitation to optimal 1 to 2 copy cloning imposed by quantum mechanics will in fact be the crucial ingredient for the protocol's security.

In what follows, we shall also restrict to individual attacks. Let us then assume that an omnipotent E can intercept the state sent by A before any noise has affected it and is capable, after manipulation, of transmitting it to B faithfully without adding any noise during the transmission. The distribution of E's input states, coherent states with random displacements modulated by a two-variable Gaussian with covariance matrix $V\mathbb{1}_2/2$ and zero mean, is described by a classical mixing channel with $Y = V\mathbb{1}_2$ acting on the covariance matrix $\boldsymbol{\sigma} = \mathbb{1}_2$ (see Section 5.3.2). The ensemble of states leaving A may hence be collectively described by the Gaussian state with zero mean and covariance matrix $\boldsymbol{\sigma}_A = (V + 1)\mathbb{1}_2$. B's homodyne measurements, in either basis, are hence Gaussian distributed with zero mean and variance $(V + 1)/2$ (see Section 5.4.1 and the footnote on page 263). In order for E to be completely undetectable, its manipulation before measuring must be a Gaussian CP-map. It is convenient here to describe such a manipulation in the Heisenberg picture. Let $\hat{\mathbf{r}}_A$ be the vector of quadrature operators associated with the mode sent by A. The most general Gaussian CP-map applied by E may be described by the following two linear relationships at the level of the canonical

---

[11] This assumption and its direct consequences (such as the restriction to Gaussian manipulations on E's part), is debatable, as the possibility of detecting alterations in the state would depend on A and B specific error analysis. However, it is not outrageous to assume it here. We use the term "completely undetectable" to refer to this type of eavesdropping in formal statements.

operators:

$$\hat{\mathbf{r}}_B = x\hat{\mathbf{r}}_A + \hat{\mathbf{n}}_B \, , \qquad (8.119)$$

$$\hat{\mathbf{r}}_E = w\hat{\mathbf{r}}_A + \hat{\mathbf{n}}_E \, , \qquad (8.120)$$

where $x, w \in \mathbb{R}$ and $\hat{\mathbf{n}}_B$ and $\hat{\mathbf{n}}_E$ are generic linear combinations of canonical operators of additional modes, all commuting with $\hat{\mathbf{r}}_A$ (and uncorrelated with it). We will not need the specific canonical structure of such linear combinations, which must correspond to the symplectic dilation of a deterministic Gaussian CP-map, other than to note that the output canonical modes satisfy $[\hat{\mathbf{r}}_B, \hat{\mathbf{r}}_E^\mathsf{T}] = 0$ (in our standard outer product notation, set out in Section 1.3.3). Then, observe that

$$[\hat{\mathbf{n}}_B, \hat{\mathbf{n}}_E^\mathsf{T}] = xw[\hat{\mathbf{r}}_A, \hat{\mathbf{r}}_A^\mathsf{T}] = ixw\Omega \, . \qquad (8.121)$$

In particular, $[\hat{n}_{B1}, \hat{n}_{E2}] = [\hat{n}_{E1}, \hat{n}_{B2}] = ixw$, from which the standard necessary implications below follow[12]

$$\Delta\hat{n}_{B1}^2\Delta\hat{n}_{E2}^2 \geq \frac{x^2 w^2}{4} \, , \qquad \Delta\hat{n}_{B2}^2\Delta\hat{n}_{E1}^2 \geq \frac{x^2 w^2}{4} \, , \qquad (8.122)$$

where $\Delta\hat{A}^2$ stands for the variance of the operator $\hat{A}$. The maps described by Eqs. (8.119) and (8.120) send the input covariance matrix $\boldsymbol{\sigma}_A$ into $\boldsymbol{\sigma}_B = x^2\boldsymbol{\sigma}_A + \{\hat{\mathbf{n}}_B, \hat{\mathbf{n}}_B^\mathsf{T}\}$ and $\boldsymbol{\sigma}_E = w^2\boldsymbol{\sigma}_A + \{\hat{\mathbf{n}}_E, \hat{\mathbf{n}}_E^\mathsf{T}\}$ (where $\hat{\mathbf{n}}_B$ and $\hat{\mathbf{n}}_E$ may be assumed to have null first moments). For E to be completely undetectable, it must hence be $x^2 = \eta$ and $\{\hat{\mathbf{n}}_B, \hat{\mathbf{n}}_B^\mathsf{T}\} = 2\mathrm{diag}(\Delta\hat{n}_{B1}^2, \Delta\hat{n}_{B2}^2) = n_{\mathsf{th}}(1-\eta)\mathbb{1}_2$ (perfectly reproducing the transmission noise). When inserted into the no-cloning relations (8.122), these values yield the following inequality on the noise affecting E's state:

$$\frac{\Delta\hat{n}_{Ej}^2}{w^2} \geq \frac{\eta}{2n_{\mathsf{th}}(1-\eta)} \, , \qquad \text{for} \quad j = 1, 2 \, . \qquad (8.123)$$

As we have seen in the discussion that led to Eq. (8.118), the ratio between added noise and attenuation is precisely what sets the attainable mutual information between two parties sharing correlated Gaussian variables.[13] Hence, by virtue of the Shannon–Hartley formula, the mutual information I_{AE} E can establish with A is bound by (8.123) as per

$$I_{AE} = \frac{1}{2}\log_2\left(1 + \frac{\frac{V}{2}}{\frac{1}{2} + \frac{\Delta\hat{n}_{Ej}^2}{w^2}}\right) \leq \frac{1}{2}\log_2\left(1 + \frac{n_{\mathsf{th}}(1-\eta)V}{\eta + n_{\mathsf{th}}(1-\eta)}\right) \, . \qquad (8.124)$$

Now, a classical analysis reveals that A and B can establish a secure key

---

[12]The Heisenberg uncertainty principle, in its customary from, which can be derived through the same methods of Section 3.4, stating that $4\Delta\hat{A}^2\Delta\hat{B}^2 \geq |\langle[\hat{A}, \hat{B}]\rangle|^2$.

[13]Here, we are implicitly assuming that E is restricted to homodyne measurements on the modes $\hat{\mathbf{r}}_E$.

after sliced reconciliation and privacy amplification if and only if their mutual information I_{AB} is greater than the mutual information I_{AE} between A and E which, using Eqs. (8.118) and (8.124), yields the following sufficient condition for security:

$$I_{AB} > I_{AE} \quad \Leftrightarrow \quad \eta > \frac{n_{\text{th}}}{n_{\text{th}} + 1} \,. \tag{8.125}$$

We have thus established what follows:

Security condition for coherent state quantum key distribution. Two parties A and B, linked by an attenuation channel with attenuation factor η and thermal noise n_{th} can distribute a classical key securely against individual, completely undetectable homodyne attacks if

$$\eta > \frac{n_{\text{th}}}{n_{\text{th}} + 1} \,. \tag{8.126}$$

In the case of negligible thermal noise, $n_{\text{th}} \gtrsim 1$, as would be the case for extremely high-frequency telecommunication or optical signals, the inequality (8.126) reduces to $\eta > 1/2$: as long as less than half the signal is lost, A and B can establish a secure key.

Problem 8.9. (*Optimal eavesdropping*). Identify an optimal attack for the case of a pure loss transmission line, $n_{\text{th}} = 1$.

Solution. In such a case, the optimal attack consists simply in mixing the intercepted mode with the vacuum at a beam splitter with transmittivity η: one output of the beam splitter, which reproduces the noise exactly, is sent to B, while the other is kept to undergo detection. In fact, such a strategy is equivalent to the Heisenberg evolution

$$\hat{\mathbf{r}}_B = \sqrt{\eta}\,\hat{\mathbf{r}}_A + \sqrt{1 - \eta}\,\hat{\mathbf{r}}_C \,, \tag{8.127}$$

$$\hat{\mathbf{r}}_E = -\sqrt{1 - \eta}\,\hat{\mathbf{r}}_A + \sqrt{\eta}\,\hat{\mathbf{r}}_C \,, \tag{8.128}$$

where $\hat{\mathbf{r}}_C$ belongs to an additional mode in the vacuum state. Thus, the noise on B is perfectly reproduced and undetectable, whilst $\Delta\hat{n}_{E_j}^2/w^2 = \eta/[2(1 - \eta)]$ saturates (8.124) and is optimal.

Notice also that the differential mutual information $I_{AB} - I_{AE}$ actually constitutes the rate at which A and B can establish a secure key. As long as the inequality (8.126) is fulfilled, any rate can be achieved, in principle, by increasing the modulation V.

This basic proof of the possibility of cryptographic exchanges with continuous variables can be greatly refined, in terms of both generalising the protocol adopted and of refining the security analysis. All the same, as it is, it already provides a spectacular demonstration of the possibility of performing quantum tasks with very ordinary Gaussian resources. Although the whole procedure could be reproduced with classical means, as is always the case for Gaussian protocols, a classical model would not be constrained by the fundamental restrictions (8.122), on which our security argument relies. Such uncertainty relations come in fact from the underlying Hilbert space structure, which classical probabilistic models cannot boast.

8.5 FURTHER READING

Most things Gaussian in theoretical quantum information are reviewed in [84], to which we refer the reader looking for an exhaustive survey, in particular concerning quantum teleportation, key distribution and distinguishability measures on the set of Gaussian states. For an overview of continuous variable protocols, including an account of early seminal experiments in the field, see [14]. The edited volumes [13, 17] also offer a broad selection covering early developments in the area, including entanglement theory, quantum computation, and implementations.

A general formula for the fidelity between two arbitrary Gaussian states was derived in [4]. The argument for the classical benchmark for quantum teleportation and storage of Section 8.1.2 is drawn from [44], while the extension for squeezed states based on symmetries may be found in [66]. A more general approach based on symmetries, encompassing broader ensembles of input states, is presented in [19]. The mathematical framework for justifying the restriction to Weyl-covariant channels is introduced in [85]. A scheme for the teleportation of a classical probability distribution that may be tested with equipment worth three pence is set forth in [21]. A confronting theoretical exchange on the feasibility of coherent state quantum teleportation, with bearing on much more general issues such as the notion of a quantum reference frame and clock synchronisation, is contained in [73, 88].

The solution to the phase-insensitive Gaussian capacity problem was first reported in [39], although the treatment we followed is taken from [60]. See also Appendix D for further, more general references on quantum channels. The existence of superadditive quantum channels was first announced in [45]. The extremality properties of Gaussian states are discussed in [92]. The connection between the additivity of the minimum output entropy and the optimality of Gaussian decompositions in evaluating Gaussian discord is made in [69].

Useful introductions to quantum estimation theory may be found in [67], as well as in [90]. The formula (8.111) for single-mode Gaussian parameter estimation first appeared in [68], in which it is derived through the Bures distance. Our multi-mode derivation of the quadratic symmetric logarithmic derivative follows the treatment of [61]. A more general study encompassing all states in exponential form with arbitrary generators is reported in [56].

The continuous variable quantum key distribution protocol we selected was put forward in [42], to which the reader is referred for further references concerning the classical portions of the protocol, such as the security condition based on the positivity of the differential mutual information, used in

Eq. (8.125). The no-cloning argument adopted to prove security was drawn from [41] where, interestingly, a malicious setting for quantum teleportation, with a coherent state threshold of $2/3$ rather than $1/2$, is also introduced. Let us note once more that a discussion of developments in continuous variable quantum key distribution (as well as the related subject of optimal cloning) may be found in [84]. A general introduction to quantum cryptography is contained in Nielsen and Chuang's classic textbook [64].

CHAPTER 9

A grand tour of continuous variable platforms

CONTENTS

With the exception of references made to optical elements and detection techniques when discussing Gaussian operations and POVMs, and in their applications, the theory developed in this book has been largely presented as an abstract mathematical construction. This chapter intends to liaise our theoretical pictures to the wide world of controllable, coherent quantum systems that share the continuous variable description. Emphasis will be on the emergence of the canonical commutation relations in different contexts and on the introduction of the relevant Hamiltonian descriptions with the understanding that, once the basic formal analogy is established, the formalisms and results of the previous chapters will be in principle applicable to the system at hand. Note, as a caveat, that some of the Hamiltonians listed in this chapter are not quadratic, notably those involving Josephson junctions and cold atoms, so that the Gaussian theory would not apply to them.

The aim of this chapter is hence not providing a thorough and exhaustive discussion of each such continuous variable system, but rather to grant the reader a degree of awareness on the possibilities offered by the diverse experimental platforms to which the theory applies. Thus, the conceptual elements will typically overshadow the (often very important when applications are concerned!) technical aspects, and glaring simplifications will be adopted.

Given the central role played by the optical embodiment of quantum continuous variables, an exception to this cursory treatment has been made for the quantisation of the electromagnetic field, carried out below in detail from the classical theory. This will also present the opportunity to provide a full derivation of the input-output formalism and the associated quantum Langevin equations, which to our knowledge were never previously dealt with at a pedagogical level in the time domain, and to discuss a bag of quantum optical tools, such as the rotating wave approximation, which, although featuring in a number of introductory quantum optics textbooks, find such wide application to deserve a place in any attempt at a self-contained treatment.

9.1 QUANTUM LIGHT

9.1.1 Classical light

Classical electromagnetic waves in a vacuum are governed by the Maxwell equations for the electric and magnetic fields $\mathbf{E}$ and $\mathbf{B}$ which, in the absence of sources and currents, take the well-known form:

$$\nabla \cdot \mathbf{E} = 0 \tag{9.1}$$

$$\nabla \times \mathbf{E} = -\partial_t \mathbf{B} \tag{9.2}$$

$$\nabla \cdot \mathbf{B} = 0 \tag{9.3}$$

$$c^2 \nabla \times \mathbf{B} = \partial_t \mathbf{E} \tag{9.4}$$

These equations imply the wave equations $\nabla^2 \mathbf{E} - \frac{1}{c^2}\partial_{tt}^2 \mathbf{E} = \nabla^2 \mathbf{B} - \frac{1}{c^2}\partial_{tt}^2 \mathbf{B} = 0$, although they are not equivalent to them, in the sense that not all solutions to the wave equations are solutions to the Maxwell equations. Specifically, the conditions on the divergences (9.1) and (9.3) imply that the fields are transverse with respect to the propagation direction of the wave solutions, while induction, expressed by Eqs. (9.2) and (9.4), interlinks the magnetic and electric fields, halving the number of independent parameters of the electromagnetic field. Besides, it should be noted that the fields $\mathbf{E}$ and $\mathbf{B}$ are always real vectors. A general solution to the Maxwell equations in a vacuum, obtained by solving the wave equations by separation of variables and by imposing all these additional constraints, may be written as follows:

$$\mathbf{E} = i \sum_{\mathbf{k}} \sum_{s} \boldsymbol{\epsilon}_s \left(\beta_{s,\mathbf{k}} \, e^{i(\mathbf{k} \cdot \mathbf{r} - \omega t)} - \beta_{s,\mathbf{k}}^* \, e^{-i(\mathbf{k}^* \cdot \mathbf{r} - \omega t)} \right) , \qquad (9.5)$$

$$\mathbf{B} = i \sum_{\mathbf{k}} \sum_{s} \frac{\boldsymbol{\kappa} \times \boldsymbol{\epsilon}_s}{c} \left(\beta_{s,\mathbf{k}} \, e^{i(\mathbf{k} \cdot \mathbf{r} - \omega t)} - \beta_{s,\mathbf{k}}^* \, e^{-i(\mathbf{k}^* \cdot \mathbf{r} - \omega t)} \right) , \qquad (9.6)$$

under the dispersion relation:

$$\omega^2 = (k_x^2 + k_y^2 + k_z^2)c^2 , \qquad (9.7)$$

that associates the wave vector $\mathbf{k} = (k_x, k_y, k_z)^{\mathsf{T}} = \boldsymbol{\kappa} \, k \, e^{i\phi}$ with the frequency ω.[1] Here, $\boldsymbol{\kappa}$ has been defined as a unit (real) vector, k is a positive real number (equal to the Euclidean length $|\mathbf{k}|$ if the vector $\mathbf{k}$ is real), while $e^{i\phi}$ accounts for imaginary wave vectors (corresponding to 'evanescent waves'): this form of $\mathbf{k}$ accounts for the most general case given the additional constraint of reality imposed on $\mathbf{E}$ and $\mathbf{B}$. The notation $\mathbf{k}^*$ stands for the complex conjugation of the vector $\mathbf{k}$.

The variable s can take only two discrete values, typically taken as -1 and $+1$ and, through the independent unit vectors $\{\boldsymbol{\epsilon}_s, s = -1, 1\}$, accounts for the two independent polarisations of the field; the transversality conditions (9.1) and (9.3) imply $\boldsymbol{\epsilon}_s \cdot \boldsymbol{\kappa} = 0$.

As already noted, the vector $\mathbf{k}$, which labels the different modes of the field in each polarisation direction, may be complex, encompassing the possibility of decaying evanescent waves, while the sum above may generalise to an integral. A real wave vector $\mathbf{k}$ pertains to a wave solution propagating along the direction $\boldsymbol{\kappa}$, with wavelength $\lambda = \frac{2\pi}{k}$.

Notice that the complex coefficients $\beta_{s,\mathbf{k}}$ are the only independent degrees of freedom of the field. The two individual solutions occurring inside the summations of Eqs. (9.5) and (9.6) are both propagating in the same direction in space (along $\boldsymbol{\kappa}$) but feature opposite field phases, which rotate in opposite senses in the complex plane. Somewhat misleadingly, and typically in the quantum field theory literature, these are sometimes distinguished as the negative and positive frequency solutions, a terminology which one should be aware of.

In order to quantise the field in the discrete paradigm of quantum optics,

---

[1] Note that all independent solutions are accounted for by setting $\omega = c\sqrt{k_x^2 + k_y^2 + k_z^2}$, where the square root is taken with positive real part (or positive imaginary part if the real part is zero), so that frequencies may always be regarded as positive if $\mathbf{k}$ is a real vector.

one still needs to constrain the solutions above by imposing spatial boundary conditions given by perfectly reflective mirrors, which is essentially how cavities are modelled. In order to impose spatial boundary conditions, it is convenient to consider the factorisation of the electric field in space- and time-dependent parts $\mathbf{E} = \boldsymbol{\chi}(\mathbf{r})T(t)$ (as we saw with the general solution, once the electric field is determined, the magnetic field will have to follow suit). The function $\boldsymbol{\chi}$ must obey the spatial part of the wave equation after separation of variables, which reads

$$\nabla^2 \boldsymbol{\chi} = -k^2 \boldsymbol{\chi} \tag{9.8}$$

(while the time dependent factor of the solution $T(t)$ obeys $\frac{\mathrm{d}^2}{\mathrm{d}t^2}T = -\omega^2 T$ with solutions $\mathrm{e}^{\mp i\omega t}$, as above), as well as the additional transversality condition

$$\nabla \cdot \boldsymbol{\chi} = 0 . \tag{9.9}$$

The boundary conditions $\mathbf{n} \cdot \mathbf{E} = \mathbf{n} \times \mathbf{B} = 0$, $\mathbf{n}$ being normal to the conducting surface, are guaranteed by imposing

$$\mathbf{n} \cdot \boldsymbol{\chi} = \mathbf{n} \times (\nabla \times \boldsymbol{\chi}) = 0 \tag{9.10}$$

on the cavity surface. In the presence of ideal conductors we can assume a real separation variable k, since the possibility of a complex wave vector is ruled out by rather general geometric arguments: as one should expect, sharp boundary conditions cannot sustain the exponentially decaying solutions to evanescent waves. Notice that we are assuming the space-dependent vector function $\boldsymbol{\chi}$ to be dimensionless, while the time dependent T has the dimensions of a field.

The solutions to Eqs. (9.8), (9.9) and (9.10) for cavities of realistic shapes are often impossible to determine analytically, although one can show, very generally, that if a cavity along any direction exists, the overall spatial solutions, which we will denote with $\boldsymbol{\chi}_{s,m}(\mathbf{r})$, form always a discrete set, each associated with a separation variable k_m. The label s can take only two values, and accounts for the fact that two independent vector solutions exist for each k_m (corresponding to the two independent polarisations of the wave solutions in free space): the third independent vector direction is ruled out by Eq. (9.9).

The 'mode functions' $\boldsymbol{\chi}_{s,n}(\mathbf{r})$ can be chosen to be real without loss of generality, and will hereafter replace the continuous set of complex plane waves $\mathrm{e}^{i\mathbf{k}\cdot\mathbf{r}}$ that were solutions to the unconstrained equation. The mode-functions always form an orthogonal set, and we will assume them to be normalised in the following sense:

$$\int_{\mathbb{R}^3} \boldsymbol{\chi}_{r,m}(\mathbf{r}) \cdot \boldsymbol{\chi}_{s,n}(\mathbf{r})\,\mathrm{d}^3\mathbf{r} = \delta_{rs}\delta_{mn}V_m , \tag{9.11}$$

which will allow us to highlight the role of the finite 'mode-volume' V_m. For future reference, note that the previous equation also implies[2]

$$\int_{\mathbb{R}^3} (\nabla \times \boldsymbol{\chi}_{r,m}(\mathbf{r})) \cdot (\nabla \times \boldsymbol{\chi}_{s,n}(\mathbf{r}))\,\mathrm{d}^3\mathbf{r} = \delta_{rs}\delta_{mn}k_m^2 V_m . \tag{9.12}$$

---

[2]The implication can be shown by resorting to the general vector identity $(\nabla \times \boldsymbol{\chi}) \cdot (\nabla \times$

Obviously, the mode-volume of infinite plane waves is undefined (infinite), a fact that, as we shall see, has significant physical implications.

The general solution to the Maxwell equations in the presence of a cavity may hence be written as follows

$$\mathbf{E} = i \sum_{s,m} \left(\alpha_{s,m} \, e^{-i\omega_m t} - \alpha^*_{s,m} \, e^{i\omega_m t} \right) \boldsymbol{\chi}_{s,m} , \tag{9.13}$$

$$\mathbf{B} = \sum_{s,m} \left(\frac{\alpha_{s,m}}{\omega_m} \, e^{-i\omega_m t} + \frac{\alpha^*_{s,m}}{\omega_m} \, e^{i\omega_m t} \right) \nabla \times \boldsymbol{\chi}_{s,m} , \tag{9.14}$$

where we set $\omega_m = ck_m$ and changed the name of the coefficients associated with each independent solution to $\alpha_{s,m}$, so as to distinguish the field in a cavity from the more general solutions (9.5) and (9.6). These solutions may be tested by direct insertion into the Maxwell equations, recalling that the $\boldsymbol{\chi}_{s,m}$ satisfy Eqs. (9.8) and (9.9), as well as the general relation $\nabla \times \nabla \times \boldsymbol{\chi}_{s,m} = \nabla(\nabla \cdot \boldsymbol{\chi}_{s,m}) - \nabla^2 \boldsymbol{\chi}_{s,m}$.

In order to introduce the notion of Hamiltonian and energy of the electromagnetic wave field, it is convenient to absorb the time-dependent exponentials into the solution coefficients by replacing $\alpha_m \, e^{-i\omega_m t}$ with $\alpha_{s,m}(t) = \frac{x_{s,m}(t) + i\, p_{s,m}(t)}{2}$ in Eqs. (9.13) and (9.14), to write

$$\mathbf{E} = i \sum_{s,m} \left(\alpha_{s,m}(t) - \alpha^*_{s,m}(t) \right) \boldsymbol{\chi}_{s,m} = - \sum_{s,m} p_{s,m}(t) \boldsymbol{\chi}_{s,m} , \tag{9.15}$$

$$\mathbf{B} = \sum_{s,m} \left(\frac{\alpha_{s,m}(t)}{\omega_m} + \frac{\alpha^*_{s,m}(t)}{\omega_m} \right) \nabla \times \boldsymbol{\chi}_{s,m} = \sum_{s,m} \frac{x_{s,m}(t)}{\omega_m} \nabla \times \boldsymbol{\chi}_{s,m} . \tag{9.16}$$

Inserting these solutions into the Maxwell equations, and applying the orthogonality condition (9.11), yields the following dynamics governing the time-dependent coefficients $x_{s,m}$ and $p_{s,m}$:

$$\dot{x}_{s,m} = \omega_m p_{s,m} , \quad \dot{p}_{s,m} = -\omega_m x_{s,m} . \tag{9.17}$$

The latter are nothing but the equations of motion of a set of decoupled harmonic oscillators described by pairs of canonically conjugate quadratures $x_{s,m}$ and $p_{s,m}$: up to physical dimensions, the field in a cavity admits a classical description in terms of a discrete set of canonical quadratures $\{(x_{s,m}, p_{s,m}), s \in \{0,1\}, n \in \mathbb{Z}\}$, each of which is ruled by the classical Hamiltonian $\omega_m(x^2_{s,m} + p^2_{s,m})$.[3] Then, by applying the orthogonality condition (9.11), it is promptly

---

$\boldsymbol{\chi}) = \nabla \cdot (\nabla \times \boldsymbol{\chi}) + \boldsymbol{\chi} \cdot \nabla(\nabla \cdot \boldsymbol{\chi}) - \boldsymbol{\chi} \cdot \nabla^2 \boldsymbol{\chi}$. The integration of the first term on the right hand side may be turned into a surface integral which vanishes under the boundary conditions (9.10). The second term is zero because of Eq. (9.9), while the third term may be simplified through Eq. (9.8), to obtain (9.12) from (9.11).

[3] Let us remind the reader that, if $H(x,p)$ is the classical Hamiltonian of a system with canonical variables x and p, the Hamilton equations of motion read $\dot{x} = \partial_p H$ and $\dot{p} = -\partial_x H$. Also note that, if the Hamiltonian is quadratic, each canonical variable may be rescaled by an arbitrary factor so that the same dynamics will be described by a properly redefined quadratic Hamiltonian. Hence, this argument does not fix the constants in front of the two terms in the Hamiltonian, but only their dependence on the variables $x_{s,m}$ and $p_{s,m}$.

seen that the total Hamiltonian of the cavity field $\sum_{s,m} \omega_m (x_{s,m}^2 + p_{s,m}^2)$ is proportional to the integral $\int_{\mathbb{R}^3} (\mathbf{E}^2 + c^2 \mathbf{B}^2) d^3 \mathbf{r}$.

In point of fact, classical electromagnetism allows one to derive from the conservation principle a general, explicit expression for the energy density H of the field in vacuum that abides by our heuristic argument and also sets the proportionality constant. One has

$$H = \frac{\varepsilon_0}{2} \left(\mathbf{E}^2 + c^2 \mathbf{B}^2 \right) , \tag{9.18}$$

where we adopted the SI units convention by the explicit inclusion of the electric permittivity in vacuum ε_0.

9.1.2 Canonical quantisation

The canonical quantisation of the light field is enacted by considering the expression (9.15) and (9.16) for the electromagnetic radiation fields and by establishing the following correspondence between classical variables and quantum operators:

$$\alpha_{s,m}(t) \longleftrightarrow E_m \hat{a}_{s,m} , \quad \alpha_{s,m}(t)^* \longleftrightarrow E_m \hat{a}_{s,m}^\dagger , \tag{9.19}$$

where $\hat{a}_{s,m}$ is a dimensionless operator on a Hilbert space while E_m is constant with the dimension of an electric field. The operators $\hat{a}_{s,m}$ and their Hermitian adjoints $\hat{a}_{s,m}^\dagger$ are then stipulated to obey the canonical commutation relations:

$$[\hat{a}_{r,m}, \hat{a}_{s,n}^\dagger] = \delta_{rs} \delta_{mn} \hat{\mathbb{1}} . \tag{9.20}$$

The electric and magnetic field operators along the polarisation direction $\boldsymbol{\epsilon}$ are thus expressed as

$$\hat{\mathbf{E}} = i \sum_{s,m} E_m \left(\hat{a}_{s,m}(t) - \hat{a}_{s,m}^\dagger(t) \right) \boldsymbol{\chi}_{s,m} , \tag{9.21}$$

$$\hat{\mathbf{B}} = \sum_{s,m} \frac{E_m}{\omega_m} \left(\hat{a}_{s,m}(t) + \hat{a}_{s,m}^\dagger(t) \right) \nabla \times \boldsymbol{\chi}_{p,m} . \tag{9.22}$$

Henceforth, we shall occasionally refer to the mode of the field with a generic annihilation operator $\hat{a}$ as 'mode a'. The quantities E_m are determined as follows. Inserting the expressions above for the field operators $\hat{\mathbf{E}}$ and $\hat{\mathbf{B}}$ into the classical density of energy (9.18), and integrating over $\mathbb{R}^3$ by utilising the orthonormality conditions (9.11) and (9.12) yields the quantum Hamiltonian $\hat{H}$ of the field:

$$\hat{H} = \sum_m \varepsilon_0 V_m E_m^2 (\hat{a}_m \hat{a}_m^\dagger + \hat{a}_m^\dagger \hat{a}_m) = \sum_m 2 \varepsilon_0 V_m E_m^2 \left(\hat{a}_m^\dagger \hat{a}_m + \frac{1}{2} \right) . \tag{9.23}$$

We want to set the constant E_m so that the Hamiltonian above corresponds to a set of decoupled quantum harmonic oscillators, each with Hamiltonian

$\omega_m \hbar \left(\hat{a}_m^\dagger \hat{a}_m + \frac{1}{2} \right)$. This is ensured by the choice

$$E_m = \sqrt{\frac{\omega_m \hbar}{2\varepsilon_0 V_m}} , \qquad (9.24)$$

which completes the quantisation procedure.

The presence of a cavity, which sustains only a specific set of spatial modes and thus coarse-grains the field degrees of freedom, has the twofold effect of discretising the theory and of determining finite mode volumes $\{V_m, m \in \mathbb{Z}\}$. A discrete, numerable set of degrees of freedom obeying the CCR is precisely what we referred to as a continuous variable system in this book and it is, in a sense, the hallmark distinguishing quantum optics from the more general quantum field theory. Note that the constant E_m is not at all conventional and, through its role in determining it as per Eq. (9.24), the mode volume V_m plays a key part in the physics of quantum light. In fact, Eqs. (9.21), (9.22) and (9.24) show that, whenever the mode n of the field couples to any other degree of freedom – $e.g.$, linearly to an electric dipole – the overall coupling constant will decrease with increasing V_m, and vanish as V_m grows to infinity. A finite mode volume is thus necessary in order to achieve any coupling between electromagnetic radiation and other degrees of freedom. In practice, experimentalists constantly strive to tailor as small mode volumes as possible, in order to achieve strong couplings. The strand of quantum optics dedicated to the study of fields in cavities and of their coupling to matter goes under the name of cavity QED ("quantum electrodynamics").

Canonical quantisation implies a huge leap of faith, in the sense that infinite alternative stipulations might have been made in order to obtain quantised eigenstates of the field free Hamiltonian. No fundamental physical principle vouches for the particular choice of $\hat{a}_m$ and $\hat{a}_m^\dagger$ as ladder operators. The simplicity of this construction, as well as the inspiring correspondence between radiation fields and harmonic oscillators, which it maintains, are its most convincing advocates. With hindsight, one may even wonder at the accuracy with which the canonical quantisation has been able to describe experiments with light in the quantum regime over the last fifty years.

9.1.3 Quantum electromagnetic fields in free space

The discussion above introduced the spatial modes of a cavity, with the precise intention of highlighting the role played by the mode volume in the quantisation programme. However, the continuum of degrees of freedom constituting the electromagnetic radiation field can be quantised directly, and continuous or discrete sets of modes can be identified in this framework too. Such 'travelling' modes are the object of quantum optics in free space or along waveguides, such as long optical fibres. In this section, we shall specialise to the case of real wave vectors, which comprise a complete set of solutions, and consider all possible freely propagating modes, so that the summations in the electric and magnetic field solutions may be replaced by an integral over $\mathbf{k} \in \mathbb{R}^3$.

The quantisation in free space is best handled by sorting the terms in the general solutions (9.5) and (9.6) by their spatial profiles. This is achieved by changing the dummy variables from $\mathbf{k}$ to $-\mathbf{k}$ in the second terms appearing inside the sums and by restricting the integrals over wave vectors to one hemisphere, which we shall denote by $\widehat{\mathbb{R}}^3$. The field solutions for propagating waves may hence be cast as

$$\mathbf{E} = i \int_{\widehat{\mathbb{R}}^3} \mathrm{dk} \sum_s \boldsymbol{\epsilon}_s \left(\beta_{s,\mathbf{k}}(t) - \beta^*_{s,-\mathbf{k}}(t) \right) e^{i\mathbf{k}\cdot\mathbf{r}} + \text{c.c.} \,, \tag{9.25}$$

$$\mathbf{B} = -i \int_{\widehat{\mathbb{R}}^3} \mathrm{dk} \sum_s \frac{\boldsymbol{\kappa} \times \boldsymbol{\epsilon}_s}{c} \left(\beta_{s,\mathbf{k}}(t) + \beta^*_{s,-\mathbf{k}}(t) \right) e^{i\mathbf{k}\cdot\mathbf{r}} + \text{c.c.} \,, \tag{9.26}$$

where c.c. stands for the complex conjugate of the preceding expression. Note that the time-dependent exponentials have been absorbed into the definition $\beta_{s,\mathbf{k}}(t) = \beta_{s,\mathbf{k}} e^{-i\omega t}$. The field quantisation is then effected by the correspondence

$$\beta_{s,\mathbf{k}}(t) \longleftrightarrow E_{\mathbf{k}} \,\hat{a}_{s,\mathbf{k}} \,, \quad \beta^*_{s,\mathbf{k}}(t) \longleftrightarrow E_{\mathbf{k}} \,\hat{a}^\dagger_{s,\mathbf{k}} \,, \tag{9.27}$$

under the continuous commutation relations

$$[\hat{a}_{r,\mathbf{k}}, \hat{a}^\dagger_{s,\mathbf{k}'}] = \delta_{rs} \, \delta^3(\mathbf{k} - \mathbf{k}') \hat{\mathbb{1}} \,. \tag{9.28}$$

Let us remark that the condition (9.28) implies that the operators $\hat{a}_{s,\mathbf{k}}$ and $\hat{a}^\dagger_{s,\mathbf{k}}$ have physical dimensions of the square root of a volume, so that the real scalar factors $E_{\mathbf{k}}$ must have dimensions of an electric field over the square root of a volume. The field operators in free space then read

$$\hat{\mathbf{E}} = i \int_{\widehat{\mathbb{R}}^3} \mathrm{dk} \sum_s \boldsymbol{\epsilon}_s E_{\mathbf{k}} \left(\hat{a}_{s,\mathbf{k}} - \hat{a}^\dagger_{s,-\mathbf{k}} \right) e^{i\mathbf{k}\cdot\mathbf{r}} + \text{h.c.} \,, \tag{9.29}$$

$$\hat{\mathbf{B}} = -i \int_{\widehat{\mathbb{R}}^3} \mathrm{dk} \sum_s \frac{\boldsymbol{\kappa} \times \boldsymbol{\epsilon}_s}{c} E_{\mathbf{k}} \left(\hat{a}_{s,\mathbf{k}} + \hat{a}^\dagger_{s,-\mathbf{k}} \right) e^{i\mathbf{k}\cdot\mathbf{r}} + \text{h.c.} \,, \tag{9.30}$$

where h.c. stands for the Hermitian conjugate of the preceding operators. Putting the operators above into Eq. (9.18), integrating over the whole space of $\mathbf{r} \in \mathbb{R}^3$, and making use of the delta-function representation (1.16), yields the free Hamiltonian operator of the field $\hat{H}$:

$$\begin{aligned}
\hat{H} &= \int_{\mathbb{R}^3} \mathrm{dr} \frac{\varepsilon_0}{2} \left(|\hat{\mathbf{E}}|^2 + c^2 |\hat{\mathbf{B}}|^2 \right) \\
&= \int_{\widehat{\mathbb{R}}^3} \mathrm{dk} \sum_s \varepsilon_0 E_{\mathbf{k}}^2 (2\pi)^3 \left(\hat{a}_{s,\mathbf{k}} \hat{a}^\dagger_{s,\mathbf{k}} + \hat{a}^\dagger_{s,\mathbf{k}} \hat{a}_{s,\mathbf{k}} + \hat{a}_{s,-\mathbf{k}} \hat{a}^\dagger_{s,-\mathbf{k}} + \hat{a}^\dagger_{s,-\mathbf{k}} \hat{a}_{s,-\mathbf{k}} \right) \\
&= \int_{\mathbb{R}^3} \mathrm{dk} \sum_s \frac{\hbar\omega}{2} \left(\hat{a}_{s,\mathbf{k}} \hat{a}^\dagger_{s,\mathbf{k}} + \hat{a}^\dagger_{s,\mathbf{k}} \hat{a}_{s,\mathbf{k}} \right)
\end{aligned} \tag{9.31}$$

where, in order to obtain the Hamiltonian operator corresponding to a continuum of decoupled harmonic oscillators, the factor $E_\mathbf{k}$ has been set as follows:

$$E_\mathbf{k} = \sqrt{\frac{\hbar\omega}{2\varepsilon_0(2\pi)^3}} \ . \tag{9.32}$$

Clearly, ω depends on $\mathbf{k}$ through the dispersion relation (9.7). Notice that, in Eq. (9.31), the integration over the whole domain of the parameter $\mathbf{k} \in \mathbb{R}^3$ has been restored by switching the sign in the integral over $-\mathbf{k}$.

The free field may hence be described as a continuum of propagating modes, labelled by the wave-vectors $\mathbf{k} \in \mathbb{R}^3$. Such a continuum may be rearranged at will to best suit the problem at issue. For instance, it is possible to switch to a position representation by taking the Fourier transform of the field operators:

$$\hat{a}_{s,\mathbf{r}} = \frac{1}{(2\pi)^{3/2}} \int_{\mathbb{R}^3} \mathrm{d}\mathbf{k} \, e^{-i\mathbf{k}\cdot\mathbf{r}} \hat{a}_{s,\mathbf{k}} \ , \tag{9.33}$$

whence, by applying Eq. (9.28),

$$[a_{r,\mathbf{r}}, a_{s,\mathbf{s}}] = \delta_{rs}\,\delta^3(\mathbf{r} - \mathbf{s}) \ . \tag{9.34}$$

The operators $a_{s,\mathbf{r}}$ and $a_{s,\mathbf{r}}^\dagger$ respectively destroy and create a photon at the point $\mathbf{r}$ in space.

More generally, given any set of functions f_t such that

$$\int_{\mathbb{R}^3} \mathrm{d}\mathbf{r} \, f_t(\mathbf{r}) f_{t'}(\mathbf{r}) = \delta(t - t') \ , \tag{9.35}$$

one can define a continuum of modes

$$\hat{a}_{s,t} = \int_{\mathbb{R}^3} \mathrm{d}\mathbf{r} \, f_t(\mathbf{r}) \hat{a}_{s,\mathbf{r}} \tag{9.36}$$

such that

$$[\hat{a}_{r,t}, \hat{a}_{s,t'}] = \delta_{rs}\,\delta(t - t') \ . \tag{9.37}$$

The functions f_t represent, in this case, the spatial profiles of a selected 'train' of travelling modes (notice that in applying this coarse graining we have lost the possibility to describe *any* field in free space). We have kept f_t real and dependent on a single real variable for simplicity, as the extension to more general parametrisations of the modes is straightforward. If relevant, Eq. (9.37) might also be modified to admit a discrete set of localised travelling modes, obeying the discrete commutation relations (9.20) analogous to those satisfied by cavity fields, although adopting such a numerable set would imply a further reduction in the possible representable solutions.

This formalism finds notable application whenever travelling modes interact with static optical elements. For instance, a simple model of a beam splitter is obtained by considering the Hamiltonian $\hat{H}_{bs}(t)$ that mixes up two trains of travelling modes at the input ports a and b:

$$\hat{H}_{bs}(t) = i \int_{-\infty}^{+\infty} dx \int_{-\infty}^{+\infty} dy \left(f_x(t) f_y^*(t) \hat{a}_x^\dagger \hat{b}_y - f_x^*(t) f_y(t) \hat{a}_x \hat{b}_y^\dagger \right) , \quad (9.38)$$

where $\hat{a}_x$ and $\hat{b}_y$ are defined, respectively, as the continuous set of modes interacting through port a and port b of the beam splitter and satisfy the commutation relations $[\hat{a}_x, \hat{a}_y^\dagger] = \delta(x-y)$ and $[\hat{b}_x, \hat{b}_y^\dagger] = \delta(x-y)$. The complex function $f_x(t)$ determines the profile of the modes that interact at the beam splitter at any given time t as well as the strength of their coupling, and make the Hamiltonian above time dependent as, in the full field theoretical picture, different sets of travelling modes interact at every time t. If, as we shall conveniently assume, x and y have the dimensions of time, then $f_x(t)$ must have the dimensions of frequency (so that the Hamiltonian itself has the dimensions of frequency, in natural units where $\hbar = 1$). Note that the Hamiltonian above is the full Hamiltonian of the system in interaction picture, where the free Hamiltonian of the field is taken into account by just letting the travelling modes propagate. This simple description highlights the practical concern of 'mode-matching': in order for optical elements involving multiple inputs to work properly, one has to take care that the relevant input signals enter them at the same time and with the right spatial profile.

Problem 9.1. (*Beam splitter on travelling wave fields*). Show that the ideal condition where only a specific pair of modes, labelled by t, interact at the beam splitter at any given time t, corresponds to the case $f_x(t) = R\,\delta(x-t)$ in the Hamiltonian $H_{bs}(t)$ given above, where R is a dimensionless real parameter. Determine R for a 50:50 beam splitter.

Solution. The Heisenberg equation of motion for $\hat{a}_x$ is readily written as

$$\dot{\hat{a}}_x = i[\hat{H}_{bs}(t), \hat{a}_x] = \int_{-\infty}^{+\infty} dx' \int_{-\infty}^{+\infty} dy f_{x'}^*(t) f_y(t) \delta(x' - x) \hat{b}_y$$

$$= \int_{-\infty}^{+\infty} dy f_x^*(t) f_y(t) \hat{b}_y = R^2 \delta(x-t) \hat{b}_t . \quad (9.39)$$

Similarly, for $\hat{b}_x$:

$$\dot{\hat{b}}_x = -R^2 \delta(x-t) \hat{a}_t \quad (9.40)$$

As an aside, note that the formal solution of the latter equation for $\hat{b}_x(t)$ reads

$$\hat{b}_x(t) = \hat{b}_x(0) - R^2 \int_0^t \delta(x - t') \hat{a}_{t'}\, dt' = \hat{b}_x(0) + \theta(x)\theta(t-x) R^2 \hat{a}_x ,$$

$$(9.41)$$

where the product of Heaviside functions $\theta(x)\theta(t-x)$ equals, respectively, 1 for $x \in (0,t)$ and 0 for $x \notin (0,t)$: quite simply, modes with x smaller than 0 will never interact, because they passed the beam splitter before it started operating, while modes with x greater than t have not yet met the beam splitter: all such modes have not evolved at all at time t ($\hat{b}_x(t) = \hat{b}_x(0)$, as one should expect).

The system of linear differential equations (9.39) and (9.40) is, up to the time-dependent delta function, nothing but a linear system corresponding to a quadratic Hamiltonian which, as we showed systematically in Chapter 3, gives rise to a symplectic transformation on the mode operators. In particular, in Section 5.1.2.1 we already dealt with a beam-splitting Hamiltonian generating a discrete beam-splitting transformation. Here, we just need to switch from quadrature to creation and annihilation operators, through the transformation $\bar{U}$ of Eq. (3.6) and to take into account the delta functions by letting the transformation parameters depend on time as Heaviside step functions. Explicitly, by recalling that $\frac{d}{dz}\theta(z) = \delta(z)$, it is immediate to check that the following operators solve the system above and meet the desired boundary conditions:

$$\hat{a}_x(t) = \cos\left(\theta(x)\theta(t-x)R^2\right)\hat{a}_x(0) + \sin\left(\theta(x)\theta(t-x)R^2\right)\hat{b}_x(0),$$
$$(9.42)$$

$$\hat{b}_x(t) = -\sin\left(\theta(x)\theta(t-x)R^2\right)\hat{a}_x(0) + \cos\left(\theta(y)\theta(t-x)R^2\right)\hat{b}_x(0).$$
$$(9.43)$$

The beam splitter's transmittance can be evaluated at $t > x > 0$, once the modes at issue have interacted, and equals $\cos(R^2)$. For a 50:50 beam splitter, one would have $\cos(R^2) = \sqrt{1/2}$, whence $R = \sqrt{\pi/4}$ (up to periodicity).

9.1.4 Input-output interfaces and quantum Langevin equations

A further, very well-known application of the description provided by Eq. (9.37) is the so called input-output formalism, which models the interaction between a set of 'static' intra-cavity modes and continuous trains of incoming and outgoing modes (at variance with the beam splitter interface above, where all the modes involved were travelling). Besides its general relevance, this formalism has been applied in the book to the description of open quantum systems in Chapter 6, whose full significance and applicability will be hence elucidated by the clarification we shall provide for the reader here.

The continuous train of modes interacting with a 'localised' mode with ladder operator $\hat{a}$ will be associated to the ladder operator $\hat{b}_t$, where the label $t \in \mathbb{R}$ has dimensions of time. Under the following Hamiltonian, the label t

identifies the mode that interacts with the system at time t:

$$\hat{H}_{IO} = i \int_{\mathbb{R}} \sqrt{\kappa} \delta(x - t)(\hat{a}^\dagger \hat{b}_x - \hat{a}\hat{b}_x^\dagger) \, dx \, , \tag{9.44}$$

where the squared coupling constant κ has the dimensions of frequency (note that $\hat{b}_x$ has the dimensions of the square root of a frequency). Each mode belonging to the continuous set only evolves at time t (if one neglects its free Hamiltonian), and is unaffected at all other times. The effect of the free field evolution on such modes is essentially space propagation: the travelling mode with ladder operator b_t approaches the input-output interface at all times smaller than t, reaches the interface and interacts with the localised mode at time t, and then is scattered off and propagates away from the interface at times greater than t. At each time, the system interacts with a different mode. Under the white noise assumption, such modes interacting at different times obey the following standard commutation relations, already introduced above:

$$[\hat{b}_x, \hat{b}_y^\dagger] = \delta(x - y) \, . \tag{9.45}$$

This relationship is not in general true for systems interacting with a continuum, though it is remarkably accurate in quantum optical set-ups. Along with an analogous condition on the covariances of operators interacting with the system at different times, it corresponds to assuming the system is Markovian, in that the environmental modes that will interact in the future are completely uncorrelated with those that interacted with the system in the past, and hence no information about the system will ever flow back into it. In mathematical terms, this ensures the corresponding evolution of the quantum state in the Schrödinger picture only depends on the state at a certain time and not on any time integral kernel. Also, this implies that the corresponding master equation for the system can be written in Lindblad form. Notice that the archetypal case of an optical cavity along a single spatial direction would require one set of interacting modes for each cavity mirror. However, unless one is interested in the detection of the outgoing modes, the partial dynamics of the system itself may in the case of multiple interfaces still be described by a single continuous set of environmental modes by redefining the latter and the coupling constants.

The Heisenberg equations of motion of the operators involved read

$$\dot{\hat{a}}(t) = i[\hat{H}_{IO}, \hat{a}] = \sqrt{\kappa} \int_{\mathbb{R}} \hat{b}_x \delta(x - t) \, dx = \sqrt{\kappa} \, \hat{b}_t \, , \tag{9.46}$$

$$\dot{\hat{b}}_x(t) = i[\hat{H}_{IO}, \hat{b}_x] = -\sqrt{\kappa} \int_{\mathbb{R}} \hat{a} \delta(x - t)\delta(x - x') \, dx' = -\sqrt{\kappa} \, \delta(x - t)\hat{a} \, . \tag{9.47}$$

The equation for $\hat{b}_x$ can be formally integrated to obtain, assuming $t, x > 0$:

$$\hat{b}_x(t) = \hat{b}_x(0) - \sqrt{\kappa} \, \theta(t - x) \, \hat{a} \, , \tag{9.48}$$

where the half maximum convention $\theta(0) = \frac{1}{2}$, corresponding to the following integration of a delta centred at the integration extremum: $\int_0^t f(x)\delta(t-x)\,\mathrm{d}x = \frac{f(t)}{2}$, is understood. Putting Eq. (9.48) into Eq. (9.46) then yields the quantum Langevin equation for $\hat{a}(t)$:

$$\dot{\hat{a}} = -\frac{\kappa}{2}\hat{a} + \sqrt{\kappa}\,\hat{b}_t(0)\,, \tag{9.49}$$

where, it should be noted, any internal dynamics has been neglected.

In the literature, it is customary to denote the annihilation operator $\hat{b}_t(t')$, for $t' > t > 0$ (of the scattered mode after the interaction at time t) with $-\hat{b}_{out}(t)$ and the annihilation operator of the same mode before the interaction $\hat{b}_t(0)$ with $\hat{b}_{in}(t)$, whence the terminology "input-output" formalism. Eq. (9.48) may thus be cast as

$$\hat{b}_{in}(t) + \hat{b}_{out}(t) = \sqrt{\kappa}\,\hat{a}\,. \tag{9.50}$$

This is at times referred to as a 'boundary condition' in analogy with classical electromagnetism although, as was just shown, it can be derived from the Hamiltonian dynamics once the initial conditions of the continuum are set. Notice that the white noise condition (9.45) may be re-written in terms of the input and output fields as

$$[\hat{b}_{in}(t), \hat{b}_{in}^\dagger(t')] = [\hat{b}_{out}(t), \hat{b}_{out}^\dagger(t')] = \delta(t - t')\,. \tag{9.51}$$

Eq. (9.49) describes the diffusive dynamics resulting from the damping of the localised mode as well as the additional input quantum noise that accounts for preserving the commutation relation and hence the uncertainty relation. Notice that the state of the continuum modes has not been specified and is completely generic: typically, it may be a finite temperature thermal Gibbs state, or the vacuum state (a situation which is often referred to as the 'pure loss' case). All such dynamics were discussed in the multi-mode case and great generality in Chapter 6: the input field $\hat{b}_{in}$ corresponds to $\hat{r}_{in}$, whilst the Langevin Equation (9.49) is but a special case of what was derived in Section 6.2.2.

Problem 9.2. (*OPO in the Langevin equation formulation*). Reconsider the single-mode optical parametric oscillator at zero temperature of Section 6.5, with Hamiltonian $-\chi\hat{x}\hat{p}$ for $\chi > 0$ (in interaction picture) and loss factor $\kappa = 2\gamma$ with $\gamma > \chi$ (that ensures stability), and determine the steady-state squeezing through the associated quantum Langevin equation.

Solution. We will only deal with the operator $\hat{x} = (\hat{a} + \hat{a}^\dagger)/\sqrt{2}$, that turns out to be the squeezed one for such a choice of χ. Adding the Hamiltonian term $-i\chi[\hat{x}\hat{p}, \hat{x}]$ to Eq. (9.49), yields the differential equation

$$d\hat{x} = -(\chi + \gamma)\,\hat{x}\,dt + \sqrt{2\gamma}\,\hat{x}_{in}(t)dt\,. \tag{9.52}$$

The zero temperature condition is reflected in the correlations

$$\langle \hat{x}_{in}(t)\hat{x}_{in}(t')\rangle = \frac{1}{2}\delta(t - t')\,. \tag{9.53}$$

Eq. (9.52) admits the following solution

$$\hat{x}(t) = e^{-(\chi+\gamma)t}\hat{x}(0) + \sqrt{2\gamma}\int_0^t e^{-(\chi+\gamma)(t-t')}\hat{x}_{in}(t')\,dt'\,, \tag{9.54}$$

with variance (the first moments are always zero)

$$\langle \hat{x}(t)^2 \rangle = e^{-2(\chi+\gamma)t}\langle \hat{x}(0)^2 \rangle + 2\gamma\int_0^t e^{-(\chi+\gamma)(2t-t'-t'')}\langle \hat{x}_{in}(t')\hat{x}_{in}(t'')\rangle dt'\,dt''$$

$$= e^{-2(\chi+\gamma)t}\langle \hat{x}(0)^2 \rangle + \gamma\frac{1 - e^{-2(\chi+\gamma)t}}{2(\gamma + \chi)}\,, \tag{9.55}$$

where (9.53) was used. The steady-state variance is simply

$$\lim_{t\to\infty}\langle \hat{x}(t)^2 \rangle = \frac{\gamma}{2(\gamma + \chi)} \tag{9.56}$$

which, recalling that our covariance matrix diagonal elements are twice the variance, coincides with the steady-state squeezing of Eq. (6.70) (we changed the sign of the Hamiltonian here, so the variances of $\hat{x}$ and $\hat{p}$ are swapped).

Problem 9.3. (*Coherent feedback*). The parametric down-conversion process discussed in the problem above, with loss rate 2γ, is equivalent to the action of two input operators, $\hat{x}_{in,1}(t)$ and $\hat{x}_{in,2}(t)$, each associated to a loss rate γ. This would describe a cavity with two lossy mirrors. Other than by measuring the output fields, as discussed in Chapter 6, the steady-state squeezing may be improved by a 'coherent feedback loop' whereby the output of one cavity, $\hat{x}_{out,2}$ is fed into the input of the other cavity. Assume the ideal case of no delay but a transmission loss η along the feedback loop, and determine the optimal steady-state squeezing achievable for all η.

Solution. The Langevin equation for $\hat{x}$ is

$$d\hat{x} = -(\chi + \gamma)\,\hat{x}\,dt + \sqrt{\gamma}(\hat{x}_{in,1} + \hat{x}_{in,2})\,dt \tag{9.57}$$

(we are dispensing with the argument (t) after input and output fields).

We are going to need now the input-output relation (9.50). The presence of the feedback loop implies that

$$\hat{x}_{in,1} = \mathcal{E}_\theta^1(\hat{x}_{out,2}) = \sqrt{\eta}\,\hat{x}_{out,2} + \sqrt{1-\eta}\,\hat{x}_{in,3}$$
$$= \sqrt{\eta\gamma}\,\hat{x} - \sqrt{\eta}\,\hat{x}_{in,2} + \sqrt{1-\eta}\,\hat{x}_{in,3} , \qquad (9.58)$$

where $\mathcal{E}_\theta^1$ is the quantum limited attenuator channel of Eq. (5.77), with $\cos^2\theta = \eta$ which, in the Heisenberg picture, has been represented as the mixing at a beam splitter with another input field, $\hat{x}_{in,3}$. Notice that all the input fields are uncorrelated with each other, and have self-correlations as in Eq. (9.53). Inserting (9.58) into (9.57) gives

$$d\hat{x} = -[\chi + \gamma(1-\sqrt{\eta})]\hat{x}\,dt + \sqrt{\gamma}[(1-\sqrt{\eta})\hat{x}_{in,2} + \sqrt{1-\eta}\,\hat{x}_{in,3}]dt . \quad (9.59)$$

From the corresponding equation for $\hat{p}$, one would see that the system is stable, in the sense of admitting a steady state, only if $\gamma(1-\sqrt{\eta}) > \chi$. Let us then assume this to be the case, and evaluate the steady-state variance (note that η may be decreased at leisure in a laboratory, by adding a controlling beam splitter along the feedback loop). Retracing the steps of the previous problem shows that each input field contributes a term to the asymptotic variance of $\hat{x}$ equal to $c^2/4|d|$, where c is the factor that multiplies them in the quantum Langevin equation, while d is the factor that multiplies $\hat{x}$ in the same equation. Thus, one gets

$$\lim_{t\to\infty} \langle\hat{x}(t)^2\rangle = \frac{\gamma(1-\sqrt{\eta})}{2[(\gamma(1-\sqrt{\eta})+\chi)]} , \qquad (9.60)$$

which is a decreasing function of η, as one should expect. The minimum steady-state value $\langle\hat{x}^2\rangle_{min}$ is obtained by replacing η with its maximum stable value, which corresponds to $(1-\sqrt{\eta}) = \chi/\gamma$:

$$\lim_{t\to\infty} \langle\hat{x}(t)^2\rangle > \langle\hat{x}^2\rangle_{min} = \frac{1}{4} . \qquad (9.61)$$

This is the notorious 3 dB of squeezing already encountered in Eq. (6.71). Quite remarkably, a properly tailored coherent feedback loop is capable of approaching this value of squeezing for any values of parametric coupling and loss rate, since the optimal value above is independent from such parameters. This bound on the steady-state (in-loop) squeezing may be beaten by monitoring the output fields, as we saw in Section 6.5.

9.1.5 Driven cavities

A most obvious, yet far-reaching, application of the input-output formalism is the case of constant laser light at a certain frequency ω_L impinging on an optical cavity, a situation which is sometimes referred to as 'linear driving' of the cavity field. Neglecting polarisation, one may assume the initial state of the light field outside the cavity, induced by the laser, to be a coherent state $|\alpha\rangle$ in the mode with fixed wave vector $k = \omega_L/c$, propagating towards the cavity mirror. The action of the free Hamiltonian of the field is then just to rotate the coherent amplitude α with frequency ω_L, as $e^{i\omega_L t}\alpha$.[4]

The laser mode will in general have a certain overlap with the input mode interacting with the cavity mirror, whose annihilation operator will be denoted with $\hat{b}_{in}(t)$, as per the previous section. While the laser is in operation, and the cavity is being driven, one may assume such an overlap to be constant in time. It is immediate to realise that the effect of such a finite overlap is just to reduce the coherent amplitude of the state: the statistics of the input mode are still those of a Gaussian state with covariance matrix equal to the identity and first moments rescaled with respect to the laser mode. We shall indicate this amplitude with $e^{i\omega_L t}\beta$, bearing in mind that this coherent amplitude rotates at the laser frequency, just like α did. The parameter β, which can be assumed to be real, depends on the power of the laser, as well as on the mode matching just discussed. It is convenient to assign to β the dimensions of a square root of a frequency, setting $\langle \hat{b}_{in}(t)\rangle = e^{i\omega_L t}\beta$ (recall that $\hat{b}_{in}(t)$ has dimensions of a square root of a frequency).

Let us now, for simplicity and to fix ideas, consider the evolution of a single cavity mode $\hat{a}$ at frequency ω_c, given by the Langevin Equation (9.49) supplemented with the free Hamiltonian of the mode:

$$\dot{\hat{a}} = -i\omega_c\hat{a} - \frac{\kappa}{2}\hat{a} + \sqrt{\kappa}\,\hat{b}_{in}(t)\,, \tag{9.62}$$

which admits the following general solution

$$\hat{a}(t) = e^{-(i\omega_c+\kappa/2)t}\hat{a}(0) + \sqrt{\kappa}\int_0^t e^{-(i\omega_c+\kappa/2)(t-t')}\hat{b}_{in}(t')\,dt'\,. \tag{9.63}$$

Whence, through knowledge of the input field statistics $\langle\hat{b}_{in}(t)\rangle = e^{i\omega_L t}\beta$, $\langle\hat{b}_{in}(s)\hat{b}_{in}(t)\rangle = 0$ and $\langle\hat{b}_{in}(s)\hat{b}_{in}^\dagger(t)\rangle = \delta(s-t)$, the first and second moments of the cavity field state are straightforward to derive:

$$\langle\hat{a}(t)\rangle = e^{-(i\omega_c+\kappa/2)t}\langle\hat{a}(0)\rangle + \sqrt{\kappa}\,\beta\int_0^t e^{(i(\omega_L-\omega_c)-\kappa/2)t'}\,dt'\,, \tag{9.64}$$

$$\langle\hat{a}(t)\hat{a}(t)\rangle = e^{-(i2\omega_c+\kappa)t}\langle\hat{a}(0)\hat{a}(0)\rangle\,, \tag{9.65}$$

$$\langle\hat{a}(t)\hat{a}^\dagger(t)\rangle = e^{-\kappa t}\langle\hat{a}(0)\hat{a}^\dagger(0)\rangle + (1 - e^{-\kappa t})\,. \tag{9.66}$$

---

[4]This results from the solution of Eq. (6.3) for the free Hamiltonian matrix $H = \mathbb{1}$. The ensuing rotation of the first moment vector corresponds to a phase factor multiplying the coherent amplitude α.

At steady state, achieved in practice for $t \gg 1/\kappa$, the second moments are those of the vacuum (or any coherent state), while the first moments are determined as the real and imaginary parts of the limit below

$$\lim_{t \to \infty} \langle \hat{a}(t) \rangle = \frac{\sqrt{\kappa}\,\beta}{\frac{\kappa}{2} - i(\omega_L - \omega_c)}. \tag{9.67}$$

The state tends to a coherent state, with amplitude depending on the laser detuning $\Delta = \omega_L - \omega_c$, as well as on the cavity loss rate κ and the laser power and specifics, contained in β. Clearly, if the laser is widely detuned, so that $|\Delta| \gg \kappa$, one has that the asymptotic coherent state is essentially the vacuum: a widely off-resonant driving has no effect on the cavity mode, as one should expect. The maximum steady-state amplitude is achieved at exact resonance, where it reaches the value $\frac{2\beta}{\sqrt{\kappa}}$.

9.1.6 Linear optical quantum computing

In the so called KLM (Knill–Laflamme–Milburn) model of quantum computation, each qubit is embodied by two continuous variable modes a and b, with annihilation operators $\hat{a}$ and $\hat{b}$, through the "dual rail" encoding described by the following correspondence:

$$|0\rangle \leftrightarrow \hat{a}^\dagger |vac\rangle\,, \quad |1\rangle \leftrightarrow \hat{b}^\dagger |vac\rangle\,, \tag{9.68}$$

where $|0\rangle$ and $|1\rangle$ stand for the two logical qubit states while $|vac\rangle$ is the two-mode vacuum state of the continuous variable system ($|0\rangle$ corresponds to one photon in mode a and zero photons in mode b, while $|1\rangle$ corresponds to zero photons in mode a and one photon in mode b).

As shown explicitly in the solution to the following problem, *compositions of phase shifters and beam splitters acting on the two modes may reproduce all single-qubit unitary operations on the qubit defined above.*

Problem 9.4. (*KLM single-qubit operations*). Show that the direct sum of phase shifter transformations $R_{\varphi_a} \oplus R_{\varphi_b}$, with each block given by Eq. (5.12), realises the phase gate $\mathrm{diag}(e^{-i\varphi_a}, e^{-i\varphi_b})$ on a dual rail qubit. Further, show that the 50:50 beam splitter transformation $R_{\pi/4}$, given by Eq. (5.13) for $\theta = \pi/4$, realises the Hadamard gate $\frac{1}{\sqrt{2}}\begin{pmatrix} 1 & 1 \\ -1 & 1 \end{pmatrix}$ on such a qubit.

Solution. The transformation $R = R_{\varphi_a} \oplus R_{\varphi_b}$ acts, in the Heisenberg picture, on the array of operators $\hat{\mathbf{r}} = (\hat{x}_a, \hat{p}_a, \hat{x}_b, \hat{p}_b)^\mathsf{T}$, with $\hat{a} = \frac{\hat{x}_a + i\hat{p}_a}{\sqrt{2}}$ and $\hat{b} = \frac{\hat{x}_b + i\hat{p}_b}{\sqrt{2}}$. The corresponding linear transformation $\bar{U} R \bar{U}^\dagger$ on the array $\hat{\mathbf{a}} = (\hat{a}, \hat{a}^\dagger, \hat{b}, \hat{b}^\dagger)^\mathsf{T}$, is hence obtained by acting on R with the unitary transformation $\bar{U}^\dagger$ such that $\hat{\mathbf{r}} = \bar{U}^\dagger \hat{\mathbf{a}}$, given by Eq. (3.6). One thus gets $\bar{U}R\bar{U}^\dagger = \mathrm{diag}(e^{i\varphi_a}, e^{-i\varphi_a}, e^{i\varphi_b}, e^{-i\varphi_b})$. That is, the phase shifters multiply the two creation operators by the phases $e^{-i\varphi_a}$ and $e^{-i\varphi_b}$: the effect on a state encoded in the qubit (9.68) is thus just the target phase gate.

As for the beam splitter, denoting it with the transformation $R_{\pi/4}$, one obtains $\bar{U}R_{\pi/4}\bar{U}^\dagger = R_{\pi/4}$ ($R_{\pi/4}$ is invariant under the change of basis $\bar{U}^\dagger$). This shows that the Hilbert space action of the 50:50 beam splitter $\hat{R}_{\pi/4}$ maps the ladder operators $\hat{a}^\dagger$ and $\hat{b}^\dagger$ as follows: $\hat{a}^\dagger \mapsto \frac{1}{\sqrt{2}}(\hat{a}^\dagger + \hat{b}^\dagger)$ and $\hat{b}^\dagger \mapsto \frac{1}{\sqrt{2}}(-\hat{a}^\dagger + \hat{b}^\dagger)$. Through the correspondence (9.68) this carries over to the single photon states, such that

$$|0\rangle \equiv \hat{a}^\dagger|vac\rangle \mapsto \frac{1}{\sqrt{2}}(\hat{a}^\dagger + \hat{b}^\dagger)|vac\rangle \equiv \frac{1}{\sqrt{2}}(|0\rangle + |1\rangle),$$

$$|0\rangle \equiv \hat{b}^\dagger|vac\rangle \mapsto \frac{1}{\sqrt{2}}(-\hat{a}^\dagger + \hat{b}^\dagger)|vac\rangle \equiv \frac{1}{\sqrt{2}}(-|0\rangle + |1\rangle), \qquad (9.69)$$

which is precisely the action of a Hadamard gate.

Hadamard gates and phase gates with arbitrary phases allow one to enact any single-qubit unitary transformation.

In order to obtain a universal set of unitary gates on any number of qubits, one needs to be able to complement single-qubit unitaries with at least one two-qubit entangling gate, such as the controlled-z gate with unitary representation $\mathrm{diag}(1, 1, 1, -1)$. The major breakthrough of the KLM proposal was proving that such a unitary operation can actually be achieved on the dual rail encoding by beam splitter transformations and photodetections alone.

As a preliminary step towards demonstrating this possibility, we will need to enlarge the state space of one mode, say a, to include the number state $2^{-1/2}\hat{a}^{\dagger 2}|vac\rangle$, and show that the following three-dimensional linear phase gate $\hat{P}$ can be implemented:

$$\hat{P}(\hat{1} + \hat{a}^\dagger + \hat{a}^{\dagger 2})|vac\rangle = (\hat{1} + \hat{a}^\dagger - \hat{a}^{\dagger 2})|vac\rangle \qquad (9.70)$$

(note that the normalisation of the input states is immaterial when defining $\hat{P}$, which is linear). In turn, this requires one to append two ancillary modes, with ladder operators $\hat{a}_1$ and $\hat{a}_2$, to a and to observe that a combination of beam splitters among these three modes exists that results in the Heisenberg picture mapping

$$\begin{pmatrix} \hat{a}^\dagger \\ \hat{a}_1^\dagger \\ \hat{a}_2^\dagger \end{pmatrix} \mapsto V \begin{pmatrix} \hat{a}^\dagger \\ \hat{a}_1^\dagger \\ \hat{a}_2^\dagger \end{pmatrix}, \qquad (9.71)$$

with real unitary V given by

$$V = \begin{pmatrix} 1 - \sqrt{2} & 1/\sqrt[4]{2} & \sqrt{3 - \sqrt{2}} \\ 1/\sqrt[4]{2} & 1/2 & 1/2 - 1/\sqrt{2} \\ \sqrt{3 - \sqrt{2}} & 1/2 - 1/\sqrt{2} & \sqrt{2} - 1/2 \end{pmatrix}. \tag{9.72}$$

Below, we shall denote with $\hat{V}$ the unitary operation that corresponds to V, when acting on states in the Schrödinger picture.

It turns out that, by preparing the initial ancilla in the state $\hat{a}_1^\dagger |0_{12}\rangle$ (where $|0_{12}\rangle$ now stands for the vacuum state of the modes a_1 and a_2), by acting with $\hat{V}$ on the whole system, and by then post-selecting the final state upon the measurement of the ancillary state $\hat{a}_1^\dagger |0_{12}\rangle$ (the same as the initial state), one implements non-deterministically the quantum gate $\hat{P}$, with success probability $1/4$. In formulae,

$$\hat{P} = \langle 0_{12} | \hat{a}_1 \hat{V} \hat{a}_1^\dagger |0_{12}\rangle. \tag{9.73}$$

This may be easily shown since the operation V, being passive, must preserve the total number of excitations, so that each system vector $\hat{a}^{\dagger j}|0_{12}\rangle$ must be sent into itself up to a multiplicative factor (this is because the number of excitations in the final, measured ancillary state is the same as the initial one, so the excitations in the system must be conserved too). Such factors may be determined explicitly from the form of P (9.73), and they are given, including normalisation (which should be reinstated after the measurement), by the quantities

$$l_j = \frac{1}{j!} \langle 0_{12} | \hat{a}_1 \hat{a}^j V \hat{a}^{\dagger j} \hat{a}_1^\dagger |0_{12}\rangle$$

$$= \frac{1}{j!} \langle 0_{12} | \hat{a}_1 \hat{a}^j (V_{00}\hat{a}^\dagger + V_{10}\hat{a}_1^\dagger + V_{20}\hat{a}_2^\dagger)^j (V_{01}\hat{a}^\dagger + V_{11}\hat{a}_1^\dagger + V_{21}\hat{a}_2^\dagger)|0_{12}\rangle. \tag{9.74}$$

Thus, $l_0 = V_{11} = 1/2$, $l_1 = V_{00}V_{11} + V_{01}V_{10} = 1/2$ and $l_2 = V_{00}(V_{00}V_{11} + 2V_{01}V_{10}) = -1/2$: this is indeed the gate we were after, which flips the sign of the number state with two excitations without affecting the other states involved. The success probability of this non-deterministic, heralded operation is $(1/2)^2 = 1/4$, corresponding to the probability of reading out the outcome associated with the state $a_1^\dagger |0_{12}\rangle$.

The ability to perform the phase gate P above allows one to implement, probabilistically, a full entangling controlled-z gate. To this aim, one should consider two dual rail qubits encoded into four modes with annihilation operators $\hat{a}$, $\hat{b}$, $\hat{c}$ and $\hat{d}$, and run the following protocol:

- Let modes b and d interact via a 50:50 beam splitter;

- apply the gate P to modes b and d, with overall success probability $1/16$;

- let modes b and d interact once more through a 50:50 beam splitter.

Problem 9.5. (*KLM probabilistic entangling gate*). Show that the procedure above implements the controlled-z gate.

Solution. Inspecting the encoding (9.68) reveals that the 50:50 beam splitter between b and d rotates the two-qubit states $|01\rangle \equiv \hat{a}^\dagger \hat{d}^\dagger |vac\rangle$ and $|10\rangle \equiv \hat{b}^\dagger \hat{c}^\dagger |vac\rangle$, leaves $|00\rangle \equiv \hat{a}^\dagger \hat{c}^\dagger |vac\rangle$ unchanged, and sends $|11\rangle \equiv \hat{b}^\dagger \hat{d}^\dagger |vac\rangle$ into the state $(\hat{d}^{\dagger 2} - \hat{b}^{\dagger 2})|vac\rangle/\sqrt{2}$, outside the computational basis. The coefficient multiplying the latter state is then flipped by applying the $\hat{P}$ gate on modes b and d. The last beam splitter then undoes the first one (since a 50:50 beam splitter squares to the identity transformation), with the net final effect of a phase in front of the coefficient of $|11\rangle$.

We have thus shown that *a scheme using only photodetectors* (capable of distinguishing between 0, 1 and more photons), *beam splitters and phase shifters may serve as a platform for universal quantum computation, at least in terms of allowing one to implement, in principle, any unitary transformation.* The KLM scheme paved the way for the development of what is now known as 'linear optical quantum computation' (where, it should be noted, "linear" here excludes squeezing transformations whose implementation, as we saw in Chapter 5, would require optical nonlinearities).

Linear optical quantum architectures typically imply micro- or nanofabricated structures, where the propagating modes are supported by optical waveguides. In such implementations, beam splitters are realised by the evanescent coupling between neighbouring waveguides (due to the overlap of evanescent modes extending outside the waveguides). Such miniaturised, integrated architectures on chips have spurred several interesting developments and hold considerable promise, for both gate- and measurement-based quantum computation (where a large initial entangled state is prepared and the information is processed through measurements and feedforward). It should be already clear from our brief discussion that the availability of reliable sources of pure single photons, accessible on demand, is critical to all linear optical schemes.

9.2 ATOM-LIGHT INTERACTIONS

The main stream of quantum optics is dedicated to the interaction of the electromagnetic field in the quantum regime with elementary constituents of matter, such as atoms and molecules, which are of interest to both fundamental research and quantum technologies. A systematic treatment of such a wide topic, which would imply the introduction of extensive notions from atomic and molecular physics, lies well outside the remit of this book. Here, we shall nonetheless provide the reader with a brief outline of the typical, most basic models of interaction between continuous variables and effective finite dimensional systems in atomic systems, and also take this as an opportunity to give a qualitative account of the well-known rotating wave approximation, which is ubiquitous in quantum optics and will be applied in the following sections.

We shall consider a cavity QED set-up, where an atom has somehow been trapped in a cavity field.[5] The internal (electronic) degrees of freedom of this generic atom, which is clearly another continuous variable, will be idealised as a two-level system since all the population of all other levels are either energetically unfavourable or altogether prohibited by selection laws. The atom, with internal energy eigenstates $|0\rangle$ and $|1\rangle$ separated by an energy gap ω_a (with dimensions of a frequency, since $\hbar = 1$), is supposed to interact with a single mode of the field at frequency ω and annihilation operator $\hat{a}$ through an electric dipole interaction. The total atom-light Hamiltonian $\hat{H}$ reads

$$\hat{H} = \frac{\omega_a}{2}\hat{\sigma}_z + \omega\hat{a}^\dagger\hat{a} + \chi(\hat{\sigma}_- + \hat{\sigma}_+)(\hat{a} + \hat{a}^\dagger) , \qquad (9.75)$$

where χ is the coupling strength and the qubit operators $\hat{\sigma}_z$, $\hat{\sigma}_-$ and $\hat{\sigma}_+$ are defined as $\hat{\sigma}_z = (|1\rangle\langle 1| - |0\rangle\langle 0|)$, $\hat{\sigma}_- = |0\rangle\langle 1|$ and $\hat{\sigma}_+ = |1\rangle\langle 0|$. In practice, the direct dipole coupling between atom and light is rather small for realistic mode volumes and values of the bare atomic dipole moments. In order to realise the coupling assumed above with reasonable, non-negligible parameters, one has typically to resort to laser driving and more sophisticated atomic level schemes. The simplest of such schemes is arguably the Raman, "lambda" transition, where a third atomic level at a higher energy is coherently coupled to the excited state $|1\rangle$ and coupled through a strong classical, slightly detuned laser to the ground state $|0\rangle$. Under the conditions where the highest level may be 'adiabatically eliminated', as this dynamics is very fast and hence 'slave' to the dynamics of the other two levels, one realises an effective coupling of the form (9.75) above. A number of quantum optics textbooks cover this as well as more refined schemes and we will hence not dwell on it here. We will, however, see a case where laser driving is instrumental in turning on an interaction later, in the case of optomechanics, with only continuous variable degrees of freedom involved.

Problem 9.6. (*Bare dipole coupling*). Give a rough estimate of the mode volume required to couple coherently the ground and first excited level of hydrogen with a resonant light field through direct electric dipole coupling.

Solution. Let a_0 be the Bohr radius, e the electron charge and ω the light frequency. The transition energy is given by $(1 - 1/4)e^2/(4\pi\varepsilon_0 a_0) = \omega\hbar$, while the dipole moment may be estimated as $a_0 e$. The typical intensity of the electric field is given by (9.24) as $\sqrt{\omega\hbar/(2\varepsilon_0 V)}$, where V is the mode volume. For the coupling to act on

---

[5]Note that a most famous, Nobel-awarded, cavity QED set-up is actually embodied by flying atoms interacting with a static cavity field at microwave frequencies.

the typical time scale of the atomic dynamics, it must be

$$\frac{3}{4}\frac{e^2}{4\pi\varepsilon_0 a_0} \approx \frac{a_0 e}{\sqrt{2\varepsilon_0 V}}\sqrt{\frac{3}{4}\frac{e^2}{4\pi\varepsilon_0 a_0}},\tag{9.76}$$

which simplifies to $V \approx \sqrt{\frac{8\pi}{3}}a_0^3$. This is ridiculously small, of course. One need not despair though: much more favourable systems exist, like atoms in "Rydberg states" with high dipole moments, where strong direct couplings may be envisaged.

9.2.1 The rotating wave approximation

Let us take the Hamiltonian $\hat{H}$ Eq. (9.75) as a starting point and analyse a very instructive facet of the ensuing dynamics. Notice that $\hat{H} = \hat{H}_{loc} + \hat{H}_{int}$ is composed of a 'local' part $\hat{H}_{loc} = \frac{\omega_a}{2}\hat{\sigma}_z + \omega\hat{a}^\dagger\hat{a}$, whose terms act separately on the local Hilbert spaces of the bosonic mode and of the qubit, and of an interaction part $\hat{H}_{int} = \chi(\hat{\sigma}_- +\hat{\sigma}_+)(\hat{a}+\hat{a}^\dagger)$. Given the simple structure of the local parts, the overall action of $\hat{H}$ may be better understood by switching to the interaction picture, where the local Hamiltonian terms are absorbed in the time dependence of the re-defined local Hilbert space bases. All operators in the interaction picture, which we shall distinguish with a tilde ~ symbol, are re-defined by the action by congruence of the unitary operator $\mathrm{e}^{i\hat{H}_{loc}t}$. In formulae, given a generic operator $\hat{O}$, we shall set $\tilde{\hat{O}} = \mathrm{e}^{i\hat{H}_{loc}t}\hat{O}\mathrm{e}^{-i\hat{H}_{loc}t}$. It is then straightforward to check that, if a generic density operator ϱ is governed by the Heisenberg equation $\dot{\varrho} = i[\varrho, \hat{H}_{loc} + \hat{H}_{int}]$, then the interaction picture operator $\tilde{\varrho}$ obeys $\dot{\tilde{\varrho}} = i[\tilde{\varrho}, \tilde{\hat{H}}_{int}(t)]$, where $\tilde{\hat{H}}_{int}(t) = \mathrm{e}^{i\hat{H}_{loc}t}\hat{H}_{int}\mathrm{e}^{-i\hat{H}_{loc}t}$.

The effect of the local Hamiltonian has been thus subsumed into a redefinition of the interaction Hamiltonian, at the price of making the latter time dependent. In our case, the exact form of $\tilde{\hat{H}}_{int}(t)$ can be easily determined, because the action by congruence of the local evolution operator $\mathrm{e}^{i\hat{H}_{loc}t}$ on the operators at play is known. The field part of such an operator is just a single-mode phase shifter which, as explained in Section 5.1.2.1, acts on $\hat{a}$ as $\mathrm{e}^{i\omega\hat{a}^\dagger\hat{a}t}\hat{a}\mathrm{e}^{-i\omega\hat{a}^\dagger\hat{a}t} = \mathrm{e}^{-i\omega t}\hat{a}$. Further, it is a straightforward exercise involving only 2×2 matrices to show that $\mathrm{e}^{i\omega_a\hat{\sigma}_z t/2}\hat{\sigma}_-\mathrm{e}^{-i\omega_a\hat{\sigma}_z t/2} = \mathrm{e}^{-i\omega_a t}\hat{\sigma}_-$. Because of these relations, the interaction picture is also known as the 'rotating frame' in this instance.

By switching to the rotating frame, we have reduced the dynamics to the action of the following time dependent Hamiltonian:

$$\begin{aligned}\tilde{\hat{H}}_{int}(t) &= \chi\left(\mathrm{e}^{-i\omega_a t}\hat{\sigma}_- + \mathrm{e}^{i\omega_a t}\hat{\sigma}_+\right)\left(\mathrm{e}^{-i\omega t}\hat{a} + \mathrm{e}^{i\omega t}\hat{a}^\dagger\right)\\ &= \chi\left(\hat{a}\hat{\sigma}_+\mathrm{e}^{i(\omega_a-\omega)t} + \mathrm{h.c.}\right) + \chi\left(\hat{a}^\dagger\hat{\sigma}_+\mathrm{e}^{i(\omega_a+\omega)t} + \mathrm{h.c.}\right).\end{aligned}\tag{9.77}$$

The operators in the first term of the Hamiltonian of Eq. (9.77) rotate in time at the frequency difference $(\omega_a - \omega)$, while the operators in the second term rotate at the frequency sum $(\omega_a + \omega)$. The former terms are said to be "rotating", whilst the latter are "counter-rotating".

Now, in a broad variety of situations, the counter-rotating terms can be neglected, for reasons that will be discussed qualitatively below, and the Hamiltonian can be approximated with the "rotating wave" Hamiltonian $\hat{H}_{rw}$:

$$\hat{H}_{rw} = \chi \left(\hat{a} \, \hat{\sigma}_+ e^{i(\omega_a - \omega)t} + \text{h.c.} \right) . \tag{9.78}$$

Notice that the surviving terms in the rotating wave approximation are the ones whereby an excitation is transferred between the field and the two-level system, while the counter-rotating terms are those where excitations are destroyed or created in pairs, in both the field and the two-level system. Also note that, in the exactly resonant case $\omega_a = \omega$, the rotating terms lose their time dependence altogether and become constant, and the interaction Hamiltonian (9.78) reduces to the "Jaynes–Cummings" Hamiltonian $\chi(\hat{a} \, \hat{\sigma}_+ + \hat{a}^\dagger \hat{\sigma}_-)$. As a side remark, let us mention that there are physical situations, typically in systems where the number of excitations (particles, in the field description) is constrained by a superselection rule, where the Jaynes–Cummings model applies exactly.

To form an appreciation of why and when the rotating wave approximation holds, it is expedient to resort to a standard approach to the study of time-dependent Hamiltonians. The Heisenberg equation for the density operator in interaction picture, $\dot{\tilde{\varrho}} = -i[\hat{\tilde{H}}_{int}(t), \tilde{\varrho}]$, can be formally solved to obtain:

$$\tilde{\varrho} = \tilde{\varrho}(0) - i \int_0^t [\hat{\tilde{H}}_{int}(t_1), \tilde{\varrho}(t_1)] \, dt_1 , \tag{9.79}$$

which can be re-inserted into the Heisenberg equation and formally solved once more, yielding

$$\tilde{\varrho} = \tilde{\varrho}(0) - i \int_0^t [\hat{\tilde{H}}_{int}(t_1), \tilde{\varrho}(0)] \, dt_1 - \int_0^t \int_0^{t_2} [\hat{\tilde{H}}_{int}(t_2), [\hat{\tilde{H}}_{int}(t_1), \tilde{\varrho}(t_1)]] \, dt_1 \, dt_2 . \tag{9.80}$$

The iteration of this process of substitution and formal solution gives rise to the celebrated Dyson series:

$$\tilde{\varrho} = \tilde{\varrho}(0) + \sum_{j=0}^{\infty} (-i)^j \int_0^t \int_0^{t_j} \cdots \int_0^{t_2} [\hat{\tilde{H}}_{int}(t_j), \ldots, [\hat{\tilde{H}}_{int}(t_1), \tilde{\varrho}(0)]] \, dt_1 \ldots dt_j , \tag{9.81}$$

where all the repeated commutators are, crucially, evaluated with respect to the density operator at time $t = 0$. In the case of the Hamiltonian $\hat{\tilde{H}}$, it is clear from Eq. (9.81) that the density operator at any time t will be a sum of terms weighted by the integrals in time of the complex exponentials $e^{i(\omega_a - \omega)t}$ and

$e^{i(\omega_a+\omega)t}$, up to complex conjugation. If the integrals of the counter-rotating terms are negligible with respect to those of the rotating terms, then the rotating wave approximation applies. Notice that, technically, this criterion depends not only on the dynamical parameters but also on the time scale one is interested in. If the exact resonance condition $\omega = \omega_a$ is met, then the criterion above reads $2\omega t \gg \sqrt{2 - 2\cos(2\omega t)}$, which is going to be guaranteed as time grows, but not at very short time scales, because of the linear growth of the left hand side of the inequality. Typically one can rely on the rotating wave Hamiltonian for $2\omega t \gg 1$ which, on the typical dynamical time-scales set by the interaction Hamiltonian strength χ, is satisfied if $\omega \gg \chi$: in this regime, the counter-rotating terms undergo a very large number of oscillations in the time taken for the interaction Hamiltonian to produce any appreciable effect.

If the resonant condition is not met, the sufficient condition above is not going to hold indefinitely, because both sides of the inequality would involve oscillating integrals (albeit possibly at vastly different frequencies). However, a more detailed analysis shows that the approximation still holds very well, at almost all times, if the conditions

$$|\omega_a - \omega| \ll \omega_a + \omega \quad \text{and} \quad \omega_a + \omega \gg \chi \qquad (9.82)$$

are met. The analysis of these exact conditions of applicability would be rather involved and would lead us astray. Let us just sum this state of affairs up by claiming that, *when the conditions* (9.82) *are satisfied,* and at times such that $t(\omega_a + \omega) \gg 1$, *one can reliably approximate the full interaction Hamiltonian* (9.77), *with the rotating wave Hamiltonian* (9.78).

Finally, notice that, although the approximation has been exemplified here for the interaction of a field with a two-level system, this would also apply to the interaction of two continuous variable degrees of freedom a and b with local Hamiltonians $\omega_a \hat{a}^\dagger \hat{a}$ and $\omega_b \hat{b}^\dagger \hat{b}$ and a bilinear coupling, such as the product of two quadratures $\chi \hat{x}_a \hat{x}_b = \chi(\hat{a} + \hat{a}^\dagger)(\hat{b} + \hat{b}^\dagger)$. Under the same conditions detailed above on the local frequencies, such a coupling can be approximated with its rotating wave version $\chi(\hat{a}\hat{b}^\dagger + \hat{a}^\dagger\hat{b})$.

9.2.2 Dispersive interactions

The rotating wave criteria (9.82) still allow for the situation where the detuning $\Delta = (\omega - \omega_a)$ is much larger than the coupling strength: $|\Delta| \gg \chi$. It is then interesting to take the Jaynes–Cummings interaction Hamiltonian $\chi(\hat{a}\,\hat{\sigma}_+ + \hat{a}^\dagger\,\hat{\sigma}_-)$ as a starting point, and consider the regime where $|\Delta| \gg \chi$. By switching to the interaction picture and considering the second-order term in the Dyson series, which we defined above to justify the rotating wave approximation, it may be shown that the effective field–qubit interaction Hamiltonian $\hat{H}_{eff}$ is then given by

$$\hat{H}_{eff} = \frac{\chi^2}{\Delta}[\hat{a}\,\hat{\sigma}_+, \hat{a}^\dagger\,\hat{\sigma}_-] = \frac{\chi^2}{\Delta}\hat{a}^\dagger\hat{a}\hat{\sigma}_z + \frac{\chi^2}{2\Delta}\hat{\sigma}_z + \frac{\chi^2}{2\Delta}\hat{\mathbb{1}}. \qquad (9.83)$$

The second term in the right hand side of Eq. (9.83) is just a Stark shift to the atomic excited level due to the interaction with the field, which may be re-absorbed and will be disregarded in what follows. One is then left with the "dispersive" interaction Hamiltonian $\frac{\chi^2}{\Delta}\hat{a}^\dagger\hat{a}\hat{\sigma}_z$.

For strong laser driving of the field with dimensionless real amplitude $|\alpha|$, the dispersive interaction may be 'linearised' by replacing $\hat{a} \mapsto |\alpha| + \hat{a}$ in the effective Hamiltonian and keeping only first-order terms in $\hat{a}$ to obtain the linearised dispersive interaction Hamiltonian $\hat{H}_d = \frac{\sqrt{2}|\alpha|\chi^2}{\Delta}\hat{x}\hat{\sigma}_z$, with $\sqrt{2}\hat{x} = \hat{a} + \hat{a}^\dagger$.[6] Tuning the optical phase of the driving, assumed to be real in the above, would allow one to couple any field quadrature to the qubit via the linearised dispersive interaction. Accessing the Hamiltonian $\hat{H}_d$ paves the way to certain non-Gaussian manipulation of the continuous variable degree of freedom, as illustrated by the iconic example below.

Problem 9.7. (*Generation of cat-like states*). An experimentalist may couple an atomic qubit to a mode of the light field through the linearised dispersive Hamiltonian $\hat{H}_d = \xi\hat{x}\hat{\sigma}_z$ and is capable of preparing the qubit in any single qubit state and to read it out, by unitary manipulation and fluorescence, in any orthonormal basis. Describe a scheme that would allow the experimentalist to create, from the field vacuum $|0\rangle$, pure cat-like states of the form $(|\alpha\rangle + |-\alpha\rangle)/\sqrt{2 + 2e^{-2|\alpha|^2}}$, where $|\alpha\rangle$ is a coherent state.

Solution. The experimentalist should follow this recipe:

- Prepare the qubit in the state $\frac{1}{\sqrt{2}}(|g\rangle + |e\rangle)$, where $|e\rangle$ and $|g\rangle$ are the eigenvectors of $\hat{\sigma}_z$.

- Turn on the dispersive coupling for a time t. Since $\hat{\sigma}_z|g\rangle = -|g\rangle$ and $\hat{\sigma}_z|e\rangle = |e\rangle$, this will act as two opposite displacement operators on the field on the two coherent branches of the quantum state: in formulae,

$$e^{-i\hat{H}_d t}\frac{1}{\sqrt{2}}(|g\rangle \otimes |0\rangle + |e\rangle \otimes |0\rangle) = \frac{1}{\sqrt{2}}(e^{i\xi t\hat{x}}|g\rangle \otimes |0\rangle + e^{-i\xi t\hat{x}}|e\rangle \otimes |0\rangle)$$

$$= \frac{1}{\sqrt{2}}(|g\rangle \otimes |\alpha\rangle + |e\rangle \otimes |-\alpha\rangle),$$

(9.84)

where the coherent amplitude has been set as $\alpha = i\xi t/\sqrt{2}$.

- Note that the state obtained above is the actual analogue of the

---

[6] The procedure of linearisation is treated in some detail in Section 9.3.1 for the case of optomechanics, where a dispersive coupling arises in the original Hamiltonian, rather than as an effective description.

cat of Schrödinger's original narrative, where the continuous variable degree of freedom is entangled with an atom. In order to get the so called 'cat-like' state we are after (a coherent superposition of coherent states with different amplitudes), one can measure the qubits in the basis of eigenvectors of $\hat{\sigma}_x$, $(|g\rangle \mp |e\rangle)/\sqrt{2}$. The field then collapses in the state $(|\alpha\rangle \mp |-\alpha\rangle)/\sqrt{2 + 2e^{-2|\alpha|^2}}$, whose relative phase depends on the qubit reading. Note that the state must necessarily be normalised, so there was no need to evaluate its normalisation from the measurement process.

This is the cat-like state the experimentalist was after, whose phase space representation was investigated in Problem 4.3.

9.3 QUANTUM OPTOMECHANICS

"Optomechanics" is a broad designation that stands for composite systems where massive mechanical degrees of freedom are coupled to light modes, ranging from gravitational wave detectors in large interferometers to micro- and nanomechanical oscillators coupled to cavity light. More specifically, cavity optomechanics refers to the coupling of a mechanical oscillator to modes of the electromagnetic field sustained by a cavity which, as we have seen above, allows for stronger coherent couplings. Several such systems have in fact by now entered, or are approaching, the quantum regime, and are a prominent example of controllable quantum continuous variables.

To fix ideas we shall focus, in this brief discussion, on the standard quantum optomechanical paradigm of a cavity whose length, and hence resonant frequencies, depends on the position of a mirror which is tethered to a substrate through a mechanical oscillator. Our treatment can be easily generalised to optomechanical couplings of a different nature, such as whispering gallery modes around a toroidal microcavity or a levitated bead, or any other set-up where localised oscillators affect the structure of the electromagnetic modes. Hereafter, we shall single out a single cavity light mode with associated annihilation operator $\hat{a}'$ and a single mechanical degree of freedom with annihilation operator $\hat{b}'$. The Hamiltonian operator of the cavity field is, up to the vacuum energy, $\hat{\omega}(\hat{x}')\hat{a}'^{\dagger}\hat{a}'$. As emphasised by the notation adopted, $\hat{\omega}(\hat{x}')$ is not a constant but an operator acting on the mechanical oscillator Hilbert space as per

$$\hat{\omega}(\hat{x}') = \frac{w}{\hat{x}'}\,, \qquad (9.85)$$

where w is a constant, with dimensions of velocity, that characterises the cavity mode in hand and $\hat{x}' = \sqrt{\frac{\hbar}{m\omega_m}}\frac{\hat{b}'+\hat{b}'^{\dagger}}{\sqrt{2}}$, with m and ω_m being, respectively, mass and bare frequency of the mechanical oscillator.

We can take into account independent Markovian environments for both light and the mechanical oscillator by introducing the respective loss rates κ and κ_m and input fields $\hat{a}_{in}$ and $\hat{b}_{in}$. The coupled quantum Langevin equations read

$$\dot{\hat{a}}' = -\frac{\kappa}{2}\hat{a}' - i\hat{\omega}(\hat{x}')\hat{a}' + \sqrt{\kappa}\,\hat{a}_{in}(t)\,, \tag{9.86}$$

$$\dot{\hat{b}}' = -\frac{\kappa_m}{2}\hat{b}' - i\omega_m\hat{b}' + i\left[\hat{\omega}(\hat{x}'),\hat{b}'\right]\hat{a}'^\dagger\hat{a}' + \sqrt{\kappa_m}\,\hat{b}_{in}(t)\,, \tag{9.87}$$

where we just added the Heisenberg Hamiltonian terms to two Langevin equations of the form (9.49), derived in Section 9.1.4. Typically, at room temperature, the second-order statistics of the light input field will be close to those of a vacuum state: $\langle\hat{a}_{in}(s)\hat{a}_{in}^\dagger(t)\rangle = \delta(s-t)$, with first moments $\langle\hat{a}_{in}(t)\rangle = \lambda\,e^{-i\omega_L t}$, where we allow for a laser driving with amplitude λ and frequency ω_L. The mechanical oscillator is instead subject to heating due to the interaction with a phonon field at finite temperature, with $\langle\hat{b}_{in}^\dagger(s)\hat{b}_{in}(t)\rangle = N\delta(s-t)$ and $\langle\hat{b}_{in}(t)\rangle = 0$. The parameter N corresponds to the number of thermal phonons that occupy the oscillator at steady state when no coupling to light is acting. The primary aim of cavity quantum optomechanics is precisely to cool down the quantum oscillator, in order to control massive oscillators in the quantum regime. This would pave the way for applications to the testing of variants to standard quantum mechanics, such as wave function collapse models (possibly induced by quantum gravity), as well as to broader quantum sensing techniques. We will show in the next pages how a canny driving of the cavity field allows one to cool down the quantum oscillator.

As apparent, the non-quadratic Hamiltonian coupling introduced above results in a quantum dynamics that is not linear, and hence difficult to deal with in full generality. It will, for this very reason, provide us with the opportunity to illustrate the procedure of linearisation of the dynamical equations of motion whereby, in the regime of small quantum fluctuations around large mean values, the dynamics can be approximated by a Gaussian one, with second-order couplings and affine equations of motion.

9.3.1 Linearised dynamics

Before proceeding with the linearisation procedure, let us notice that the addition of c-numbers, even if time dependent, to the field operators does not alter their commutation relations. One can hence define standard bosonic annihilation operators $\hat{a}$ and $\hat{b}$ as $\hat{a} = \hat{a}' - \alpha(t)$ and $\hat{b} = \hat{b}' - \beta(t)$ (where $\alpha(t)$ and $\beta(t)$ are understood to multiply the identity operator). Linearisation will rest on the assumption of small fluctuations of the bosonic operators around $\alpha(t)$ and $\beta(t)$: $|\langle\hat{a}\rangle| \ll |\alpha(t)|$ and $|\langle\hat{b}\rangle| \ll |\beta(t)|$,[7] whereby we will Taylor

---

[7]These conditions, which can be checked a posteriori, only compare the first statistical moment of the fluctuations with the large moduli of $\alpha(t)$ and $\beta(t)$. It should be noted that this necessary condition is not enough to guarantee that all the relevant matrix elements of the operators $\hat{a}$ and $\hat{b}$ are small with respect to $\alpha(t)$ and $\beta(t)$. A rigorous analysis of

expand the cavity frequency operator as $\hat{\omega}(\hat{x}') \approx \frac{w}{l} - \frac{w}{l^2}\hat{x}'$; for convenience, we have defined the oscillator 'reference length' $l = \sqrt{\frac{\hbar}{m\omega_m}} \frac{\beta(t)+\beta(t)^*}{\sqrt{2}}$ (which is time dependent). In turn, this allows one to simplify the commutator in Eq. (9.87) as $\left[\hat{\omega}(\hat{x}'), \hat{b}'\right] \approx \frac{w}{l^2}\sqrt{\frac{\hbar}{2m\omega_m}} = g$, where the optomechanical coupling g has been defined. The dynamical equation thus obtained for $\hat{b}'$ is not yet linear, because of the presence of $\hat{a}'^\dagger\hat{a}'$, but the latter may also be expanded to the first-order in $\hat{a} = \hat{a}' - \alpha(t)$ to obtain the linearised quantum Langevin equations

$$\dot{\hat{a}}' \approx -\frac{\kappa}{2}\hat{a}' - i\omega(l)\hat{a}' - ig\alpha(t)(\hat{b} + \hat{b}^\dagger) + \sqrt{\kappa}\,\hat{a}_{in}(t)\,, \tag{9.88}$$

$$\dot{\hat{b}}' \approx -\frac{\kappa_m}{2}\hat{b}' - i\omega_m\hat{b}' - ig|\alpha(t)|^2 - ig(\alpha(t)\hat{a}^\dagger + \alpha^*(t)\hat{a}) + \sqrt{\kappa_m}\,\hat{b}_{in}(t)\,. \tag{9.89}$$

Notice that, as part of the linearisation procedure, we have also substituted the operator $\hat{\omega}(\hat{x})\hat{a}'$ in Eq. (9.86) with its linearised version $\omega(l) + g\alpha(t)(\hat{b} + \hat{b}^\dagger)$. Generally, the linearised dynamics is obtained for any number of variables by determining the Jacobian of the vector function that expresses the time derivative $\dot{\hat{\boldsymbol{a}}}$ of the vector of operators $\hat{\boldsymbol{a}}$ (which contains all the ladder operators of the system).

We are now free to set $\alpha(t)$ and $\beta(t)$ as

$$\alpha(t) = e^{-i\omega_L t}\frac{\sqrt{\kappa}\,\lambda}{\frac{\kappa}{2} + i\omega(l) - i\omega_L}\,, \tag{9.90}$$

$$\beta(t) = -\frac{ig|\alpha(t)|^2}{\frac{\kappa_m}{2} + i\omega_m}\,, \tag{9.91}$$

which solve the following pair of coupled differential equations,[8]

$$\dot{\alpha}(t) = -\left(\frac{\kappa}{2} + i\omega(l)\right)\alpha(t) + \sqrt{\kappa}\,\lambda\,e^{-i\omega_L t}\,, \tag{9.92}$$

$$\dot{\beta}(t) = -\left(\frac{\kappa_m}{2} + i\omega_m\right)\beta(t) - ig|\alpha(t)|^2 \tag{9.93}$$

that, if inserted into Eqs. (9.88) and (9.89), recalling that $\hat{a}' = \alpha(t) + \hat{a}$ and

---

the applicability of the linearised dynamics would hinge on the analytical details of the multivariable function that is Taylor-expanded, and should possibly involve the introduction of more sophisticated notions of distance between operators. This major burden seems unnecessary in such a well-trodden scenario. In the end, although we shall provide later on a heuristic justification for its validity for strongly driven systems, the most convincing argument for the linearisation is its long-standing success in explaining observed dynamics in the appropriate regimes.

[8]Note that $\beta(t)$, as determined by Eqs. (9.90) and (9.91), is actually a constant in time, since $\alpha(t)$ depends on time only through its phase. Also note that the two equations are coupled not only through the presence of $|\alpha(t)|^2$, but also through the dependence of l on β. This results in a third-order algebraic equation to determine $\alpha(t)$.

$\hat{b}' = \beta(t) + \hat{b}$, yield the linearised equations for the fluctuations $\hat{a}$ and $\hat{b}$

$$\dot{\hat{a}} \approx -\left(\frac{\kappa}{2} + i\omega(l)\right)\hat{a} - ig\alpha(t)(\hat{b} + \hat{b}^\dagger) + \sqrt{\kappa}\left(\hat{a}_{in}(t) - \lambda e^{-i\omega_L t}\right), \quad (9.94)$$

$$\dot{\hat{b}} \approx -\left(\frac{\kappa_m}{2} + i\omega_m\right)\hat{b} - ig(\alpha(t)\hat{a}^\dagger + \alpha^*(t)\hat{a}) + \sqrt{\kappa_m}\,\hat{b}_{in}(t). \quad (9.95)$$

The very last term in Eq. (9.94) cancels the laser driving contribution to the first moments of the linearised dynamics (let us remind the reader that $\lambda e^{-i\omega_L t}$ is the input laser amplitude): it is promptly seen that the classical equation for the first moments after linearisation admits a zero constant solution. If one then assumes, as reasonable for a very strong laser driving λ, that the steady state of the cavity mode a' is close to what it would be in the absence of the optomechanical coupling, and that at the beginning of the relevant dynamics the mode is already at steady state, then the initial first moments of the linearised mode a must be very small [this can be inferred by noting that that the steady state's first moments of a driven empty cavity, determined in Eq. (9.67), are the same as $\alpha(0)$]. Together with the increase of $|\alpha(t)|$ and $|\beta|$ with λ, apparent from Eqs. (9.90) and (9.91), this ensures that the necessary linearity conditions of the small quantum fluctuations' first moment are satisfied for a strong enough laser driving. We shall not, here, concern ourselves with a detailed analysis of the applicability of the linear regime, but rather illustrate some general, consequential features of the linearised dynamics.

9.3.2 Sideband driving

It is now convenient to adopt the notation $\alpha(t) = \alpha e^{-i\omega_L t}$, by defining the constant α, which is determined by Eq. (9.92) above, and express the linearised quantum Langevin equations for the fluctuations as

$$\dot{\hat{a}} = -\left(\frac{\kappa}{2} + i\omega(l)\right)\hat{a} - ig\alpha e^{-i\omega_L t}(\hat{b} + \hat{b}^\dagger) + \sqrt{\kappa}\left(\hat{a}_{in}(t) - \lambda e^{-i\omega_L t}\right), \quad (9.96)$$

$$\dot{\hat{b}} = -\left(\frac{\kappa_m}{2} + i\omega_m\right)\hat{b} - ig(\alpha e^{-i\omega_L t}\hat{a}^\dagger + \alpha^* e^{i\omega_L t}\hat{a}) + \sqrt{\kappa_m}\,\hat{b}_{in}(t). \quad (9.97)$$

Notwithstanding linearisation, the equations above correspond to a time dependent, open dynamics whose exact solution would prove rather technical. Here, we will content ourselves with a qualitative, yet very revealing, treatment of this dynamics.

The unitary part of the dynamics above can be identified by setting $\kappa = 0$. Inspection of such a case shows that it corresponds to the following time dependent Hamiltonian:

$$\hat{H}(t) = \omega(l)\hat{a}^\dagger\hat{a} + \omega_m\hat{b}^\dagger\hat{b} + g(\alpha e^{-i\omega_L t}\hat{a}^\dagger + \alpha^* e^{i\omega_L t}\hat{a})(\hat{b} + \hat{b}^\dagger). \quad (9.98)$$

The local Hamiltonian terms can be dispensed with by moving to the interaction picture, that is, by letting the unitary transformation $e^{i(\omega(l)\hat{a}^\dagger\hat{a} + \omega_m\hat{b}^\dagger\hat{b})t}$ act by congruence on all operators at play. As we saw in Section 9.2.1, this transforms the ladder operators as $\hat{a} \mapsto e^{-i\omega(l)t}\hat{a}$ and $\hat{b} \mapsto e^{-i\omega_m t}\hat{b}$ (whereby this time dependent change of basis is also referred to as the "rotating frame") and yields the rotating interaction Hamiltonian $\hat{H}_{int}(t)$

$$\hat{H}_{int}(t) = g(\alpha e^{-i\Delta t}\hat{a}^\dagger + \alpha^* e^{i\Delta t}\hat{a})(e^{-i\omega_m t}\hat{b} + e^{i\omega_m t}\hat{b}^\dagger) , \qquad (9.99)$$

where the laser detuning with respect to the cavity frequency $\Delta = (\omega_L - \omega(l))$ was introduced.

Now, if one picks $\Delta = -\omega_m$, a driving regime referred to as the "red sideband" since the laser frequency is lower than the cavity one (notice that the optical frequency is typically several orders of magnitude larger than the mechanical one), then, by virtue of the rotating wave approximation discussed in Section 9.2.1, the dynamics will be well approximated by the following time independent Hamiltonian:

$$\hat{H}_{red} = g(\alpha \hat{a}^\dagger \hat{b} + \alpha^* \hat{a}\hat{b}^\dagger) . \qquad (9.100)$$

But, as we saw, this is nothing but a beam-splitting Hamiltonian, whereby excitations are not created but keep swapping between the optical and the mechanical mode. Since the higher-frequency optical mode has very low thermal noise even at room temperature, and since the cavity damping rate κ is typically comparable to the dynamical frequencies, this creates the conditions to cool the mechanical oscillator, by swapping its thermal excitations to the cavity and letting them leak out before they get back to the oscillator. A more accurate analysis, including noise in the quantum Langevin equations derived above, shows that this driving frequency ($\Delta = -\omega_m$) is typically optimal, in the sense of attaining mechanical steady states with the smallest mean number of phonons (excitations) and also, more importantly, the smallest entropy. This cooling technique, which is widely applied in atomic and molecular physics too, and generally wherever cavities with high enough finesse to 'resolve' the sidebands are available, is termed 'sideband cooling'.

If one instead were to drive the optomechanical system on the blue sideband, i.e., with $\Delta = \omega_m$, then the effective time-independent Hamiltonian, determined once again through the same argument that underlies the rotating wave approximation, would be

$$\hat{H}_{blu} = g(\alpha \hat{a}^\dagger \hat{b}^\dagger + \alpha^* \hat{a}\hat{b}) . \qquad (9.101)$$

As we saw in Chapter 5, this Hamiltonian generates two-mode squeezing, and thus entangles the light field with the mechanical oscillator. This heats up the oscillator's local state and increases its entropy because of the correlations established with the optical mode. Typically, the blue sideband regime is unstable and does not admit a steady state.

It is straightforward to verify that, in the switch to the interaction picture, the Hamiltonian is changed but the terms responsible for dissipation and input noise in the Langevin equation are not affected. To be clear, the quantum

Langevin equation for the operators $\hat{a}$ and $\hat{b}$ in the red sideband regime read

$$\dot{\hat{a}} = -\frac{\kappa}{2}\hat{a} - i\omega(l)\hat{l} - ig\alpha\hat{b} + \sqrt{\kappa}\,\hat{a}_{in}(t)\,, \tag{9.102}$$

$$\dot{\hat{b}} = -\frac{\kappa_m}{2}\hat{b} - i\omega_m\hat{b} - ig\alpha^*\hat{a} + \sqrt{\kappa_m}\,\hat{b}_{in}(t)\,, \tag{9.103}$$

and the corresponding equations for for the blue sideband dynamics are obtained by substituting $\alpha\hat{b}$ with $\alpha\hat{b}^\dagger$ in Eq. (9.102) and $\alpha^*\hat{a}$ with $\alpha\hat{a}^\dagger$ in Eq. (9.103).

9.4 TRAPPED IONS

The motional degrees of freedom of trapped ions, i.e., the positions of the centre of mass of each ion, distinguished from the 'internal' degrees of freedom which refer to the discrete electronic structure, are another glaring example of quantum continuous variables on which a high degree of experimental control can be exerted. It is well known that, essentially because of the Laplace equation for the electric field in the vacuum, no system of charged particles can be confined in three dimensions through external static electric fields alone, a result known as Earnshaw's theorem. Over the years, two main approaches have been put forward and developed to achieve the trapping, and subsequent control, of arrays of ions: the Paul trap, based on a quadrupole electric field oscillating at radio frequencies (and hence called the rf field), and the Penning trap, which employs a magnetic field along with an electric quadrupole field. We shall not enter here into the details of the trapping mechanisms, but will just assume a harmonic trapping potential acting on all ions in all spatial directions on a 1-d chain of ions, and briefly see how the Coulomb repulsion between the ions results in a coupling between their positions, with promise for far-reaching consequences.[9] It is also worth mentioning that, although, as we will see, trapped ions play a prominent part as prototypes for quantum computing, the interest in controlling trapped ions pre-dates quantum computation, being mainly related to mass spectroscopy.

In what follows, we shall distinguish the bare longitudinal trapping frequency ω, that is the trapping frequency in the direction of the linear ions' array, set as the x direction, from the transverse trapping frequency ω_T along y and z.[10] For simplicity, such frequencies will be assumed to be the same for all the n ions, although more daring configurations might be envisaged in certain set-ups. The mass m and charge q will also be assumed to be the same

---

[9] Let us just briefly mention that, in the case of a Paul trap, such a trapping potential is actually a "pseudo-potential", arising from the time average of the oscillating trapping potential, which furnishes an appropriate description on long enough time scales with respect to the radio-frequency field oscillations.

[10] In the literature, the longitudinal and transverse directions are also referred to as "axial" and "radial", respectively.

for each ion. The overall potential energy of the array of n ions is

$$V = \frac{1}{2}m \sum_{j=1}^{n} \left(\omega^2 x_j^2 + \omega_T^2 (y_j^2 + z_j^2) \right) \tag{9.104}$$

$$+ \sum_{j<k} \frac{q^2}{4\pi\epsilon_0 \sqrt{(x_j - x_k)^2 + (y_j - y_k)^2 + (z_j - z_k)^2}}, \tag{9.105}$$

where (x_j, y_j, z_j) denote the position in space of the jth ion. The equilibrium positions $\bar{x}_j$ of the ions along the axial trapping direction are determined by setting to zero the partial derivatives of the potential above (the force acting on each particle), which is equivalent to solving the coupled algebraic system

$$x_j = \sum_{k \neq j} \frac{\text{sgn}(j-k)l^3}{(x_j - x_k)^2} \quad \text{for} \quad j \in [1, \ldots, n], \tag{9.106}$$

where $\text{sgn}(j-k)$ equals 1 if $j > k$ and -1 otherwise and $l^3 = \frac{q^2}{4\pi\epsilon_0 m\omega^2}$ sets the length scale of the system. For $n = 2$ and $n = 3$, this system may be solved analytically, yielding the equilibrium positions $[-2^{-2/3}l, 2^{-2/3}l]$ and $[-(5/4)^{1/3}l, 0, (5/4)^{1/3}l]$ respectively, while for $n > 3$ it must be solved numerically. Note that the values $y_j = z_j = 0$ for all $j \in [1, \ldots, n]$ are always stationary for the transverse coordinates y and z once the x_j's are set as above.

In typical experimental conditions, the Coulomb potential may be reliably approximated by expanding it to the second order around the equilibrium positions $\bar{x}_j$, thus obtaining a quadratic Hamiltonian, following a procedure analogous to that described in detail in Section 9.3.1 for linearised optomechanical couplings.[11] This yields a quadratic Hamiltonian acting on the ions' positions and momenta, with couplings given by the Hessian matrix C of the second derivatives of the potential V. Let us focus, at first, on the longitudinal displacements along the x direction alone. In order to fit our description into the general framework we developed in the previous chapters, it is convenient to rescale the ions' continuous variables by introducing a $2n$-dimensional vector $\hat{\mathbf{r}}$ of dimensionless positions and momenta: $\hat{r}_j = \hat{x}_j \sqrt{\frac{m\omega}{\hbar}}$ and $\hat{r}_{n+j} = \hat{p}_j / \sqrt{m\hbar\omega}$ for $j \in [1, \ldots, n]$, so that the quadratic Hamiltonian $\hat{H}$ in the linearised regime can be written as

$$\hat{H} = \frac{\hbar\omega}{2} \hat{\mathbf{r}}^\mathsf{T} H \hat{\mathbf{r}}, \quad \text{with} \quad H = \begin{pmatrix} C & 0 \\ 0 & \mathbb{1} \end{pmatrix}, \tag{9.107}$$

where

$$C_{jk} = \frac{1}{m\omega^2} \frac{\partial^2 V}{\partial x_j \partial x_k} \bigg|_{x_j = \bar{x}_j} = \begin{cases} 1 + \sum_{l \neq j} \frac{2l^3}{|\bar{x}_j - \bar{x}_l|^3} & \text{if} \quad j = k \\ -\frac{2l^3}{|\bar{x}_j - \bar{x}_k|^3} & \text{if} \quad j \neq k \end{cases}. \tag{9.108}$$

---

[11] In order to obtain a linear dynamics for the operators, in Section 9.3.1 we expanded the Langevin equations to the first order, whereas here we expand the Hamiltonian operator to the second order. Since the nonlinear terms of the original dynamics are in both cases Hamiltonian terms, the two procedures are completely equivalent.

The transverse displacements along y and z are subject to a quadratic Hamiltonian of the same form, with ω replaced by ω_T both in the formulae above and in the definition of the dimensionless variables.[12] Longitudinal and transverse modes are decoupled when the full Hamiltonian is linearised around the equilibrium position.

In the linearised regime, the ions behave like a harmonic chain and interact with each other through quadratic interactions, with local trapping frequencies modified by the Coulomb repulsion. The Hamiltonian matrix H is always positive, and hence admits a symplectic diagonalisation that corresponds to a decomposition in decoupled, collective normal modes with eigenfrequencies corresponding to the symplectic eigenvalues of H. One such mode is the centre of mass mode, whereby all the ions with equal mass oscillate collectively with the same velocity and the same displacement from the equilibrium position. The frequency of such a mode is just the bare trapping frequency ω (we are referring here, to fix ideas, to the longitudinal modes, though the same could be claimed for the transverse ones).

Problem 9.8. (*Centre of mass mode*). Show that the centre of mass mode is a normal mode of the quadratic Hamiltonian of Eqs. (9.107) and (9.108), and that its associated frequency is ω.

Solution. The Hamiltonian matrix above is written in xp-ordering, with symplectic form given by J as in (3.4). Under such an ordering, any direct sum of equal $n \times n$ orthogonal matrices $R \oplus R$ is symplectic. One such matrix will diagonalise H, since the identity block is invariant under orthogonal transformations and C is symmetric (notice that, in general, H will not be put in normal form by such a rotation, since the x and p components pertaining to each decoupled mode may still be unbalanced, and some single-mode squeezing might be in order). Observe now that the sum of the elements on each row of C is the same and equal to 1 – which is a reflection at the second order of the fact that the potential V is central. Hence, the vector associated to the centre of mass mode, $(1,\ldots,1)^\mathsf{T}$ is an eigenvector of C with eigenvalue 1. Since this eigenvalue, 1, is the same as the eigenvalue associated with the corresponding momentum vector – the Hamiltonian matrix is trivially the identity in the momentum subspace – this is already a normal mode (there is no need of applying further squeezing as mentioned above), with eigenfrequency given by the factor that multiplies the Hamiltonian matrix over $\hbar$, which is ω.

The quantum continuous variables embodied by the ions' motion can be

---

[12] Notice that the longitudinal trapping frequency ω still plays a role in the dynamics of the transverse modes as it enters the definition of l. This is as expected since the longitudinal trap sets the spacings between the ions and hence the strength of the Coulomb interaction (and, hence, the 'speed of sound' through the harmonic chain, if you will).

manipulated directly, for instance, by controlling the trapping frequencies in time (see the problem set out below). Thanks to the Coulomb coupling, this allows for the generation of entanglement and squeezing provided, of course, that the ions can be previously cooled down and kept operating in a coherent regime. Arguably more interesting, however, are the possibilities allowed by the interplay between such motional degrees of freedom and the internal electronic levels of the ions, which can be coupled to each other by shining a standing wave laser on the ions. The driving of each ion's internal transitions depends clearly on the field amplitude, and hence on the ion's position in the standing wave: this effect results in a coupling between light and motion. Besides, by tuning the laser frequency, coherent interactions between specific energy levels and motional degrees of freedom may be turned on, leaving the remainder of the system unaffected. We shall illustrate the possibilities offered by the coherent coupling of the internal and motional degrees of freedom by addressing one of its best-known suggested applications in the next section.

Problem 9.9. (*Squeezing by trap modulation*). Consider a single particle of mass m held in a harmonic potential with frequency ω_0, initially found in the ground state of such a potential. Show that a sudden change in frequency, from ω_0 to ω_1, along one direction, enacts a squeezing operation on the particle's motion.

Solution. Let us consider only one spatial direction, say x. The particle starts in the ground state at frequency ω_0, with covariance matrix $\mathbb{1}$ with respect to the dimensionless operators $\hat{r}_1 = \hat{x}\sqrt{\frac{m\omega_0}{\hbar}}$ and $\hat{r}_2 = \frac{\hat{p}}{\sqrt{m\hbar\omega_0}}$, which is obviously stationary with respect to the initial Hamiltonian $\frac{1}{2}(m\omega^2\hat{x}^2 + \hat{p}^2/m)$, with Hamiltonian matrix $H_0 = \mathbb{1}$. Let $\iota = \omega_1^2/\omega_0^2$; the new Hamiltonian matrix after the frequency change is given by $H_1 = \mathrm{diag}(\sqrt{\iota}, 1)$ (with respect to the old coordinates, with new Hamiltonian operator $\frac{\hbar\omega_0}{2}\hat{r}^\mathsf{T} H_1 \hat{r}$). This Hamiltonian matrix is such that $(\Omega H_1)^2 = -\iota\mathbb{1}$, so its associated generator is straightforward to diagonalise and yields (let us also rescale time with $1/\omega_0$, for convenience)

$$e^{\Omega H_1 t} = \begin{pmatrix} \cos(\sqrt{\iota}t) & \sin(\sqrt{\iota}t)/\sqrt{\iota} \\ -\sqrt{\iota}\sin(\sqrt{\iota}t) & \cos(\sqrt{\iota}t) \end{pmatrix}, \qquad (9.109)$$

which can also be verified by differentiation in $t = 0$. At $t = \frac{\pi}{2\sqrt{\iota}}$, the transformation above becomes $e^{\Omega H_1 \frac{\pi}{2\sqrt{\iota}}} = \Omega\,\mathrm{diag}(\frac{\omega_1}{\omega_0}, \frac{\omega_0}{\omega_1})$, which is the product of a passive phase shifter and of a diagonal squeezer. The squeezing grows with the ratio $\frac{\omega_1}{\omega_0}$, or the inverse thereof, depending on whether $\omega_1 > \omega_0$ or the other way around. Notice that, if $\omega_1 > \omega_0$, then the $\hat{x}$ quadrature is squeezed (consider in fact that the action of Ω swaps the quadratures), as one should expect since the potential becomes steeper, and the particle is 'literally' squeezed. On the other hand, if $\omega_0 > \omega_1$, then the momentum is squeezed, as a larger spread in position localises the momentum around lower values. Finally, let us point out that the squeezing is transient, as after the maximum squeezing flagged up above the state rotates back to the vacuum state, under the action of the original, 'free' Hamiltonian (note however that this recurrence takes place on the time scale $1/\omega_1$, and may hence be in principle delayed by an arbitrarily long time).

9.4.1 The Cirac–Zoller quantum computer

Let us now review a celebrated proposal for a quantum computing architecture with trapped ions, which hinges on a continuous variable bus linking together internal electronic levels of separate ions, where digital quantum information is stored. As usual, we intend to focus here on the bare conceptual elements underpinning the scheme, stripped of all technical detail. This will imply the adoption of certain glaring simplifications, which have been confronted over the years in a number of more specialised accounts.

Let us consider a linear array of trapped ions as a quantum processor, with quantum information stored in two specific energy levels of each ion, denoted with $|0\rangle_j$ and $|1\rangle_j$, where the label j stands for the jth ion. We shall assume each ion may be individually addressed with laser light and cooled down to the ground state of both motional and internal degrees of freedom. By driving the internal transition through the laser, any single-qubit unitary quantum gate on the internal levels of any ion may in principle be implemented. It is well known that, in order to obtain a universal quantum processor, which is a device that allows one to implement any unitary gate on the global qubit register, one only needs to supplement single-qubit gates with the ability to perform a specific entangling gate on each pair of qubits.

The first proposal on how to achieve this with trapped ions was put forward by Juan Ignacio Cirac and Peter Zoller in 1995. The Cirac–Zoller entangling gate employs tilted laser standing waves with respect to the trap's main axis, which allow one to address the ions individually whilst coupling to the motional mode along the axis. If the frequency of the laser shining on the jth ion matches the difference between the internal, electronic energy gap and the centre of mass longitudinal mode, then, by the very same reasoning that led to the rotating wave approximation and the sideband driving in optomechanics above (see Sections 9.2.1 and 9.3.2), the time-independent Hamiltonian $\hat{H}_j$ is enacted:

$$\hat{H}_j = |1\rangle_{jj}\langle 0|\hat{a}_{cm} + |0\rangle_{jj}\langle 1|\hat{a}_{cm}^\dagger \,, \tag{9.110}$$

where $\hat{a}_{cm}$ stands for the annihilation operator of the centre of mass mode, and we have neglected an overall factor, which just amounts to appropriately rescaling time in what follows. Notice that the transitions involving all other motional normal modes are negligible if the frequency spacing between the different modes is large enough.

The Hamiltonian $\hat{H}_j$ preserves the total number of excitations, and is hence easily diagonalised in each of the two-dimensional subspaces in which the Hilbert space may be decomposed as a direct sum. If one assumes the centre of mass mode to be initialised in the vacuum state, the solution in the single excitation sector will suffice to describe the dynamics induced by the Hamiltonian $\hat{H}_j$, which acts as follows

$$|10\rangle \mapsto \cos(t)|10\rangle - i\sin(t)|01\rangle , \qquad (9.111)$$

$$|01\rangle \mapsto \cos(t)|01\rangle - i\sin(t)|10\rangle , \qquad (9.112)$$

where t is the rescaled time and $|ab\rangle = |a\rangle_j \otimes |b\rangle_{cm}$ for $a, b \in \{0, 1\}$ ($|b\rangle_{cm}$ is the state with b phonons in the centre of mass mode). Notice that the Hamiltonian above obviously leaves the state $|00\rangle$ unchanged. We shall assume that, as is rather common, a laser with different, orthogonal polarisation to the one considered so far can be shined on the ion to couple the same ground state $|0\rangle_j$ to a different excited state $|2\rangle_j$, through a Hamiltonian of the same form as (9.110). This dynamics does not affect the state $|10\rangle$, and acts as follows on $|01\rangle$:

$$|01\rangle \mapsto \cos(t)|01\rangle - i\sin(t)|20\rangle . \qquad (9.113)$$

The key idea behind the Cirac–Zoller gate is letting any pair of separate ions interact sequentially with the centre of mass mode through the unitary dynamics described above, in such a way that the phonon mode will be unentangled from the ion qubits at the end of the operation, and yet the latter will have undergone a non-trivial, entangling unitary operation. This is achieved by first acting on ion j through a π pulse which, in our notation, amounts to setting $t = \pi/2$ in the evolution (9.111, 9.112) above. By adopting the notation $|abc\rangle = |a\rangle_j \otimes |b\rangle_k \otimes |c\rangle_{cm}$ for $a, b, c \in \{0, 1\}$, this first step corresponds to

$$|000\rangle \mapsto |000\rangle , \qquad (9.114)$$

$$|010\rangle \mapsto |010\rangle , \qquad (9.115)$$

$$|100\rangle \mapsto -i|001\rangle , \qquad (9.116)$$

$$|110\rangle \mapsto -i|011\rangle . \qquad (9.117)$$

Next, a 2π pulse – i.e., one with $t = \pi$ in our notation – in the orthogonal polarisation is shone on the k-th ion. This leaves all the final states in the last array of equations unaffected except for $|001\rangle$ which, as shown by Eq. (9.113), acquires a minus sign, so that the total unitary evolution at this step reads (notice that this step of the evolution involves the second and third systems, and that the state $|1\rangle_k$ of the second system is preserved by the Hamiltonian)

$$|000\rangle \mapsto |000\rangle \,, \tag{9.118}$$
$$|010\rangle \mapsto |010\rangle \,, \tag{9.119}$$
$$|100\rangle \mapsto i|001\rangle \,, \tag{9.120}$$
$$|110\rangle \mapsto -i|011\rangle \,. \tag{9.121}$$

Note in passing that the state $|20\rangle$ never enters the discrete gate operations: its only role is letting the state $|01\rangle$ rotate and acquire a relative phase according to Eq. (9.113). In order to complete the operation, the initial π pulse on the j-th ion is applied again, resulting in the total evolution:

$$|000\rangle \mapsto |000\rangle \,, \tag{9.122}$$
$$|010\rangle \mapsto |010\rangle \,, \tag{9.123}$$
$$|100\rangle \mapsto |100\rangle \,, \tag{9.124}$$
$$|110\rangle \mapsto -|110\rangle \,. \tag{9.125}$$

The centre of mass mode is hence obviously decoupled from the two qubits: it acts as a continuous variable link that mediates the interaction between the internal levels of the ions. The minus sign affecting the last computational basis state (for the two qubits alone) corresponds to the action of a controlled-z gate, which is equivalent to the so called controlled not gate, and is an entangling gate, as can be easily verified by a proper choice of initial states. The possibility of enacting such a gate between any two qubits in the ion chains, along with the ability to perform any single-qubit unitary, which may be realised by coherent laser driving of the internal levels, allows for the realisation of any unitary operation and hence, in principle, of any coherent quantum computation.

The Cirac–Zoller scenario, where continuous variables mediate coherent interactions between "discrete", digital quantum variables where the quantum information is stored and processed, has been, and still is, a classic paradigm in the design and development of schemes for the implementation of controlled, coherent quantum dynamics.

9.5   ATOMIC ENSEMBLES

Another class of systems amenable to the bosonic continuous variable description is comprised of ensembles of N finite-dimensional, non-interacting quantum degrees of freedom, with associated individual spin operators $\hat{J}_{a,k}$, with $a \in \{x, y, z\}$ and $k \in [1, \ldots, N]$, such that $[\hat{J}_{x,j}, \hat{J}_{y,k}] = i\delta_{jk}\hat{J}_{z,j}$ (in natural units). As well known from the theory of angular momentum, each such representation of the $SU(2)$ algebra is defined on a Hilbert space of dimension $2s + 1$, in terms of the 'spin' s, which can be integer or half-integer: for simplicity, we shall assume all such individual values of the spin to be the same for all degrees of freedom. We shall also assume familiarity with the basics of the quantum theory of angular momentum. One may then define the collective operators $\hat{J}_a = \sum_{k=1}^{N} \hat{J}_{a,k}$ for all $a \in \{x, y, z\}$, which also represent the angular momentum algebra:

$$\left[\hat{J}_x, \hat{J}_y\right] = i\hat{J}_z . \tag{9.126}$$

Let us now consider a macroscopic ensemble, i.e., one with $N \gg 1$, which is strongly 'polarised' along, say, the z direction, in the sense that the dynamics is restricted to a subspace of the Hilbert space where $\hat{J}_z$ is at all times close to the maximum value Ns. We shall not introduce here the mathematics of the Holstein–Primakoff transformation, that would provide our approximation with a rigorous footing, but just heuristically note that, under such conditions, one may disregard deviations from the large expectation value $\langle \hat{J}_z \rangle \approx Ns$ and replace the operator $\hat{J}_z$ with the c-number $\langle \hat{J}_z \rangle$ (which we have assumed to be positive, close to $+Ns$, though obviously the opposite value would do too). Then, the rescaled operators $\hat{x}_a = \frac{\hat{J}_x}{\sqrt{\langle \hat{J}_z \rangle}}$ and $\hat{p}_a = \frac{\hat{J}_y}{\sqrt{\langle \hat{J}_z \rangle}}$ approximately satisfy the CCR:

$$[\hat{x}_a, \hat{p}_a] = i\frac{\hat{J}_z}{\langle \hat{J}_z \rangle} \approx i . \tag{9.127}$$

Hence, strongly polarised ensembles of spins behave like bosonic continuous variables, and may be described through the formalism we developed in this book.

In practice, the most prominent example of such systems is represented by atomic clouds of two-level atoms, where near-maximal polarisation is maintained by applying a strong magnetic field along a given spatial direction. Each atom in the ensemble is also referred to as a 'pseudo-spin', hinting at the fact that, strictly speaking, it does not embody a spin degree of freedom, although the span of the two internal atomic eigenstates which are relevant to this case form a two-dimensional Hilbert space which is completely equivalent to a spin 1/2 Hilbert space. By exploiting auxiliary atomic levels in different ways, such systems may be coupled coherently to a travelling light mode a_l through either a beam splitting $i(\hat{a}_a^\dagger \hat{a}_l - \hat{a}_a \hat{a}_l^\dagger)$, a two-mode squeezing $(\hat{a}_a \hat{a}_l + \hat{a}_a^\dagger \hat{a}_l^\dagger)$ or a 'quantum non-demolition' $(\hat{p}_a \hat{p}_l)$ Hamiltonian (also known as QND or 'Faraday' in this context). Along with the efficient homodyne detection of the light modes, these couplings have allowed for the realisation of continuous variable quantum memories, where the state of the light field may be mapped and then retrieved after storage, as well as for quantum teleportation and entanglement generation between light and atoms. A major advantage of such atomic interfaces is that they operate coherently, though not ideally, even at room temperature. Note that the development of reliable quantum memories is instrumental to the advent of quantum technologies, in particular because of their importance to the design of quantum repeaters that, as already mentioned in Section 8.1, would in turn pave the way for long-distance quantum communication.

9.6 INTEGRATED QUANTUM CIRCUITS AND JOSEPHSON JUNCTIONS

Micro- and nano-fabricated superconducting electrical circuits at low temperatures (around 1 K) feature very low dissipation and thermal noise, and may hence enter the quantum regime, in the sense that certain of their collective, electronic observables must be treated as quantum operators.

The basic case of a quantum LC oscillator is modelled by stipulating that the charge on the capacitor $\hat{C}$, and the flux through the inductor $\hat{\Phi}$ are conjugated canonical operators satisfying the CCR

$$[\hat{\Phi}, \hat{Q}] = i\hbar \,, \tag{9.128}$$

whose dynamics is governed by the Hamiltonian:

$$\hat{H}_{LC} = \frac{1}{2}\frac{\hat{\Phi}^2}{L} + \frac{1}{2}\frac{\hat{Q}^2}{C} \,, \tag{9.129}$$

where L and C are, respectively, the inductance and the capacitance of the circuit.[13] This is yet another example of a quantum harmonic oscillator, that can in principle also be coupled to the continuous variables of light.

In the context of quantum technologies, there is a strong interest in modifying the linear circuitry above, governed by a Gaussian dynamics and featuring an evenly spaced energy spectrum, by the introduction of controlled non-linear elements. This is achieved by employing Josephson junctions: pairs of thin superconducting electrodes separated by a thin insulating barrier (about 1 nm wide, typically fabricated by oxidation of the electrodes), through which Cooper pairs may tunnel. In the RF-SQUID (radio frequency superconducting quantum interference device) architecture, the two electrodes forming the junction, with capacitance C, are linked through a superconducting loop with inductance L. This results in the following anharmonic (non-quadratic) Hamiltonian:

$$\hat{H}_{SQ} = \frac{1}{2}\frac{\hat{\Phi}^2}{L} + \frac{1}{2}\frac{\hat{Q}^2}{C} - E_J \cos\left[\frac{2e}{\hbar}(\hat{\Phi} - \Phi_{ext})\right] \,, \tag{9.130}$$

---

[13]Note that this description does not account for the electric charge granularity, as the canonical operator $\hat{Q}$ has a continuous spectrum. A different prescription is often put forward that stipulates canonical commutation relations between a number operator $\hat{N}$ with discrete spectrum and a phase operator $\hat{\theta}$ with eigenvalues in $[0, 2\pi[$. These operators are reminiscent of discretised quasiposition and periodic quasimomentum in translationally invariant quantum systems, with the difference that $\hat{N}$ is usually assumed to have a positive semidefinite spectrum, since its expectation value corresponds to the number of excess, elementary (negative) charges on the capacitor. Although it may produce good predictions in certain situations (such as for the superconducting charge qubits embodied in the so called Cooper pair boxes), we shall not adopt this formalism, since it is fraught with formal difficulties and would require a lengthy discussion and particular care. It will suffice to say here that the relationship $[\hat{N}, \hat{\theta}] = i$, often introduced, is untenable for self-adjoint operators with the properties listed above.

where E_J is the adjustable Josephson energy of the junction, Φ_{ext} is an external flux that may be exerted through an auxiliary coil and e is the electron charge.

The applicative interest in non-linear Hamiltonians such as (9.130) lies mainly in the associated unevenly spaced energy spectrum: this allows one to selectively address transitions between only the two lowest lying energy levels of the system, since higher transitions will be sufficiently detuned. Along with the ability to cool the system down to the ground state, this property allows one to restrict the system dynamics to coherent manipulations in a two-level Hilbert space, and hence to obtain a qubit ready for applications in quantum computation and information processing. Qubits embodied in RF-SQUIDs are usually referred to as 'flux qubits' and are, for instance, the elementary constituents of the well-known D-Wave quantum processors. Different architectures and variations of the Josephson dynamics lead to quantum degrees of freedom characterised by different internal dynamics, such as the 'charge qubits' in Cooper pair boxes, where the Josephson electrodes are connected through a voltage source and an additional gate capacitor. Although their main interest so far has been in their exploitation as qubits, effective finite dimensional systems, integrated quantum circuits may also be employed in their whole Hilbert space, as full-blown continuous variables, with the internal Hamiltonian given above. It should be noted that the control of such systems is particularly handy, since it can be applied through wired electric signals. Screening the quantum degrees of freedom from their control infrastructure has, on the other hand, been one of the main technological challenges in the development of the field.

It is also worth noting that systems of coupled Josephson junctions can be fabricated in arrays, thus realising a coherent many-body quantum system. The Hamiltonian modelling of such arrays will be very briefly touched upon in the next section.

9.7 COLD BOSONIC ATOMS IN OPTICAL LATTICES

Cold bosonic atoms in optical lattices also display coherent quantum features and abide by the continuous variable description. Through magnetic and optical trapping techniques, atoms can be loaded on an optical lattice realised by counter-propagating laser beams, which corresponds to a standing external sinusoidal potential acting on each atom along two dimensions. The whole system of many atoms, accounting for interactions between them, can be described by a continuous bosonic field operator (assuming the atoms in question have integer spin), which can be expanded in a orthonormal set of Wannier functions, as customary in translationally invariant systems. If the external fields are weak enough, and no band crossing occurs, then one can restrict to the lowest Wannier band, and reduce the continuous system to a discrete bosonic field governed by the following Hamiltonian:

$$\hat{H}_{BH} = \frac{1}{2}\sum_{j,k} \hat{a}_j^\dagger \hat{a}_j \hat{a}_k^\dagger \hat{a}_k U_{jk} - \mu \sum_j \hat{a}_j^\dagger \hat{a}_j - t \sum_{j,k:\,\mathrm{nn}} (\hat{a}_j \hat{a}_k^\dagger + \hat{a}_j^\dagger \hat{a}_k), \quad (9.131)$$

where each annihilation operator $\hat{a}_j$ destroys an atom at lattice site j, the summations extend over all the sites of the two-dimensional lattice, and the notation $j, k :$ nn indicates that only nearest neighbour sites are considered. The density–density interaction matrix U_{jk}, the 'chemical potential' μ (as one refers to a Hamiltonian term proportional to the number of particles, whose canonical statistical treatment is equivalent to introducing a grand canonical ensemble with the same chemical potential) and the hopping strength t may all be obtained as integrals over Wannier functions of the lowest band. The bosonic atoms are clearly modelled as a many-body continuous variable system as they satisfy the CCR:

$$[\hat{a}_j, \hat{a}_k^\dagger] = \delta_{jk} . \quad (9.132)$$

The Hamiltonian $\hat{H}_{BH}$ is a generalisation of the celebrated Bose–Hubbard Hamiltonian, which is obtained in the specific case $U_{jk} = \delta_{jk}U$, where only onsite repulsion (assuming a positive U) is present.

The Bose–Hubbard model and its variations are of very broad interest in condensed matter physics and a paradigm to investigate coherent quantum many-body phenomena. In the standard Bose–Hubbard form, the ground state of $\hat{H}_{HB}$ displays the very well-known Mott insulator to superfluid quantum phase transition, where the field switches from a separable state of localised bosons at each lattice site to a delocalised state featuring entanglement and long-range correlations. The Hamiltonian above may be applied reliably to several systems, such as granular superconductors and ultra-thin films and, interestingly, its one-dimensional version may also be applied to model arrays of Josephson junctions (see the previous section) in the limit where the charge operator may be approximated by a large c-number (its average). In the context of cold atoms, it should also be mentioned that Josephson junctions may also be mimicked in the laboratory by achieving Bose–Einstein condensation in each of the wells of an optical double-well potential. Such condensates allow for an approximated canonical phase-number of particles description analogous to the one involving phase and number of Cooper pairs in a superconducting Josephson junction, and hence represent yet another case of controllable quantum system amenable to a continuous variable description.

9.8 FURTHER READING

For a derivation of energy and Hamiltonian of the classical electromagnetic field, the reader may consult the classic treatment by Jackson [52]. An exhaustive, and truly passionate, account of cavity quantum electrodynamics, including the details of the field quantisation in the presence of a cavity, may be found in [24]. The rotating wave approximation is handled in detail and cast in a wider framework in a very useful paper by James and Jerke [55], where Eq. (9.83) is justified too. The input-output formalism was first introduced by Gardiner and Collett [33], and is included in the book by Walls and Milburn [83]. Theory and practice of coherent feedback control of squeezing were reported in [51], including the analysis of finite delays, while a formal treatment of cascaded input-output interfaces was introduced in [40]. The original KLM paper on linear optical quantum computation [57] contains several refinements (such as quantum gate teleportation) over the crude, basic analysis reported here. Unitary gate decompositions and universal sets are discussed in detail in [64].

Quantum optomechanics was reviewed by Bowen and Milburn in the monograph [12]. Reference [54], by James, is a very clear introduction to the motional degrees of freedom of trapped ions. The original proposal by Cirac and Zoller for a quantum processor with trapped ions was published in [20]. A systematic, specialist reference on atomic ensembles as an interface between light and matter is the review paper [43]. A clear account of the Holstein–Primakoff transformation is given in this set of lecture notes [18]. Zagoskin [93] gives a systematic account of superconducting degrees of freedom, comprising an accurate discussion of the controversial status of phase and number as conjugate canonical variables. An overview of the possibilities offered by cold atoms in optical lattices is given in [10].

More details on the fundamentals of experimental implementations of quantum protocols with continuous variables may also be found in the collection of articles [17].

A note on fermions

CONTENTS

The fermionic algebra can be expressed in terms of anti-commutation relations between annihilation and creation operators of n modes as

$$\{\hat{a}_j, \hat{a}_k^\dagger\} = \delta_{jk}\hat{\mathbb{1}}, \quad \{\hat{a}_j, \hat{a}_k\} = 0, \quad \text{for } j \in [1, \ldots, n], \tag{A.1}$$

which may be recast in terms of the self-adjoint 'Majorana operators' $\hat{x}_j = (\hat{a}_j + \hat{a}_j^\dagger)/\sqrt{2}$ and $\hat{p}_j = i(\hat{a}_j^\dagger - \hat{a}_j)/\sqrt{2}$ as

$$\{\hat{x}_j, \hat{x}_k\} = \{\hat{p}_j, \hat{p}_k\} = \delta_{jk}\hat{\mathbb{1}}, \quad \{\hat{x}_j, \hat{p}_k\} = 0. \tag{A.2}$$

Note that, if the modes of the field were extended to a continuum, the annihilation and creation operators above would describe fermions, such as electrons, in free space, on an infinite-dimensional Hilbert space. In first quantisation, which applies in the non-relativistic limit, this would correspond to a standard continuous variable system (constrained by anti-symmetry).

Here, let us instead consider a finite number of fermionic modes – as could be the case in practice for fermionic atoms populating an optical lattice, where only a discrete set of sites may be occupied – and analyse the action of quadratic Hamiltonians on them, in analogy with what was done with bosonic fields in Chapter 3. To this aim, let us define the vector of self-adjoint fermionic operators $\hat{\mathbf{r}} = (\hat{x}_1, \hat{p}_1, \ldots, \hat{x}_n, \hat{p}_n)^\mathsf{T}$ and the purely quadratic, Hermitian Hamiltonian operator $\hat{H} = \frac{i}{2}\hat{\mathbf{r}}^\mathsf{T} H \hat{\mathbf{r}}$, in terms of an as yet unqualified, real Hamiltonian matrix H. Disregarding the identity operator $\hat{\mathbb{1}}$, the fermionic algebra is compactly rendered as

$$\{\hat{\mathbf{r}}, \hat{\mathbf{r}}^\mathsf{T}\} = \mathbb{1}_{2n}, \tag{A.3}$$

which is to be read, in components, as $\{\hat{r}_j, \hat{r}_k\} = \delta_{jk}$. The fermionic analogue of the Heisenberg evolution (3.18) is then written as

$$\dot{\hat{r}}_j = i[\hat{H}, \hat{r}_j] = -\frac{1}{2}\sum_{kl}[\hat{r}_k H_{kl} \hat{r}_l, \hat{r}_j]$$

$$= -\frac{1}{2}\sum_{kl} H_{kl} (\hat{r}_k\{\hat{r}_l, \hat{r}_j\} - \{\hat{r}_k, \hat{r}_j\}\hat{r}_l) = -\frac{1}{2}\sum_k (H_{jk}^\mathsf{T} - H_{jk})\hat{r}_k, \tag{A.4}$$

where we used the Leibniz relation $[\hat{A}\hat{B}, \hat{C}] = \hat{A}\{\hat{B}, \hat{C}\} - \{\hat{A}, \hat{C}\}\hat{B}$. Only the antisymmetric part of H hence contributes to the evolution here, and we can hence assume $H = -H^\mathsf{T}$. The possible linear evolutions of the field operators at finite times are thus given by the transformations e^H (time has been absorbed in H), for skew-symmetric H. Such transformations form the (compact) special orthogonal group $SO(2n)$, and clearly preserve the anti-commutation relations (A.3).

The same programme carried out for bosons in this book might in principle apply to fermions, for which one may define Gaussian states as ground and thermal states of the quadratic Hamiltonians defined above, with the group $SO(2n)$ replacing the symplectic group in the characterisation of Gaussian unitary operations. However, certain subtleties proper to fermions emerge if one pushes this approach beyond dynamics, into the characterisation of entropies and quantum correlations of fermionic fields. For instance, it turns out that the partial transpositions of fermionic Gaussian states are not Gaussian, so that their spectrum is not easily accessible. Hence, inquiries into quantum entanglement based on the PPT criterion, which were so successful for bosons, do not quite carry over to fermions, where more sophisticated techniques are in order. A systematic approach to the entanglement of fermionic modes, highlighting some of its striking peculiarities, is presented in [29].

Some notable facts about the symplectic group

CONTENTS

In this appendix, we shall prove useful notions concerning the real symplectic group $Sp_{2n,\mathbb{R}} = \{S \in \mathcal{M}(2n, \mathbb{R}) \mid SJS^{\mathsf{T}} = J\}$, with

$$J = \begin{pmatrix} 0_n & \mathbb{1}_n \\ -\mathbb{1}_n & 0_n \end{pmatrix} \qquad (\text{B.1})$$

(throughout this appendix, it will be convenient to adopt the xp-ordering introduced in (3.4), by rearranging the vector of operators as an array of $\hat{x}$'s followed by an array of $\hat{p}$'s). Notice that all the statements found below apply to the standard ordering with symplectic form Ω of Eq. (3.2).

Although all the mathematical properties we made use of in the book will be derived here, some readers may be interested in a more comprehensive coverage of the group, which may be found in [3, 22].

First off, let us list a few basic statements:

$$J = \in Sp_{2n,\mathbb{R}} , \qquad (\text{B.2})$$

$$J^{-1} = J^{\mathsf{T}} = -J \in Sp_{2n,\mathbb{R}} , \qquad (\text{B.3})$$

$$S^T = J^{\mathsf{T}} S^{-1} J \in Sp_{2n,\mathbb{R}} . \qquad (\text{B.4})$$

Notice that Eq. (B.4) explicitly shows that $S^{\mathsf{T}} \in Sp_{2n,\mathbb{R}}$.

Any symplectic matrix that can be written as a 'one-shot' matrix exponential $S = e^A$ may be written as $S = e^{JHt}$, where H is real and symmetric and t is a real parameter. This can be seen by setting $S = e^{JLt}$ and taking the first derivative with respect to t of $SJS^\mathsf{T} = J$, which defines the group. Thus, one obtains

$$\frac{\mathrm{d}}{\mathrm{d}t}\left(e^{JLt}Je^{-L^\mathsf{T}Jt}\right)\bigg|_{t=0} = JLJ - JL^\mathsf{T}J = 0 \quad \Rightarrow \quad L = L^\mathsf{T}, \qquad \text{(B.5)}$$

which shows that all generators of the symplectic group must be of the form JH, with symmetric H. Not all symplectic transformations admit a representation as the one-shot exponential of a real generator.[1] At the very end of this appendix, we will show that any symplectic can however be given as the product of two matrix exponentials. Notice also that, to obtain the whole symplectic group, one does not restrict to positive definite Hamiltonian matrices (such a restriction was only imposed in defining Gaussian states).

B.1   THE ORTHOGONAL COMPACT SUBGROUP

Next, let us consider the intersection $K(n)$ between the real symplectic group and the orthogonal group: $K(n) = Sp_{2n,\mathbb{R}} \cap O(2n)$. As we shall see, the subgroup $K(n)$ constitutes the maximal compact subgroup of $Sp_{2n,\mathbb{R}}$. We will prove now that $K(n)$ is isomorphic to $U(n)$, and derive a particularly simple form for any element of $K(n)$ in the process.

Write a generic $2n \times 2n$ real matrix S in terms of $n \times n$ blocks:

$$S = \begin{pmatrix} X & Y \\ W & Z \end{pmatrix}. \qquad \text{(B.6)}$$

The equivalent conditions $SJS^\mathsf{T} = J$ and $S^\mathsf{T}JS = J$ read

$$XY^\mathsf{T} - YX^\mathsf{T} = WZ^\mathsf{T} - ZW^\mathsf{T} = 0_n, \qquad \text{(B.7)}$$

$$XZ^\mathsf{T} - YW^\mathsf{T} = \mathbb{1}_n, \qquad \text{(B.8)}$$

$$X^\mathsf{T}W - W^\mathsf{T}X = Y^\mathsf{T}Z - Z^\mathsf{T}Y = 0_n, \qquad \text{(B.9)}$$

$$X^\mathsf{T}Z - W^\mathsf{T}Y = \mathbb{1}_n, \qquad \text{(B.10)}$$

while orthogonality $(S^\mathsf{T}S = S^\mathsf{T}S = \mathbb{1}_{2n})$ implies

$$XX^\mathsf{T} + YY^\mathsf{T} = WW^\mathsf{T} + ZZ^\mathsf{T} = \mathbb{1}_n, \qquad \text{(B.11)}$$

$$XW^\mathsf{T} + YZ^\mathsf{T} = 0_n, \qquad \text{(B.12)}$$

$$X^\mathsf{T}X + W^\mathsf{T}W = Y^\mathsf{T}Y + Z^\mathsf{T}Z = \mathbb{1}_n, \qquad \text{(B.13)}$$

$$X^\mathsf{T}Y + W^\mathsf{T}Z = 0_n. \qquad \text{(B.14)}$$

Multiplying Eq. (B.10) by X on the left, and then inserting, in order,

---

[1] For instance, the transformation $\mathrm{diag}(-z, -1/z)$ for $z > 0$, clearly a symplectic, does not.

Eqs. (B.12), (B.11) and (B.9), one obtains

$$XX^\mathsf{T}Z - XW^\mathsf{T}Y = X \quad \Rightarrow \quad XX^\mathsf{T}Z + YZ^\mathsf{T}Y = X$$
$$\Rightarrow \quad Z - YY^\mathsf{T}Z + YZ^\mathsf{T}Y = X \quad \Rightarrow \quad Z = X \,. \tag{B.15}$$

Inserting $X = Z$ and Eq. (B.13) into Eq. (B.10) yields $W^\mathsf{T}Y + W^\mathsf{T}W = 0$, which can be multiplied by W on the left and then simplified by using Eqs. (B.11), (B.14) and (B.9) to obtain

$$\begin{aligned}
0 = WW^\mathsf{T}(Y + W) &= (Y + W) - (XX^\mathsf{T}Y + XX^\mathsf{T}W) \\
&= (Y + W) - (-XW^\mathsf{T}X + XX^\mathsf{T}W) \\
&= (Y + W) - (-XW^\mathsf{T}X + XW^\mathsf{T}X) = Y + W \,.
\end{aligned} \tag{B.16}$$

Hence the most general $2n \times 2n$ orthogonal symplectic matrix can be written as

$$S = \begin{pmatrix} X & Y \\ -Y & X \end{pmatrix} \quad \text{with} \quad \begin{aligned} XY^\mathsf{T} - YX^\mathsf{T} &= 0_n \,, \\ XX^\mathsf{T} + YY^\mathsf{T} &= \mathbb{1}_n \,. \end{aligned} \tag{B.17}$$

Note that the condition for simplecticity and orthogonality coincide for block matrices in this form.

On the other hand, splitting an $n \times n$ unitary U in real and imaginary parts as per $U = X + iY$, with X and Y real, implies the following conditions on X and Y:

$$UU^\dagger = XX^\mathsf{T} + YY^\mathsf{T} + i(YX^\mathsf{T} - XY^\mathsf{T}) = \mathbb{1}_n \,, \tag{B.18}$$

which are identical to the conditions of Eq. (B.17), thus proving the isomorphism relating $K(n)$ to $U(n)$.

Furthermore, notice that any $S \in K(n)$ can be block-diagonalised by applying the unitary $\bar{U}'$ that describes the passage from canonical to annihilation and creation operators [$\bar{U}'$ is the re-ordered version of $\bar{U}$ of Eq. (3.6)]:

$$\bar{U}' \begin{pmatrix} X & Y \\ -Y & X \end{pmatrix} \bar{U}'^\dagger = \begin{pmatrix} X - iY & 0_n \\ 0_n & X + iY \end{pmatrix} = \begin{pmatrix} U^* & 0_n \\ 0_n & U \end{pmatrix}, \tag{B.19}$$

for

$$\bar{U}' = \frac{1}{\sqrt{2}} \begin{pmatrix} \mathbb{1}_n & i\mathbb{1}_n \\ \mathbb{1}_n & -i\mathbb{1}_n \end{pmatrix}. \tag{B.20}$$

Besides providing a very compact way of expressing an orthogonal symplectic matrix, Eq. (B.19) also shows that the determinant of any $S \in K(n)$ is equal to $+1$. Note that the change to complex variables enacted by $\bar{U}'$ simply corresponds to a switch from canonical to ladder operators, with symplectic form $\bar{U}'J\bar{U}'^\dagger$. Symplectic transformations written in the new basis are commonly known, in the field theoretical and condensed matter literature, as 'Bogoliubov' transformations. Let us emphasise that the latter are hence just symplectic transformations written down in a specific basis.

B.2 THE SINGULAR VALUE DECOMPOSITION

We are now in a position to introduce and prove the decomposition (5.10) of a symplectic transformation, which is used extensively throughout the volume. We shall maintain the xp-ordering here, and it should be clear that, once re-arranged, the statement below is the same as Eq. (5.10).

Singular value decomposition of a symplectic transformation. Any symplectic matrix $S \in Sp_{2n,\mathbb{R}}$ can be decomposed as

$$S = O_1 Z O_2 , \tag{B.21}$$

with O_1, $O_2 \in K(n)$ and

$$Z = Z_x \oplus Z_x^{-1} , \tag{B.22}$$

with $Z_x = \mathrm{diag}(z_1, \ldots, z_n)$ and $z_j > 0 \; \forall \; j \in [1, \ldots, n]$.

Proof. Let us notice that any matrix S allows for a singular value decomposition where $S = O_1 Z O_2$ for some O_1, $O_2 \in K(n)$ and Z diagonal and positive semidefinite. We just need to prove that the singular value decomposition of a symplectic operation may always be given in terms of *symplectic* orthogonal O_1 and O_2 and of Z of the form above.

As for Z note that, by Eq. (B.4), the inverse of any symplectic S^{-1} is orthogonally equivalent to its transpose S^{T}. Since the set of singular values is preserved by orthogonal transformations and transposition, it follows that the set of singular values of S must be the same as the set of singular values of S^{-1}, which are obviously just the inverse of the singular values of S. Hence, the set of $2n$ singular values of S must be invariant under element-wise inversion, i.e., it must be made up of pairs z_j and z_j^{-1}. This is consistent with the statement (B.21). Also note that the singular value decomposition is determined up to the ordering of the singular values, which may be varied by redefining O_1 and O_2 and inserting (orthogonal) permutations acting on the central diagonal matrix. We can hence adopt an ordering such that $Z' = D \oplus D^{-1}$, with $D = \mathrm{diag}(z_1, \ldots, z_n)$ and $z_j \geq 1 \; \forall \; j$, in agreement with the statement above (either z_j or $1/z_j$ must in fact be greater than 1). The matrix Z' is symplectic with respect to the form J, as can be straightforwardly verified. The singular value decomposition of S is hence determined up to an orthogonal transformation $R \in O(2n)$:

$$S = O_1 R Z' R^{\mathsf{T}} O_2 , \tag{B.23}$$

where $R = R_1 \oplus R_2$ and R_1 and R_2 are any two $n \times n$ orthogonal transformations that act on the subspaces pertaining to degenerate singular values of S (that is, R_1 and R_2 act as the identity on basis vectors corresponding to non-degenerate singular values and are block-diagonal orthogonal on any degenerate subspace). Also note that R_1 may be picked different from R_2. No further ambiguity is allowed by the singular value decomposition, other than the trivial ordering of the z_j, which we shall disregard.

Now, we set to prove that a choice of R_1 and R_2 exists such that $O_1 R$ and $R^\mathsf{T} O_2$ are symplectic. Notice that

$$ SS^\mathsf{T} \in Sp_{2n,\mathbb{R}} \quad \Rightarrow \quad O_1 R Z'^2 R^\mathsf{T} O_1^\mathsf{T} J O_1 R Z'^2 R^\mathsf{T} O_1^\mathsf{T} = J' , \qquad (\text{B.24}) $$

which implies

$$ Z'^2 J' Z'^2 = J' \qquad (\text{B.25}) $$

for the anti-symmetric form

$$ J' = R^\mathsf{T} O_1^\mathsf{T} J O_1 R = \begin{pmatrix} A & B \\ -B^\mathsf{T} & C \end{pmatrix} , \qquad (\text{B.26}) $$

where the $n \times n$ blocks A, B and C are implicitly defined, with $A = -A^\mathsf{T}$, $C = -C^\mathsf{T}$ and B arbitrary. We will now show that Eq. (B.25) allows one to choose R such that $J' = J$. Then, Eq. (B.26) is equivalent to stating that RO_1 is symplectic. In components, and adopting the summation convention for repeated indexes, Eq. (B.25) reads

$$ A_{jk} z_j^2 z_k^2 = A_{jk} \quad \forall \ \ j, k \in [1, \ldots, n] , \qquad (\text{B.27}) $$

$$ C_{jk} z_j^{-2} z_k^{-2} = C_{jk} \quad \forall \ \ j, k \in [1, \ldots, n] , \qquad (\text{B.28}) $$

$$ B_{jk} z_j^2 z_k^{-2} = B_{jk} \quad \forall \ \ j, k \in [1, \ldots, n] . \qquad (\text{B.29}) $$

Since A and C are anti-symmetric and $z_j \geq 1 \ \forall \ j$, the first two equations above show that, unless $z_j = z_k = 1$ for $j \neq k$, then $A_{jk} = C_{jk} = 0$. Moreover, $B_{jk} = 0$ for $j \neq k$, unless $z_j = z_k$. Now, the subspace described by the vectors corresponding to $z_j = 1$, can be handled by choosing the arbitrary R to put the antisymmetric form defined on it in canonical form, as detailed in the footnote on page 43. This will render $A_{jk} = C_{jk} = 0$, and $B_{jk} = 0$ for $j \neq k$. Finally, the matrix B_{jk} may be diagonalised on the subspaces with degenerate $z_j \neq 1$ too, by choosing R_1 and R_2 that enact the singular value decomposition of the restriction of B on such subspaces. Note in fact that

$$ RJ'R^\mathsf{T} = \begin{pmatrix} R_1 A R_1^\mathsf{T} & R_1 B R_2^\mathsf{T} \\ -R_2 B^\mathsf{T} R_1^\mathsf{T} & R_2 C R_2^\mathsf{T} \end{pmatrix} , \qquad (\text{B.30}) $$

so that the freedom of R_1 and R_2 is enough to singular value decompose on the degenerate subspaces. We have hence shown that a choice of R

exists that puts the anti-symmetric form of Eq. (B.26) in canonical form, with $A = B = 0$ and B diagonal and positive semi-definite. Inspection of the reduction to canonical form (see the footnote on page 43), reveals that the diagonal entries of B are the square roots of the eigenvalues of $-J'^2$ and therefore do not depend on the orthogonal transformation acting on the anti-symmetric form. But, by its definition given in Eq. (B.26), $-J'^2 = \mathbb{1}_{2n}$, so that the diagonal entries of B in canonical form must all be equal to 1. This shows that R may always be chosen such that $J' = J$, and hence that $O_1 R$ is symplectic. Since S, Z' and $O_1 R$ are all symplectic, then $R^\mathsf{T} O_2$ must also be symplectic, which proves the validity of the singular value decomposition in the form given above. $\qquad\square$

Note that, since any matrix in the compact subgroup $K(n)$ has determinant 1, the Euler decomposition implies that any symplectic matrix has determinant 1. Moreover, it shows that $K(n) = Sp_{2n,\mathbb{R}} \cap O(2n)$ is the maximal compact subgroup of $Sp_{2n,\mathbb{R}}$ (as any element $S = O_1 Z O_2$ with any singular value different from 1 may not belong to a compact subgroup that includes $K(n)$, because taking repeated multiplications of $O_1^\mathsf{T} S O_2{}^\mathsf{T}$ by itself would see certain singular values grow indefinitely).

Furthermore, the singular value decomposition implies that any symplectic may be written as $S = O_1 Z O_2 = (O_1 O_2) O_2^\mathsf{T} Z O_2 = OP$, where $O \in K(n)$, $P = P^\mathsf{T}$ and $P > 0$. But orthogonal and positive definite matrices do admit representations as single exponentials of real matrices, so that one may conclude that any symplectic may be obtained as the product of at most two matrix exponentials, as reported in the main text.

The Wiener process

CONTENTS

Given their rather abrupt insertion into the quantum stochastic formalism of Chapter 6, we felt obliged to add this explanatory note about the very basic definition of a Wiener process, as well as a justification of the Itô rule it abides by.

To this aim, we will reproduce faithfully the treatment of Jacob and Steck, published in [53] (notwithstanding this appendix, that covers the bare essentials, the reader is strongly encouraged to look up the original article). A standard, comprehensive reference on stochastic calculus, where most of these mysteries are fully unravelled, is Gardiner's volume [32], extended to the quantum domain in [31].

To begin with, let us define w as a normally distributed random variable with zero mean and variance equal to the time parameter t, whose probability density reads

$$p_w(t) = \frac{e^{-\frac{w^2}{2t}}}{\sqrt{2\pi t}} \, . \tag{C.1}$$

For a given Δt and $w(t)$, let us then define the finite increment Δw as $\Delta w = w(t + \Delta t) - w(t)$, which is clearly also a normal random variable with zero mean, and variance Δt, such that

$$p_{\Delta w}(\Delta t) = \frac{e^{-\frac{(\Delta w)^2}{2\Delta t}}}{\sqrt{2\pi \Delta t}} \, . \tag{C.2}$$

Let us now change the variable Δw into $(\Delta w)^2$ in the probability density above getting (notice that $(\Delta w)^2$ is always positive):

$$p_{(\Delta w)^2}(\Delta t) = \frac{e^{-\frac{(\Delta w)^2}{2\Delta t}}}{\sqrt{2\pi \Delta t (\Delta w)^2}} \, . \tag{C.3}$$

Note that the mean and variance of this distribution (resulting from averages over all possible realisation of the process, which we shall denote with $\langle \cdot \rangle$) are

$$\langle (\Delta w)^2 \rangle = \Delta t \, , \quad \langle [(\Delta w)^2 - \langle (\Delta w)^2 \rangle]^2 \rangle = 2(\Delta t)^2 \, . \tag{C.4}$$

The continuum limit may now be taken by partitioning the time interval between 0 and t in N interval of duration $\Delta t_N = t/N$, with associated discrete Wiener increments $\Delta w_j = w((j+1)\Delta t_N) - w(j\Delta t_N)$. Let us consider the variable given by the sum of squared increments $\sum_{j=0}^{N-1}(\Delta w_j)^2$, each of which, as we saw above, have mean t/N and variance $2t^2/N^2$. Each increment is an independent random variable, so that the central limit theorem[1] applies, and guarantees that in the continuum limit $N \to \infty$, where $(\Delta w_j)^2$ may be identified with $dw^2(t')$ and the sum turns into an integral, the sum of squared increments tends to a normal distribution with average t and variance $\lim_{N\to\infty} 2t^2/N = 0$. Hence, one may write the equality

$$\lim_{N\to\infty} \sum_{j=0}^{N-1}(\Delta w_j)^2 = \int_0^t dw^2(t') = t = \int_0^t dt' \qquad (C.5)$$

which holds *deterministically*, not merely in terms of an average over realisations, because the variance of the distribution vanishes. But, since the equation above must hold for any time interval, one is led to the deterministic Itô rule:

$$dw^2 = dt . \qquad (C.6)$$

This reasoning clarifies how it is possible that dw is a stochastic quantity, while dw^2 is not.

We have seen realisations of Wiener processes in the book as the outcomes of time-continuous, general-dyne quantum measurements of an environment, which in turn drive the stochastic update of the system state. Note that the central limit argument employed above did not depend on the normal nature of the starting distribution. Therefore, more mundanely, any classical random walk (resulting from independent and identically distributed steps with finite variance) will result in a Wiener process in the continuous limit.

---

[1] Whose simplest formulation states that, in the limit of infinite sample size N, the sum of independent and identically distributed variables with mean $\bar{x}$ and variance Δx is a normal distribution with mean $\lim_{N\to\infty} N\bar{x}$ and variance $\lim_{N\to\infty} N\Delta x$.

Selected mathematical lore on quantum channels

CONTENTS

This appendix collects all the mathematical statements about quantum channels (Gaussian and not) used in the book, as well as their proofs.

D.1 THE HOLEVO BOUND

Here, we report the statement of the Holevo bound along with a proof that makes use of the strong subadditivity of the von Neumann entropy, following [71]. A proof of the bound may also be found in [64].

Holevo bound. Let A prepare an alphabet of quantum states $\{\varrho_x\}$, each with probability p_x, with associated classical statistical ensemble X for the random variable x. Let B measure each of the states with a POVM $\{\hat{K}_y\}$, whose outcomes y define the classical statistical ensemble Y. Then, one has

$$I(X;Y) \leq \chi(p_x, \varrho_x) = S_V\left(\sum_x p_x \varrho_x\right) - \sum_x p_x S_V(\varrho_x), \qquad \text{(D.1)}$$

where $I(X;Y)$ is the mutual information between X and Y.

Proof. Note that, for a given POVM, the joint probability distribution $p_{x,y}$ is given by $p_{x,y} = p_x \text{Tr}[\hat{K}_y \varrho_x \hat{K}_y]$. Beside the system S, which is actually prepared and measured, let us define two further auxiliary systems, X, where A keeps track of the sampled classical variable x by preparing pure states in an orthogonal basis $\{|x\rangle\}$, and Y, where B will keep track of the measurement outcomes, initialised in an arbitrary pure state $|0\rangle\langle 0|$.

The global system XSY sets out in the state:

$$\varrho_{XSY} = \sum_x p_x (|x\rangle\langle x| \otimes \varrho_x \otimes |0\rangle\langle 0|) , \tag{D.2}$$

whence the local states and von Neumann entropies $\varrho_x = \sum_x p_x |x\rangle\langle x|$, with $S_V(\varrho_x) = S(X)$ (S denoting the classical Shannon entropy), $\varrho_S = \sum_x p_x \varrho_x \equiv \varrho$, with $S_V(\varrho_S) = S_V(\varrho)$, may be easily derived. Besides, since the $\{|x\rangle\}$ have been chosen mutually orthogonal, $S_V(\varrho_{XSY}) = S_V(\varrho_{XS}) = S(X) + \sum_x p_x S(\varrho_x)$.

Let us now further enlarge the system with an ancillary system Z, initialised in an arbitrary pure state $|0\rangle_Z$, which, by virtue of Naimark's theorem, will allow us to describe any POVM carried out by B as a projective measurement. The key realisation now is that B's act of measurement is always equivalent to a unitary action on the subsystem SYZ. In fact, one may define a unitary $\hat{U}_{SYZ}$ acting according to

$$\hat{U}_{SYZ}(|\psi\rangle_S \otimes |0\rangle \otimes |0\rangle_Z) = \sum_y \hat{K}_y |\psi\rangle_S \otimes |y\rangle \otimes |y\rangle_Z , \tag{D.3}$$

where the $\{|y\rangle\}$ and $\{|y\rangle_Z\}$ are all mutually orthogonal.

Let a prime $'$ denote states after the application of the unitary defined above. The initial state ϱ_{XSYZ} is sent into

$$\varrho'_{XSYZ} = \sum_{x,y,w} p_x |x\rangle\langle x| \otimes \hat{K}_y \varrho_x \hat{K}_w^\dagger \otimes |y\rangle\langle w| \otimes |y\rangle_{ZZ}\langle w| , \tag{D.4}$$

whose partial trace over Z yields

$$\varrho'_{XSY} = \sum_{x,y} p_x |x\rangle\langle x| \otimes \hat{K}_y \varrho_x \hat{K}_y^\dagger \otimes |y\rangle\langle y| . \tag{D.5}$$

Further partial tracing, over S, leads to

$$\varrho'_{XY} = \sum_{x,y} p_x \text{Tr}(\hat{K}_y \varrho_x \hat{K}_y^\dagger) |x\rangle\langle x| \otimes |y\rangle\langle y| = \sum_{x,y} p_{x,y} |x,y\rangle\langle x,y| . \tag{D.6}$$

Thus, one gets

$$S_V(\varrho'_{XY}) = S(X,Y) . \tag{D.7}$$

Also, yet more partial tracing, over X, yields $\varrho'_Y = \sum_y p_y |y\rangle\langle y|$ and

$$S_V(\varrho'_Y) = S(Y) . \tag{D.8}$$

Note now that a global unitary transformation does not affect entropies, such that

$$S_V(\varrho'_{XSYZ}) = S_V(\varrho_{XSYZ}) = S_V(\varrho_{XSY}) = S(X) + \sum_x p_x S_V(\varrho_x), \tag{D.9}$$

$$S_V(\varrho'_{SYZ}) = S_V(\varrho_{SYZ}) = S_V(\varrho_{SY}) = S_V(\varrho) . \tag{D.10}$$

We can now apply the strong subadditivity of the von Neumann entropy (2.24), to claim

$$S_V(\varrho'_{XSYZ}) + S_V(\varrho'_Y) \le S_V(\varrho'_{XY}) + S_V(\varrho'_{SYZ}) = S(X,Y) + S_V(\varrho), \tag{D.11}$$

where we also used (D.7) and (D.10). Eqs. (D.8) and (D.9) may be inserted into the left hand side to get, after re-arranging the terms,

$$I(X;Y) = S(X) + S(Y) - S(X,Y) \le S_V(\varrho) - \sum_x p_x S_V(\varrho_x) = \chi(p_x, \varrho_x) , \tag{D.12}$$

which is the Holevo bound (recall that we defined $\varrho = \sum_x p_x \varrho_x$). □

The achievability of the bound in the asymptotic limit of an infinite number of uses is a classic result derived independently by Holevo [46] and Schumacher and Westmoreland [76].

D.2   ENTANGLEMENT BREAKING CHANNELS

An entanglement breaking channel Φ is a CP-map that, when acting as $(\mathcal{I} \otimes \Phi)$ on a subsystem of a composite quantum state (recall that $\mathcal{I}$ stands for the identity superoperator), yields a separable output. Entanglement breaking channels admit a notable set of equivalent characterisations, which we shall justify here extending the treatment of [50] to the Hilbert space $L^2(\mathbb{R})$, through the limiting procedure associated with the infinite dimensional Choi-Jamiolkowski isomorphism (see Chapter 5). A more rigorous approach to this characterisation may be found in [47].

Equivalent characterisations of entanglement breaking channels.
A trace-preserving CP-map $\Phi : \mathcal{B}(L^2(\mathbb{R})) \mapsto \mathcal{B}(L^2(\mathbb{R}))$ is an entanglement breaking channel if any of the following equivalent conditions is met:

(i) $(\mathcal{I} \otimes \Phi)(\varrho) = \sum_j p_j(\varrho_{A,j} \otimes \varrho_{B,j})$ for some probability distribution p_j (with $\sum p_j = 1$) and local quantum states $\varrho_{A,j}$ and $\varrho_{B,j}$.

(ii) $\lim_{r \to \infty}(\mathcal{I} \otimes \Phi)|\psi_r\rangle\langle\psi_r| = \sum_j p_j(\varrho_{A,j} \otimes \varrho_{B,j})$, where $|\psi_r\rangle = \cosh(r)^{-1} \sum_{j=0}^{\infty} \tanh(r)^j |j,j\rangle$, which aligns with the maximally entangled unnormalised vector in the Fock basis $|\xi\rangle = \sum_{j=0}^{\infty} |j,j\rangle$ in the limit $r \to \infty$. In words, a channel is entanglement breaking if and only if its Choi state is separable.

(iii) $\Phi(\varrho) = \sum_\mu \varrho_\mu \mathrm{Tr}(\hat{K}_\mu \varrho \hat{K}_\mu^\dagger)$, for some POVM with $\sum_\mu K_\mu^\dagger K_\mu = \hat{\mathbb{1}}$ and quantum states ϱ_μ: an entanglement breaking channel is a measure-and-prepare channel.

(iv) $\Theta \circ \Phi$ is completely positive for all positive maps Θ.

(v) $\Phi \circ \Theta$ is completely positive for all positive maps Θ.

Proof. (i) $\Rightarrow$ (ii) by continuity, which holds in the weak topology (element-wise convergence of density matrices in the Fock basis).

Let us now prove (ii) $\Rightarrow$ (iii): by invoking the separability of the space $L^2(\mathbb{R})$, one can recast a separable state $\sum_j p_j(\varrho_{A,j} \otimes \varrho_{B,j})$ as $\sum_l p_l(|v_l\rangle\langle v_l| \otimes |w_l\rangle\langle w_l|)$, where $\{|v_l\rangle\}$ and $\{|w_l\rangle\}$ are generic sets of normalised – but not necessarily orthogonal – vectors belonging to the local Hilbert spaces. Thus, (ii) may be rewritten as

$$\lim_{r \to \infty}(\mathcal{I} \otimes \Phi)|\psi_r\rangle\langle\psi_r| = \sum_l p_l(|v_l\rangle\langle v_l| \otimes |w_l\rangle\langle w_l|) . \tag{D.13}$$

· Now, let Υ_r be the CP-map defined by

$$\Upsilon_r(\varrho) = \cosh(r)^2 \sum_l p_l |w_l\rangle\langle w_l| \langle v_l|\varrho|v_l\rangle . \tag{D.14}$$

Then, one has

$$\lim_{r \to \infty}(\mathcal{I} \otimes \Upsilon_r)|\psi_r\rangle\langle\psi_r| = \lim_{r \to \infty}\sum_{jkl} \tanh(r)^{j+k}(|j\rangle\langle k| \otimes |w_l\rangle\langle w_l|)p_l\langle j|v_l\rangle\langle v_l|k\rangle$$

$$= \sum_l p_l(|v_l\rangle\langle v_l| \otimes |w_l\rangle\langle w_l|) , \tag{D.15}$$

where we used the resolution of the identity $\lim_{r\to\infty}\sum_j \tanh(r)^j|j\rangle\langle j| = \hat{\mathbb{1}}$. Because of the unicity of the Choi representation, the comparison between (D.13) and (D.15) implies $\lim_{r\to\infty}\Upsilon_r = \Phi$. In order to infer from Eq. (D.14) that Φ is measure-and-prepare, it is left to be shown that the POVM property $\lim_{r\to\infty}\cosh(r)^2\sum_l p_l|v_l\rangle\langle v_l| = \hat{\mathbb{1}}$ is satisfied. This can be shown by taking the partial trace of Eq. (D.13) on the second subsystem which, recalling that the trace is linear and that Φ is trace preserving, leads to

$$\lim_{r\to\infty}\cosh(r)^2\sum_l p_l|v_l\rangle\langle v_l| = \lim_{r\to\infty}\sum_{jk}\tanh(r)^{j+k}|j\rangle\langle k|\mathrm{Tr}\left[\Phi(|j\rangle\langle k|)\right]$$

$$= \lim_{r\to\infty}\sum_j \tanh(r)^{2j}|j\rangle\langle j| = \hat{\mathbb{1}}\,, \qquad \text{(D.16)}$$

which wraps up the proof of the implication.

The implication (iii) $\Rightarrow$ (i) is straightforward: it suffices to notice that if Φ is measure-and-prepare, then

$$(\mathcal{I}\otimes\Phi)(\varrho) = \sum_\mu \mathrm{Tr}_B(\hat{K}_\mu\varrho\hat{K}_\mu^\dagger)\otimes\varrho_\mu = \sum_\mu p_\mu\mathrm{Tr}_B(\varrho)\otimes\varrho_\mu\,, \qquad \text{(D.17)}$$

where Tr_B denotes the partial trace over the second subsystem and $p_\mu = \mathrm{Tr}(\hat{K}_\mu\varrho\hat{K}_\mu)$.

The equivalence (iv) $\Leftrightarrow$ (i) holds because a state ϱ_{sep} is separable if and only if $(\mathcal{I}\otimes\Theta)\varrho_{sep}\geq 0$ for all positive – not necessarily completely positive – maps Θ. Then, $(\mathcal{I}\otimes\Theta\circ\Phi)\varrho = (\mathcal{I}\otimes\Theta)\circ(\mathcal{I}\otimes\Phi)\varrho\geq 0$ for all Θ if and only if $(\mathcal{I}\otimes\Phi)\varrho$ is separable for all bipartite input states ϱ.

Finally, (iv) $\Leftrightarrow$ (v) since a map is positive if and only if its dual is positive,[1] and the dual of an entanglement breaking channel is always entanglement breaking, as is apparent from the characterisation (iii). The condition (iv) on $\Theta^*\circ\Phi^*$ then co-implies that the dual $(\Theta^*\circ\Phi^*)^* = \Phi\circ\Theta$ is completely positive. $\qquad\square$

Let us add that an analogous characterisation holds for the broader class of "NPT-breaking" (also known as "NPT-entanglement breaking") channels whose partial action sends any input state into a – possibly entangled – state with positive partial transpose. These may be defined as channels with a PPT Choi state.

We will now reproduce an argument due to Shor [79], and show that the minimum output entropy of an entanglement breaking channel is additive. The additivity would actually extend to the classical capacity of general entanglement breaking channels, but here we will just report on the result concerning

---

[1]This follows directly from our definition of the dual as the adjoint under the Hilbert-Schmidt inner product: $\mathrm{Tr}[\Phi^*(\hat{A})^\dagger\hat{B}] = \mathrm{Tr}[\hat{A}^\dagger\Phi(\hat{B})]$ for all trace-class $\hat{A}$ and $\hat{B}$.

the output entropy, which is the only property we utilised in determining the classical capacity of bosonic channels in Chapter 8. Let us then prove

---

Additivity of the minimal output entropy of entanglement breaking channels. Let Φ be an entanglement breaking channel, then

$$\inf_{\varrho_{AB}} S_V(\Phi^{\otimes 2}(\varrho_{AB})) = 2 \inf_{\varrho_A} S_V(\Phi(\varrho_A)) . \qquad (D.18)$$

Proof. Clearly, the left hand side is upper bounded by the right hand side, since the choice $\varrho_{AB} = \varrho_A \times \varrho_A$ is included in the minimisation on the left hand side. In order to prove the equality above, we will employ the strong subadditivity of the von Neumann entropy to show that the right hand side is also a lower bound for the left hand side.

First, note that, by property (i), one has that

$$\Phi^{\otimes 2}(\varrho_{AB}) = \sum_j p_j \Phi(|v_j\rangle\langle v_j|) \otimes |w_j\rangle\langle w_j| . \qquad (D.19)$$

An orthonormal set $\{|j\rangle\}$ in a third, ancillary Hilbert space may always be found to extend the state above to

$$\tau_{ABC} = \sum_j p_j \Phi(|v_j\rangle\langle v_j|) \otimes |w_j\rangle\langle w_j| \otimes |j\rangle\langle j| , \qquad (D.20)$$

with local reductions

$$\tau_{AB} = \sum_j p_j \Phi(|v_j\rangle\langle v_j|) \otimes |w_j\rangle\langle w_j| = \Phi^{\otimes 2}(\varrho_{AB}) , \qquad (D.21)$$

$$\tau_{AC} = \sum_j p_j \Phi(|v_j\rangle\langle v_j|) \otimes |j\rangle\langle j| , \qquad (D.22)$$

$$\tau_{BC} = \sum_j p_j |w_j\rangle\langle w_j| \otimes |j\rangle\langle j| , \quad \tau_C = \sum_j p_j |j\rangle\langle j| , \qquad (D.23)$$

$$\tau_B = \sum_j p_j |w_j\rangle\langle w_j| = \mathrm{Tr}_A[(\mathbb{1} \otimes \Phi)\varrho_{AB}] = \Phi(\mathrm{Tr}_A \varrho_{AB}) . \qquad (D.24)$$

We can then apply strong subadditivity, in the form $S_V(\tau_{AB}) \geq S_V(\tau_{ABC}) - S_V(\tau_{BC}) + S_V(\tau_B)$ noting that, because of the orthonormality of the $\{|j\rangle\}$, one has $S_V(\tau_{ABC}) = S_V(\tau_{AC})$ and $S_V(\tau_{BC}) = S_V(\tau_C)$, which gives

$$S_V(\Phi^{\otimes 2}(\varrho_{AB})) \geq S_V(\tau_{AC}) - S_V(\tau_C) + S_V(\Phi(\mathrm{Tr}_A \varrho_{AB})) . \qquad (D.25)$$

Notice that the left hand side of this inequality is precisely the output entropy we would like to bound from below. The term $S_V(\tau_{AC}) - S_V(\tau_C)$ can be simplified through the so called "chain rule":

$$S_V(\tau_{AC}) - S_V(\tau_C) = S_V\left(\sum_j p_j \Phi(|v_j\rangle\langle v_j|) \otimes |j\rangle\langle j|\right) - S_V\left(\sum_j p_j |j\rangle\langle j|\right)$$

$$= \sum_j p_j S_V(\Phi(|v_j\rangle\langle v_j|)) \tag{D.26}$$

which, inserted into Eq. (D.25), yields

$$S_V(\Phi^{\otimes 2}(\varrho_{AB})) \geq \sum_j p_j S_V(\Phi(|v_j\rangle\langle v_j|)) + S_V(\Phi(\mathrm{Tr}_A \varrho_{AB}))$$

$$\geq 2 \inf_{\varrho_A} S_V(\Phi(\varrho_A)) . \tag{D.27}$$

Along with the obvious converse inequality, this proves the additivity of the minimum output entropy. □

D.3 LINEAR BOSONIC CHANNELS

Let us a define a *linear* bosonic channel Φ as one whose dual Φ^* as defined in Section 5.3.1 acts on a Weyl operator $\hat{D}_{\mathbf{r}}$ as per

$$\Phi^*(\hat{D}_{\mathbf{r}}) = f(\mathbf{r})\hat{D}_{V\mathbf{r}} , \tag{D.28}$$

for some $2n \times 2n$ real matrix V and $f : \mathbb{R}^{2n} \to \mathbb{C}$. As apparent from Eq. (5.55), all deterministic Gaussian CP-maps are linear according to this definition.

A channel of the form (D.28) is a CP-map if and only if the function $f(\mathbf{r})$ is continuous, $f(\mathbf{0}) = 1$ and if the matrix

$$L_{jk} = f(\mathbf{r}_j - \mathbf{r}_k)e^{i\mathbf{r}_k^\mathsf{T}(\Omega - V^\mathsf{T}\Omega V)\mathbf{r}_j/2} \tag{D.29}$$

is positive semi-definite for all sets of vectors $\mathcal{R} = \{\mathbf{r}_j : \mathbf{r}_j \in \mathbb{R}^{2n}, j \in [1, \ldots, j_{max}]\}$.

These conditions on $f(\mathbf{r})$ may be proven by imposing that a physical input characteristic function $\chi(\mathbf{r})$ – which, as discussed in Section 4.3.1, has to satisfy the conditions of continuity, normalisation (that is, $\chi(\mathbf{0}) = 1$), as well as the positivity condition (4.31) – is mapped into a physical characteristic function. In particular, the prescription (D.29) above ensures that the transformed characteristic function still preserves (4.31).[2] It should be noted that, per se, this condition is only necessary and sufficient for the positivity (rather than complete positivity) of the map. However, it is rather straightforward to

---

[2] This may be shown by exploiting the fact that (4.31) must hold for *all* sets of vectors of the final characteristic function, and by using the fact that the Hadamard product of two positive matrices is always a positive matrix. The Hadamard product of two matrices with entries A_{jk} and B_{jk} is the 'wrong', element-wise product, with entries $A_{jk}B_{jk}$.

show that maps that take the simple form (D.28) and satisfy (D.29) cannot make any initial characteristic function unphysical, even if they only act on a subset of the phase space variables.

Observe that, if $V = \mathbb{1}$, then the exponential factor in Eq. (D.29) is always one and $f(\mathbf{r})$ must be a classical characteristic function (the Fourier transform of a multivariate classical probability distribution).

D.3.1 Weyl-covariant channels

A Weyl-covariant CP-map Υ is defined as a bosonic channel such that

$$\Upsilon(\hat{D}_{\mathbf{r}}^{\dagger}\hat{O}\hat{D}_{\mathbf{r}}) = \hat{D}_{\mathbf{r}}^{\dagger}\Upsilon(\hat{O})\hat{D}_{\mathbf{r}} \,, \quad \forall \hat{O} \in \mathcal{B}(L^2(\mathbb{R}^{2n})) \,, \; \mathbf{r} \in \mathbb{R}^{2n} \,. \tag{D.30}$$

Notice that such channels are often referred to as "phase-space covariant" channels. We have chosen to tweak this terminology slightly here in order to better distinguish them from the "phase-covariant" and "phase-contravariant" channels (see the next section), which occupy much of our discussion of channels' information capacities. There, the notion of covariance is relative to phase space rotations, and not displacements.

The dual Υ^* of a Weyl-covariant channel is Weyl-covariant too:

$$\begin{aligned}
\mathrm{Tr}[\hat{A}\Upsilon^*(\hat{D}_{\mathbf{r}}^{\dagger}\hat{B}\hat{D}_{\mathbf{r}})] &= \mathrm{Tr}[\Upsilon(\hat{A})\hat{D}_{\mathbf{r}}^{\dagger}\hat{B}\hat{D}_{\mathbf{r}}] = \mathrm{Tr}[\hat{D}_{\mathbf{r}}\Upsilon(\hat{A})\hat{D}_{\mathbf{r}}^{\dagger}\hat{B}] \\
&= \mathrm{Tr}[\Upsilon(\hat{D}_{\mathbf{r}}\hat{A}\hat{D}_{\mathbf{r}}^{\dagger})\hat{B}] = \mathrm{Tr}[\hat{A}\hat{D}_{\mathbf{r}}^{\dagger}\Upsilon^*(\hat{B})\hat{D}_{\mathbf{r}}] \,,
\end{aligned} \tag{D.31}$$

which must hold for all trace class $\hat{A}$ and $\hat{B}$, and thus implies $\Upsilon^*(\hat{D}_{\mathbf{r}}^{\dagger}\hat{B}\hat{D}_{\mathbf{r}}) = \hat{D}_{\mathbf{r}}^{\dagger}\Upsilon^*(\hat{B})\hat{D}_{\mathbf{r}}$.

The action of the dual of a Weyl-covariant channel on Weyl operators is very revealing. The Baker-Campbell-Hausdorff relation (3.11) yields, due the Weyl-covariance of Υ^*,

$$[\Upsilon^*(\hat{D}_{\mathbf{r}})\hat{D}_{\mathbf{r}}^{\dagger}, \hat{D}_{\mathbf{s}}] = 0 \,, \quad \forall \mathbf{r}, \mathbf{s} \in \mathbb{R}^{2n} \,. \tag{D.32}$$

This is a very powerful relationship. In point of fact, the irreducibility of the Weyl set of displacement operators implies that the only operator that commutes with all Weyl operators is the identity, whence

$$\Upsilon^*(\hat{D}_{\mathbf{r}}) = f(\mathbf{r})\hat{D}_{\mathbf{r}} \,, \tag{D.33}$$

for some $f : \mathbb{R}^{2n} \to \mathbb{C}$. In order for the original map Υ to be a physical CP-map, $f(\mathbf{r})$ has to comply with the condition of positivity (D.29) for $V = \mathbb{1}$, that is, it must be a classical characteristic function. This completely characterises all Weyl-covariant CP-maps.

A quantum CP-map Φ is said to be NPT-breaking if and only if the compositions $\Phi \circ T$ and $T \circ \Phi$, where T denotes the transposition map, are also CP-maps. This definition amounts to stating that the output of $\mathcal{I} \otimes \Phi$ has always a positive partial transpose.[3] It is clear that all entanglement breaking channels must be NPT-breaking (since a positive partial transpose is in general necessary for separability). Note also that, since a total transposition does not affect the trace, one has $\mathrm{Tr}[\hat{A}T(\hat{B})] = \mathrm{Tr}[T(\hat{A})\hat{B}]$ for all trace class A and B: the transposition map is its own dual ($T^* = T$).

In Chapter 7 we also saw that transposition in the Fock basis (all these statements do not depend on the transposition basis) is described in the phase space of n modes by flipping the sign of all the p variables, through the linear operator $\Sigma_n = \sigma_z^{\oplus n}$.

By applying the characterisation of Weyl-covariant channels (D.33), as well as the CP-map condition (D.29) to $\Upsilon \circ T$, one obtains the following characterisation of Weyl-covariant entanglement-breaking channels:

Weyl-covariant NPT-breaking channels. A Weyl-covariant channel Υ is NPT entanglement breaking if and only if

$$\Upsilon^*(\hat{D}_{\mathbf{r}}) = f(\mathbf{r})\hat{D}_{\mathbf{r}} , \qquad (D.34)$$

where Υ^* denotes the dual of Υ and $f(\mathbf{r}/\sqrt{2})$ is a quantum characteristic function: a density matrix τ exists such that $f(\mathbf{r}) = \mathrm{Tr}(\hat{D}_{-\sqrt{2}\mathbf{r}}\tau)$.

Proof. A Weyl-covariant channel Υ must be linear, and therefore satisfy (D.33). The dual of $\Upsilon \circ T$ is just $T \circ \Upsilon^*$. The complete positivity of $\Upsilon \circ T$ then amounts to setting $V = \Sigma_n$ in (D.29), and hence to the positivity of the matrix

$$L_{jk} = f(\mathbf{r}_j - \mathbf{r}_k)e^{i\mathbf{r}_k^{\mathsf{T}}(2\Omega)\mathbf{r}_j/2} \qquad (D.35)$$

for all sets of vectors $\mathcal{R} = \{\mathbf{r}_j : \mathbf{r}_j \in \mathbb{R}^{2n}, j \in [1, \ldots, j_{max}]\}$ (note that we used the property $\Sigma_n \Omega_n \Sigma_n^{\mathsf{T}} = -\Omega_n$). Switching to the variable $\mathbf{r}/\sqrt{2}$ shows that the requirement above is equivalent to the positivity of

$$M_{jk} = f(\mathbf{r}_j/\sqrt{2} - \mathbf{r}_k/\sqrt{2})e^{i\mathbf{r}_k^{\mathsf{T}}\Omega\mathbf{r}_j/2} , \qquad (D.36)$$

which is exactly the condition (4.31) that $f(\mathbf{r}/\sqrt{2})$ must satisfy in order to be a quantum characteristic function (since it must also be $f(0) = 1$). This completes the proof. Notice that our statement is equivalent to Eq. (8.49), where $\mathbf{r}$ was rescaled by a factor $\sqrt{2}$. □

D.4 PHASE-CONTRAVARIANT GAUSSIAN CHANNELS

Single-mode phase-contravariant bosonic Gaussian channels are defined as deterministic Gaussian CP-maps $\bar{\Phi}$ such that $\bar{\Phi}(\hat{R}_\varphi \varrho \hat{R}_\varphi^\dagger) = \hat{R}_\varphi^\dagger \bar{\Phi}(\varrho)\hat{R}_\varphi$ for all

---

[3]This condition is analogous to the requirements (iv) and (v) that characterise the smaller set of entanglement breaking CP-maps (see page 324).

unitary $\hat{R}_\varphi = e^{i\varphi \hat{a}^\dagger \hat{a}}$ corresponding to a phase shifter transformation. A characterisation of such channels may be obtained proceeding along the lines of Problem 5.16: since a bosonic Gaussian channel is entirely characterised by its action on Gaussian states, it is sufficient to consider a generic input Gaussian state with covariance matrix σ [the action of the channel may then be carried over to generic states through the relation (5.69)]. Denoting with R_φ all matrices of the form of Eq. (5.12), phase-contravariance amounts then to

$$X R_\varphi \sigma R_\varphi^\mathsf{T} X^\mathsf{T} + Y = R_\varphi^\mathsf{T} X \sigma X^\mathsf{T} R_\varphi + R_\varphi^\mathsf{T} Y R_\varphi \quad \forall \, \sigma : \sigma + i\Omega \geq 0 \quad \text{(D.37)}$$

and for all $\varphi \in [0, 2\pi[$. As the equality above must hold for all σ, it is clear that the linear and constant parts of the equation above must equate separately and independently. This immediately implies that Y must be invariant under any rotation, which means it must be proportional to the identity. As for X, one is left with

$$(R_\varphi X R_\varphi)\sigma(R_\varphi^\mathsf{T} X^\mathsf{T} R_\varphi^\mathsf{T}) = X \sigma X^\mathsf{T}, \quad \text{(D.38)}$$

where one can insert $\sigma = \mathbb{1}$ to get $R_\varphi^\mathsf{T} X X^\mathsf{T} R_\varphi = X X^\mathsf{T}$, which implies the necessary condition $X X^\mathsf{T} \propto \mathbb{1}$. Thus, it must be $X = xQ$ for some transformation $Q \in O(2)$ and $x \in \mathbb{R}$. Two-dimensional orthogonal transformations come in only two flavours: either $Q = R_\theta$ or $Q = \sigma_z R_\theta$, where σ_z is the Pauli z matrix and $R \in SO(2)$ with $\theta \in [0, 2\pi[$. Now, a major simplification to our task comes from the abelian nature of $SO(2)$, such that $R_\varphi R_\theta = R(\varphi + \theta)$. If $Q = R_\theta$ were a special orthogonal, then it can be shown, in the very same manner we discarded the option $\sigma_z R_\theta$ for phase-covariant channels, that Eq. (D.38) could not be satisfied. If, on the other hand, $Q = \sigma_z R_\theta$, then Eq. (D.38) turns into

$$R_\varphi \sigma_z R_{\theta+\varphi} \sigma R_{-\theta-\varphi} \sigma_z R_{-\varphi} = \sigma_z R_\theta \sigma R_{-\theta} \sigma_z = \sigma_z R_\theta \sigma R_{-\theta} \sigma_z, \quad \text{(D.39)}$$

where we used the property $R_\varphi \sigma_z = \sigma_z R_{-\varphi}$. Hence, it must be $X = x\sigma_z R_\theta = R_{-\theta}\sigma_z$.

Up to a phase rotation, contravariant channels are thus characterised by $X = x\sigma_z$ and $Y = y\mathbb{1}$. Imposing the Heisenberg condition (5.37) leads to the necessary condition

$$y \geq x^2 + 1. \quad \text{(D.40)}$$

Notice that this differs from the phase-covariant analogue (5.90).

Since $\sigma_z^2 = \mathbb{1}$, the action of the channel on a generic covariance matrix σ may be written as $\sigma_z(x^2\sigma + y\mathbb{1})\sigma_z$. In dealing with partial transposition (Chapter 7), we have learned that the inversion of the canonical quadrature $\hat{p}$ corresponds, at the Hilbert space level, to transposition in the Fock basis. We have also seen that transposition is not, on its own, a physical quantum channel, as it is not completely positive. However, (D.40) implies that, at variance with the phase-covariant case, a minimum, strictly positive amount of additive noise always needs to be present: that comes to the rescue and makes the map completely positive. It is hence the case that any phase-contravariant

Gaussian bosonic channel can be obtained by transposition from a phase-covariant channel for which $y \geq |x^2 + 1|$.

The union of the sets of phase-covariant and contravariant channels is referred to as the class of 'phase-insensitive' channels.

D.4.1 Conjugate channels

We saw that any quantum CP-map Φ admits a representation in terms of a Stinespring dilation: $\Phi(\varrho) = \mathrm{Tr}_B(\hat{U}\varrho \otimes |\varphi\rangle\langle\varphi|\hat{U}^\dagger)$ where $\varrho \in \mathcal{B}(\mathcal{H}_A)$, $|\varphi\rangle \in \mathcal{H}_B$ and $\hat{U}$ is a unitary acting on the global Hilbert space $\mathcal{H}_A \otimes \mathcal{H}_B$.

The quantum channel $\bar{\Phi} = \mathrm{Tr}_A(\hat{U}\varrho \otimes |\varphi\rangle\langle\varphi|\hat{U}^\dagger)$ is said to be 'conjugate' to Φ. Pairs of conjugate channels share the same minimum output entropy: this is promptly seen by noting that, due to the concavity of entropy and to the linearity of a quantum CP-map, there must always exist a pure input state – an extreme element in the convex set of density matrices – that attains the minimum input entropy. But the spectra of the output of a channel and its conjugate under pure inputs are clearly the same (this just follows from the definitions above and the unicity of the Schmidt coefficients of pure quantum states). Hence, the minimum output entropies must be the same. The kinship among conjugate channels extends, in general, to their classical capacities, although in the book we only require the identity of the minimum output entropies, and hence we shall skip the capacity proof.

Clearly, it is easy to determine the conjugate of a Gaussian bosonic channel once its Stinespring dilation is known. In particular, for our proof of the classical capacity of phase-insensitive channels, we require the form of the conjugate of a quantum limited amplifier channel $\mathcal{A}_r^0$, given by Eq. (5.87) for $n_{\text{th}} = 1$. Such a channel is the outcome of the action on the input, with covariance matrix $\boldsymbol{\sigma}$, and an appended vacuum state of a two-mode squeezing operation:

$$\begin{pmatrix} c\mathbb{1} & s\sigma_z \\ s\sigma_z & c\mathbb{1} \end{pmatrix} \begin{pmatrix} \boldsymbol{\sigma} & 0 \\ 0 & \mathbb{1} \end{pmatrix} \begin{pmatrix} c\mathbb{1} & s\sigma_z \\ s\sigma_z & c\mathbb{1} \end{pmatrix} = \begin{pmatrix} c^2\boldsymbol{\sigma} + s^2\mathbb{1} & cs(\boldsymbol{\sigma} + \mathbb{1})\sigma_z \\ cs(\boldsymbol{\sigma} + \mathbb{1})\sigma_z & s^2\sigma_z\boldsymbol{\sigma}\sigma_z + c^2\mathbb{1} \end{pmatrix},$$
(D.41)

where c and s were shorthand notation for $\cosh(r)$ and $\sinh(r)$. If, instead of pinching out the upper left sub-block, one takes the lower-right one (which corresponds to partial tracing over A rather than B), one obtains the conjugate channel $\bar{\mathcal{A}}_r^0$, defined by the matrices:

$$\bar{X} = \sinh(r)\sigma_z, \quad \bar{Y} = \cosh(r)^2.$$
(D.42)

This is a phase-contravariant channel, obtained by transposition times the phase-covariant channel with $X = \sinh(r)\mathbb{1}$ and $\cosh(r)^2\bar{Y}$ (*note that this is not the original quantum limited amplifier*, since $\cosh(r)$ and $\sinh(r)$ are swapped). The latter, phase-covariant channel, can be in turn decomposed, along the lines of Problem 5.17, as the product of channels $\mathcal{A}_r^0 \circ \mathcal{E}_\theta^0$, for $\theta = \arccos(\tanh(r))$. Putting everything together, we have shown that the conjugate channel $\bar{\mathcal{A}}_r^0$ to the quantum limited amplifier $\mathcal{A}_r^0$ may be written as

$$\bar{\mathcal{A}}_r^0 = T \circ \mathcal{A}_r^0 \circ \mathcal{E}_\theta^0 , \quad \text{with} \quad \theta = \arccos\left(\tanh(r)\right) . \tag{D.43}$$

This equation was key, in Section 8.2.2, to determining the classical capacity of generic phase-insensitive channels.

Moreover, we can proceed as in Problem 5.17 to obtain a decomposition of a generic phase-contravariant channel. The composition of a pure loss channel $\mathcal{E}_\theta^0$ followed by a phase-contravariant amplifier $\bar{\mathcal{A}}_r^0$ (the conjugate to a quantum limited amplifier) yields a channel with $X = (\sinh r \cos\theta)\sigma_z$ and $Y = [(\sinh r)^2(\sin\theta)^2 + (\cosh r)^2]\mathbb{1}_2$. Such matrices span the whole set of phase-contravariant channels, parametrised by $x = \cos\theta \sinh r$ and $y = [(\sin\theta)^2(\sinh r)^2 + (\cosh r)^2]$, as may be shown by obtaining the inverse solutions

$$(\cos\theta)^2 = \frac{2x^2}{x^2 - 1 + y} \quad \text{and} \quad (\cosh r)^2 = \frac{x^2 + 1 + y}{2} , \tag{D.44}$$

and proving that they are consistent with the bound (D.40). A generic phase-contravariant channel $\bar{\Phi}$ may therefore be decomposed as

$$\bar{\Phi} = \bar{\mathcal{A}}_r^0 \circ \mathcal{E}_\theta^0 , \tag{D.45}$$

with r and θ determined by Eqs. (D.44).

Classical and quantum estimation bounds

CONTENTS

As a bare introduction to quantum estimation, let us include the basic proofs of the classical and quantum Cramér–Rao bounds, associated, respectively, with the classical and quantum Fisher information, in the vanilla case of an unbiased estimator of a single parameter.

E.1   CLASSICAL FISHER INFORMATION

Let ϑ be the parameter to be estimated, on which a probability density $p(x|\vartheta)$ governing a random variable x depends. An 'unbiased estimator' $\tilde{\vartheta}(x)$ of ϑ is a function that yields an estimated value of ϑ from the sampling of the probability $p(x|\vartheta)$, and such that

$$\langle \tilde{\vartheta} \rangle = \vartheta \, , \tag{E.1}$$

where $\langle \cdot \rangle$ indicates, in this whole section, a sampling average (*i.e.*, an average over x). The mean of an unbiased estimator thus gives the accurate value of the deterministic parameter ϑ.

Let us then define the quantity $\zeta = \partial_\vartheta \ln p(x|\vartheta) = \partial_\vartheta p(x|\vartheta)/p(x|\vartheta)$, for which one has[1]

$$\langle \zeta \rangle = \int \partial_\vartheta p(x|\vartheta) \, \mathrm{d}x = \partial_\vartheta \int p(x|\vartheta) \, \mathrm{d}x = \partial_\vartheta 1 = 0 \, , \tag{E.2}$$

---

[1]We can swap partial differentiation and integration here because of the properties of $p(x|\vartheta)$, which has either a bounded support independent from ϑ, or has infinite support but is continuously differentiable with uniformly convergent integral.

as well as

$$\Delta\zeta^2 = \langle\zeta^2\rangle = \int \frac{(\partial_\vartheta p(x|\vartheta))^2}{p(x|\vartheta)}\,\mathrm{d}x = \int p(x|\vartheta)\,(\partial_\vartheta \ln p(x|\vartheta))^2\,\mathrm{d}x = I_\vartheta\;, \quad \text{(E.3)}$$

where we have defined the classical Fisher information I_ϑ associated with ϑ.

Notice also that the covariance $\Delta(\tilde\vartheta\zeta)$, between the quantities ζ and the estimator $\tilde\vartheta$ is just given by $\langle\tilde\vartheta\zeta\rangle$ (since $\langle\zeta\rangle = 0$, which cancels all the contribution from the first moments), and may be explicitly evaluated as

$$\Delta(\tilde\vartheta\zeta) = \int \tilde\vartheta(x)\partial_\vartheta p(x|\vartheta)\,\mathrm{d}x = \partial_\vartheta \int \tilde\vartheta(x)p(x|\vartheta)\,\mathrm{d}x = \partial_\vartheta\vartheta = 1\;. \quad \text{(E.4)}$$

One can now invoke the Cauchy-Schwarz inequality in the form:[2]

$$\Delta(\tilde\vartheta\zeta)^2 \le \Delta\tilde\vartheta^2 \Delta\zeta^2\;, \quad \text{(E.5)}$$

where $\Delta\tilde\vartheta^2$ and $\Delta\zeta^2$ are the variances associated with $\tilde\vartheta$ and ζ. Inequality (E.5), together with (E.3) and (E.4), yields

$$\Delta\tilde\vartheta^2 \ge \frac{1}{\Delta\zeta^2} = \frac{1}{I_\vartheta}\;, \quad \text{(E.6)}$$

which is nothing but the Cramér–Rao bound over one sampling of the distribution.

If N independent samplings $\mathbf{x} = (x_1, \ldots, x_N)$ are considered (essentially, N 'measurements'), the total probability factorises as $p(\mathbf{x}) = \prod_{j=1}^N p(x_j)$, and the total Fisher information is trivially additive, because of the logarithm in its definition. The classical Cramér–Rao bound thus takes the form (8.81) given in the main text (notice also that we have avoided the introduction of an estimator in Section 8.3 by just giving the standard deviation of the parameter θ, instead of its estimator's, with a slight abuse of notation).

E.2   QUANTUM FISHER INFORMATION

In quantum mechanics, in the scenario described on page 251 of Section 8.3, the classical probability distribution is given as $p(\mu|\vartheta) = \mathrm{Tr}(\hat K_\mu \varrho_\vartheta \hat K_\mu^\dagger)$, with the outcome μ playing the role of the random variable x of the previous section, and the dependence on ϑ contained in the dependence of an ensemble of quantum states ϱ_ϑ.

Then, the question in hand becomes the optimisation of the Cramér–Rao bound (E.6) over all possible POVMs $\{\hat K_\mu\}$. We can assume, without loss of generality, positive, self-adjoint $\hat K_\mu$'s such that $\hat K_\mu = \hat K_\mu^\dagger \ge 0$.[3] Let us

---

[2] This just follows from the fact that $\int \mathrm{d}x[f(x) - \langle f\rangle][g(x) - \langle g\rangle]$ acts as scalar product on well-behaved functions $f(x)$ and $g(x)$.

[3] Since, given the elements $\{\hat K_\mu\}$, the same statistics may always be reproduced by the POVM $\{(\hat K_\mu^\dagger \hat K_\mu)^{1/2}\}$.

then define the symmetric logarithmic derivative of ϱ_ϑ, $\hat{\mathcal{L}}_\vartheta$, as the self-adjoint operator that satisfies $(\hat{\mathcal{L}}_\vartheta \varrho_\vartheta + \varrho_\vartheta \hat{\mathcal{L}}_\vartheta) = 2\partial_\vartheta \varrho_\vartheta$, such that one has

$$\partial_\vartheta p(\mu|\vartheta) = \mathrm{Tr}[\partial_\vartheta \varrho_\vartheta \hat{K}_\mu^2] = \mathrm{Re}(\mathrm{Tr}[\varrho_\vartheta \hat{K}_\mu^2 \hat{\mathcal{L}}_\vartheta]). \tag{E.7}$$

Therefore, the classical Fisher information for this POVM is given, as per Eq. (E.3), by

$$I_{\hat{K}_\mu,\vartheta} = \sum_\mu \frac{\left(\mathrm{Re}(\mathrm{Tr}[\varrho_\vartheta \hat{K}_\mu^2 \hat{\mathcal{L}}_\vartheta])\right)^2}{\mathrm{Tr}[\varrho_\vartheta \hat{K}_\mu^2]}. \tag{E.8}$$

This quantity may be upper bounded to obtain the quantum Cramér–Rao bound, as per

$$I_{\hat{K}_\mu,\vartheta} \leq \sum_\mu \left| \frac{\mathrm{Tr}[\varrho_\vartheta \hat{K}_\mu^2 \hat{\mathcal{L}}_\vartheta]}{\sqrt{\mathrm{Tr}[\varrho_\vartheta \hat{K}_\mu^2]}} \right|^2 = \sum_\mu \left| \mathrm{Tr}\left[\frac{\sqrt{\varrho_\vartheta}\hat{K}_\mu}{\sqrt{\mathrm{Tr}[\varrho_\vartheta \hat{K}_\mu^2]}} \hat{K}_\mu \hat{\mathcal{L}}_\vartheta \sqrt{\varrho_\vartheta} \right] \right|^2$$

$$\leq \sum_\mu \mathrm{Tr}[\hat{K}_\mu^2 \hat{\mathcal{L}}_\vartheta \varrho_\vartheta \hat{\mathcal{L}}_\vartheta] = \mathrm{Tr}[\hat{\mathcal{L}}_\vartheta^2 \varrho_\vartheta], \tag{E.9}$$

where the Cauchy–Schwarz inequality with respect to the Hilbert–Schmidt inner product and the POVM normalisation $\sum_\mu \hat{K}_\mu^2 = \hat{\mathbb{1}}$ were employed.

This bound is achievable for a projective measurement over the eigenstates of the symmetric logarithmic derivative $\hat{\mathcal{L}}_\vartheta$, which offers further valuable insight into the optimal estimation process and, upon definition of the quantum Fisher information $I_\vartheta = \mathrm{Tr}[\hat{\mathcal{L}}_\vartheta^2 \varrho_\vartheta]$, completes the proof of Inequality (8.83), with I_ϑ that indeed corresponds to the definition (8.82) (the case of N measurements is handled as in the classical case at the end of the preceding section). For a more exhaustive discussion of this and related matters, the reader is referred to [67], and references therein.

References

[1] G. Adesso and A. Datta. Quantum versus Classical Correlations in Gaussian States. *Phys. Rev. Lett.*, 105:030501, 2010.

[2] V.I. Arnold. *Mathematical Methods of Classical Mechanics.* Springer, 1978.

[3] Arvind, B. Dutta, N. Mukunda, and R. Simon. The Real Symplectic Groups in Quantum Mechanics and Optics. *Pramana*, 45:471, 1995.

[4] L. Banchi, S. Pirandola, and S.L. Braunstein. Quantum Fidelity for Arbitrary Gaussian States. *Phys. Rev. Lett.*, 115:260501, 2015.

[5] S.M. Barnett and P.M. Radmore. *Methods in Theoretical Quantum Optics.* Clarendon Press, Oxford, 1997.

[6] V.P. Belavkin. Quantum continual measurements and a posteriori collapse on CCR. *Commun. Math. Phys.*, 146:611, 1992.

[7] V.P. Belavkin. Measurement, Filtering and Control in Quantum Open Dynamical Systems. *Rep. Math. Phys.*, 43:A405, 1999.

[8] J.S. Bell. *Speakable and Unspeakable in Quantum Mechanics: Collected Papers on Quantum Philosophy.* Cambridge University Press, 2004.

[9] R. Bhatia. *Matrix Analysis.* Springer, New York, 1996.

[10] I. Bloch. Ultracold quantum gases in optical lattices. *Nat. Phys.*, 1:23, 2005.

[11] A. Botero and B. Reznik. Mode-wise Entanglement of Gaussian States. *Phys. Rev. A*, 67:052311, 2003.

[12] W.P. Bowen and G.J. Milburn. *Quantum Optomechanics.* CRC Press, 2015.

[13] S.L. Braunstein and A.K. Pati, editors. *Quantum Information with Continuous Variables.* Springer, 2003.

[14] S.L. Braunstein and P. van Look. Quantum information with continuous variables. *Rev. Mod. Phys.*, 77:513, 2005.

[15] D.E. Browne, J. Eisert, S. Scheel, and M.B. Plenio. Driving non-Gaussian to Gaussian states with linear optics. *Phys. Rev. A*, 67:062320, 2003.

[16] H.J. Carmichael. *Statistical Methods in Quantum Optics 1.* Springer (Berlin, Heidelberg), 1991.

[17] N.J. Cerf, G. Leuchs, and E.S. Polzik, editors. *Quantum Information with Continuous Variables of Atoms and Light.* Imperial College Press, 2007.

[18] J. Chalker. Quantum Theory of Condensed Matter. https://www-thphys.physics.ox.ac.uk/people/JohnChalker/qtcm/lecture-notes.pdf.

[19] G. Chiribella and G. Adesso. Quantum Benchmarks for Pure Single-Mode Gaussian States. *Phys. Rev. Lett.*, 112:010501, 2014.

[20] J.I. Cirac and P. Zoller. Quantum Computations with Cold Trapped Ions. *Phys. Rev. Lett.*, 74:4091, 1995.

[21] O. Cohen. Classical Teleportation of Classical States. arXiv:quant-ph/0310017, 2003.

[22] M. de Gosson. *Symplectic Geometry and Quantum Mechanics.* Birkhäuser, Basel, 2006.

[23] G. De Palma, A. Mari, V. Giovannetti, and A.S. Holevo. Normal form decomposition for Gaussian-to-Gaussian superoperators. *J. Math. Phys.*, 56:052202, 2015.

[24] S.M. Dutra. *Cavity Quantum Electrodynamics: The Strange Theory of Light in a Box.* Wiley, 2004.

[25] J. Eisert, S. Scheel, and M.B. Plenio. Distilling Gaussian States with Gaussian Operations is Impossible. *Phys. Rev. Lett.*, 89:137903, 2002.

[26] A. Ferraro, S. Olivares, and M.G.A. Paris. *Gaussian States in Quantum Information.* Bibliopolis, Napoli, 2005.

[27] J. Fiurášek. Gaussian Transformations and Distillation of Entangled Gaussian States. *Phys. Rev. Lett.*, 89:137904, 2002.

[28] G.B. Folland. *Harmonic Analysis in Phase Space.* Princeton University Press, 1989.

[29] N. Friis, A.R. Lee, and D.E. Bruschi. Fermionic-mode entanglement in quantum information. *Physical Review A*, 87:022338, 2013.

[30] R. García-Patrón, J. Fiurášek, N.J. Cerf, J. Wenger, R. Tualle-Brouri, and Ph. Grangier. Proposal for a Loophole-Free Bell Test Using Homodyne Detection. *Phys. Rev. Lett.*, 93:130409, 2004.

[31] C. Gardiner and P. Zoller. *Quantum Noise.* Springer, Heidelberg, 2010.

[32] C.W. Gardiner. *Stochastic Methods: A Handbook for the Natural and Social Sciences.* Springer, 2009.

[33] C.W. Gardiner and M.J. Collett. Input and output in damped quantum systems: Quantum stochastic differential equations and the master equation. *Phys. Rev. A*, 31:3761, 1985.

[34] M.G. Genoni, S. Mancini, H.M. Wiseman, and A. Serafini. Quantum filtering of a thermal master equation with a purified reservoir. *Phys. Rev. A*, 90:063826, 2014.

[35] G. Giedke and J.I. Cirac. Characterization of Gaussian operations and distillation of Gaussian states. *Phys. Rev. A*, 66:032316, 2002.

[36] G. Giedke, J. Eisert, J.I. Cirac, and M.B. Plenio. Entanglement transformations of pure Gaussian states. *Quant. Inf. Comp.*, 3:211, 2003.

[37] G. Giedke, M.M. Wolf, O. Krüger, R.F. Werner, and J.I. Cirac. Entanglement of Formation for Symmetric Gaussian States. *Phys. Rev. Lett.*, 91:107901, 2003.

[38] P. Giorda and M.G.A. Paris. Gaussian Quantum Discord. *Phys. Rev. Lett.*, 105:020503, 2010.

[39] V. Giovannetti, A.S. Holevo, and R. García-Patrón. A solution of the Gaussian optimizer conjecture. *Comm. Math. Phys.*, 334(1553), 2015.

[40] J.E. Gough, M.R. James, and H.I. Nurdin. Squeezing components in linear quantum feedback networks. *Phys. Rev. A*, 81:023804, 2010.

[41] F. Grosshans and Ph. Grangier. Quantum cloning and teleportation criteria for continuous quantum variables. *Phys. Rev. A*, 64:010301, 2001.

[42] F. Grosshans and Ph. Grangier. Continuous variable quantum cryptography using coherent states. *Phys. Rev. Lett.*, 88:057902, 2002.

[43] K. Hammerer, A.S. Sørensen, and E.S. Polzik. Quantum interface between light and atomic ensembles. *Rev. Mod. Phys.*, 82:1041, 2010.

[44] K. Hammerer, M. M. Wolf, E.S. Polzik, and J.I. Cirac. Quantum benchmark for storage and transmission of coherent states. *Phys. Rev. Lett.*, 94:150503, 2005.

[45] M.B. Hastings. Superadditivity of communication capacity using entangled inputs. *Nat. Phys.*, 5:255, 2009.

[46] A.S. Holevo. The capacity of the quantum channel with general signal states. *IEEE Trans. Inf. Theory*, 44:269, 1998.

[47] A.S. Holevo. Entanglement-breaking channels in infinite dimensions. *Problems of Information Transmission*, 44:3, 2008.

[48] A.S. Holevo. On the Choi–Jamiolkowski Correspondence in Infinite Dimensions. *J. Math. Phys.*, 52:042202, 2011.

[49] A.S. Holevo. *Probabilistic and statistical aspects of quantum mechanics.* Edizioni della Normale, 2011.

[50] M. Horodecki, P.W. Shor, and M.B. Ruskai. General Entanglement Breaking Channels. *Rev. Math. Phys.*, 15(629), 2003.

[51] S. Iida, M. Yukawa, H. Yonezawa, N. Yamamoto, and A. Furusawa. Experimental Demonstration of Coherent Feedback Control on Optical Field Squeezing. *IEEE Trans. Automat. Contr.*, 57:2045, 2012.

[52] J.D. Jackson. *Classical Electrodynamics.* Wiley (New York), 1998.

[53] K. Jacobs and D.A. Steck. A straightforward introduction to continuous quantum measurements. *Contemp. Phys.*, 47:279, 2006.

[54] D.F.V. James. Quantum dynamics of cold trapped ions with application to quantum computation. *Appl. Phys. B*, 66:181, 1998.

[55] D.F.V. James and J. Jerke. Effective Hamiltonian theory and its applications in quantum information. *Canadian Journal of Physics*, 85:625, 2007.

[56] Z. Jiang. Quantum Fisher information for states in exponential form. *Phys. Rev. A*, 89:032128, 2014.

[57] E. Knill, R. Laflamme, and G.J. Milburn. A scheme for efficient quantum computation with linear optics. *Nature*, 409:46, 2001.

[58] L. Lami, A. Serafini, and G. Adesso. Gaussian entanglement revisited. arXiv:1612.05215, 2016.

[59] E.H. Lieb and M.B. Ruskai. Proof of the Strong Subadditivity of Quantum Mechanical Entropy. *J. Math. Phys.*, 14:1938, 1973.

[60] A. Mari, V. Giovannetti, and A.S. Holevo. Quantum state majorization at the output of bosonic Gaussian channels. *Nature Communications*, 5(3826), 2014.

[61] A. Monras. Phase space formalism for quantum estimation of Gaussian states. arXiv:1303.3682, 2013.

[62] C. Navarrete-Benlloch, R. García-Patrón, J.H. Shapiro, and N.J. Cerf. Enhancing quantum entanglement by photon addition and subtraction. *Phys. Rev. A*, 86:012328, 2012.

[63] H. Nha and H.J. Carmichael. Proposed Test of Quantum Nonlocality for Continuous Variables. *Phys. Rev. Lett.*, 93:020401, 2004.

[64] M.A. Nielsen and I.L. Chuang. *Quantum Computation and Information.* Cambridge University Press, 2000.

[65] H.I. Nurdin and N. Yamamoto. Distributed entanglement generation between continuous-mode Gaussian fields with measurement-feedback enhancement. *Physical Review A*, 86:022337, 2012.

[66] M. Owari, M.B. Plenio, E.S. Polzik, A. Serafini, and M.M. Wolf. Squeezing the limit: Quantum benchmarks for the teleportation and storage of squeezed states. *New. J. Phys.*, 10:113014, 2008.

[67] M.G.A. Paris. Quantum estimation for quantum technology. *Int. J. Quantum Inf.*, 7:125, 2009.

[68] O. Pinel, P. Jian, C. Fabre, N. Treps, and D. Braun. Quantum parameter estimation using general single-mode Gaussian states. *Phys. Rev. A*, 88:040102, 2013.

[69] S. Pirandola, G. Spedalieri, S.L. Braunstein, N.J. Cerf, and S. Lloyd. Optimality of Gaussian Discord. *Phys. Rev. Lett.*, 113:140405, 2014.

[70] M.B. Plenio. The Logarithmic Negativity: A Full Entanglement Monotone That is not Convex. *Phys. Rev. Lett.*, 95:090503, 2005.

[71] J. Preskill. Course Information for Physics 219/Computer Science 219 – Quantum Computation (Formerly Physics 229). http://www.theory.caltech.edu/people/preskill/ph219/.

[72] R.R. Puri. *Mathematical Methods of Quantum Optics.* Springer, 2001.

[73] T. Rudolph and B.C. Sanders. Requirement of Optical Coherence for Continuous-Variable Quantum Teleportation. *Phys. Rev. Lett.*, 87:077903, 2001.

[74] J.J. Sakurai. *Modern Quantum Mechanics.* Addison Wesley, 1993.

[75] W.P. Schleich. *Quantum Optics in Phase Space.* Wiley-VCH Verlag, Berlin, 2001.

[76] B. Schumacher and M. Westmoreland. Sending classical information via noisy quantum channels. *Phys. Rev. A*, 56:131, 1997.

[77] B. Schumacher and M. Westmoreland. *Quantum Processes, Systems, and Information.* Cambridge University Press, 2010.

[78] E. Shchukin and W. Vogel. Inseparability Criteria for Bipartite Quantum States. *Phys. Rev. Lett.*, 95:230502, 2005.

[79] P.W. Shor. Additivity of the Classical Capacity of Entanglement-Breaking Quantum Channels. *J. Math Phys.*, 43:4334, 2002.

[80] R. Simon. Peres-Horodecki Separability Criterion for Continuous Variable Systems. *Phys. Rev. Lett.*, 84:2726, 2000.

[81] R. Simon, V. Srinivasan, and S.K. Chaturvedi. Congruences and Canonical Forms for a Positive Matrix: Application to the Schweinler-Wigner Extremum Principle. *J. Math. Phys.*, 40:7, 1998.

[82] G. Vidal and R.F. Werner. A computable measure of entanglement. *Phys. Rev. A*, 65:032314, 2002.

[83] D.F. Walls and G.J. Milburn. *Quantum Optics*. Springer, Berlin, 2008.

[84] C. Weedbrook, S. Pirandola, R. García Patrón, T.C. Ralph, J.H. Shapiro, and S. Lloyd. Gaussian quantum information. *Rev. Mod. Phys.*, 84:621, 2012.

[85] R. F. Werner. The Uncertainty Relation for Joint Measurement of Position and Momentum. In O. Hirota, editor, *Quantum Information, Statistics, Probability*, page 153. Paramus, NJ: Rinton, 2004.

[86] R.F. Werner and M.M. Wolf. Bound Entangled Gaussian States. *Phys. Rev. Lett.*, 86:3658, 2001.

[87] J. Williamson. On the Algebraic Problem Concerning the Normal Forms of Linear Dynamical Systems. *Am. J. Math.*, 58:141, 1936.

[88] H.M Wiseman. Optical coherence and teleportation: Why a laser is a clock, and not a quantum channel. In *Proc. SPIE 5111, Fluctuations and Noise in Photonics and Quantum Optics*, volume 78, 2003.

[89] H.M Wiseman and A.C. Doherty. Optimal Unravellings for Feedback Control in Linear Quantum Systems. *Phys.Rev. Lett.*, 94:070405, 2005.

[90] H.M. Wiseman and G.J. Milburn. *Quantum Measurement and Control*. Cambridge University Press, 2010.

[91] M.M. Wolf, J. Eisert, and M.B. Plenio. Entangling Power of Passive Optical Elements. *Phys. Rev. Lett.*, 90:047904, 2003.

[92] M.M. Wolf, G. Giedke, and J.I. Cirac. Extremality of Gaussian Quantum States. *Phys. Rev. Lett.*, 96:080502, 2006.

[93] A.M. Zagoskin. *Quantum Engineering: Theory and Design of Quantum Coherent Structures*. Cambridge University Press, 2011.

Index

Printed in the United States
by Baker & Taylor Publisher Services